# *Pleurotus*

Long ago, *Pleurotus* mushrooms (commonly known as oyster mushrooms) were discovered growing naturally on decaying logs across Asia and Europe. Valued for their delicate flavor and revered in traditional medicine, they were believed to boost immunity and promote heart health. As the demand for sustainable food sources grew, it was found that *Pleurotus* species could be cultivated easily on agricultural wastes like wheat straw and cotton stalks. This made them an ideal candidate for low-cost, eco-friendly farming, supporting food security, and rural livelihoods. Further, modern scientific research validated their status as a superfood, revealing a rich nutritional profile, including high-quality protein, dietary fiber, B-complex vitamins, essential minerals, and a range of bioactive compounds such as β-glucans and phenolics. These compounds impart a wide range of therapeutic properties, including antioxidant, anticancer, antidiabetic, immunomodulatory, and other health-promoting effects. Today, *Pleurotus* mushrooms have transcended their traditional roles to become key ingredients in functional foods, dietary supplements, and even green nanotechnology, symbolizing the fusion of ancient wisdom with modern science.

**Key Features:**

- The book explores the historical and cultural significance of *Pleurotus* mushrooms in traditional medicine, aligns it with current scientific evidence, and provides detailed descriptions of their morphological characteristics.
- It offers a comprehensive guide to sustainable oyster mushroom farming, from mycelial culture and spawn production to agro-waste-based cultivation and postharvest management, promoting waste-to-wealth models and eco-friendly agriculture.
- The book highlights the rich nutritional profile and therapeutic potential of *Pleurotus* mushrooms, with practical sections on culinary applications and traditional recipes. The book also discusses biofortification strategies to enhance micronutrient content, helping combat hidden hunger and promote dietary wellness.
- It illustrates the emerging role of *Pleurotus* mushrooms in green nanotechnology, detailing eco-friendly nanoparticle synthesis methods and their promising medicinal applications in antioxidant, antimicrobial, and anticancer therapies.
- The book promotes low-cost cultivation and value-added product development, supporting food security, rural income generation, and women-led entrepreneurship. It includes a market analysis of the growing global demand for *Pleurotus* products and potential avenues for small-scale businesses.

# *Pleurotus*

## From Farm to Pharmacy

Edited by
Somanjana Khatua and Krishnendu Acharya

CRC Press
Taylor & Francis Group
Boca Raton London New York

CRC Press is an imprint of the
Taylor & Francis Group, an **informa** business

Designed cover image: Shutterstock

First edition published 2026
by CRC Press
2385 NW Executive Center Drive, Suite 320, Boca Raton FL 33431

and by CRC Press
4 Park Square, Milton Park, Abingdon, Oxon, OX14 4RN

*CRC Press is an imprint of Taylor & Francis Group, LLC*



ISBN: 978-1-032-86449-5 (hbk)
ISBN: 978-1-032-86450-1 (pbk)
ISBN: 978-1-003-52757-2 (ebk)

DOI: 10.1201/9781003527572

Typeset in Times
by Apex CoVantage, LLC

# Contents

# Preface

*Pleurotus* species, commonly known as oyster mushrooms, have long held an esteemed place in traditional food systems and medicinal practices across cultures. Celebrated for their delicate flavor and tender texture, they have been staples in culinary traditions and featured in herbal tonics and soups believed to boost vitality, strengthen immunity, and support overall well-being. The early 20th century marked a pivotal moment for oyster mushrooms, as their nutritional and therapeutic potential inspired systematic cultivation efforts. Notably, large-scale production began in Germany during World War I in response to food shortages and the search for alternative protein sources. Since then, cultivation practices have evolved rapidly, spreading across continents and establishing *Pleurotus* as one of the most widely cultivated edible mushrooms globally. Their rapid growth, adaptability, and ability to thrive on a wide range of lignocellulosic agricultural wastes have made them a sustainable, low-cost solution for agro-residue recycling. Today, they rank as the third most cultivated mushroom worldwide, with China leading global production.

Beyond their culinary appeal, oyster mushrooms stand out as a nutritional powerhouse. They provide high-quality protein, essential dietary fiber, and an array of B-complex vitamins, making them particularly valuable in vegetarian diets and protein-deficient populations. Recent advances in biofortification have further enhanced their nutritional profile by enriching mushrooms with critical micronutrients such as iron, zinc, selenium, and calcium through substrate supplementation. These biofortified mushrooms act as functional foods, helping to combat malnutrition and micronutrient deficiencies. Given their high moisture content and delicate nature, effective postharvest strategies such as refrigeration, drying, and breathable packaging are vital for preserving quality and extending shelf life. Their versatility in the kitchen, from stir-fries and grills to soups and stews, has secured their place in global cuisines.

Medicinally, oyster mushrooms are a reservoir of bioactive compounds with wide-ranging health benefits. Rich in β-glucans, they modulate immune responses by activating macrophages, enhancing dendritic cell activity, and regulating cytokines. Extensive studies have confirmed their antioxidant, antimicrobial, anti-inflammatory, anti-aging, neuroprotective, hepatoprotective, prebiotic, hypocholesterolemic, and cardioprotective effects. Additionally, their antidiabetic potential, manifested through α-glucosidase inhibition, improved insulin sensitivity, and postprandial glucose regulation, underscores their functional value. The mushrooms have also demonstrated hepatoprotective effects and anticancer activity in various *in vitro* and *in vivo* models, owing to their unique secondary metabolites.

An exciting frontier in *Pleurotus* research lies in nanotechnology. Oyster mushrooms have emerged as eco-friendly biofactories for the green synthesis of metal nanoparticles, including silver, gold, and zinc oxide. Utilizing mushroom extracts as natural reducing and stabilizing agents, these nanoparticles exhibit significant antimicrobial and anticancer properties, opening up new possibilities in drug delivery and biomedical applications. This confluence of mycology and nanotechnology exemplifies innovation rooted in nature.

Complementing all these advances, the global market for oyster mushrooms is experiencing remarkable growth, driven by increased demand for plant-based nutrition, heightened awareness of their health benefits, and the sustainability of mushroom farming. This expansion is further supported by advancements in biotechnology and cultivation technologies, which improve yields. Moreover, the mushroom industry is playing a vital role in promoting entrepreneurship, rural start-ups, and women's empowerment. By addressing critical issues such as unemployment, gender inequality, and malnutrition, the oyster mushroom-based industry aligns with several Sustainable Development Goals, offering a powerful model for integrated socio-economic development.

Building on this momentum, *Pleurotus: From Farm to Pharmacy* offers a comprehensive overview of the cultural legacy, nutritional composition, medicinal potential, technological innovations, and market prospects of oyster mushrooms. Designed for researchers, cultivators, nutritionists, entrepreneurs, and students, it seeks to integrate traditional wisdom with contemporary scientific insights, highlighting the transformative power of this extraordinary fungus in shaping a healthier and better future.

This volume is the result of collective effort and shared vision. We extend our deepest gratitude to the many researchers, scholars, and contributors whose timely support and invaluable insights have made this work possible. We hope that students, educators, health professionals, and stakeholders in mushroom-based industries will find this book both informative and inspiring.

# Editor Biographies

**Dr. Somanjana Khatua** is an assistant professor in the Department of Botany, University of Allahabad, Uttar Pradesh, India. She pursued a master's in Botany and Ph.D. from the University of Calcutta. Earlier, she was appointed by the West Bengal Education Service and posted at Krishnagar Government College. The areas of her research interest include drug development, immunology, carbohydrate biology, natural products, medical mycology, and functional food. To date, she has published 93 papers in different peer-reviewed journals such as *Current Research in Biotechnology, Phytotherapy Research, Biodegradation, Chemico-Biological Interactions, Food and Function, Frontiers in Pharmacology, Scientific Reports, PLOS ONE*, the *Journal of Pharmacy and Pharmacology, PeerJ, Carbohydrate Polymers, Carbohydrate Research*, and the *International Journal of Biological Macromolecules.* She has 83 papers in Scopus and 1,922 Scopus citations, as well as 2,426 Google Scholar citations. Her h-index is 26 (Scopus) and 30 (Google Scholar), while her Google Scholar i10 index is 54. She has received many awards, such as the Woman Botanist Award 2020 from the Indian Botanical Society and the Outstanding Paper Award from the Government of West Bengal, India. She has been included in the top 2% of scientists list in the Elsevier (Stanford University) ranking in 2023. She delivers many invited lectures in different colleges across India.

**Prof. Krishnendu Acharya** pursued a master's in Botany from the University of Calcutta. He then completed M.Tech. and Ph.D. from Jadavpur University. Prof. Acharya joined the University of Calcutta as Lecturer in Botany in 2004 and was promoted to Professor in 2012. To date, he has supervised 24 students for Ph.D. and published 520 research articles and 10 books. He has 12,955 Scopus citations and 17,945 Google Scholar citations. His h-index is 59 (Scopus) and 66 (Google Scholar), while his Google Scholar i10 index is 308. He was ranked first among the top 100 authors of state universities based on the Indian Citation Index 2016 by the Confederation of Indian Industry (CII). He has been listed on Elsevier's (Stanford University) list of the world's top 2% of scientists for the last four years. Prof. Acharya has been elected as Fellow of the West Bengal Academy of Science and Technology, the Linnean Society of London, the Mycological Society of India, the Indian Mycological Society, and the International College of Nutrition, Canada. He has received many awards, including the Sir Edwin John Butler Memorial Award, the Patel Memorial Award, the Global Achievement in the Field of Phyllosphere Biology, and the Prof. K. Natarajan Memorial Award. He is on the editorial boards of many national and international journals. Prof. Acharya actively and regularly engages in lecturing at different seminars, universities, and colleges to boost young minds and inspire students. Beyond his academic roles and various administrative responsibilities, Prof. Acharya is actively promoting self-sustainability through entrepreneurship. His low-cost organic cultivation technology development has directly benefited over 1,800 farmers, further demonstrating his commitment to improving lives through sustainable practices.

# 1 The Oyster Mushroom
## *History, Science, and Biology*

*Entaj Tarafder, Maqsood Ahmed Khaskheli, Yong Wang, and Mir Muhammad Nizamani**
Department of Plant Pathology, College of Agriculture, Guizhou University, Guiyang, China

### 1.1 INTRODUCTION

Oyster mushrooms (*Pleurotus* spp.) have long been valued for their multifaceted roles in human history and continue to be of significant relevance in contemporary times. Bridging the domains of nutrition, medicine, and environmental sustainability, these fungi have emerged as vital biological resources with wide-ranging applications (Raman et al., 2021; Sangeeta et al., 2024). Renowned for their delicate texture and mild, savory flavor, oyster mushrooms have been consumed for centuries as a highly nutritious food. They are rich in essential proteins, dietary fiber, vitamins (notably B-complex and D), minerals such as iron, zinc, and potassium, and potent antioxidants. These attributes have cemented their place in traditional diets and contributed to their expanding role in modern health-conscious culinary practices (Khatua et al., 2013; Jana and Acharya, 2022).

Beyond their nutritional value, oyster mushrooms possess significant medicinal potential. Traditionally used in folk medicine, they are now scientifically recognized for their immunomodulatory, anti-inflammatory, antioxidant, and antitumor properties (Thakur, 2020; Kumar et al., 2021). A growing body of research is uncovering diverse bioactive compounds in *Pleurotus* species, offering promising avenues for the development of functional foods and natural therapeutics (Devi et al., 2024). From an ecological perspective, oyster mushrooms play a critical role as saprotrophs in natural ecosystems. Their ability to degrade complex organic materials makes them key agents in nutrient cycling and organic waste decomposition. As such, they contribute to environmental sustainability by enhancing soil fertility and supporting waste-to-wealth initiatives (Sarkar et al., 2024). Their cultivation supports sustainable agriculture by converting agro-industrial residues into valuable biomass, thus mitigating environmental pollution (Malik et al., 2021). In recent decades, the commercial cultivation of *Pleurotus* species has witnessed rapid global expansion. Advancements in cultivation techniques, coupled with rising consumer awareness of health and sustainability, have made these mushrooms more accessible and economically viable (Jayaraman et al., 2024). Today, they are a staple in cuisines worldwide and play a growing role in rural livelihoods, food security, and entrepreneurship development (Ghazaryan et al., 2024).

This chapter provides a comprehensive overview of oyster mushrooms, exploring their historical significance, taxonomy, anatomical features, life cycle, and reproductive strategies. It also highlights their nutritional and medicinal properties, ecological contributions, and relevance to sustainable development. Furthermore, it reviews current trends in cultivation, global market dynamics, and emerging research directions, underscoring the versatile importance of *Pleurotus* mushrooms in food systems, health sciences, and environmental management.

### 1.2 SECTION 1: HISTORY OF OYSTER MUSHROOMS AS FOOD

#### 1.2.1 Ancient Uses

The history of oyster mushrooms as a food source dates back centuries, with some of the earliest records found in ancient China and Japan, where they were not only consumed as a dietary staple but also valued for their health benefits (Singh et al., 2018; Ba et al., 2023). Historical texts from the Han Dynasty in China (206 BCE–220 CE) mention the collection and consumption of wild mushrooms, including oyster macrofungi, underscoring their importance in traditional Chinese cuisine and medicine (Niego et al., 2021). In Japan, *Pleurotus* spp. were similarly documented as a valuable food source, particularly during times of scarcity (Chiu et al., 2000). These early uses were primarily based on their availability in the wild, with people gathering them from forests due to their ease of identification, palatable flavor, and versatility in cooking (Khan et al., 2024). Additionally, ancient healers and herbalists used oyster mushrooms in various medicinal practices to treat ailments ranging from respiratory conditions to immune system support. This dual role as both food and medicine highlights the enduring appreciation for oyster mushrooms, which have evolved from wild-harvested delicacies to widely cultivated and commercially available food sources today (Chugh et al., 2022).

Oyster mushrooms have played a significant role in the culinary and medicinal traditions of various cultures throughout history (Rathod et al., 2021). In ancient China, their use dates back to the Han Dynasty (206 BCE–220 CE), where they were valued not only for their nutritional

* Corresponding author: mirmohammadnizamani@outlook.com

DOI: 10.1201/9781003527572-1

benefits but also for their medicinal properties, such as boosting the immune system and improving overall health (Hobbs, 2023). In Japan, oyster mushrooms have long been a staple in traditional dishes, appreciated for their delicate flavor and texture, and were particularly crucial during periods of food scarcity (Thakur, 2020). In Europe, oyster mushrooms were less commonly known until the 18th century, when they began to be cultivated and recognized for their culinary potential (Jongman et al., 2018). Indigenous cultures in North America also utilized oyster mushrooms, incorporating them into their diets and traditional medicinal practices. Across these diverse cultures, the historical significance of oyster mushrooms is evident in their enduring popularity and their versatile applications in both food and medicine (Pánek et al., 2019).

### 1.2.2 Oyster Mushrooms in Traditional Medicine

Oyster mushrooms have been used in traditional medicine for centuries, recognized for their potential health benefits and therapeutic properties (Oloke, 2017; Elkhateeb et al., 2019). In ancient China, these mushrooms were incorporated into herbal remedies to boost the immune system, improve respiratory health, and promote overall vitality (Mingyi et al., 2019; Ray et al., 2024). Traditional Chinese Medicine (TCM) texts often highlighted their use in treating ailments such as coughs, colds, and asthma due to their anti-inflammatory and antioxidant properties (Chugh et al., 2022). Similarly, in Japan, oyster mushrooms were utilized in Kampo medicine, where they were believed to support liver health and aid in digestion (Gepshtein et al., 2008). In Europe, the medicinal uses of oyster mushrooms became more recognized during the 18th century, with folk medicine practitioners using them to treat wounds and skin infections, leveraging their antibacterial properties (González and Vallejo, 2023). In Russia and other parts of Eastern Europe, oyster mushrooms were employed to combat infections and boost immunity, particularly during harsh winters when illnesses were more prevalent (Stajić et al., 2023).

The integration of oyster mushrooms into traditional healing practices varied across cultures but consistently underscored their perceived medicinal value. In TCM, oyster mushrooms were often dried and used in decoctions, infusions, and teas, sometimes combined with other medicinal herbs to enhance their efficacy (Mingyi et al., 2019; Sangeeta et al., 2024). This practice continues today, with many modern TCM formulations including oyster mushrooms as a key ingredient. In Japan, Kampo medicine incorporated oyster mushrooms into various preparations, such as soups and broths, consumed to alleviate digestive issues and strengthen the body (Aditya et al., 2024; Tarafder et al., 2024). The use of mushrooms in food as medicine was a common theme, reflecting the holistic approach of Eastern medical traditions (Austin-Asomeji et al., 2024).

In Native American healing practices, oyster mushrooms were sometimes used in poultices applied to wounds to prevent infection and promote healing (Anil Kumar et al., 2023). The knowledge of mushroom foraging and their medicinal uses was passed down through generations, integrating seamlessly into their broader understanding of natural remedies (Ioffe, 2024). In Europe, particularly in Eastern European countries, the knowledge of mushroom foraging included medicinal uses of oyster mushrooms (Sudheer et al., 2023; Chauvin-Payan, 2024). They were integrated into home remedies, often combined with other local herbs and plants to treat common ailments. This practice persists in some rural areas today, where traditional knowledge of natural remedies is still valued (Schulp et al., 2014; Jalinik et al., 2024).

## 1.3 MODERN CULINARY USE

### 1.3.1 Popularity in Contemporary Cuisines

In contemporary cuisines worldwide, oyster mushrooms have gained significant popularity due to their versatile flavor profile, nutritional benefits, and ease of preparation (Aditya et al., 2024). They are widely appreciated for their delicate texture and subtle, slightly sweet taste, which make them suitable for a variety of dishes (Verma, 2023). In Western cuisine, oyster mushrooms are often featured in gourmet dishes, stir-fries, soups, and pasta. They are particularly favored in vegetarian and vegan diets as a meat substitute due to their umami-rich flavor and satisfying texture. In Asian cuisine, oyster mushrooms continue to be a staple ingredient, found in dishes ranging from hot pots and noodle soups to stir-fried and tempura preparations (Grigson, 2008; Romine, 2018).

### 1.3.2 Culinary Trends and Innovations

Culinary trends and innovations have further propelled the popularity of oyster mushrooms in modern kitchens. Chefs and home cooks alike are experimenting with new ways to incorporate these mushrooms into their recipes, pushing the boundaries of traditional cooking methods (Aditya et al., 2024). For instance, oyster mushrooms are now being used in plant-based burgers and other meat alternatives, capitalizing on their texture and flavor to create satisfying and sustainable food options (Sangeeta et al., 2024). They are also being utilized in fermentation processes to develop unique flavors and textures, adding a new dimension to traditional mushroom preparations (Bierend, 2021; Khan et al., 2024).

In addition to these innovations, there has been a rise in the use of oyster mushrooms in health-focused culinary trends. As consumers become more health-conscious, the demand for nutrient-dense, low-calorie ingredients has increased (Sangeeta et al., 2024). Oyster mushrooms fit this profile perfectly, leading to their inclusion in a variety of health-oriented dishes, such as salads, grain bowls, and low-carb meals (Ali et al., 2024). Moreover, the farm-to-table movement has embraced oyster mushrooms due to their relatively easy cultivation and minimal environmental impact, making them a preferred choice for sustainable and locally sourced ingredients (Barritt, 2015).

The culinary versatility of oyster mushrooms is also reflected in their use in global fusion cuisine, where they are paired with a wide array of ingredients and cooking styles to create innovative and exciting new dishes. This adaptability ensures that oyster mushrooms remain a beloved and dynamic component of modern culinary arts, continuing to inspire chefs and food enthusiasts around the world (Chang et al., 2012; Adetunji et al., 2022).

## 1.4 COMMERCIAL CULTIVATION

### 1.4.1 Development of Cultivation Techniques

The commercial cultivation of oyster mushrooms has evolved significantly over the past century, transforming from small-scale operations to sophisticated, large-scale enterprises (Thakur, 2020; Carrasco et al., 2021). Early cultivation methods were often rudimentary, involving simple inoculation of logs or stumps with mushroom spores (Rathod, 2023). However, as the demand for oyster mushrooms grew, more efficient and controlled techniques were developed. The introduction of substrate cultivation, using materials like straw, sawdust, and agricultural waste marked a major advancement (Mahari et al., 2020). These substrates are sterilized or pasteurized to eliminate competing organisms and then inoculated with oyster mushroom spawn, creating an optimal environment for mycelium growth and fruiting body formation (Jongman et al., 2018; Jarial et al., 2024).

Advancements in technology and agricultural science have further refined these methods (Tarafder et al., 2024). Climate-controlled growing rooms now allow for year-round production, ensuring consistent quality and yield. Innovations such as the use of spawn bags, automated misting systems, and precise temperature and humidity controls have streamlined the cultivation process (Ahmed, 2024). Additionally, research into substrate composition has led to the development of specialized blends that enhance mushroom growth and productivity. These technological advancements have not only increased the efficiency of oyster mushroom cultivation but have also made it accessible to a wider range of growers, from small urban farms to large agricultural operations (Guragain et al., 2024).

### 1.4.2 Expansion of Global Production

The expansion of global production has made oyster mushrooms one of the most widely cultivated and consumed mushrooms worldwide (Singh et al., 2020). Countries like China, the United States, and India have become major producers, supplying both domestic markets and international exports. In China, which leads the world in mushroom production, extensive commercial operations utilize state-of-the-art facilities to meet the high demand (Li and Xu, 2022). The United States and Europe have also seen significant growth in oyster mushroom farming, driven by increasing consumer interest in healthy and sustainable food sources (Pandey et al., 2018; Mahari et al., 2020).

The global expansion has been facilitated by the adaptability of oyster mushrooms to various climatic conditions and substrates, allowing them to be grown in diverse regions (Boukary et al., 2024). This adaptability has also spurred innovation in local cultivation practices, with farmers experimenting with different agricultural by-products as substrates, thereby reducing waste and promoting sustainability (Jayaraman et al., 2024). In developing countries, oyster mushroom cultivation has become a valuable source of income and nutrition, contributing to food security and economic development (Gao et al., 2023). Programs and initiatives by governments and non-profit organizations have supported small-scale farmers in establishing mushroom farms, providing training and resources to enhance production (Martín et al., 2023; Loc et al., 2024).

Furthermore, the international trade of oyster mushrooms has flourished, with fresh, dried, and processed mushrooms being exported to meet the growing global demand. This trade is supported by advancements in logistics and supply chain management, ensuring that high-quality mushrooms reach consumers worldwide (De Cianni et al., 2023; Silva et al., 2024). The popularity of oyster mushrooms in diverse culinary traditions has also driven their global market expansion, with a steady increase in consumption across Asia, Europe, North America, and beyond (Zięba et al., 2020).

## 1.5 SECTION 2: SCIENTIFIC CLASSIFICATION

### 1.5.1 Taxonomy

Oyster mushrooms belong to the kingdom Fungi, a diverse and widespread group of organisms that play crucial roles in ecosystems as decomposers, symbionts, and pathogens (Gupta et al., 2017; Bahram and Netherway, 2022). Fungi are distinct from plants, animals, and bacteria, with unique cellular structures and reproductive methods. This kingdom includes a vast array of species, ranging from microscopic yeasts to large, complex mushrooms (Wallen et al., 2018; Naranjo-Ortiz and Gabaldón, 2019, 2020).

Within the kingdom Fungi, oyster mushrooms are classified under the phylum Basidiomycota. Basidiomycota is one of the most diverse fungal phyla and is characterized by the production of spores on specialized cells called basidia (Oberwinkler and Bauer, 2018; Shreejana et al., 2022). This group includes not only mushrooms but also other forms such as puffballs, shelf fungi, and rusts. The defining feature of Basidiomycota is the presence of basidia, which are typically found on the gills or pores of mushrooms, where sexual spores called basidiospores are produced (Halbwachs and Bässler, 2015; Coelho et al., 2017; Khan et al., 2024).

Oyster mushrooms fall into the class Agaricomycetes, which encompasses most of the familiar mushroom-forming fungi (Nagy et al., 2023; Földi et al., 2024). This class is notable for its members' ability to form large, fleshy fruiting bodies. Agaricomycetes play significant ecological roles in decomposing organic matter and forming mycorrhizal associations with plants. The class includes a wide range of mushrooms, from the edible to the poisonous, and displays a variety of morphological forms (Eduardo et al., 2018; Virágh et al., 2022; Yan et al., 2023).

Within Agaricomycetes, oyster mushrooms are further classified in the order Agaricales (Merényi et al., 2022; Bhatia et al., 2024). The Agaricales order is commonly known as the gilled mushrooms, as most species in this group have gills on the underside of their caps, where spores are produced (Atri et al., 2017). This order is incredibly diverse, containing many well-known edible mushrooms, such as the button mushroom (*Agaricus bisporus*), as well as many toxic species (Zafar et al., 2020; Usman et al., 2021). Members of Agaricales are found in various habitats, contributing to the decomposition of organic material and nutrient cycling in ecosystems (Birkebak et al., 2013; Sharma, 2017).

Oyster mushrooms are members of the family Pleurotaceae, which includes mushrooms that typically grow on wood and have a laterally attached stem or no stem at all (Jarial et al., 2024). This family is characterized by its members' saprophytic lifestyle, breaking down lignin and cellulose in dead wood and contributing to forest nutrient cycles. Pleurotaceae mushrooms are often found on decaying logs, stumps, and other wood debris, playing a vital role in forest ecosystems (Singh et al., 2021).

The genus *Pleurotus*, to which oyster mushrooms belong, is distinguished by its oyster-shaped caps and ability to grow on a variety of substrates (Silva et al., 2024). *Pleurotus ostreatus* is the most well-known and widely grown oyster mushroom. It has a broad, fan-shaped cap that can range in color from white to gray to brown, and it is prized for its mild flavor and tender texture (Phan et al., 2019; Raman et al., 2021). *Pleurotus pulmonarius*, also known as the Phoenix oyster mushroom, has a lighter cap and is more heat-tolerant, making it suitable for cultivation in warmer climates (Parkes et al., 2003). *Pleurotus eryngii*, known as the king oyster mushroom, is characterized by its thick stem and smaller cap, valued for its firm texture and savory flavor (Su et al., 2020). Another notable species is *Pleurotus djamor*, the pink oyster mushroom, which is remarkable for its vibrant pink color and rapid growth rate (Siti-Nuramira et al., 2022; Valdez-Solana et al., 2023).

### 1.5.2 Morphological Characteristics

Oyster mushrooms are easily recognizable due to their unique and distinctive morphological characteristics (Pánek et al., 2019; Li et al., 2020). One of the most notable features of oyster mushrooms is their cap, which is broad, fan-shaped, and resembles an oyster shell, hence their common name (Thakur, 2014). The cap can vary in color, ranging from white and gray to brown and even pink, depending on the species and environmental conditions. The cap surface is typically smooth and slightly moist when fresh, often becoming more leathery and dry as it ages (Młodzińska, 2009; Bhatt et al., 2016).

Another distinguishing feature is the gills, which are located on the underside of the cap. The gills of oyster mushrooms are decurrent, meaning they extend down the stem, giving them a distinctive appearance (Kües et al., 2015; de Mattos-Shipley et al., 2016). These gills are closely spaced and can range in color from white to cream or yellowish as the mushroom matures. The gills are an essential part of the mushroom's reproductive system, producing and releasing spores into the environment (Atri et al., 2017; Li et al., 2023).

Oyster mushrooms usually have a short, stubby stem (stipe) or sometimes lack a stem altogether (Hanlin, 2004). When present, the stem is often off-center (eccentric) or lateral, attaching to the side of the cap rather than the center (Geikowsky and Aksak, 2020). This lateral attachment is a key characteristic that helps differentiate oyster mushrooms from other gilled fungi (Singh et al., 2021; Money, 2023).

The flesh of oyster mushrooms is white, firm, and slightly brittle, breaking cleanly when snapped (Dufossé et al., 2021; Berger et al., 2022). They have a mild, pleasant aroma and flavor that is often described as subtly sweet and nutty. The texture is tender yet meaty, making them a popular choice for various culinary applications (Sungsinchai et al., 2019; Belova, 2024).

### 1.5.3 Comparison with Other Fungi

When comparing oyster mushrooms to other fungi, several morphological differences become apparent. One major difference is the shape and attachment of the cap (Jarial et al., 2024). While many gilled mushrooms, such as those in the genus *Agaricus* (which includes the common button mushroom), have a centrally attached stem with a convex to flat cap, oyster mushrooms have a more eccentric or lateral stem attachment and a fan-shaped cap. This distinctive shape helps in the easy identification of oyster mushrooms in the wild and in markets (Tadiosa and Lubos, 2020; Yan et al., 2023; Jarial et al., 2024).

Another key difference is the habitat. Oyster mushrooms are typically saprophytic, growing on decaying wood, such as logs, stumps, and fallen branches (Wendiro et al., 2019). This contrasts with many other mushrooms, such as mycorrhizal fungi, which form symbiotic relationships with living trees (Pérez-Moreno et al., 2021). For example, mushrooms in the genus *Amanita*, known for their often striking appearance and some highly toxic species, form ectomycorrhizal relationships with tree roots, whereas oyster mushrooms are primarily decomposers (de Mattos-Shipley et al., 2016; Kües et al., 2018).

The gill structure and spore print color can also be distinguishing factors. Oyster mushrooms have decurrent gills that run down the stem, and their spore print is typically white to lilac-gray (Mihilall et al., 2011; Mohapatra et al., 2015). In contrast, mushrooms like those in the genus

*Psilocybe* (which includes some psychoactive species) have adnate to adnexed gills that do not run down the stem, and their spore prints are often dark purple to black (Gotvaldová et al., 2021; Rahman et al., 2022).

Furthermore, the growth patterns and fruiting bodies of oyster mushrooms set them apart from polypores, another group of wood-decaying fungi (Schmidt, 2007; Kües et al., 2018). Polypores, such as those in the genus *Ganoderma* (reishi mushrooms), have a porous underside instead of gills and often form large, woody fruiting bodies that persist for years (Seethapathy et al., 2023; Keller et al., 2024). In contrast, oyster mushrooms produce more delicate, short-lived fruiting bodies that decompose quickly after releasing their spores (Spatafora et al., 2017).

### 1.5.4 Phylogenetic Relationships

Oyster mushrooms (*Pleurotus* spp.) belong to the order Agaricales within the class Agaricomycetes, a group known for its diversity and ecological significance (Jarial et al., 2024). The evolutionary lineage of oyster mushrooms traces back to the early diversification of the Basidiomycota, one of the major phyla of fungi (de Mattos-Shipley et al., 2016; Yu et al., 2024). Basidiomycota is characterized by the production of spores on basidia, which are specialized, club-shaped cells (Chowdhary et al., 2014). This group includes not only mushrooms but also rusts, smuts, and shelf fungi, showcasing a wide range of morphologies and ecological roles (Manoharachary et al., 2005; Raghukumar and Raghukumar, 2017).

The genus *Pleurotus* is part of the family Pleurotaceae, which is distinguished by its saprophytic lifestyle, primarily decomposing wood (Téllez-Téllez and Díaz-Godínez, 2019). This family evolved mechanisms to break down complex organic materials like lignin and cellulose, enabling them to thrive on decaying wood and contribute to nutrient cycling in forest ecosystems (Horwath and Paul, 2015; Martín and López, 2023). The Pleurotaceae family is closely related to other families within the Agaricales that share similar ecological niches and morphological traits, such as the Marasmiaceae and the Tricholomataceae (Psurtseva et al., 2023).

The diversification of *Pleurotus* species is thought to be driven by their adaptability to various substrates and environmental conditions (Li et al., 2020). This adaptability has allowed them to colonize a wide range of habitats, from temperate to tropical regions (Linder et al., 2018). Their evolutionary success is also attributed to their efficient reproductive strategies, which include the production of large numbers of spores that can disperse widely and colonize new substrates quickly (Naranjo-Ortiz and Gabaldón, 2019).

### 1.5.5 Genetic Studies and Molecular Analysis

Genetic studies and molecular analyses have provided significant insights into the phylogenetic relationships within the genus *Pleurotus* and between *Pleurotus* and other fungal genera (Li et al., 2024). Molecular techniques, such as DNA sequencing of ribosomal RNA (rRNA) genes, particularly the internal transcribed spacer (ITS) region, have been instrumental in elucidating these relationships (Furan et al., 2024). The ITS region is highly variable and serves as a reliable genetic marker for distinguishing between closely related species (Chen et al., 2010).

Phylogenetic analyses based on ITS sequences have revealed that the genus *Pleurotus* is monophyletic, meaning that all species within the genus share a common ancestor and form a distinct clade. These analyses have also helped clarify the relationships among different *Pleurotus* species (Li et al., 2020; Xu et al., 2021). For example, *P. ostreatus* and *P. pulmonarius* are closely related and often cluster together in phylogenetic trees, reflecting their similar morphological and ecological characteristics (Knop et al., 2015).

Further molecular studies have used additional genetic markers, such as mitochondrial DNA and protein-coding genes, to provide a more comprehensive understanding of the evolutionary history of *Pleurotus*. These studies have supported the monophyly of the genus and have helped identify several cryptic species, species that are morphologically similar but genetically distinct (Quattrini et al., 2019; Zeb et al., 2024).

Comparative genomics has also shed light on the genetic basis of key traits in *Pleurotus* species, such as lignin degradation and substrate utilization (Sun et al., 2023). The genomes of several *Pleurotus* species have been sequenced, revealing a repertoire of genes involved in the production of ligninolytic enzymes (Gupta et al., 2018). These enzymes, including laccases and peroxidases, enable *Pleurotus* species to efficiently break down lignin and other complex organic molecules in wood, giving them a competitive advantage in their ecological niche (Fernández-Fueyo et al., 2016).

Molecular analysis has also been used to investigate the genetic diversity within *Pleurotus* populations, revealing patterns of genetic variation and gene flow that reflect their ecological and geographical distribution. Such studies are essential for understanding the adaptive strategies of *Pleurotus* species and their potential responses to environmental changes (Zhao et al., 2016; Dai et al., 2019).

## 1.6 SECTION 3: ANATOMY OF OYSTER MUSHROOMS

### 1.6.1 Macroscopic Features

#### 1.6.1.1 Cap (Pileus): Shape, Size, and Color Variations

The cap, or pileus, is one of the most distinctive and identifiable features of oyster mushrooms (Jarial et al., 2024). The shape is usually broad and laterally attached, giving it a resemblance to an oyster shell (Bhambri et al., 2022; Darshan et al., 2024). The shape of the cap can vary somewhat depending on the specific species and growing conditions (Harris and Theriot, 2018). Generally, it starts off convex when young and flattens out as it matures, sometimes developing a wavy or undulating edge. The size of the cap can also vary widely (Ovrebo et al., 2021). In *P. ostreatus*, caps typically range from 5 to 25 cm in diameter, though they can occasionally grow larger under optimal conditions. The cap's surface is smooth to slightly rough and may become somewhat leathery with age (Sibounnavong et al., 2008; Rodriguez Estrada and Pecchia, 2017).

The color of the cap is another key characteristic that can help in identifying different species of oyster mushrooms (Zhang et al., 2021; Yang et al., 2023). The color varies significantly across species and can also be influenced by environmental factors such as light, temperature, and humidity (Koski and Galloway, 2018). For instance, *P. ostreatus* generally has a cap that ranges from light to dark gray, sometimes exhibiting a bluish tint when young, transitioning to brownish or whitish tones as it ages (Adebiyi and Yakubu, 2016).

*Pleurotus pulmonarius* often displays lighter colors, such as cream, pale gray, or light brown (Islam et al., 2009; Mshandete, 2011). This species is also known to have a slightly more elongated cap compared to *P. ostreatus*. *Pleurotus djamor*, the pink oyster mushroom, stands out with its striking pink to salmon-colored caps, which can add a vibrant splash of color to culinary dishes and cultivation setups (Shepard et al., 2008; Castanera et al., 2014; Ryu et al., 2024). *Pleurotus eryngii*, known as the king oyster mushroom, features a more robust and thicker cap that is usually tan to brown and often has a more centrally attached stem, setting it apart from other *Pleurotus* species (Seid, 2023; Tijani et al., 2024).

The cap's surface is usually smooth but can sometimes have a slightly velvety or hairy texture depending on the species and environmental conditions (Buyck et al., 2017). When fresh, the cap is somewhat moist to the touch, but it dries out and becomes more leathery as the mushroom ages. The margin of the cap is initially inrolled and becomes wavy or lobed as it matures, contributing to the characteristic appearance of the oyster mushroom (Semwal et al., 2014; Borah et al., 2019; Bhambri et al., 2022; Shuting, 2024).

Oyster mushrooms typically grow in clusters, with multiple caps often overlapping or growing in a shelf-like arrangement (Diro et al., 2021). This growth habit is especially prominent in wild settings, where they are commonly found growing on decaying hardwood trees, stumps, or logs (Kõrkjas et al., 2021). The clustered growth pattern not only aids in their identification but also maximizes spore dispersal, as the overlapping caps create a large surface area for spore release (Heinz et al., 2012).

#### 1.6.1.2 Gills (Lamellae): Arrangement and Function

The gills, or lamellae, are located on the underside of the cap (pileus) and are easily visible when the mushroom is picked or examined from below (Frank et al., 2010; Alim et al., 2023). The arrangement of the gills in oyster mushrooms is known as decurrent, meaning that the gills extend down the stem (stipe), often running a considerable distance downward (Semwal et al., 2014). This decurrent arrangement is one of the defining characteristics of the *Pleurotus* genus and helps distinguish oyster mushrooms from other gilled fungi. The gills are typically closely spaced and can be quite crowded, maximizing the surface area available for spore production (Samantha et al., 2016; Iapichino et al., 2021).

In terms of physical appearance, the gills are usually broad and somewhat thick compared to those of other mushrooms (Singh et al., 2021). Their color can vary depending on the species and age of the mushroom (Bazzicalupo et al., 2017). When young, the gills are generally white to cream-colored, and as the mushroom matures, they may take on a slightly yellowish or grayish hue. This color change can be an indicator of the mushroom's age and spore maturity (Semwal et al., 2014; Castaño et al., 2017).

The primary function of the gills is to produce and release spores, which are the reproductive units of the mushroom (Li et al., 2023). The gills provide a large surface area for the formation of basidia, which are microscopic, club-shaped cells where spores are produced. Each basidium typically produces four spores, which are then released into the environment to propagate the species (Halbwachs and Bässler, 2015; Kües and Navarro-González, 2015).

The spores are ejected from the basidia with considerable force, allowing them to fall into the air currents below the cap (Iapichino et al., 2021). The close spacing and orientation of the gills ensure that a high number of spores can be produced and dispersed efficiently (Gill et al., 2022). This efficient spore release mechanism is essential for the wide distribution and colonization capabilities of oyster mushrooms (Jarial et al., 2024).

The gill structure also plays a role in the mushroom's overall aerodynamics and spore dispersal strategy (Wang et al., 2015). The decurrent arrangement of the gills helps guide the falling spores away from the mushroom body, increasing the likelihood that they will be carried away by air currents and find a suitable substrate for germination (Halbwachs et al., 2016; Lamichhane et al., 2018).

In addition to their reproductive role, the gills of oyster mushrooms can also be used as a means of identifying different species within the *Pleurotus* genus. Variations in gill spacing, thickness, and color can provide clues to the specific species and aid in accurate identification (Kohda et al., 2015; Rodríguez-Barrera et al., 2021).

#### 1.6.1.3 Stem (Stipe): Structure and Texture

The stem, or stipe, of oyster mushrooms is another important macroscopic feature that contributes to their identification and overall morphology. While the stem of oyster mushrooms is typically less prominent than the cap, it has distinctive structural and textural characteristics that are significant for both identification and culinary use (Wong and Cheung, 2008; Ozaki et al., 2022).

Oyster mushrooms are known for their lateral or eccentric stem attachment, meaning the stem is often attached to the side of the cap rather than the center (Mishra, 2003). In some species, the stem may be very short or nearly absent, giving the mushroom a more shelf-like appearance as it grows out from the substrate. This growth habit is particularly adapted to the mushroom's natural habitat, where it typically grows on the sides of trees or logs (de Mattos-Shipley et al., 2016; Halbwachs et al., 2016).

When present, the stem of an oyster mushroom is generally cylindrical and can vary in length from a few millimeters to several centimeters, depending on the species and growing conditions (Emiru et al., 2016). For example, *P. ostreatus* often has a short, stout stem, while *P. eryngii* features a much thicker and longer stem that is more centrally attached (Chamutpong et al., 2019). This robust stem of *P. eryngii* is a notable exception within the *Pleurotus* genus and is highly prized for its culinary uses (Zervakis et al., 2014).

The texture of the oyster mushroom stem can also vary. In general, the stem is firm and fibrous, providing structural support to the mushroom (Chang et al., 2021; Ibrahim and Huda-Faujan, 2024). This fibrous nature makes the stem more resistant to breaking compared to the cap. However, the texture can change with age; young stems are typically tender and easier to chew, while older stems may become tougher and more fibrous (Shah et al., 2017; Nishinari et al., 2024).

In species like *P. ostreatus*, the stem is usually white to off-white and smooth to the touch, often matching the color of the gills and cap (Prakasam et al., 2013). Flesh inside the stem is dense and somewhat spongy, contributing to its firm texture. In contrast, *Pleurotus eryngii* has a more substantial, meaty stem that remains tender and flavorful even as it ages, making it a popular choice for culinary applications where the stem itself is a main ingredient (Leighton et al., 2010; Sangeeta et al., 2024).

The texture and structure of the stem can also impact the way the mushroom is prepared and cooked (Ketnawa and Rawdkuen, 2023). In many culinary traditions, the tender stems of young oyster mushrooms are used along with the caps in a variety of dishes (Obodai et al., 2014). However, in the case of tougher stems from older mushrooms, they may be used differently, such as being finely chopped or cooked longer to soften their texture (Zafar et al., 2020).

## 1.6.2 Microscopic Features

### 1.6.2.1 Spore-producing Structures

Oyster mushrooms, like other members of the phylum Basidiomycota, produce spores in specialized structures known as basidia (Lu et al., 2024). The basidia are club-shaped cells located on the surface of the gills (lamellae) on the underside of the mushroom cap (Kües and Navarro-González, 2015). Each basidium typically produces four spores, called basidiospores, which are borne on small projections called sterigmata (Borman and Johnson, 2023). The basidiospores are forcibly discharged from the basidia and then dispersed by air currents, aiding in the propagation of the species (Halbwachs and Bässler, 2015).

Under the microscope, the gill surface of oyster mushrooms appears densely packed with basidia, each bearing spores. The basidiospores are usually elliptical or cylindrical in shape and can vary in size depending on the species (Garnica et al., 2007; Singh et al., 2021). For instance, *P. ostreatus* spores are generally smooth, elongated, and measure approximately 8–12 μm in length and 3–4 μm in width. The spores are also characterized by their color when viewed en masse in a spore print, which for oyster mushrooms is typically white to lilac-gray (Mihilall et al., 2011; Barh et al., 2023).

### 1.6.2.2 Hyphal Network and Cellular Composition

The hyphal network is the fundamental structural component of oyster mushrooms, forming both the vegetative and reproductive parts of the fungus (Barh et al., 2022; Porter and Naleway, 2022). Hyphae are long, thread-like cells that interconnect to form a dense network known as mycelium. The mycelium spreads through the substrate, breaking down organic matter and absorbing nutrients, which fuel the growth of the mushroom (Houette et al., 2022; Huang et al., 2024).

Oyster mushroom hyphae are septate, meaning they have cross-walls called septa that divide the hyphae into individual cells (Fazenda et al., 2008; Tadiosa and Lubos, 2020). Each septum has a central pore that allows for the movement of cytoplasm, organelles, and nutrients between cells, facilitating coordinated growth and function (Das and Kunwar, 2024). The cell walls of hyphae are primarily composed of chitin, a strong and flexible polysaccharide that provides structural support (Gow et al., 2017; Cheng et al., 2024).

One of the distinctive features of the hyphal network in oyster mushrooms is the presence of clamp connections (de Mattos-Shipley et al., 2016). These small, bridge-like structures form at the septa are involved in the process of nuclear division during hyphal growth (Springstein et al., 2020). Clamp connections help ensure the even distribution of nuclei between the divided cells, maintaining the dikaryotic state (where each cell contains two genetically distinct nuclei) essential for sexual reproduction in Basidiomycota (Aanen et al., 2023).

The ability of mycelium to secrete enzymes is crucial for the decomposition of organic material (De Beeck et al., 2021). Oyster mushrooms produce a range of extracellular enzymes, including cellulases, ligninases, and hemicellulases, which break down complex polymers such as cellulose, lignin, and hemicellulose into simpler compounds that can be absorbed and utilized as nutrients (Lim et al., 2013; Abe et al., 2021; de Souza and Kawaguti, 2021). This enzymatic activity not only supports the growth of the mushroom but also plays a vital role in nutrient cycling in ecosystems (Pintarič et al., 2024).

At the cellular level, the hyphae contain typical eukaryotic organelles, including nuclei, mitochondria, endoplasmic reticulum, and Golgi apparatus, all of which are involved in cellular metabolism, energy production, and the synthesis of proteins and other molecules (Riquelme et al., 2018; Mohan et al., 2023). The cytoplasm of hyphal cells is highly dynamic, with continuous movement of organelles and vesicles facilitating growth and the secretion of enzymes (Shoji et al., 2014; Riquelme et al., 2018).

#### 1.6.2.3 Nutritional Composition

Oyster mushrooms are a nutritionally rich food source, offering a well-rounded profile of macro- and micronutrients along with various bioactive compounds (Shukla and Shrivastava, 2024). They are a good source of protein, containing around 2–3 g/100 g of fresh weight, which includes essential amino acids, making them particularly valuable for vegetarian and vegan diets (Gorissen et al., 2018). The carbohydrate content is moderate, at approximately 4–5 g/100 g, including dietary fiber that aids in digestive health. Additionally, they are low in fat, with less than 0.5 g/100 g, and the fats present are primarily unsaturated, including beneficial linoleic acid (Barceló-Coblijn and Murphy, 2009; Reynolds et al., 2019).

Oyster mushrooms are also rich in vitamins, notably the B vitamins such as niacin (B3), riboflavin (B2), and pantothenic acid (B5), and they have significant vitamin D content, especially when exposed to sunlight or UV light (Hrubša et al., 2022; Ali et al., 2024). The mineral content is also impressive, with high levels of potassium, phosphorus, iron, zinc, selenium, and copper, all of which play crucial roles in maintaining various bodily functions such as cardiovascular health, bone health, immune function, and neurological function (Godswill et al., 2020; Binesh and Venkatachalam, 2024).

Beyond these nutrients, oyster mushrooms contain several bioactive compounds that contribute to their health benefits (Yadav and Negi, 2021). These include polysaccharides like β-glucans, which enhance the immune response and have potential antitumor effects, and pleuran, known for its immunostimulatory and cholesterol-lowering properties (Khan et al., 2018; Murphy et al., 2020). Phenolic compounds, which act as antioxidants, help combat oxidative stress and inflammation. Ergothioneine, another potent antioxidant present in high levels, offers cellular protection and potential neuroprotective benefits (Zhang and Tsao, 2016; Wijesekara and Xu, 2024). Additionally, oyster mushrooms produce lovastatin, a compound that can lower cholesterol levels by inhibiting a key enzyme in cholesterol synthesis, and terpenoids, which have anti-inflammatory, anti-cancer, and antimicrobial properties (Golak-Siwulska et al., 2018; Devi et al., 2024). Lectins found in these mushrooms show promise in cancer treatment by inducing apoptosis in cancer cells, and natural antimicrobial compounds produced by oyster mushrooms can inhibit the growth of various bacteria and fungi, enhancing food safety and preservation (Mohanta et al., 2018; Kurl et al., 2024). Overall, the combination of these nutritional and bioactive components makes oyster mushrooms a highly beneficial addition to the diet (Kumar et al., 2021; Devi et al., 2024).

## 1.7 SECTION 4: LIFE CYCLE OF OYSTER MUSHROOMS

### 1.7.1 Spore Germination

#### 1.7.1.1 Conditions for Spore Viability

Oyster mushrooms begin their life cycle as spores, which are microscopic reproductive units released from the gills of mature mushrooms. For these spores to be viable and germinate, specific conditions must be met (Gupta et al., 2011; Hou et al., 2024). First, they require a suitable substrate rich in nutrients, such as decaying wood, straw, or composted agricultural waste. The substrate must be adequately moist, as water is essential for spore activation and metabolic processes. Optimal temperature ranges between 20 and 25 °C (68°F–77°F), though this can vary slightly depending on the oyster mushroom species (Elisashvili et al., 2019; Guo et al., 2024; Jarial et al., 2024). Additionally, adequate oxygen levels and a pH level between 5.5 and 6.5 are critical for successful spore germination. The environment should also be free from contaminants and competing microorganisms to ensure the spores can develop without being outcompeted or inhibited (Pinto et al., 2020; Allwood et al., 2023).

#### 1.7.1.2 Initial Stages of Growth

Once the spores are in a conducive environment, they begin the initial stages of growth. The first step is the germination of spores, which involves the absorption of water and swelling of the spore (Bressuire-Isoard et al., 2018). The spore then breaks its dormancy and starts to grow hyphae, which are thread-like structures. These hyphae extend out from the spore, seeking nutrients in the substrate (Schürg et al., 2012; Riquelme et al., 2018). As the hyphae grow and branch, they form a network called mycelium. The mycelium is the vegetative part of the fungus and can spread extensively through the substrate, absorbing nutrients and breaking down complex organic materials into simpler compounds that the fungus can use (Jones et al., 2017; Du et al., 2018).

During this initial growth phase, the mycelium must establish itself strongly in the substrate to ensure the survival and continued development of the fungus (Boddy and Hiscox, 2016; Berger et al., 2022). This phase is crucial, as a robust mycelial network will support the later stages of growth, leading to the formation of fruiting bodies (mushrooms) (Li et al., 2023). Environmental factors such as temperature, humidity, and substrate composition must be carefully controlled to support optimal mycelial growth (Aiduang et al., 2024). Once the mycelia sufficiently colonizes the substrate, it will enter the next phase of development, preparing to produce fruiting bodies under the right conditions (Gonkhom et al., 2022).

### 1.7.2 Mycelium Development

#### 1.7.2.1 Hyphal Growth and Network Formation

After spore germination, the hyphae continue to grow and extend into the substrate, creating a complex and extensive network known as mycelium (Fischer and Glass, 2019). Hyphal growth involves the elongation of the filamentous structures as they branch out and explore the substrate for nutrients (Riquelme et al., 2018). The mycelium serves as the vegetative part of the fungus, effectively acting as its root system. This network is crucial for nutrient absorption, as the mycelium secretes enzymes that break down complex organic materials in the substrate into simpler compounds (De Beeck et al., 2021; Manan et al., 2021). These compounds are then absorbed by the hyphae to fuel further growth and development. As the mycelium spreads, it forms a dense and interconnected web that optimizes the fungus's ability to colonize the substrate and gather nutrients (Klein et al., 2014; Shakir and Ahmad, 2024; Voutetaki and Mpalaskas, 2024).

#### 1.7.2.2 Role of Substrate

The substrate plays a pivotal role in the development of the mycelium. It serves as the primary source of nutrients required for the growth and proliferation of the fungus (Fleck et al., 2011; Devi et al., 2020). For oyster mushrooms, ideal substrates are those rich in lignocellulosic materials, such as hardwood sawdust, straw, or agricultural waste like cottonseed hulls and coffee grounds. These materials provide the necessary carbon, nitrogen, and other nutrients essential for fungal growth (Itoo and Reshi, 2014; Doroški et al., 2022).

The substrate must be properly prepared to support mycelial development. This often involves pasteurization or sterilization to eliminate competing microorganisms and contaminants that could inhibit mycelial growth (Aiduang et al., 2024; Grimm et al., 2024). Once the substrate is prepared, it is inoculated with mushroom spawn, which contains living mycelium. Under optimal conditions of temperature, humidity, and aeration, the mycelium will rapidly colonize the substrate. Moisture content is particularly important, as the mycelium requires a humid environment to grow effectively (Zhang et al., 2019; Landingin et al., 2020; Tacer-Caba et al., 2020).

As the mycelium develops, it continues to digest and decompose the substrate, converting it into biomass and preparing for the next stage of the mushroom life cycle: the formation of fruiting bodies (Sonnenberg et al., 2022; Silva et al., 2024). The health and density of the mycelial network directly influence the quality and yield of the mushrooms that will eventually be produced. Thus, understanding and managing the conditions for mycelium development is crucial for successful oyster mushroom cultivation (Sattar et al., 2021; Suwannarach et al., 2022).

### 1.7.3 Fruiting Body Formation

#### 1.7.3.1 Environmental Triggers

The transition from mycelial growth to fruiting body formation in oyster mushrooms is influenced by several environmental triggers (Sakamoto, 2018). Key factors include changes in temperature, humidity, light, and the availability of fresh air. Typically, a drop in temperature and an increase in humidity signal the mycelium to start forming fruiting bodies (Pelkmans et al., 2016; Hao et al., 2022). For many oyster mushroom species, optimal fruiting temperatures range from 15 to 20 °C (59 to 68 °F), with humidity levels between 85% and 95%. Light, particularly natural or artificial light with a wavelength similar to daylight, also plays a crucial role in initiating fruiting (Paradiso and Proietti, 2022; Jarial et al., 2024). Additionally, the introduction of fresh air, which increases the level of $O_2$ and reduces $CO_2$ concentration around the mycelium, is essential for triggering mushroom formation (González-Fandos et al., 2000; Yang et al., 2024).

### 1.7.4 Developmental Stages from Primordia to Mature Mushrooms

#### 1.7.4.1 Primordia Formation

Once the environmental conditions are favorable, the mycelium starts to form tiny, pinhead-sized structures known as primordia. These are the initial indicators of fruiting body development. The formation of primordia is a critical stage, as it sets the foundation for the subsequent growth of mushrooms (Kim et al., 2005; Straatsma et al., 2013).

#### 1.7.4.2 Pinhead Stage

The primordia develop into small, pinhead-sized structures called pins. These pins continue to grow as they start to differentiate into the distinct parts of the mushroom. At this stage, maintaining high humidity and adequate fresh air is crucial to support proper development (Guerin-Laguette et al., 2014; Tsikouras et al., 2016; Shamshiri et al., 2018).

#### 1.7.4.3 Baby Mushroom Stage

The pins enlarge and begin to take on the recognizable shape of mushrooms. The caps start to expand, and the stems elongate (Lu and Liaw, 2020). This stage is marked by rapid growth, and the baby mushrooms become more defined in their structure. Consistent environmental conditions are necessary to avoid deformities and ensure healthy growth (Zhang et al., 2013; Hanson and Gluckman, 2014).

#### 1.7.4.4 Mature Mushroom Stage

The mushrooms reach full maturity, with fully developed caps and stems. At this point, they are ready for harvest (Pilz and Molina, 2002). The caps of mature oyster mushrooms typically take on a distinctive, oyster-like shape, and their size can vary depending on the species and growing conditions (Akhil et al., 2023). Mature mushrooms should be harvested promptly to maintain their quality and prevent spore release, which can affect the growing environment and future yields (Zhang et al., 2018).

## 1.8 REPRODUCTIVE CYCLE

### 1.8.1 Spore Production and Dispersal

Oyster mushrooms reproduce primarily through the production and dispersal of spores (James, 2015). These spores are generated in the gills located on the underside of the mushroom cap. As the mushroom matures, millions of microscopic spores are formed and eventually released into the air (Stamets and Zwickey, 2014; Castaño et al., 2017). This process typically occurs when the environmental conditions, such as humidity and temperature, are optimal for spore survival and dispersal (Zanatta et al., 2018). The spores are dispersed by air currents and can travel considerable distances. When they land in a suitable environment with the right substrate and conditions, they can germinate and start a new life cycle (Money and Fischer, 2009; Warren et al., 2019).

### 1.8.2 Sexual versus Asexual Reproduction Mechanisms

#### 1.8.2.1 Sexual Reproduction

Sexual reproduction in oyster mushrooms involves the fusion of compatible hyphae, which are the thread-like structures that make up the mycelium (Barh et al., 2019). When two spores of opposite mating types land in proximity and germinate, their hyphae grow toward each other and fuse in a process called plasmogamy (Halbwachs and Bässler, 2015). This results in the formation of dikaryotic mycelium, containing two distinct nuclei per cell, one from each parent spore (Aanen et al., 2023). The dikaryotic mycelium grows and eventually forms fruiting bodies. Inside the fruiting bodies, karyogamy occurs, where the two nuclei fuse to form a diploid nucleus (Kües, 2000; Shin et al., 2021). This diploid nucleus undergoes meiosis to produce haploid spores, completing the sexual reproductive cycle. This genetic recombination allows for greater genetic diversity, which can enhance adaptability and resilience in changing environments (Wallen and Perlin, 2018; Garland and Curry, 2022).

#### 1.8.2.2 Asexual Reproduction

Oyster mushrooms can also reproduce asexually through a process called budding or the formation of conidia (Nmom et al., 2021). In asexual reproduction, the mycelium can produce spores without the fusion of hyphae from different mating types (Francisco et al., 2020). These asexual spores, known as conidia, are produced directly by the mycelium and are genetically identical to the parent fungus (Ojeda-López et al., 2018). Conidia can also disperse through the air and germinate in suitable conditions to form new mycelium. While asexual reproduction does not generate genetic diversity, it allows for rapid colonization and proliferation in favorable environments (Rivas et al., 2014; Yang and Kim, 2016).

## 1.9 CONCLUSION

Oyster mushrooms (*Pleurotus* spp.) exhibit a fascinating life cycle marked by distinct stages, each critical for their growth, development, and reproduction. Beginning with spore germination, these fungi require specific environmental conditions to initiate the growth of hyphae and establish a robust mycelial network. The substrate plays a pivotal role in this stage, providing the necessary nutrients and conditions for the mycelium to thrive. As the mycelium matures, environmental triggers such as changes in temperature, humidity, light, and fresh air initiate the formation of fruiting bodies. These bodies go through several developmental stages, from primordia to mature mushrooms, each requiring careful environmental management to ensure healthy growth. The reproductive cycle of oyster mushrooms is equally complex, involving both sexual and asexual reproduction mechanisms. Sexual reproduction, involving the fusion of hyphae from compatible spores, introduces genetic diversity, which enhances the adaptability and resilience of the species. Asexual reproduction, through the formation of conidia, allows for rapid colonization and propagation in favorable conditions. This dual approach ensures the widespread dispersal and survival of oyster mushrooms across various environments. In addition to their biological intricacies, oyster mushrooms are highly valued for their nutritional composition. They are rich in proteins, vitamins, minerals, and bioactive compounds, making them a nutritious and health-promoting addition to diets. The presence of compounds like β-glucans, phenolics, ergothioneine, and lovastatin contributes to their medicinal properties, including immune modulation, antioxidant protection, and cholesterol reduction. Understanding the life cycle, nutritional composition, and reproductive strategies of oyster mushrooms provides valuable insights for their cultivation and utilization. These fungi not only offer significant health benefits but also play an important role in ecological sustainability by decomposing organic matter and enriching soil health. As such, oyster mushrooms continue to be an important subject of study and cultivation in both scientific and culinary fields.

## REFERENCES

Aanen, D. K., A. van't Padje, and B. Auxier. 2023. Longevity of fungal mycelia and nuclear quality checks: a new hypothesis for the role of clamp connections in dikaryons. *Microbiology and Molecular Biology Reviews* 87(3): e00022–21.

Abe, M. M., M. C. Branciforti, and M. Brienzo. 2021. Biodegradation of hemicellulose-cellulose-starch-based bioplastics and microbial polyesters. *Recycling* 6(1): 22.

Adebiyi, A. O., and H. O. Yakubu. 2016. A survey of mushrooms in two local government areas of Ekiti state, Nigeria. *Donnish Journal of Agricultural Research* 3(2): 13–16.

Adetunji, C. O., O. T. Olaniyan, J. B. Adetunji, O. O. Osemwegie, and B. E. Ubi. 2022. African mushrooms as functional foods and nutraceuticals. In: *Fermentation and Algal Biotechnologies for the Food, Beverage and Other Bioproduct Industries*, ed. J. C. Ogbonna, S. Uzochukwu, E. G. Y. Nwoba, et al., 233–251. CRC Press, Boca Raton.

Aditya, D. K., S. Kashyap, and S. Gupta. 2024. Mushrooms as functional foods: trends and innovations. In: *Mushroom Magic: Biochemistry and Nutritional Value of Fungi*, ed. D. Kumar, S. Gupta, M. J. Ansari, et al., 16 pages. CRC Press, London.

Ahmed, N., B. Zhang, L. Deng, et al. 2024. Advancing horizons in vegetable cultivation: a journey from ageold practices to high-tech greenhouse cultivation—a review. *Frontiers in Plant Science* 15: 1357153.

Aiduang, W., K. Jatuwong, T. Luangharn, et al. 2024. A review delving into the factors influencing mycelium-based green composites (MBCs) production and their properties for long-term sustainability targets. *Biomimetics* 9(6): 337.

Akhil, G. L., S. S. Thara, N. S. Radhika, P. K. Sajeesh, and N. K. Binitha. 2023. Performance of different oyster mushrooms in five agro-ecological zones of Kerala. *Journal of Tropical Agriculture* 61(2): 207–217.

Ali, S., N. Yousaf, M. Usman, et al. 2024. *Volvariella volvacea* (paddy straw mushroom): a mushroom with exceptional medicinal and nutritional properties. *Heliyon* 10(21): e39747.

Alim, Y., W. Sidhoum, and S. Dib. 2023. First record of the edible mushroom *Lepista sordida* in Western Algerian forest: nutritional value and physicochemical parameters of mycelial culture. *Journal of Fungi* 9(8): 858.

Allwood, J. G., L. T. Wakeling, L. S. Post, and D. C. Bean. 2023. Food safety considerations in the production of traditional fermented products: Japanese rice koji and miso. *Journal of Food Safety* 43(4): e13048.

Anil Kumar, N. V., C. Quispe, J. Herrera-Bravo, et al. 2023. Potential of mushrooms bioactive for the treatment of skin diseases and disorders. *Journal of Food Biochemistry* 2023(1): 5915769.

Atri, N. S., M. Kaur, and S. Sharma. 2017. Characterization of lamellate mushrooms—an appraisal. In: *Developments in Fungal Biology and Applied Mycology*, ed. T. Satyanarayana, S. Deshmukh, and B. Johri, 471–500. Springer, Singapore.

Austin-Asomeji, I., M. O. Raimi, E. O. Mbong, A. B. Ariyo, F. G. Evans, E. G. Ekine, . . . and E. O. Akhigbe. 2024. Evaluating the anti-cancer potential of ganoderma lucidum: A critical analysis of clinical evidence and methodological advances. *Preprints*. 2024070119

Ba, D. M., S. Zhang, Y. Nishita, et al. 2023. Mushroom consumption and hyperuricemia: results from the National Institute for Longevity Sciences-Longitudinal Study of Aging and the National Health and Nutrition Examination Survey (2007–2018). *Nutrition Journal* 22(1): 62.

Bahram, M., and T. Netherway. 2022. Fungi as mediators linking organisms and ecosystems. *FEMS Microbiology Reviews* 46(2): fuab058.

Barceló-Coblijn, G., and E. J. Murphy. 2009. Alpha-linolenic acid and its conversion to longer chain n− 3 fatty acids: benefits for human health and a role in maintaining tissue n− 3 fatty acid levels. *Progress in Lipid Research* 48(6): 355–374.

Barh, A., K. Sharma, P. Bhatt, et al. 2022. Identification of key regulatory pathways of basidiocarp formation in *Pleurotus* spp. using modeling, simulation and system biology studies. *Journal of Fungi* 8(10): 1073.

Barh, A., S. Kamal, V. P. Sharma, K. Sharma, B. Kumari, and M. Nath. 2023. Identification and morpho-molecular characterization of low spore strain in oyster mushroom. *Molecular Biology Reports* 50(6): 5029–5038.

Barh, A., V. P. Sharma, S. K. Annepu, S. Kamal, S. Sharma, and P. Bhatt. 2019. Genetic improvement in *Pleurotus* (oyster mushroom): a review. *3 Biotech* 9(9): 322.

Barritt, T. W. 2015. *Long Island Food: A History from Family Farms and Oysters to Craft Spirits*. History Press Library Editions.

Bazzicalupo, A. L., B. Buyck, I. Saar, J. Vauras, D. Carmean, and M. L. Berbee. 2017. Troubles with mycorrhizal mushroom identification where morphological differentiation lags behind barcode sequence divergence. *Taxon* 66(4): 791–810.

Belova, A. 2024. Multisensory perception of exotic fruits and flavour conveyability in English. *Cognition, Communication, Discourse* 28: 7–23.

Berger, R. G., S. Bordewick, N. K. Krahe, and F. Ersoy. 2022. Mycelium vs. fruiting bodies of edible fungi—a comparison of metabolites. *Microorganisms* 10(7): 1379.

Bhambri, A., M. Srivastava, V. G. Mahale, S. Mahale, and S. K. Karn. 2022. Mushrooms as potential sources of active metabolites and medicines. *Frontiers in Microbiology* 13: 837266.

Bhatia, J. N., and A. N. Yadav. 2024. Characterization and yield performance of spawn prepared from *Hypsizygus ulmarius* (Bull.) Redhead and some *Pleurotus* species (Agaricomycetes). *Biocatalysis and Agricultural Biotechnology* 56: 103047.

Bhatt, R., U. Singh, and S. Stephenson. 2016. Wild edible mushrooms from high elevations in the Garhwal Himalaya-I. *Current Research in Environmental and Applied Mycology* 6(2): 118–131.

Bierend, D. 2021. *In Search of Mycotopia: Citizen Science, Fungi Fanatics, and the Untapped Potential of Mushrooms*. Chelsea Green Publishing Co.

Binesh, A., and K. Venkatachalam. 2024. Copper in human health and disease: a comprehensive review. *Journal of Biochemical and Molecular Toxicology* 38(11): e70052.

Birkebak, J. M., J. R. Mayor, K. M. Ryberg, and P. B. Matheny. 2013. A systematic, morphological and ecological overview of the Clavariaceae (Agaricales). *Mycologia* 105(4): 896–911.

Boddy, L., and J. Hiscox. 2016. Fungal ecology: principles and mechanisms of colonization and competition by saprotrophic fungi. *Microbiology Spectrum* 4(6): FUNK-0019–2016.

Borah, T. R., A. R. Singh, P. Paul, H. Talang, B. Kumar, and S. Hazarika. 2019. Spawn production and mushroom cultivation technology. *ICAR research complex for NEH region* 46.

Borman, A. M., and E. M. Johnson. 2023. Changes in fungal taxonomy: mycological rationale and clinical implications. *Clinical Microbiology Reviews* 36(4): e0009922.

Boukary, A. A., A. B. Olou, M. Piepenbring, and N. S. Yorou. 2024. Mushroom cultivation in tropical Africa: successes, challenges, and opportunities. *Journal of Agriculture and Food Research* 18: 101264.

Bressuire-Isoard, C., V. Broussolle, and F. Carlin. 2018. Sporulation environment influences spore properties in Bacillus: evidence and insights on underlying molecular and physiological mechanisms. *FEMS Microbiology Reviews* 42(5): 614–626.

Buyck, B., B. Duhem, K. Das, et al. 2017. Fungal biodiversity profiles 21–30. *Cryptogamie, Mycologie* 38(1): 101–146.

Carrasco, J., D. C. Zied, M. J. Navarro, F. J. Gea, and A. Pardo-Giménez. 2021. Commercial cultivation techniques of mushrooms. In: *Advances in Macrofungi: Industrial Avenues and Prospects*, ed. K. R. Sridhar, and S. K. Deshmukh, 11–40. CRC Press, Boca Raton.

Castanera, R., G. Pérez, L. López, et al. 2014. Highly expressed captured genes and cross-kingdom domains present in Helitrons create novel diversity in *Pleurotus ostreatus* and other fungi. *BMC Genomics* 15: 1–17.

Castaño, C., J. Oliva, J. Martinez de Aragon, et al. 2017. Mushroom emergence detected by combining spore trapping with molecular techniques. *Applied and Environmental Microbiology* 83(13): e0060017.

Chamutpong, S., T. Pummarin, and C. Noysang. 2019. Nutritional properties, antioxidant and anti-acetylcholinesterase activities of *Pleurotus ostreatus*. *Applied Mechanics and Materials* 891: 14–20.

Chang, S. T., and S. P. Wasser. 2012. The role of culinary-medicinal mushrooms on human welfare with a pyramid model for human health. *International Journal of Medicinal Mushrooms* 14(2): 95–134.

Chang, X., A. Yang, X. Bao, et al. 2021. An innovative structured fruit (SF) product made from litchi juice, king oyster mushroom (*Pleurotus eryngii*) and gellan gum: nutritional, textural, sensorial properties. *LWT* 152: 112344.

Chauvin-Payan, C. 2024. Wild mushrooms. Uses, consumption and popular beliefs from Europe to Eastern Siberia. *Revue d'ethnoécologie* 25.

Chen, S., H. Yao, J. Han, et al. 2010. Validation of the ITS2 region as a novel DNA barcode for identifying medicinal plant species. *PloS ONE* 5(1): e8613.

Cheng, Q., M. C. Dickwella Widanage, J. R. Yarava, et al. 2024. Molecular architecture of chitin and chitosan-dominated cell walls in zygomycetous fungal pathogens by solid-state NMR. *Nature Communications* 15(1): 8295.

Chiu, S. W., S. C. Law, M. L. Ching, K. W. Cheung, and M. J. Chen. 2000. Themes for mushroom exploitation in the 21st century: sustainability, waste management, and conservation. *The Journal of General and Applied Microbiology* 46(6): 269–282.

Chowdhary, A., S. Kathuria, K. Agarwal, and J. F. Meis. 2014. Recognizing filamentous basidiomycetes as agents of human disease: a review. *Medical Mycology* 52(8): 782–797.

Chugh, R. M., P. Mittal, N. Mp, et al. 2022. Fungal mushrooms: a natural compound with therapeutic applications. *Frontiers in Pharmacology* 13: 925387.

Coelho, M. A., G. Bakkeren, S. Sun, M. E. Hood, and T. Giraud. 2017. Fungal sex: the Basidiomycota. *Microbiology Spectrum* 5(3). https://doi.org/10.1128/microbiolspec.funk-0046-2016.

Dai, Y., Sun, L., Yin, X., et al. 2019. *Pleurotus eryngii* genomes reveal evolution and adaptation to the Gobi desert environment. *Frontiers in Microbiology* 10: 2024.

Darshan, K., S. P. Sagar, B. Vajramma, et al. 2024. Medicinal potential of Turkey tail mushroom (*Trametes versicolor*): a comprehensive review. *South African Journal of Botany* 172: 254–266.

Das, A., and A. Kunwar. 2024. Septins: structural insights, functional dynamics, and implications in health and disease. *Journal of Cellular Biochemistry* 126(1): e30660.

De Beeck, M. O., P. Persson, and A. Tunlid. 2021. Fungal extracellular polymeric substance matrices–highly specialized microenvironments that allow fungi to control soil organic matter decomposition reactions. *Soil Biology and Biochemistry* 159: 108304.

De Cianni, R., G. C. Varese, and T. Mancuso. 2023. A further step toward sustainable development: the case of the edible mushroom supply chain. *Foods* 12(18): 3433.

de Mattos-Shipley, K. M., K. L. Ford, F. Alberti, A. M. Banks, A. M. Bailey, and G. D. Foster. 2016. The good, the bad and the tasty: the many roles of mushrooms. *Studies in Mycology* 85(1): 125–157.

de Souza, T. S., and H. Y. Kawaguti. 2021. Cellulases, hemicellulases, and pectinases: applications in the food and beverage industry. *Food and Bioprocess Technology* 14(8): 1446–1477.

Devi, P. V., J. Islam, P. Narzary, D. Sharma, and F. Sultana. 2024. Bioactive compounds, nutraceutical values and its application in food product development of oyster mushroom. *Journal of Future Foods* 4(4): 335–342.

Devi, R., T. Kaur, D. Kour, K. L. Rana, A. Yadav, and A. N. Yadav. 2020. Beneficial fungal communities from different habitats and their roles in plant growth promotion and soil health. *Microbial Biosystems* 5(1): 21–47.

Diro, B. B., T. Daba, and T. G. Genemo. 2021. Production and characterization of cellulase from mushroom (*Pleurotus ostreatus*) for effective degradation of cellulose. *International Journal of Biological and Pharmaceutical Sciences Archive* 2(1): 135–150.

Doroški, A., A. Klaus, A. Režek Jambrak, and I. Djekic. 2022. Food waste originated material as an alternative substrate used for the cultivation of oyster mushroom (*Pleurotus ostreatus*): a review. *Sustainability* 14(19): 12509.

Du, H., M. Ayouz, P. Lv, and P. Perré. 2018. A lattice-based system for modeling fungal mycelial growth in complex environments. *Physica A: Statistical Mechanics and its Applications* 511: 191–206.

Dufossé, L., M. Fouillaud, and Y. Caro. 2021. Fungi and fungal metabolites for the improvement of human and animal nutrition and health. *Journal of Fungi* 7(4): 274.

Eduardo, N., S. Florencia, P. Nicolás, and G. József. 2018. Richness, species composition and functional groups in Agaricomycetes communities along a vegetation and elevational gradient in the Andean Yungas of Argentina. *Biodiversity and Conservation* 27: 1849–1871.

Elisashvili, V., E. Kachlishvili, and M. L. Chikindas. 2019. Recent advances in the physiology of spore formation for Bacillus probiotic production. *Probiotics and Antimicrobial Proteins* 11: 731–747.

Elkhateeb, W. A., G. M. Daba, P. W. Thomas, and T. C. Wen. 2019. Medicinal mushrooms as a new source of natural therapeutic bioactive compounds. *Egyptian Pharmaceutical Journal* 18(2): 88–101.

Emiru, B., K. Zenebech, and F. Kebede. 2016. Effect of substrates on the yield, yield attribute and dietary values of oyster mushroom (*Pleurotus ostreatus*) in the pastoral regions of northern Ethiopia. *African Journal of Food, Agriculture, Nutrition and Development* 16(4): 11198–11218.

Fazenda, M. L., R. Seviour, B. McNeil, and L. M. Harvey. 2008. Submerged culture fermentation of "higher fungi": the macrofungi. *Advances in Applied Microbiology* 63: 33–103.

Fernández-Fueyo, E., F. J. Ruiz-Dueñas, M. F. López-Lucendo, et al. 2016. A secretomic view of woody and nonwoody lignocellulose degradation by *Pleurotus ostreatus*. *Biotechnology for Biofuels* 9: 1–18.

Fischer, M. S., and N. L. Glass. 2019. Communicate and fuse: how filamentous fungi establish and maintain an interconnected mycelial network. *Frontiers in Microbiology* 10: 619.

Fleck, C. B., F. Schöbel, and M. Brock. 2011. Nutrient acquisition by pathogenic fungi: nutrient availability, pathway regulation, and differences in substrate utilization. *International Journal of Medical Microbiology* 301(5): 400–407.

Földi, C., Z. Merényi, B. Balázs, et al. 2024. Snowball: a novel gene family required for developmental patterning of fruiting bodies of mushroom-forming fungi (Agaricomycetes). *mSystems* 9(3): e0120823.

Francisco, C. S., M. M. Zwyssig, and J. Palma-Guerrero. 2020. The role of vegetative cell fusions in the development and asexual reproduction of the wheat fungal pathogen Zymoseptoria tritici. *BMC Biology* 18: 1–16.

Frank, J. L., R. A. Coffan, and D. Southworth. 2010. Aquatic gilled mushrooms: *Psathyrella* fruiting in the Rogue River in southern Oregon. *Mycologia* 102(1): 93–107.

Furan, M. A., G. Genli, M. Yıldız, and H. J. Sun. 2024. Exploring secondary structures within the nuclear ribosomal DNA internal transcribed spacers (nr DNA ITS) for enhanced phylogenetic insights in *Medicago sativa* (alfalfa). *Plant Biotechnology Reports* 18(4): 551–561.

Gao, Y., Z. Wu, W. Li, et al. 2023. Expanding the valorization of waste mushroom substrates in agricultural production: progress and challenges. *Environmental Science and Pollution Research* 30(2): 2355–2373.

Garland, S., and H. A. Curry. 2022. Turning promise into practice: crop biotechnology for increasing genetic diversity and climate resilience. *PLoS Biology* 20(7): e3001716.

Garnica, S., M. Weiss, G. Walther, and F. Oberwinkler. 2007. Reconstructing the evolution of agarics from nuclear gene sequences and basidiospore ultrastructure. *Mycological Research* 111(9): 1019–1029.

Geikowsky, E., and B. Aksak. 2020. Bioinspired fibrillar adhesives with shape-controlled off-center caps for switchable and directional adhesion. *Bioinspiration and Biomimetics* 15(5): 056007.

Gepshtein, Y., G. A. Plotnikoff, and K. Watanabe. 2008. Kampo in women's health: Japan's traditional approach to premenstrual symptoms. *The Journal of Alternative and Complementary Medicine* 14(4): 427–435.

Ghazaryan, K., A. Singh, V. D. Rajput, et al. 2024. Mushrooms in food security and environmental management: sustainable solutions for a changing world. In: *Sustainable Agriculture: Nanotechnology, Biotechnology, Management and Food Security*, ed. V. D. Rajput, A. Singh, K. Ghazaryan, T. M. Minkina, and A. R. M. Al-Tawaha, 233. De Gruyter.

Gill, N. S., M. G. Turner, C. D. Brown, et al. 2022. Limitations to propagule dispersal will constrain postfire recovery of plants and fungi in western coniferous forests. *BioScience* 72(4): 347–364.

Godswill, A. G., I. V. Somtochukwu, A. O. Ikechukwu, and E. C. Kate. 2020. Health benefits of micronutrients (vitamins and minerals) and their associated deficiency diseases: a systematic review. *International Journal of Food Sciences* 3(1): 1–32.

Golak-Siwulska, I., A. Kałużewicz, T. Spiżewski, M. Siwulski, and K. Sobieralski. 2018. Bioactive compounds and medicinal properties of Oyster mushrooms (*Pleurotus* sp.). *Folia Horticulturae* 30(2): 191–201.

Gonkhom, D., T. Luangharn, K. D. Hyde, M. Stadler, and N. Thongklang. 2022. Optimal conditions for mycelial growth of medicinal mushrooms belonging to the genus *Hericium*. *Mycological Progress* 21(9): 82.

González, J. A., and J. R. Vallejo. 2023. The use of shells of marine molluscs in Spanish ethnomedicine: a historical approach and present and future perspectives. *Pharmaceuticals* 16(10): 1503.

González-Fandos, E., M. Giménez, C. Olarte, S. Sanz, and A. Simón. 2000. Effect of packaging conditions on the growth of micro-organisms and the quality characteristics of fresh mushrooms (*Agaricus bisporus*) stored at inadequate temperatures. *Journal of Applied Microbiology* 89(4): 624–632.

Gorissen, S. H., J. J. Crombag, J. M. Senden, et al. 2018. Protein content and amino acid composition of commercially available plant-based protein isolates. *Amino Acids* 50: 1685–1695.

Gotvaldová, K., K. Hájková, J. Borovička, R. Jurok, P. Cihlářová, and M. Kuchař. 2021. Stability of psilocybin and its four analogs in the biomass of the psychotropic mushroom Psilocybe cubensis. *Drug Testing and Analysis* 13(2): 439–446.

Gow, N. A., J. P. Latge, and C. A. Munro. 2017. The fungal cell wall: structure, biosynthesis, and function. *Microbiology Spectrum* 5(3): 10–1128.

Grigson, J. 2008. *The Mushroom Feast*. Grub Street Cookery.

Grimm, D., E. Sonntag, and G. Rahmann. 2024. Evaluation of different pasteurization and sterilization methods for oyster mushroom substrates. *Journal of Microbiology, Biotechnology and Food Sciences* 13(5): e10428.

Guerin-Laguette, A., N. Cummings, R. C. Butler, et al. 2014. *Lactarius deliciosus* and *Pinus radiata* in New Zealand: towards the development of innovative gourmet mushroom orchards. *Mycorrhiza* 24(7): 511–523.

Guo, J., Y. Zhang, J. Fang, et al. 2024. Reduction and reuse of forestry and agricultural bio-waste through innovative green utilization approaches: a review. *Forests* 15(8): 1372.

Gupta, B., B. P. Niranjan Reddy, and A. S. Kotasthane. 2011. Molecular characterization and mating type analysis of oyster mushroom (*Pleurotus* spp.) using single basidiospores for strain improvement. *World Journal of Microbiology and Biotechnology* 27: 1–9.

Gupta, D. K., M. Rühl, B. Mishra, et al. 2018. The genome sequence of the commercially cultivated mushroom *Agrocybe aegerita* reveals a conserved repertoire of fruiting-related genes and a versatile suite of biopolymer-degrading enzymes. *BMC Genomics* 19: 1–13.

Gupta, K. K., S. Agarwal, A. Kushwaha, et al. 2017. Oyster mushroom: a rich source of antioxidants. *Incredible World of Biotechnology* 1: 43–57.

Guragain, D. P., B. Shrestha, and I. Bajracharya. 2024. A low-cost centralized IoT ecosystem for enhancing oyster mushroom cultivation. *Journal of Agriculture and Food Research* 15: 100952.

Halbwachs, H., and C. Bässler. 2015. Gone with the wind–a review on basidiospores of lamellate agarics. *Mycosphere* 6(1): 78–112.

Halbwachs, H., J. Simmel, and C. Bässler. 2016. Tales and mysteries of fungal fruiting: how morphological and physiological traits affect a pileate lifestyle. *Fungal Biology Reviews* 30(2): 36–61.

Hanlin, R. T. 2004. Mushrooms of West Virginia and the Central Appalachians. *Native Plants Journal* 5(2): 196–197.

Hanson, M. A., and P. D. Gluckman. 2014. Early developmental conditioning of later health and disease: physiology or pathophysiology? *Physiological Reviews* 94(4): 1027–1076.

Hao, H., J. Zhang, Q. Wang, et al. 2022. Transcriptome and differentially expressed gene profiles in mycelium, primordium and fruiting body development in *Stropharia rugosoannulata*. *Genes* 13(6): 1080.

Harris, L. K., and J. A. Theriot. 2018. Surface area to volume ratio: a natural variable for bacterial morphogenesis. *Trends in Microbiology* 26(10): 815–832.

Heinz, E., T. A. Williams, S. Nakjang, et al. 2012. The genome of the obligate intracellular parasite *Trachipleistophora hominis*: new insights into microsporidian genome dynamics and reductive evolution. *PLoS Pathogens* 8(10): e1002979.

Hobbs, C. 2023. The health and clinical benefits of medicinal fungi. In: *Biochemical Engineering and Biotechnology of Medicinal Mushrooms*, ed. M. Berovic, and J.-J. Zhong, 285–356. Springer, Cham.

Horwath, W., and E. A. Paul. 2015. Carbon cycling: the dynamics and formation of organic matter. In: *Soil Microbiology, Ecology and Biochemistry*, ed. E. A. Paul, 339–382. Academic Press.

Hou, Z., Z. Meretenyi, Y. Yang, et al. 2024. A new regulator of sporulation sheds light on spore morphogenesis and ballistospory in mushroom-forming fungi. *bioRxiv* 2024–07.

Houette, T., C. Maurer, R. Niewiarowski, and P. Gruber. 2022. Growth and mechanical characterization of mycelium-based composites towards future bioremediation and food production in the material manufacturing cycle. *Biomimetics* 7(3): 103.

Hrubša, M., T. Siatka, I. Nejmanová, et al. 2022. Biological properties of vitamins of the B-complex, part 1: vitamins B1, B2, B3, and B5. *Nutrients* 14(3): 484.

Huang, Z., Y. Wei, and S. A. Hadigheh. 2024. Variations in the properties of engineered mycelium-bound composites (MBCs) under different manufacturing conditions. *Buildings* 14(1): 155.

Iapichino, M., Y. W. Wang, S. Gentry, A. Pringle, and A. Seminara. 2021. A precise relationship among Buller's drop, ballistospore, and gill morphologies enables maximum packing of spores within gilled mushrooms. *Mycologia* 113(2): 300–311.

Ibrahim, H. S. S., and N. Huda-Faujan. 2024. Physicochemical properties and nutritional composition of chicken patties with grey Oyster mushroom stems. *Malaysian Journal of Science Health and Technology* 10(1): 50–57.

Ioffe, D. 2024. Griby i Mukhi: a historical contextualization of the esoteric mushroom religion of Moscow conceptualism: fungal erotic imagery of entheogens and insects. *Religions* 15(7): 777.

Islam, M. Z., M. H. Rahman, and F. Hafiz. 2009. Cultivation of oyster mushroom (*Pleurotus flabellatus*) on different substrates. *International Journal of Sustainable Crop Production* 4(1): 45–48.

Itoo, Z. A., and Z. A. Reshi. 2014. Effect of different nitrogen and carbon sources and concentrations on the mycelial growth of ectomycorrhizal fungi under *in-vitro* conditions. *Scandinavian Journal of Forest Research* 29(7): 619–628.

Jalinik, M., T. Pawłowicz, P. Borowik, and T. Oszako. 2024. Mushroom picking as a special form of recreation and tourism in woodland areas—a case study of Poland. *Forests* 15(3): 573.

James, T. Y. 2015. Why mushrooms have evolved to be so promiscuous: insights from evolutionary and ecological patterns. *Fungal Biology Reviews* 29(3–4): 167–178.

Jana, P., and K. Acharya. 2022. Mushroom: a new resource for anti-angiogenic therapeutics. *Food Reviews International* 38(1): 88–109.

Jarial, R. S., K. Jarial, and J. N. Bhatia. 2024. Comprehensive review on oyster mushroom species (Agaricomycetes): morphology, nutrition, cultivation and future aspects. *Heliyon* 10(5): e26539.

Jayaraman, S., B. Yadav, R. C. Dalal, et al. 2024. Mushroom farming: a review focusing on soil health, nutritional security and environmental sustainability. *Farming System* 2(3): 100098.

Jones, M., T. Huynh, C. Dekiwadia, F. Daver, and S. John. 2017. Mycelium composites: a review of engineering characteristics and growth kinetics. *Journal of Bionanoscience* 11(4): 241–257.

Jongman, M., K. B. Khare, and D. Loeto. 2018. Oyster mushroom cultivation at different production systems: a review. *European Journal of Biomedical and Pharmaceutical Sciences* 5(5): 72–79.

Keller, H. W., S. E. Everhart, and C. M. Kilgore. 2024. The Myxomycetes: nature's quick-change artists. *American Scientist* 112(6): 352–359.

Ketnawa, S., and S. Rawdkuen. 2023. Properties of texturized vegetable proteins from edible mushrooms by using single-screw extruder. *Foods* 12(6): 1269.

Khan, A., W. Murad, S. Ali, et al. 2024. Contribution of mushroom farming to mitigating food scarcity: current status, challenges and potential future prospects in Pakistan. *Heliyon* 10(23): e40362.

Khan, A. A., A. Gani, F. A. Khanday, and F. A. Masoodi. 2018. Biological and pharmaceutical activities of mushroom β-glucan discussed as a potential functional food ingredient. *Bioactive Carbohydrates and Dietary Fibre* 16: 1–13.

Khatua, S., S. Paul, and K. Acharya. 2013. Mushroom as the potential source of new generation of antioxidant: a review. *Research Journal of Pharmacy and Technology* 6(5): 496–505.

Kim, Y. K., C. L. Xiao, and J. D. Rogers. 2005. Influence of culture media and environmental factors on mycelial growth and pycnidial production of Sphaeropsis pyriputrescens. *Mycologia* 97(1): 25–32.

Klein, O. I., E. P. Isakova, Y. I. Deryabina, et al. 2014. Humic substances enhance growth and respiration in the basidiomycetes Trametes maxima under carbon limited conditions. *Journal of Chemical Ecology* 40: 643–652.

Knop, D., O. Yarden, and Y. Hadar. 2015. The ligninolytic peroxidases in the genus *Pleurotus*: divergence in activities, expression, and potential applications. *Applied Microbiology and Biotechnology* 99: 1025–1038.

Kohda, M., L. A. Jordan, T. Hotta, et al. 2015. Facial recognition in a group-living cichlid fish. *PLoS ONE* 10(11): e0142552.

Kõrkjas, M., L. Remm, and A. Lõhmus. 2021. Development rates and persistence of the microhabitats initiated by disease and injuries in live trees: a review. *Forest Ecology and Management* 482: 118833.

Koski, M. H., and L. F. Galloway. 2018. Geographic variation in pollen color is associated with temperature stress. *New Phytologist* 218(1): 370–379.

Kües, U. 2000. Life history and developmental processes in the basidiomycete *Coprinus cinereus*. *Microbiology and Molecular Biology Reviews* 64(2): 316–353.

Kües, U., W. Khonsuntia, S. Subba, and B. Dörnte. 2018. Volatiles in communication of Agaricomycetes. *Physiology and Genetics* 149–212.

Kües, U., and M. Navarro-González. 2015. How do Agaricomycetes shape their fruiting bodies? 1. Morphological aspects of development. *Fungal Biology Reviews* 29(2): 63–97.

Kumar, K., R. Mehra, R. P. Guiné, et al. 2021. Edible mushrooms: a comprehensive review on bioactive compounds with health benefits and processing aspects. *Foods* 10(12): 2996.

Kurl, S., S. Kaur, N. Mittal, and G. Kaur. 2024. Mushrooms and colorectal cancer: unveiling mechanistic insights and therapeutic innovations. *Phytotherapy Research* 39(1): 480–493.

Lamichhane, J. R., P. Debaeke, C. Steinberg, M. P. You, M. J. Barbetti, and J. N. Aubertot. 2018. Abiotic and biotic factors affecting crop seed germination and seedling emergence: a conceptual framework. *Plant and Soil* 432: 1–28.

Landingin, H. R. R., B. E. Francisco, R. M. R. Dulay, S. Kalaw, and R. Reyes. 2020. Optimization of culture conditions for mycelial growth and basidiocarp production of Cyclocybe cylindracea (Maire). *CLSU International Journal of Science & Technology* 4(1): 1–17.

Leighton, C. S., H. C. Schönfeldt, and R. Kruger. 2010. Quantitative descriptive sensory analysis of five different cultivars of sweet potato to determine sensory and textural profiles. *Journal of Sensory Studies* 25(1): 2–18.

Li, C., and S. Xu. 2022. Edible mushroom industry in China: current state and perspectives. *Applied Microbiology and Biotechnology* 106(11): 3949–3955.

Li, J., L. H. Han, X. B. Liu, Z. W. Zhao, and Z. L. Yang. 2020. The saprotrophic *Pleurotus ostreatus* species complex: late Eocene origin in East Asia, multiple dispersal, and complex speciation. *IMA Fungus* 11: 1–21.

Li, Q., X. Ying, Y. Yang, and W. Gao. 2024. Genetic diversity and genome-wide association study of *Pleurotus pulmonarius* Germplasm. *Agriculture* 14(11): 2023.

Li, Y., Y. Feng, Y. Shang, H. Xu, R. Xia, Z. Hou, S. Pan, L. Li, Y. Bian, J. Zhu, and Z. Wang. 2023. Sexual spores in edible mushroom: bioactive components, discharge mechanisms and effects on fruiting bodies quality. *Food Science and Human Wellness* 12(6): 2111–2123.

Lim, S. H., Y. H. Lee, and H. W. Kang. 2013. Efficient recovery of lignocellulolytic enzymes of spent mushroom compost from oyster mushrooms, *Pleurotus* spp., and potential use in dye decolorization. *Mycobiology* 41(4): 214–220.

Linder, H. P., C. E. Lehmann, S. Archibald, C. P. Osborne, and D. M. Richardson. 2018. Global grass (Poaceae) success underpinned by traits facilitating colonization, persistence and habitat transformation. *Biological Reviews* 93(2): 1125–1144.

Loc, N. Q., T. K. L. Bui, N. M. Tu, et al. 2024. Enhancing circular economy in the mushroom production chain: systematic literature review and field study in the Central Highlands of Vietnam. *The Journal of Solid Waste Technology and Management* 50(4): 689–710.

Lu, C. P., and J. J. Liaw. 2020. A novel image measurement algorithm for common mushroom caps based on convolutional neural network. *Computers and Electronics in Agriculture* 171: 105336.

Lu, M. L., G. H. Yuan, H. Rehemujiang, et al. 2024. Effects of spent substrate of oyster mushroom (*Pleurotus ostreatus*) on ruminal fermentation, microbial community and growth performance in Hu sheep. *Frontiers in Microbiology* 15: 1425218.

Mahari, W. A. W., W. Peng, W. L. Nam, et al. 2020. A review on valorization of oyster mushroom and waste generated in the mushroom cultivation industry. *Journal of Hazardous Materials* 400: 123156.

Malik, N. A., J. Kumar, M. S. Wani, Y. R. Tantray, and T. Ahmad. 2021. Role of mushrooms in the bioremediation of soil. Microbiota and biofertilizers, Vol 2. *Ecofriendly Tools for Reclamation of Degraded Soil Environs* 77–102.

Manan, S., M. W. Ullah, M. Ul-Islam, O. M. Atta, and G. Yang. 2021. Synthesis and applications of fungal mycelium-based advanced functional materials. *Journal of Bioresources and Bioproducts* 6(1): 1–10.

Manoharachary, C., K. Sridhar, R. Singh, et al. 2005. Fungal biodiversity: distribution, conservation and prospecting of fungi from India. *Current Science* 58–71.

Martín, C., G. I. Zervakis, S. Xiong, G. Koutrotsios, and K. O. Strætkvern. 2023. Spent substrate from mushroom cultivation: exploitation potential toward various applications and value-added products. *Bioengineered* 14(1): 2252138.

Martín, J. A., and R. López. 2023. Biological deterioration and natural durability of wood in Europe. *Forests* 14(2): 283.

Merényi, Z., M. Virágh, E. Gluck-Thaler, et al. 2022. Gene age shapes the transcriptional landscape of sexual morphogenesis in mushroom-forming fungi (Agaricomycetes). *Elife* 11: e71348.

Mihilall, Y., A. Mudhoo, and R. Mohee. 2011. Development of a new substrate for the cultivation of the *Pleurotus sajor-caju* mushroom through controlled composting. *Dynamic Soil Dynamic Plant* 5: 82–92.

Mingyi, Y., T. Belwal, H. P. Devkota, L. Li, and Z. Luo. 2019. Trends of utilizing mushroom polysaccharides (MPs) as potent nutraceutical components in food and medicine: a comprehensive review. *Trends in Food Science & Technology* 92: 94–110.

Mishra, R. 2003. Some edible oyster mushrooms of Arunachal Pradesh. *Environment, Pollution and Management* 349.

Młodzińska, E. 2009. Survey of plant pigments: molecular and environmental determinants of plant colors. *Acta Biologica Cracovienca Series Botanica* 51(1): 7–16.

Mohan, A. G., B. Calenic, N. A. Ghiurau, R. M. Duncea-Borca, A. E. Constantinescu, and I. Constantinescu. 2023. The Golgi apparatus: a voyage through time, structure, function and implication in neurodegenerative disorders. *Cells* 12(15): 1972.

Mohanta, Y. K., D. Nayak, K. Biswas, et al. 2018. Silver nanoparticles synthesized using wild mushroom show potential antimicrobial activities against food borne pathogens. *Molecules* 23(3): 655.

Mohapatra, D., N. R. Singh, and S. K. Rath. 2015. Methods for identification and conservation of macro fungus: a probable global food. *Methods* 16–23.

Money, N. P. 2023. Goldilocks mushrooms: how ballistospory has shaped basidiomycete evolution. *Fungal Biology* 127(4): 975–984.

Money, N. P., and M. W. Fischer. 2009. Biomechanics of spore release in phytopathogens. *The Mycota* 5: 115–133.

Mshandete, A. M. 2011. Cultivation of *Pleurotus* HK-37 and *Pleurotus sapidus* (oyster mushrooms) on cattail weed (*Typha domingesis*) substrate in Tanzania. *International Journal of Research in Biological Sciences* 1(3): 35–44.

Murphy, E. J., E. Rezoagli, I. Major, N. J. Rowan, and J. G. Laffey. 2020. β-glucan metabolic and immunomodulatory properties and potential for clinical application. *Journal of Fungi* 6(4): 356.

Nagy, L. G., P. J. Vonk, M. Künzler, et al. 2023. Lessons on fruiting body morphogenesis from genomes and transcriptomes of Agaricomycetes. *Studies in Mycology* 104(1): 1–85.

Naranjo-Ortiz, M. A., and T. Gabaldón. 2019. Fungal evolution: major ecological adaptations and evolutionary transitions. *Biological Reviews* 94(4): 1443–1476.

Naranjo-Ortiz, M. A., and T. Gabaldón. 2020. Fungal evolution: cellular, genomic and metabolic complexity. *Biological Reviews* 95(5): 1198–1232.

Niego, A. G., S. Rapior, N. Thongklang, et al. 2021. Macrofungi as a nutraceutical source: promising bioactive compounds and market value. *Journal of Fungi* 7(5): 397.

Nishinari, K., M. A. Peyron, N. Yang, et al. 2024. The role of texture in the palatability and food oral processing. *Food Hydrocolloids* 147: 109095.

Nmom, F. W., L. O. Amadi, and N. N. Ngerebara. 2021. Influences of light regimes on reproduction, germination, pigmentation, pathogenesis and overall development of a variety of filamentous fungi–a review. *Asian Journal of Biology* 11(3): 25–34.

Oberwinkler, F., and R. Bauer. 2018. Ultrastructure in basidiomycetes–requirement for function. *Biodiversity and Ecology of Fungi, Lichens, and Mosses* 381–418.

Obodai, M., I. C. Ferreira, A. Fernandes, et al. 2014. Evaluation of the chemical and antioxidant properties of wild and cultivated mushrooms of Ghana. *Molecules* 19(12): 19532–19548.

Ojeda-López, M., W. Chen, C. E. Eagle, et al. 2018. Evolution of asexual and sexual reproduction in the aspergilli. *Studies in Mycology* 91(1): 37–59.

Oloke, J. K. 2017. Oyster mushroom (*Pleurotus* species); a natural functional food. *The Journal of Microbiology, Biotechnology and Food Sciences* 7(3): 254.

Ovrebo, C. L., R. E. Halling, K. W. Hughes, and M. Kuo. 2021. New species and records of *Tricholoma* (Agaricales: Tricholomataceae) sections Genuina and Megatricholoma from Costa Rica and United States. *Journal of the Botanical Research Institute of Texas* 15(2): 545–557.

Ozaki, Y., T. Aimi, and N. Shimomura. 2022. Detection of autophagy-related structures in fruiting bodies of edible mushroom, *Pleurotus ostreatus*. *Microscopy* 71(4): 222–230.

Pandey, V. V., A. Kumari, M. Kumar, J. Saxena, C. Kainthola, and A. Pandey. 2018. Mushroom cultivation: substantial key to food security. *Journal of Applied and Natural Science* 10(4): 1325–1331.

Pánek, M., L. Wiesnerová, I. Jablonský, D. Novotný, and M. Tomšovský. 2019. What is cultivated oyster mushroom? Phylogenetic and physiological study of *Pleurotus ostreatus* and related taxa. *Mycological Progress* 18: 1173–1186.

Paradiso, R., and S. Proietti. 2022. Light-quality manipulation to control plant growth and photomorphogenesis in greenhouse horticulture: the state of the art and the opportunities of modern LED systems. *Journal of Plant Growth Regulation* 41(2): 742–780.

Parkes, M., R. Panelli, and P. Weinstein. 2003. Converging paradigms for environmental health theory and practice. *Environmental Health Perspectives* 111(5): 669–675.

Pelkmans, J. F., L. G. Lugones, and H. A. Wösten. 2016. 15 fruiting body formation in basidiomycetes. *Growth, Differentiation and Sexuality* 387–405.

Pérez-Moreno, J., A. Guerin-Laguette, A. C. Rinaldi, et al. 2021. Edible mycorrhizal fungi of the world: what is their role in forest sustainability, food security, biocultural conservation and climate change? *Plants, People, Planet* 3(5): 471–490.

Phan, C. W., J. K. Wang, E. Y. Y. Tan, Yet al. 2019. Giant oyster mushroom, *Pleurotus giganteus* (Agaricomycetes): current status of the cultivation methods, chemical composition, biological, and health-promoting properties. *Food Reviews International* 35(4): 324–341.

Pilz, D., and R. Molina. 2002. Commercial harvests of edible mushrooms from the forests of the Pacific Northwest United States: issues, management, and monitoring for sustainability. *Forest Ecology and Management* 155(1–3): 3–16.

Pintarič, M., A. Štuhec, E. Tratnik, and T. Langerholc. 2024. Spent mushroom substrate improves microbial quantities and enzymatic activity in soils of different farming systems. *Microorganisms* 12(8): 1521.

Pinto, C. A., S. A. Moreira, L. G. Fidalgo, R. S. Inácio, F. J. Barba, and J. A. Saraiva. 2020. Effects of high-pressure processing on fungi spores: factors affecting spore germination and inactivation and impact on ultrastructure. *Comprehensive Reviews in Food Science and Food Safety* 19(2): 553–573.

Porter, D. L., and S. E. Naleway. 2022. Hyphal systems and their effect on the mechanical properties of fungal sporocarps. *Acta Biomaterialia* 145: 272–282.

Prakasam, V., G. Thiribhuvanamala, and P. A. Devi. 2013. Open field cultivation of oyster mushroom in bamboo plantations. *Mushroom Research* 22(1).

Psurtseva, N. V., A. A. Kiyashko, S. V. Senik, N. V. Shakhova, and N. V. Belova. 2023. The conservation and study of Macromycetes in the Komarov Botanical Institute Basidiomycetes Culture Collection—their taxonomical diversity and biotechnological prospects. *Journal of Fungi* 9(12): 1196.

Quattrini, A. M., T. Wu, K. Soong, M. S. Jeng, Y. Benayahu, and C. S. McFadden. 2019. A next generation approach to species delimitation reveals the role of hybridization in a cryptic species complex of corals. *BMC Evolutionary Biology* 19: 1–19.

Raghukumar, S., and S. Raghukumar. 2017. Fungi: characteristics and classification. Fungi in coastal and oceanic marine ecosystems: *Marine Fungi* 1–15.

Rahman, H., M. O. Faruq, T. B. A. Hai, et al. 2022. IoT enabled mushroom farm automation with machine learning to classify toxic mushrooms in Bangladesh. *Journal of Agriculture and Food Research* 7: 100267.

Raman, J., K. Y. Jang, Y. L. Oh, et al. 2021. Cultivation and nutritional value of prominent *Pleurotus* spp.: an overview. *Mycobiology* 49(1): 1–14.

Rathod, M. G. 2023. Mushroom farming: exploring varieties, cultivation strategies, and endless possibilities. *Research & Reviews in Biotechnology & Biosciences* 10(2): 1–11.

Rathod, M. G., R. B. Gadade, G. M. Thakur, and A. P. Pathak. 2021. Oyster mushroom: cultivation, bioactive significance and commercial status. *Frontiers in Life Science* 2: 21.

Ray, P., S. Kundu, and D. Paul. 2024. Exploring the therapeutic properties of Chinese mushrooms with a focus on their anti-cancer effects: a systemic review. *Pharmacological Research-Modern Chinese Medicine* 100433.

Reynolds, A., J. Mann, J. Cummings, N. Winter, E. Mete, and L. Te Morenga. 2019. Carbohydrate quality and human health: a series of systematic reviews and meta-analyses. *The Lancet* 393(10170): 434–445.

Riquelme, M., J. Aguirre, S. Bartnicki-García, et al. 2018. Fungal morphogenesis, from the polarized growth of hyphae to complex reproduction and infection structures. *Microbiology and Molecular Biology Reviews* 82(2): 10–1128.

Rivas, F., P. Nuñez, T. Jackson, and N. Altier. 2014. Effect of temperature and water activity on mycelia radial growth, conidial production and germination of *Lecanicillium* spp. isolates and their virulence against *Trialeurodes vaporariorum* on tomato plants. *BioControl* 59(1): 99–109.

Rodriguez Estrada, A. E., and J. Pecchia. 2017. Cultivation of *Pleurotus ostreatus*. In: *Edible and Medicinal Mushrooms: Technology and Applications*, ed. C. Z. Diego, and A. Pardo-Giménez, 339–360. Wiley-Blackwell.

Rodríguez-Barrera, T. M., M. Téllez-Téllez, J. E. Sánchez, et al. 2021. Edible mushrooms of the genus *Pleurotus* as biocontrol agents of parasites of importance for livestock. *Scientia Fungorum* 52: e1375.

Romine, S. 2018. *Cooking with Healing Mushrooms: 150 Delicious Adaptogen-Rich Recipes that Boost Immunity, Reduce Inflammation & Promote Whole Body Health*. Simon and Schuster.

Ryu, J. S., B. Park, A. F. van Peer, K. S. Na, and S. H. Lee. 2024. Quantitative trait loci analysis for molecular markers linked to agricultural traits of *Pleurotus ostreatus*. *Plos ONE* 19(8): e0308832.

Sakamoto, Y. 2018. Influences of environmental factors on fruiting body induction, development and maturation in mushroom-forming fungi. *Fungal Biology Reviews* 32(4): 236–248.

Samantha, C. K., E. Peter, J. Chen, L. M. Guojie, and R. Zhao. 2016. Correct names of two cultivated mushrooms from the genus *Pleurotus* in China. *Phytotaxa* 260(1): 036–046.

Sangeeta, S., D. Sharma, S. Ramniwas, R. Mugabi, J. Uddin, and G. A. Nayik. 2024. Revolutionizing mushroom processing: innovative techniques and technologies. *Food Chemistry: X* 101774.

Sarkar, J., A. Naskar, A. Nath, B. Gangopadhyay, E. Tarafdar, D. Das, and K. Acharya. 2024. Innovative utilization of harvested mushroom substrate for green synthesis of silver nanoparticles: a multi–response optimization approach. *Environmental Research* 248: 118297.

Sattar, M. N., S. Iftikhar, F. Saeed, et al. 2021. Mushroom farm design, management of pests, and control of diseases. In: *Mushrooms: Agaricus Bisporus*, ed. Y. N. Sassine, 297. CABI Publishing.

Schmidt, O. 2007. Indoor wood-decay basidiomycetes: damage, causal fungi, physiology, identification and characterization, prevention and control. *Mycological Progress* 6(4): 261–279.

Schulp, C. J., W. Thuiller, and P. H. Verburg. 2014. Wild food in Europe: a synthesis of knowledge and data of terrestrial wild food as an ecosystem service. *Ecological Economics* 105: 292–305.

Schürg, T., U. Brandt, C. Adis, and A. Fleißner. 2012. The *Saccharomyces cerevisiae* BEM 1 homologue in *Neurospora crassa* promotes co-ordinated cell behaviour resulting in cell fusion. *Molecular Microbiology* 86(2): 349–366.

Seethapathy, P., S. Sankaralingam, I. K. Muniraj, M. Perumal, and N. Pandurangan. 2023. Mass multiplication, economic analysis, and marketing of *Ganoderma* sp. (Reishi Mushroom). *Food Microbiology Based Entrepreneurship* 89–113.

Seid, A. E. 2023. Mushroom supplementation in promoting health and performance of poultry. *The Open Agriculture Journal* 17(1).

Semwal, K. C., S. L. Stephenson, V. K. Bhatt, and R. P. Bhatt. 2014. Edible mushrooms of the Northwestern Himalaya, India: a study of indigenous knowledge, distribution and diversity. *Mycosphere* 5(3): 440–461.

Shah, D. U., T. P. Reynolds, and M. H. Ramage. 2017. The strength of plants: theory and experimental methods to measure the mechanical properties of stems. *Journal of Experimental Botany* 68(16): 4497–4516.

Shakir, M. A., and M. I. Ahmad. 2024. Bioproduct advances: insight into failure factors in mycelium composite fabrication. *Biofuels, Bioproducts and Biorefining* 18(5): 1739–1754.

Shamshiri, R. R., J. W. Jones, K. R. Thorp, D. Ahmad, H. C. Man, and S. Taheri. 2018. Review of optimum temperature, humidity, and vapour pressure deficit for microclimate evaluation and control in greenhouse cultivation of tomato: a review. *International Agrophysics* 32(2): 287–302.

Sharma, R. 2017. Ectomycorrhizal mushrooms: their diversity, ecology and practical applications. *Mycorrhiza-Function, Diversity, State of the Art* 99–131.

Shepard, G. H., D. Arora, and A. Lampman. 2008. The grace of the flood: classification and use of wild mushrooms among the highland Maya of Chiapas. *Economic Botany* 62: 437–470.

Shin, J. M., L. Yuan, M. Ohme-Takagi, and T. Kawashima. 2021. Cellular dynamics of double fertilization and early embryogenesis in flowering plants. *Journal of Experimental Zoology Part B: Molecular and Developmental Evolution* 336(8): 642–651.

Shoji, J. Y., T. Kikuma, and K. Kitamoto. 2014. Vesicle trafficking, organelle functions, and unconventional secretion in fungal physiology and pathogenicity. *Current Opinion in Microbiology* 20: 1–9.

Shreejana, K. C., K. Kurunju, R. Thapa, S. Ghimire, and P. Shrestha. 2022. Analysis of primordia and fruiting forming days on using different sterilization techniques of straw and its effect on yield of blue oyster mushroom. *Bhartiya Krishi Anusandhan Patrika* 37(1): 80–83.

Shukla, S., and D. Shrivastava. 2024. Nutritional deficiencies and subfertility: a comprehensive review of current evidence. *Cureus* 16(8): e66477.

Shuting, C. H. A. N. G. 2024. Mushrooms: fungi for all seasons. *Acta Edulis Fungi* 31(1).

Sibounnavong, P., C. D. Cynthia, S. P. Kalaw, R. G. Reyes, and K. Soytong. 2008. Some species of macrofungi at Puncan, Carranglan, Nueva Ecija in the Philippines. *Journal of Agricultural Technology* 4(2): 105–115.

Silva, M., A. C. Ramos, F. J. Lidon, F. H. Reboredo, and E. M. Gonçalves. 2024. Pre-and postharvest strategies for *Pleurotus ostreatus* mushroom in a circular economy approach. *Foods* 13(10): 1464.

Singh, M., S. Kamal, and V. P. Sharma. 2018. Status and trends in world mushroom production-II-Mushroom production in Japan and China. *Mushroom Research* 27(1).

Singh, M., S. Kamal, and V. P. Sharma. 2020. Status and trends in world mushroom production-III-World Production of different mushroom species in 21st century. *Mushroom Research* 29(2).

Singh, S., C. Raj, H. K. Singh, R. K. Avasthe, P. Said, A. Balusamy, S. K. Sharma, S. C. Lepcha, and V. Kerketta. 2021. Characterization and development of cultivation technology of wild split gill *Schizophyllum commune* mushroom in India. *Scientia Horticulturae* 289: 110399.

Siti-Nuramira, J., R. Farhana, S. Nabil, S. M. Jafari, and S. Raseetha. 2022. Impact of drying methods on the quality of grey (*Pleurotus sajor-caju*) and pink (*Pleurotus djamor*) oyster mushrooms. *Journal of Food Measurement and Characterization* 16(5): 3331–3343.

Sonnenberg, A. S., J. J. Baars, G. Straatsma, P. M. Hendrickx, E. Hendrix, C. Blok, and A. van Peer. 2022. Feeding growing button mushrooms: the role of substrate mycelium to feed the first two flushes. *Plos One* 17(7): e0270633.

Spatafora, J. W., M. C. Aime, I. V. Grigoriev, F. Martin, J. E. Stajich, and M. Blackwell. 2017. The fungal tree of life: from molecular systematics to genome-scale phylogenies. *The Fungal Kingdom* 1–34.

Springstein, B. L., C. Woehle, J. Weissenbach, A. O. Helbig, T. Dagan, and K. Stucken. 2020. Identification and characterization of novel filament-forming proteins in cyanobacteria. *Scientific Reports* 10(1): 1894.

Stajić, M., M. Galić, and J. Ćilerdžić. 2023. Mushrooms—from traditional remedies to the modern therapeutics. *Mushrooms with Therapeutic Potentials* 147–188.

Stamets, P., and H. Zwickey. 2014. Medicinal mushrooms: ancient remedies meet modern science. *Integrative Medicine: A Clinician's Journal* 13(1): 46.

Straatsma, G., A. S. Sonnenberg, and L. J. Van Griensven. 2013. Development and growth of fruit bodies and crops of the button mushroom, *Agaricus bisporus*. *Fungal Biology* 117(10): 697–707.

Su, D., W. Lv, Y. Wang, D. Li, and L. Wang. 2020. Drying characteristics and water dynamics during microwave hot-air flow rolling drying of *Pleurotus eryngii*. *Drying Technology* 38(11): 1493–1504.

Sudheer, S., A. Raal, A. Jürgenson, and R. Bhat. 2023. Wild edible mushroom trade in Europe. *Wild Mushrooms and Health* 194–209.

Sun, L., X. Yin, F. L. Sossah, X. Han, and Y. Li. 2023. Comparative genomic analysis of pleurotus species reveals insights into

the evolution and coniferous utilization of *Pleurotus placentodes*. *Frontiers in Molecular Biosciences* 10: 1292556.
Sungsinchai, S., C. Niamnuy, P. Wattanapan, M. Charoenchaitrakool, and S. Devahastin. 2019. Texture modification technologies and their opportunities for the production of dysphagia foods: a review. *Comprehensive Reviews in Food Science and Food Safety* 18(6): 1898–1912.
Suwannarach, N., J. Kumla, Y. Zhao, and P. Kakumyan. 2022. Impact of cultivation substrate and microbial community on improving mushroom productivity: a review. *Biology* 11(4): 569.
Tacer-Caba, Z., J. J. Varis, P. Lankinen, and K. S. Mikkonen. 2020. Comparison of novel fungal mycelia strains and sustainable growth substrates to produce humidity-resistant biocomposites. *Materials & Design* 192: 108728.
Tadiosa, E. R., and L. C. Lubos. 2020. Common macroscopic fungi and mosses of North Central Mindanao. *Asian Journal of Biodiversity* 10(1).
Tarafder, E., M. M. Nizamani, S. C. Karunarathna, D. Das, X. Zeng, R. A. Rind, and F. Tian. 2024. Advancements in genetic studies of mushrooms: a comprehensive review. *World Journal of Microbiology and Biotechnology* 40(9): 275.
Téllez-Téllez, M., and G. Díaz-Godínez. 2019. Omic tools to study enzyme production from fungi in the *Pleurotus* genus. *BioResources* 14(1).
Thakur, M. P. 2014. Present status and future prospects of tropical mushroom cultivation in India: a review. *Indian Phytopathology* 67(2): 113–125.
Thakur, M. P. 2020. Advances in mushroom production: key to food, nutritional and employment security: a review. *Indian Phytopathology* 73: 377–395.
Tijani, N. A., J. Hokello, K. O. Awojobi, et al. 2024. Recent advances in Mushroom-mediated nanoparticles: a critical review of mushroom biology, nanoparticles synthesis, types, characteristics and applications. *Journal of Drug Delivery Science and Technology* 105695.
Tsikouras, P., G. Galazios, K. Romanidis, et al. 2016. Brenner tumour of the ovary-an incidental histological finding. *European Journal of Gynaecological Oncology* 37(2): 267–269.
Usman, M., G. Murtaza, and A. Ditta. 2021. Nutritional, medicinal, and cosmetic value of bioactive compounds in button mushroom (*Agaricus bisporus*): a review. *Applied Sciences* 11(13): 5943.
Valdez-Solana, M. A., E. K. Ventura-García, I. A. Corral-Guerrero, et al. 2023. *In silico* characterization of the physicochemical and biological properties of the pink (*Pleurotus djamor* var. *salmoneostramineus*) oyster mushroom chromoprotein. *Bioinformatics and Biology Insights* 17: 11779322231154139.
Verma, S. 2023. Mushroom cultivation: a sustainable approach to future agriculture to ensure quality food and nutritional security of current population in India.
Virágh, M., Z. Merényi, A. Csernetics, et al. 2022. Evolutionary morphogenesis of sexual fruiting bodies in basidiomycota: toward a new evo-devo synthesis. *Microbiology and Molecular Biology Reviews* 86(1): e0001921.
Voutetaki, M. E., and A. C. Mpalaskas. 2024. Natural fiber-reinforced mycelium composite for innovative and sustainable construction materials. *Fibers* 12(7): 57.
Wallen, R. M., and M. H. Perlin. 2018. An overview of the function and maintenance of sexual reproduction in dikaryotic fungi. *Frontiers in Microbiology* 9: 503.
Wang, P., Y. Zhang, F. Mi, et al. 2015. Recent advances in population genetics of ectomycorrhizal mushrooms *Russula* spp. *Mycology* 6(2): 110–120.
Warren, S. D., L. R. Stark, L. A. Lewis, et al. 2019. Reproduction and dispersal of biological soil crust organisms. *Frontiers in Ecology and Evolution* 7: 344.
Wendiro, D., A. P. Wacoo, and G. Wise. 2019. Identifying indigenous practices for cultivation of wild saprophytic mushrooms: responding to the need for sustainable utilization of natural resources. *Journal of Ethnobiology and Ethnomedicine* 15: 1–15.
Wijesekara, T., and B. Xu. 2024. Occurrence, dietary sources, quantification and bioactivities of natural antioxidant ergothioneine—a longavity vitamin? *International Journal of Food Science & Technology* 59(9): 5951–5963.
Wong, K. H., and P. C. K. Cheung. 2008. Sclerotium of culinary-medicinal king tuber oyster mushroom, *Pleurotus tuberregium* (Fr.) Singer (Agaricomycetideae): its cultivation, biochemical composition, and biopharmacological effects. *International Journal of Medicinal Mushrooms* 10(4): 303–313.
Xu, F., P. Chen, H. Li, et al. 2021. Comparative transcriptome analysis reveals the differential response to cadmium stress of two *Pleurotus* fungi: *Pleurotus cornucopiae* and *Pleurotus ostreatus*. *Journal of Hazardous Materials* 416: 125814.
Yadav, D., and P. S. Negi. 2021. Bioactive components of mushrooms: processing effects and health benefits. *Food Research International* 148: 110599.
Yan, Z., H. Liu, J. Li, and Y. Wang. 2023. Application of identification and evaluation techniques for edible mushrooms: a review. *Critical Reviews in Analytical Chemistry* 53(3): 634–654.
Yang, Y., O. Jia, Y. Li, et al. 2024. Effect of high $CO_2$ controlled atmosphere storage on postharvest quality of button mushroom (*Agaricus bisporus*). *Foods* 13(21): 3486.
Yang, Y., Y. Zhang, C. Huang, Q. Chen, and W. Gao. 2023. Development of cleaved amplified polymorphic sequence marker for cap color identification in *Pleurotus cornucopiae*. *Horticulturae* 9(11): 1238.
Yang, Y. Y., and J. G. Kim. 2016. The optimal balance between sexual and asexual reproduction in variable environments: a systematic review. *Journal of Ecology and Environment* 40: 1–18.
Yu, Y., T. Liu, Y. Wang, L. Liu, X. He, J. Li, F. M. Martin, W. Peng, and H. Tan. 2024. Comparative analyses of *Pleurotus pulmonarius* mitochondrial genomes reveal two major lineages of mini oyster mushroom cultivars. *Computational and Structural Biotechnology Journal* 23: 905–917.
Zafar, S., F. Jabeen, M. Akram, Z. Riaz, and N. Munir. 2020. Mushroom species and classification: bioactives in poisonous and edible mushrooms. In: *Poisonous Plants and Phytochemicals in Drug Discovery,* ed. A. G. Mtewa, C. Egbuna, and G. M. Narasimha Rao, 163–188. Wiley.
Zanatta, F., A. Vanderpoorten, L. Hedenäs, et al. 2018. Under which humidity conditions are moss spores released? A comparison between species with perfect and specialized peristomes. *Ecology and Evolution* 8(23): 11484–11491.
Zeb, U., T. Aziz, A. Azizullah, et al. 2024. Complete mitochondrial genomes of edible mushrooms: features, evolution, and phylogeny. *Physiologia Plantarum* 176(3): e14363.
Zervakis, G. I., S. Ntougias, M. L. Gargano, et al. 2014. A reappraisal of the *Pleurotus eryngii* complex—new species and taxonomic combinations based on the application of a polyphasic approach, and an identification key to *Pleurotus* taxa associated with Apiaceae plants. *Fungal Biology* 118(9–10): 814–834.
Zhang, H., and R. Tsao. 2016. Dietary polyphenols, oxidative stress and antioxidant and anti-inflammatory effects. *Current Opinion in Food Science* 8: 33–42.

Zhang, K., Y. Y. Pu, and D. W. Sun. 2018. Recent advances in quality preservation of postharvest mushrooms (*Agaricus bisporus*): a review. *Trends in Food Science & Technology* 78: 72–82.

Zhang, W. R., S. R. Liu, Y. B. Kuang, and S. Z. Zheng. 2019. Development of a novel spawn (block spawn) of an edible mushroom, *Pleurotus ostreatus*, in liquid culture and its cultivation evaluation. *Mycobiology* 47(1): 97–104.

Zhang, Y., W. Gao, A. Sonnenberg, Q. Chen, J. Zhang, and C. Huang. 2021. Genetic linkage and physical mapping for an oyster mushroom (*Pleurotus cornucopiae*) and quantitative trait locus analysis for cap color. *Applied and Environmental Microbiology* 87(21): e0095321.

Zhang, Y., C. Venkitasamy, Z. Pan, and W. Wang. 2013. Recent developments on umami ingredients of edible mushrooms—a review. *Trends in Food Science & Technology* 33(2): 78–92.

Zhao, M. R., C. Y. Huang, X. L. Wu, et al. 2016. Genetic variation and population structure of the mushroom *Pleurotus ferulae* in China inferred from nuclear DNA analysis. *Journal of Integrative Agriculture* 15(10): 2237–2246.

Zięba, P., A. Sękara, K. Sułkowska-Ziaja, and B. Muszyńska. 2020. Culinary and medicinal mushrooms: insight into growing technologies. *Acta Mycologica* 55(2): 5526.

# 2 A Guide to Cultivating *Pleurotus ostreatus* and Other Edible Oyster Mushrooms

*Priti Yadav and Somanjana Khatua**
Molecular and Applied Mycology Laboratory, Department of Botany, Faculty of Science, University of Allahabad, Prayagraj, India

## 2.1 INTRODUCTION

The global population is projected to rise from 8.2 billion in 2024 to an estimated peak of 10.3 billion by 2084. This demographic expansion will impose unprecedented demands on global food systems, compelling the agricultural sector to intensify production. However, such intensification often leads to adverse environmental impacts, including excessive reliance on chemical pesticides and the unsustainable accumulation of agricultural residues. Globally, agriculture generates approximately 1,300 million tons of wastes annually, nearly half of which remains untreated (Liu et al., 2024). In India alone, agricultural residues contribute an estimated 350–990 million tons of solid waste each year, including over 130 million tons of paddy straw. A significant proportion of this straw is openly burned, exacerbating air pollution and posing serious public health risks (Koul et al., 2022).

These pressing challenges, however, can be effectively addressed through the cultivation of mushrooms, particularly oyster mushrooms (*Pleurotus* spp.), which represent a promising low-input biotechnology for the sustainable management and valorization of agricultural residues. Oyster mushrooms have gained significant attention in the 21st century due to their rapid mycelial growth, adaptability to a wide range of agro-industrial byproducts, relatively short cropping cycle (28–35 days), multiple flushes (5–6 fruiting cycles within 120 days), resistance to pathogens, high biological efficiency (BE), ease of cultivation, minimal financial investment, and well-documented nutraceutical properties (Vieira and de Andrade, 2016; Zhang et al., 2019; Mahari et al., 2020; Ban et al., 2022; De Mastro et al., 2023). These natural saprophytes can efficiently colonize a wide range of lignocellulosic substrates, including wheat straw (commonly used as the primary base) (De Mastro et al., 2023), as well as corncobs, cottonseed hulls, sawdust, sugarcane bagasse, waste tea leaves, banana pseudostems, maize cobs, spent coffee grounds, waste paper, and various other agro-industrial residues. On many of these substrates, as well as on mixtures of different substrates in various combinations, BE values exceeding 100% have been consistently reported (Rajarathnam and Bano, 1987; Carrasco-Cabrera et al., 2019; Economou et al., 2020; Tesfay et al., 2020; Ban et al., 2022). Importantly, their cultivation typically excludes synthetic inputs, thereby reducing the ecological footprint and mitigating pesticide-related health concerns (Carrasco-Cabrera et al., 2019; Dissasa, 2022).

At present, *Pleurotus ostreatus* ranks as the second most widely cultivated edible mushroom globally, following *Agaricus bisporus*, and accounts for over 16% of total global mushroom production (Sánchez, 2010; Subedi et al., 2023). While China leads global output, countries such as Malaysia report that oyster mushrooms constitute over 90% of national mushroom production, with projected yields reaching 65,000 tons by 2020 (Mahari et al., 2020). The genus *Pleurotus* encompasses more than 40 recognized species, at least 26 of which are commercially cultivated. Apart from *P. ostreatus*, other notable examples include the grey oyster or phoenix mushroom (*Pleurotus pulmonarius*), king oyster or Cardoncello (*Pleurotus eryngii*), white oyster (*Pleurotus florida*), maple oyster (*Pleurotus cystidiosus*), red oyster (*Pleurotus flabellatus*), yellow oyster (*Pleurotus citrinopileatus*), and pink oyster (*Pleurotus djamor* var. *roseus*) (Raman et al., 2020; Alananbeh et al., 2014). Most of these species thrive under moderate climatic conditions, typically between 20 and 35 °C, with relative humidity (RH) of 65%–70% (Mahari et al., 2020).

Nevertheless, successful oyster mushroom cultivation is highly sensitive to a range of environmental and nutritional factors. The cultivation process consists of three primary stages, each influenced by specific extrinsic and intrinsic parameters. The first stage, involving mycelial culture preparation, is regulated by factors such as composition of the culture medium, carbon and nitrogen sources, incubation temperature, and pH, which collectively affect the vigor and purity of the mycelial growth (Hoa and Wang, 2015). The second stage is spawn production, where the choice of grain substrate and the nature of nutritional supplementation are critical for developing healthy and productive spawn (Sánchez, 2010). The third and most intensive phase involves substrate preparation and inoculation with spawn. This step is influenced by several variables, including the method of sterilization or pasteurization, the substrate's chemical composition (particularly the carbon-to-nitrogen ratio and lignocellulose content), the amount of spawn used,

* Corresponding author: somanjana@allduniv.ac.in


DOI: 10.1201/9781003527572-2

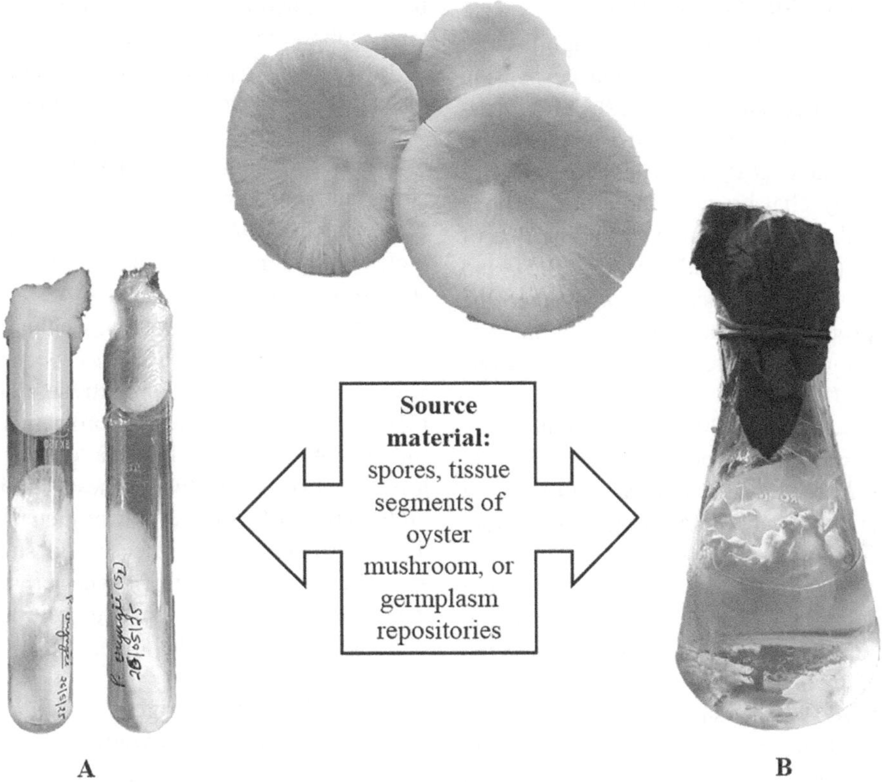


**FIGURE 2.1** Oyster mushroom pure culture grown on different media. (A) Solid medium: potato dextrose agar (PDA). (B) Liquid medium: potato dextrose broth (PDB).

and the incubation conditions (such as temperature, humidity, $CO_2$ concentration, light exposure, and pH) along with the overall nutrient availability within the substrate (Hu et al., 2023). Each of these parameters plays a vital role in determining BE, yield, and quality of the mushroom.

Therefore, systematic optimization of these cultivation parameters is essential to enhance productivity, improve product quality, and ensure economic feasibility. In this context, a comprehensive review is urgently needed to explore and consolidate advancements in oyster mushroom cultivation systems. Such an evaluation would emphasize strategies for boosting yield, maximizing the use of agro-waste substrates, and developing sustainable, low-cost production technologies. The present article aims to address this critical need by offering an in-depth examination of current practices and innovations, ultimately contributing to the development of climate-resilient and resource-efficient mushroom farming systems.

## 2.2 CULTIVATION TECHNIQUES OF OYSTER MUSHROOMS

### 2.2.1 Pure Culture Preparation

The initiation of oyster mushroom cultivation begins with the establishment of a pure mycelial culture, which forms the cornerstone of high-quality spawn production (Hoa and Wang, 2015). Mycelial cultures may be initiated from basidiospores, tissue segments of healthy fruiting bodies, or obtained from germplasm repositories and commercial culture banks (Sánchez, 2010). Once the mycelium begins to develop, a small fragment from the actively growing, uncontaminated edge of the colony is excised (6 mm in diameter) and transferred onto fresh sterile agar medium. The culture is then incubated at 25 °C to ensure the development of a genetically stable and pure culture. This step is crucial for producing disease-free and high-performing inoculum for subsequent cultivation stages. Pure cultures are typically stored at 4 °C for short-term maintenance and are subcultured at regular intervals, usually every four weeks, to maintain viability and vigor (Chang and Miles, 2004; Zhang et al., 2014; Dissasa, 2022; Pradhan et al., 2024).

In addition to solid agar media, liquid media is also employed for spawn production (Figure 2.1). For preparing liquid inoculum, a fully colonized slant culture is cut into soybean-sized agar blocks containing mycelia and transferred aseptically into sterilized liquid medium. The inoculated flask is then cultured on a rotary shaker at 150 rpm and 28 °C for four days to facilitate uniform mycelial proliferation (Liu et al., 2018). Liquid culture techniques offer advantages such as faster colonization and reduced contamination risk, making them increasingly popular for large-scale spawn production. However, to achieve optimal mycelial growth and morphology, several parameters including the composition of the culture medium, incubation temperature, pH, and light conditions must be carefully considered and optimized.

#### 2.2.1.1 Suitable Media for Mycelia Culture

The selection of a suitable culture medium significantly influences the vigor, uniformity, and radial expansion of *Pleurotus* mycelium (Lenka et al., 2022). Early studies by Suharban and Nair (1994) identified oatmeal agar and potato dextrose agar (PDA) as particularly effective in supporting robust mycelial growth. These observations were reinforced by subsequent research, including that of Mshandete and Mgonja (2009), Chuku et al. (2015), and Sardar et al. (2015), which consistently demonstrated the efficacy of PDA in promoting biomass accumulation, exopolysaccharide secretion, and protein synthesis in various *Pleurotus* species.

A recent study by Lenka et al. (2022) assessed the radial mycelial growth of five *Pleurotus* species (*P. sajor-caju*, *P. florida*, *P. ostreatus*, *P. eous*, and *P. eryngii*) on four solid media: PDA, malt extract agar (MEA), yeast malt extract agar (YME), and Sabouraud dextrose agar (SDA). The results revealed significant interspecific and intermedium variability, with radial growth diameters ranging from 6.60 cm to 9.53 cm at eight days post-inoculation. Among the tested media, PDA demonstrated the highest efficacy, followed by SDA and MEA. This superior performance is likely attributed to its balanced nutrient composition and high compatibility across *Pleurotus* strains. For liquid cultures, Gapiński et al. (2007) observed that *P. citrinopileatus*, *P. djamor*, *P. eryngii*, and *Pleurotus precocmore* exhibited greater dry biomass accumulation in potato-based media compared to Hansen medium.

In contrast, Hoa and Wang (2015) reported that yam dextrose agar (YDA) outperformed PDA in supporting the mycelial growth of *P. ostreatus* and *P. cystidiosus*, while MEA and sweet potato dextrose agar were suboptimal. Overall, these findings underscore the critical role of selecting an appropriate and cost-effective culture medium such as PDA to ensure vigorous and uniform mycelial development. Optimizing media formulations is essential for enhancing spawn quality and ensuring the scalability and economic viability of oyster mushroom cultivation systems.

#### 2.2.1.2 Optimal Carbon Source for Mycelial Growth

Carbon sources play a fundamental role in fungal metabolism, functioning as both energy-yielding substrates and structural components that support mycelial proliferation and secondary metabolite synthesis (Maseko et al., 2024). The mycelial growth rate of *Pleurotus* species is notably influenced by the carbon-to-nitrogen (C:N) ratio of the growth medium (Chanprapai et al., 2022). In an early investigation, Neelam et al. (2013) identified dextrose as the most effective carbon source for *P. ostreatus* and *P. florida*, followed closely by fructose, sucrose, and galactose. Hoa and Wang (2015) studied the effects of various carbon sources including glucose, fructose, maltose, sucrose, molasses, and lactose (each at 2%) on the mycelial development of *P. ostreatus* and *P. cystidiosus* grown on potato agar medium. Among these, *P. ostreatus* exhibited the highest radial growth with glucose (8.60 cm), sucrose (8.54 cm), and molasses (8.34 cm), while *P. cystidiosus* also performed best with glucose and sucrose. Conversely, lactose consistently supported the least growth in both species.

Interestingly, species-specific tolerance to sucrose was observed: *P. ostreatus* maintained uniform growth across a wide concentration range (1%–5%), whereas *P. cystidiosus* exhibited optimal growth at 1%–3%, beyond which osmotic stress and imbalanced C:N ratios hindered development. Maseko et al. (2024) further affirmed that 2% glucose promotes optimal mycelial expansion in *P. ostreatus*, and that sucrose concentrations between 4% and 6% still support satisfactory growth. Similarly, Lenka et al. (2022) evaluated the growth of five *Pleurotus* species on PDA amended with 2% concentrations of dextrose, maltose, and sucrose. Dextrose yielded the most vigorous mycelial development, followed by sucrose, while maltose proved comparatively less effective. Collectively, these studies underscore the preferential utilization of easily metabolizable mono- and disaccharides such as glucose and sucrose by *Pleurotus* spp. Their consistent performance across species and media makes them ideal carbon sources for large-scale, cost-effective mushroom cultivation.

#### 2.2.1.3 Optimal Nitrogen Source for Mycelial Growth

Nitrogen, an essential macronutrient in fungal physiology, plays a pivotal role in the biosynthesis of amino acids, proteins, nucleotides, and chitin. In the context of oyster mushroom cultivation, nitrogen supplementation significantly influences the vigor, morphology, and metabolic activity of the mycelium. Both inorganic and organic nitrogen sources have been widely utilized in culture media formulations, including ammonium salts [e.g., $NH_4Cl$, $(NH_4)_2SO_4$, $NH_4NO_3$], nitrates (e.g., $NaNO_3$, $KNO_3$), urea, amino acids, casein hydrolysate, peptone, and yeast extract, typically ranging between 0.012% and 0.11% nitrogen content (Lima et al., 1975; Hoa and Wang, 2015). Growth can also be promoted through the addition of ascorbic acid, wood vinegar, and ethanolic extracts of *Agave potatorum* leaves into the cultivation media. However, excessive nitrogen levels have been reported to suppress ligninolytic enzyme expression, thereby adversely affecting fruiting body formation (Chanprapai et al., 2022).

Lenka et al. (2022) systematically assessed the influence of diverse nitrogen sources on radial mycelial growth across five *Pleurotus* species. All nitrogen-supplemented media enhanced growth compared to the control, with inorganic nitrogen sources generally outperforming organic ones. Among these, $NH_4Cl$ supported the highest colony diameters, particularly in *P. eryngii* (9.39 cm) and *P. florida* (9.26 cm), followed by $(NH_4)_2SO_4$. This trend highlights the superior bioavailability and rapid assimilation of ammonium ions over other nitrogen forms. While other inorganic sources yielded moderate improvements, organic nitrogen substrates supported comparatively lower radial expansion.

These results are in concordance with the findings of Hoa and Wang (2015), who evaluated the response of *P. ostreatus* and *P. cystidiosus* to multiple nitrogen sources. $NH_4Cl$ again emerged as the most effective, with optimal mycelial diameters recorded at concentrations ranging from 0.03% to 0.09% for *P. ostreatus* and 0.03% to 0.05% for *P. cystidiosus*. Although $(NH_4)_2SO_4$ stimulated significant growth in *P. cystidiosus*, its impact on *P. ostreatus* was statistically comparable to the control. Organic sources such as urea and soybean meal exhibited limited efficacy, though some organic treatments, despite supporting smaller colony diameters, resulted in denser and more compact mycelial mats, suggesting variations in biomass architecture rather than total growth rate.

In another study, Maseko et al. (2024) identified $NH_4Cl$ (0.1%) as the most suitable nitrogen source for *P. ostreatus*, followed by urea (0.01%) and $NH_4H_2PO_4$ (0.05%). Additionally, calcium carbonate (0.1%) and ascorbic acid (0.01%) also supported favorable growth, albeit to a lesser extent, indicating potential auxiliary roles in enhancing physiological performance under specific culture conditions.

Supporting this general trend, Neelam et al. (2013) demonstrated that $NH_4Cl$ significantly outperformed nitrate-based nitrogen sources such as sodium nitrate and calcium nitrate in promoting mycelial development in *P. ostreatus* and *P. florida*. The reduced performance of nitrate sources is often attributed to their relatively lower uptake efficiency and possible inhibitory effects on nitrogen metabolism compared to the more readily assimilable ammonium forms. Taken together, the collective body of evidence underscores the critical importance of nitrogen source selection during the vegetative phase of oyster mushroom cultivation. Among the tested sources, inorganic nitrogen compounds especially $NH_4Cl$ and $(NH_4)_2SO_4$ consistently facilitate robust and uniform mycelial expansion.

#### 2.2.1.4 Optimal pH Condition for Mycelial Growth

Hydrogen ion concentration (pH) of the growth medium plays a pivotal role in determining fungal physiology. The pH sensitivity of *Pleurotus* species is well established, though the optimal range varies depending on the species, cultivation system, and substrate composition. Experimental studies evaluating pH values from 3.0 to 8.0 consistently demonstrate that both highly acidic and alkaline environments adversely affect mycelial development. For instance, Rawte (2011) reported that *Pleurotus* species exhibited maximum mycelial growth at pH 5.0, while deviations on either side significantly inhibited development. Similarly, Sardar et al. (2015) observed that slightly acidic conditions, particularly pH 6.0, supported optimal radial growth across six *Pleurotus* strains, including *P. ostreatus*, *P. sajor-caju*, *P. eryngii*, *P. columbinus*, and *P. sapidus*. These findings were further supported by Velioglu and Urek (2015), who reported that *P. djamor* produced the highest mycelial biomass in solid-state fermentation at an initial pH of 6.0.

Despite this general preference for acidic conditions, some studies have reported different optima. For instance, Sarker et al. (2007) identified pH 7.0 as the most favorable for mycelial expansion and yield in *P. ostreatus*. Such variability suggests that the pH response in *Pleurotus* spp. is highly species specific and may be influenced by the buffering capacity of the substrate, physiological adaptability, and interactions with other environmental parameters. From a cultivation perspective, maintaining the initial pH of the substrate within a moderately acidic range, typically between 5.5 and 6.5, is advisable. This range not only supports vigorous and uniform mycelial growth but also facilitates favorable enzymatic activity and nutrient uptake, thus enhancing the overall yield and quality of *Pleurotus* mushrooms.

#### 2.2.1.5 Optimal Temperature for Mycelial Growth

Temperature is a critical environmental variable that profoundly affects fungal physiology, enzymatic efficiency, and metabolic processes. Each fungal species, including those within the genus *Pleurotus*, exhibits a defined thermal tolerance range beyond which mycelial growth is markedly impaired (Hoa and Wang, 2015). Zharare et al. (2010), Neelam et al. (2013), and Hoa and Wang (2015) collectively reported that most *Pleurotus* strains achieve maximal radial expansion between 25 °C and 28 °C (Table 2.1). Growth outside this window, particularly at lower or higher extremes, results in significant inhibition of hyphal elongation and biomass accumulation. More recently, Lenka et al. (2022) confirmed that the optimal temperature range for mycelial growth across five *Pleurotus* species lies at 25 ± 2 °C. Specifically, *P. florida* and *P. sajor-caju* exhibited their highest radial growth rates within the 25–30 °C interval. However, temperatures exceeding 30 °C often induce thermal stress, leading to slower hyphal extension, morphological deformities, or even complete growth inhibition. At 35 °C and above, most *Pleurotus* strains exhibit sharply reduced growth rates or fail to grow altogether, with structural abnormalities frequently observed in the mycelial network (Choi et al., 2003; Hoa and Wang, 2015). These thermophysiological responses underscore the ecological plasticity of *Pleurotus* species, particularly their adaptability to subtropical and tropical climates.

### 2.2.2 Spawn Production

Spawn refers to the vegetative mycelium of a specific mushroom species that has been cultivated on a nutrient-rich substrate, most commonly cereal grains such as wheat, pearl millet, or sorghum. This grain-based carrier material, often supplemented with chopped straw, acts as the primary inoculum for initiating mushroom cultivation (Sánchez, 2010; Singh et al., 2022). The production process begins with the development of a pure, contaminant-free culture of the target fungus. This culture is subsequently transferred onto sterilized grains, where the fungal mycelium colonizes and proliferates, resulting in a viable spawn mixture composed of both actively growing mycelia and supportive substrate materials (Waghmare, 2016). Currently, mushroom growers utilize two principal forms of spawn: solid spawn and liquid spawn. Solid spawn is further classified into several subtypes based on the substrate used—namely grain spawn, sawdust spawn, and stick spawn—each with specific applications and advantages depending on the cultivation method and mushroom species (Liu et al., 2018; Zhang et al., 2019).

**TABLE 2.1**
**Summary of the Most Suitable Culture Media, Temperature, Carbon Sources, and Nitrogen Sources for the Mycelia Growth of Some *Pleurotus* Species**

| *Pleurotus* spp. | Culture Media | Temperature | Carbon Source (Concentration) | Nitrogen Source | Reference |
|---|---|---|---|---|---|
| *P. citrinopileatus* | PDA | 20–25 °C | Glucose | Yeast powder | Xiao et al., 2024 |
| *P. cystidiosus* | YDA followed by MEA, PDA | 25–28 °C | Glucose, sucrose (1%–3%) and dextrose | Ammonium chloride | Zharare et al., 2010; Hoa and Wang, 2015; Bellettini et al., 2019 |
| *P. florida* | SDA followed by PDA | 25 °C | Dextrose followed by glucose | Ammonium chloride followed by ammonium sulfate | Zharare et al., 2010; Bellettini et al., 2019; Lenka et al., 2022 |
| *P. eryngii* | SDA followed by PDA | 25–27 °C | Dextrose followed by glucose | Ammonium chloride followed by ammonium sulfate | Nguyen et al., 2020; Lenka et al., 2022 |
| *P. eous* | PDA | 27 °C | Dextrose followed by glucose | Ammonium sulfate followed by ammonium chloride | Lenka et al., 2022 |
| *P. ostreatus* | PDA | 24–28 °C | Glucose, sucrose (1%–3%) and molasses | Ammonium chloride | Hoa and Wang, 2015; Bellettini et al., 2019; Nguyen et al., 2020; Werghemmi et al., 2022 |
| *P. sajor-caju* | PDA | 25–27 °C | Dextrose followed by glucose | Ammonium sulfate followed by ammonium chloride | Zharare et al., 2010; Lenka et al., 2022 |
| *P. sapidus* | PDA followed by MEA, YMA | 20–28 °C | - | - | Mahadevan and Shanmugasundaram, 2018; Irshad et al., 2023 |

#### 2.2.2.1 Solid Spawn

Solid spawn remains the most commonly utilized form in oyster mushroom cultivation due to its ease of production and minimal equipment requirements. Unlike liquid fermentation systems, solid spawn preparation does not necessitate specialized fermenters, making it suitable for small- to medium-scale growers. Its quality can be assessed through simple visual inspection to ensure purity and absence of contamination prior to use. Typically, solid spawn is prepared using cereal grains, often supplemented with gypsum to enhance structure and nutrient availability. Alternatively, lignocellulosic substrates like sawdust and cottonseed hulls, enriched with additives such as wheat bran and lime, may also be used. This method continues to dominate in commercial production settings, particularly across mushroom farms in China (Zhang et al., 2014).

##### *2.2.2.1.1 Sawdust Spawn*

Sawdust-based spawn is a widely adopted method in the cultivation of wood-decaying mushrooms, including various *Pleurotus* species. Oh et al. (2003) detailed a method for producing sawdust spawn of *P. ostreatus* using an equal mixture of poplar and pine sawdust, supplemented with rice bran. The mixture underwent fermentation for 16 days, during which it was regularly moistened to maintain optimal conditions. After fermentation, the substrate was transferred into polypropylene (PP) bags, sterilized by autoclaving for 90 minutes, and cooled. Following a 12-hour rest period, the bags were inoculated under sterile conditions and incubated at 25 °C for about 20 days to allow full mycelial colonization.

In another study, researchers described the preparation of *P. eryngii* spawn using PP bags (15 cm wide and 35 cm long) filled with a substrate composed of 80 kg broadleaf sawdust, 1 kg lime, 19 kg wheat bran, and 120 L of water. The filled bags, with substrate heights of 15–16 cm, were sealed using specially designed plastic rings and vented caps fitted with microbial filters. These were sterilized for 60 minutes at 121 °C, cooled, and inoculated with a 1 $cm^2$ plug of mycelium from PDA slants. Incubation took place in dark conditions at 25 °C until full colonization (Zhang et al., 2014).

In a subsequent protocol by Zhang et al. (2019), a substrate consisting of 78% sawdust, 20% wheat bran, and 2% gypsum (on a dry weight basis) adjusted to 63% moisture was used. Each 30 cm × 15 cm PP bag was filled with 750 g of this hydrated mix. The bags were sterilized at 121 °C for two hours, cooled, inoculated with a 1 $cm^2$ agar plug from a 7-day-old PDA culture, and incubated in the dark at 25 °C with 60% RH until mycelial growth was complete. These studies collectively highlight the adaptability of sawdust spawn formulations through the utilization of locally available agricultural residues.

##### *2.2.2.1.2 Grain Spawn*

Grain spawn is a commonly used method in mushroom cultivation, favored for its efficiency and capacity to support vigorous mycelial growth (Figure 2.2). The process begins

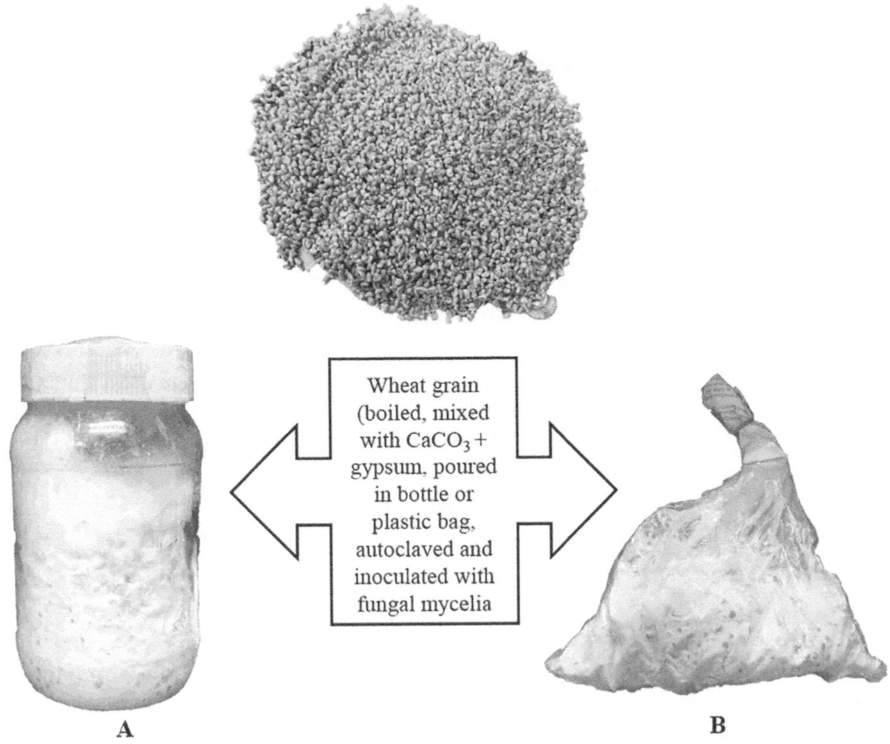


**FIGURE 2.2** Fully colonized spawn prepared using wheat grains for oyster mushroom cultivation: (A) Grain spawn in a glass bottle and (B) Grain spawn in a polypropylene packet.

with thorough cleaning of cereal grains (such as wheat, sorghum, pearl millet, or maize) by washing them 3–4 times to eliminate dust, soil particles, and other debris. The grains are then either soaked for 20–30 minutes or directly boiled in a water-to-grain ratio of 1.75 L/kg for 20–35 minutes until they reach a semi-cooked state. After boiling, they are spread over a clean muslin cloth or mesh to air dry until the surface moisture is removed. To promote optimal colonization and prevent clumping, additives such as gypsum (calcium sulfate) and chalk powder (calcium carbonate) are incorporated. A common formulation includes 20 g gypsum and 5 g chalk/kg of dry grain, or 10 g of each per kilogram of wet grain. Gypsum functions primarily as an anti-caking agent, while $CaCO_3$ helps maintain the pH in the favorable range of 7.0–8.0 (Kumar et al., 2020). Mixing is carried out under aseptic conditions to reduce contamination risks, with the final substrate adjusted to a moisture content of 55%–60% (Sharma and Kumar, 2011; Sahu et al., 2014; Hoa and Wang, 2015; Tesfay et al., 2020).

The treated grains are packed into heat-resistant PP bags or bottles, typically containing between 250 g and 1 kg of substrate, occupying approximately two-thirds of the container's volume. These containers are sterilized in an autoclave at 121 °C for durations ranging from 30 minutes to 4 hours depending on the volume (Tesfay et al., 2020; Deora et al., 2021). Once cooled, inoculation is carried out under sterile conditions, typically in a laminar airflow cabinet illuminated with UV light for 20–30 minutes. Inoculum (usually 10 agar plugs from a pure mycelial culture or from liquid culture) is introduced, and the containers are then sealed with breathable closures, followed by incubation in darkness at 20–32 °C (Bhatti et al., 2007; Vieira and de Andrade, 2016; Deora et al., 2021). Gentle shaking on the 5th and 10th days helps distribute the mycelium uniformly throughout the substrate. Within 2–3 weeks, full colonization is achieved, rendering the grain spawn ready for use in commercial-scale cultivation (Sharma and Kumar, 2011; El-Sayed et al., 2024; Sahu et al., 2014; Sofi et al., 2014; Hoa and Wang, 2015; Waghmare, 2016; Prasad et al., 2024).

Several factors, such as grain type, size, porosity, and water-holding capacity, significantly influence the efficiency of colonization (Thakur and Rathod, 2021). Among various grains studied, wheat (*Triticum aestivum*) has consistently shown excellent performance. Its balanced nutrient profile and favorable physical properties support strong mycelial growth, leading to high BE across multiple species such as *P. florida*, *P. djamor*, and *P. eous* (Sharma, 2003; Pal et al., 2008). However, it sometimes requires a longer colonization period compared to other grains, such as sorghum (Kerketta et al., 2019). Sorghum (*Sorghum bicolor*) has also emerged as a superior grain for faster spawn development. It supports rapid colonization, with studies noting colonization times as short as 7–8 days in *P. eous* and *P. columbinus* (Table 2.2) (Pal et al., 2008; Sahu et al., 2014). Sorghum's structure and nutrient content make it an ideal substrate for high-quality spawn production across a range of *Pleurotus* species. Maize (*Zea mays*) has shown promise as well, particularly in white and yellow varieties. White

**TABLE 2.2**
**Use of Different Grain Substrates for Spawn Production of *Pleurotus* spp.**

| *Pleurotus* spp. | Grain | Reference |
|---|---|---|
| *P. columbinus* | Bajra | Irshad et al., 2023 |
| *P. cystidiosus* | Brown rice | Hoa and Wang, 2015 |
| *P. djamor* | Wheat | Sharma, 2003 |
| *P. eous* | Wheat | Pal et al., 2008; Bhatia and Yadav, 2024 |
| *P. eryngii* | Wheat | Philippoussis et al., 2001; Sharma, 2003; Deora et al., 2021; Bhatia and Yadav, 2024 |
| *P. flabellatus* | Wheat | Pandey et al., 2017 |
| *P. florida* | Wheat | Sharma, 2003 |
| *P. ostreatus* | Brown rice, sorghum, wheat | Philippoussis et al., 2001; Obodai et al., 2003; Hoa and Wang, 2015; Girmay et al., 2016; Tesfay et al., 2020 |
| *P. pulmonarious* | White maize wheat | Tinoco et al., 2001; Philippoussis et al., 2001; Stanley and Awi-Waadu, 2010 |
| *P. sajor-caju* | Sorghum | Chauhan and Pant, 1988 |
| *P. sapidus* | Sorghum | Sahu et al., 2014; Borah et al., 2019 |
| *P. tuber-regium* | White maize | Tinoco et al., 2001; Stanley and Awi-Waadu, 2010 |

maize supported full colonization of *P. pulmonarius* and *P. tuber-regium* within 10 days (Stanley et al., 2010), while yellow maize yielded robust growth in *P. ostreatus* and *P. cystidiosus* (Hoa and Wang, 2015). Brown rice (*Oryza sativa*) has shown high compatibility with certain species. Hoa and Wang (2015) observed the largest mycelial colonies on brown rice in *P. ostreatus* and *P. cystidiosus*, while Sahu et al. (2014) noted quick colonization in paddy-based spawn (8.66 days), slightly slower than sorghum. Other less commonly used grains, such as barley, oats, rye, and chickpeas, have also been studied (Singh et al., 2022; Ufitinema et al., 2023); but more data are needed to validate their performance across different strains.

The choice of grain also intersects with the temperature preferences of various *Pleurotus* species. For instance, *P. eous* thrives in warmer environments (30–35 °C), while *P. eryngii* exhibits a broader temperature tolerance (10–35 °C), making it adaptable to diverse climates (Oei and Nieuwenhuijzen, 2005; Telang et al., 2010). Hence, tailoring incubation temperature to the species being cultivated is crucial for achieving optimal mycelial growth. Post-colonization, spawn can be stored at 0–2 °C for up to a week or refrigerated at 4 °C for about a month without significant viability loss. When stored at ambient temperature (~25 °C), the usable life is generally 30 days. Although refrigeration can prolong shelf life up to three months, viability and performance typically decline after two months (Borah et al., 2019; Bandura et al., 2022).

##### *2.2.2.1.3 Stick Spawn*

Stick spawn has recently been adopted in mushroom cultivation for its ability to shorten spawn running time and boost yields. In *P. eryngii*, rough-surfaced wooden sticks (15 cm × 0.6 cm × 0.8 cm) are soaked, layered with sawdust, packed in bags, autoclaved, and inoculated, then incubated at 25 °C in darkness (Zhang et al., 2014). Similarly, corn stalk chips (12 cm × 1 cm × 2 mm) have been used as carriers for *P. ostreatus* in liquid cultures (Liu et al., 2018).

#### 2.2.2.2 Liquid Spawn

Liquid spawn allows for more uniform distribution within the substrate and has demonstrated faster colonization and higher mushroom yields compared to solid spawn. However, its adoption is limited in less developed regions due to the need for costly equipment and strict sterile conditions (Zhang et al., 2014). A typical liquid culture medium for *P. pulmonarius* includes 20 g/L brown sugar, 4 g/L each of rice bran, malt extract, and yeast extract, adjusted to pH 5.5, and incubated at 28 °C with 250 rpm agitation and 30%–40% oxygen partial pressure. Under these conditions, a maximum dry biomass of 11.72 g/L was achieved after three days of fermentation. This liquid spawn effectively colonized sterile rubber wood sawdust more rapidly than grain spawn and yielded superior sporophore production (Abdullah et al., 2013).

### 2.2.3 Substrate Preparation and Inoculation With Spawn

The historical roots of oyster mushroom cultivation trace back to 1917, when Flack in Germany first experimented with growing *Pleurotus* species on natural wooden substrates such as logs and stumps (Muswati et al., 2021). Although hardwood species like beech, chestnut, and poplar have proven effective for such traditional methods, their labor-intensive nature and slow yield cycles render them less favorable for present-day commercial-scale operations. Instead, artificially prepared substrates have become the cornerstone of modern mushroom production (Mahari et al., 2020; Sariyer et al., 2023). A typical cultivation cycle begins with substrate selection and physical processing. Straw (preferably aged for around two months postharvest) is either chopped manually or reduced via mechanical milling to 3–5 cm segments. This size optimization increases surface area, facilitating better substrate colonization by fungal mycelia (Chang and Miles, 2004). Subsequently, the processed substrate undergoes thermal or chemical treatment to eliminate unwanted microorganisms. Steam pasteurization at 80–90 °C for 5 to 9 hours is widely practiced, often using heat-resistant PP or polyethylene (PE) bags (Mahari et al., 2020). Traditionally, the substrate for oyster mushrooms needs to undergo high-pressure sterilization (121–123 °C for 2–3 h) or atmospheric pressure sterilization (95–100 °C for 8–12 h) (Yang et al., 2022).

In contexts where fuel or equipment is limited, growers often turn to chemical or alkaline treatments. For example, treating straw with a combination of 2% formalin and 0.1% potassium permanganate in hot water significantly improves yields by suppressing competitive microbes (Banik and Nandi, 2004). Another low-cost yet efficient method involves immersing the substrate in lime water (prepared using calcium carbonate or calcium hydroxide), which not only suppresses microbial competitors but also enhances BE, promotes faster spawn run, and improves final productivity (Iossi et al., 2018). Alternatives such as chlorine disinfection have also gained traction. In this approach, straw is soaked in a 0.3% chlorine bleach solution for durations ranging from 10 minutes to overnight. This achieves both sterilization and optimal moisture content. After treatment, the straw is spread out on sanitized surfaces to air dry briefly, allowing excess chlorine to dissipate and moisture to stabilize (Zhang et al., 2002; Deora et al., 2021). In advanced systems, such as those optimized for *Pleurotus tuoliensis*, cultivation bags containing 2 kg of substrate are autoclaved at 110 °C for six hours before inoculation (Zou et al., 2020).

In recent years, short-term composting methods have gained popularity for oyster mushroom cultivation. Composting is an eco-friendly biological process wherein native microorganisms break down organic material into simpler, more stable compounds. Following this process, the composted substrate typically achieves a C/N ratio of 28:1 to 30:1, which is considered optimal for *Pleurotus* species cultivation. One major advantage of using composted substrates is the reduction in sterilization requirements. In many cases, the substrate may only require heating at 100 °C for about 4 hours, and in some situations, sterilization may be omitted entirely (Yang et al., 2022). For example, Hernández et al. (2003) reported successful cultivation of *P. ostreatus* on a lignocellulosic substrate with an initial pH of 8.7, even without prior pasteurization or composting. However, they also observed that a brief composting period of 2–3 days led to a notable improvement in BE. Similarly, Vieira and de Andrade (2016) found that seven-day composting with conditioning significantly enhanced both yield and BE of the mushroom.

The critical step post-treatment is the adjustment of substrate moisture to an optimal 60–65%. Moisture content below this range may inhibit fungal metabolism, while oversaturation can restrict aeration and favor contaminants. Once cooled, the substrate is inoculated with mushroom spawn, usually at 5%–18% of the wet weight. This can be done either by evenly mixing the spawn into the substrate or layering it strategically (Deora et al., 2021; Royse et al., 2004).

For inoculated substrate containment, cylindrical baglogs remain the standard, especially for *P. ostreatus*. Growers typically use plastic bags (PP, PE, or polyvinyl chloride) measuring 15 × 30 cm or 40 × 50 cm. These are sealed using plastic neck rings and breathable sponge plugs to permit gas exchange while minimizing contamination risks. Alternatively, rubber bands may be used to close the bag's mouth, with perforations added to facilitate aeration and fruiting body emergence (Zhang et al., 2002). The substrate should have a pH that is slightly acidic to slightly basic, ideally close to neutral (Chukwu et al., 2022).

Spawn run representing mycelial colonization of the substrate takes place in a dark, humidity- and temperature-controlled chamber. Conditions are typically maintained at 24–28 °C with 80%–90% RH. Proper aeration during this phase is vital, and care must be taken not to over-compress the substrate, which could restrict oxygen flow and hinder mycelial growth (Oei, 2003). Upon completion of the spawn run, which usually takes two to three weeks, bags are transferred to fruiting rooms where environmental conditions are adjusted depending on the species cultivated. If the bags have perforations, fruiting bodies will naturally emerge from these sites. Otherwise, the bags are opened to allow fruit development. Watering, typically via fine mist sprays in the morning and evening, helps maintain substrate moisture at optimal levels (Zhang et al., 2002; Baysal et al., 2003). Fruiting typically begins within 25 to 30 days post-inoculation, with all the flushes of mushrooms harvested manually. Timely harvesting is crucial to ensure peak quality and to trigger subsequent flushes (Zhang et al., 2002).

#### 2.2.3.1 Substrate Selection for Optimized Oyster Mushroom Cultivation

The efficiency of oyster mushroom cultivation is largely dependent on the appropriate selection and preparation of substrate. While wood-based materials such as sawdust were among the earliest substrates utilized, their application often leads to reduced mushroom yield and BE due to their inherently low protein content, which limits fungal biomass development (Akcay et al., 2023). To overcome this limitation, researchers have explored various supplementation strategies to improve nutritional balance and enhance productivity. For instance, rubber wood sawdust amended with rice bran, magnesium sulfate ($MgSO_4$), and $CaCO_3$, adjusted to a moisture content of 60%–65% using water or diluted lemon straw extract, has been effectively employed for the cultivation of *P. sajor-caju* under controlled baglog conditions. Notably, the inclusion of lemon straw extract significantly enhanced BE, increasing it by approximately 1.5-fold compared to water supplementation alone (Chanprapai et al., 2022).

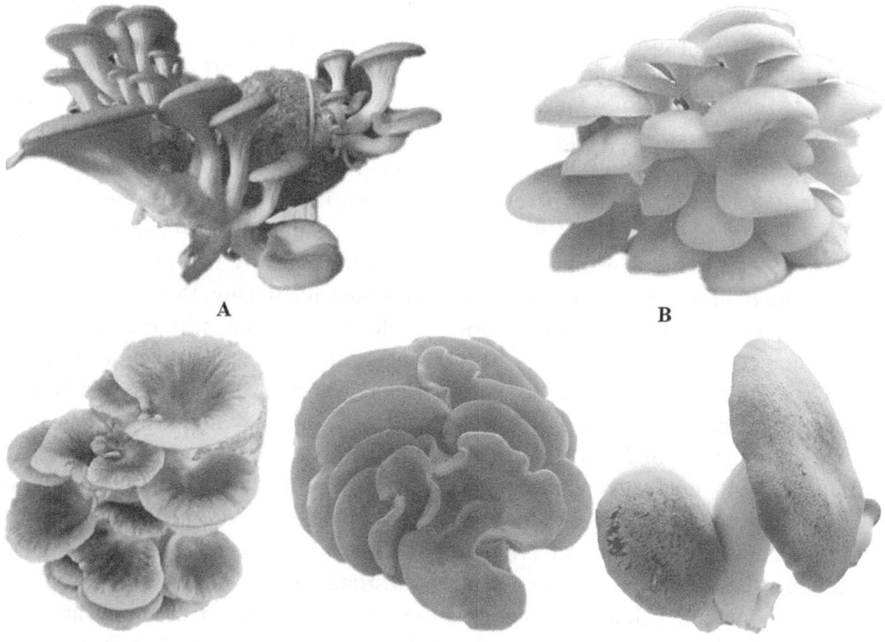


**FIGURE 2.3** Oyster mushroom cultivation using wheat grain spawn on wheat substrate: (A) *Pleurotus ostreatus*, (B) *Pleurotus florida*, (C) *Pleurotus sajor-caju*, (D) *Pleurotus djamor*, (E) *Pleurotus eryngii*.

Among various agro-waste substrates, wheat straw stands out as one of the most dependable and widely adopted plant-based materials for oyster mushroom cultivation (Figures 2.3 and 2.4). Its lignocellulosic structure, enriched with cellulose (39%–51%) and lignin (~18%), offers an ideal matrix for fungal colonization, while its moderate protein content (3.5%–5%) supports vigorous mycelial proliferation and fruit body development (Akcay et al., 2023). When used alone, wheat straw has demonstrated promising BE, reaching up to 79.4% in certain optimized cultivation conditions (Iossi et al., 2018; Zou et al., 2020). To further boost productivity and mushroom quality, supplementation with nitrogen-rich amendments such as wheat bran, corncob, peach sawdust, and even unconventional inputs like broken egg mixtures has proven effective (Wei et al., 2024). A notable example includes the use of a 4:1 cottonseed hull to wheat bran ratio, which resulted in significantly larger cap diameters (up to 10.4 cm) alongside improved yield (Yang et al., 2013). Furthermore, combining wheat straw with other biodegradable wastes such as waste paper or hazelnut leaves has been shown to enhance fruiting body production and structural integrity (Yildiz et al., 2002). Incorporating nutrient-enriched amendments, like anaerobically digested distillery effluent, has also shown promise when co-composted with wheat straw, yielding improved oyster mushroom growth and productivity (Pant et al., 2006). Oyster mushrooms can also be cultivated on wheat straw substrates enriched with up to 50% olive mill waste without significantly affecting cultivation efficiency or the overall quality of the fruiting bodies. However, a distinct change in the coloration of the mushrooms has been observed under these conditions (Ruiz-Rodríguez et al., 2010).

Rice straw, another readily available agricultural residue, serves as a highly effective basal substrate for oyster mushroom cultivation. In the case of *P. sajor-caju*, rice straw outperformed wheat straw by yielding approximately 10% more mushrooms under identical environmental conditions (Zhang et al., 2002). Deora et al. (2021) evaluated multiple substrates for the cultivation of *P. eryngii* and reported that paddy straw exhibited the highest BE and supported the maximum number of fruiting bodies. Beyond rice, other fibrous crop residues such as *Brachiaria brizantha* and *Brachiaria decumbens* grasses, as well as sugarcane straw, have shown considerable potential. Notably, a composted substrate comprising *B. brizantha* supplemented with wheat bran achieved a remarkable BE of 123.95%, underscoring the importance of proper substrate preconditioning and optimized carbon-to-nitrogen (C/N) ratios (Vieira and de Andrade, 2016). Furthermore, rice straw demonstrated enhanced performance when enriched with biomanure sources like poultry litter or green weeds such as *Leonotis* and *Cassia sophera*. These combinations not only improved yield but also elevated the protein content of the resulting fruiting bodies, contributing to the nutritional value of the harvest (Banik and Nandi, 2004; Das and Mukherjee, 2007).

In recent years, researchers have diversified their substrate palette by incorporating agricultural and urban waste materials. Waste paper, when finely chopped and supplemented with rice husk, chicken manure, or peat, offers a sustainable and effective base for cultivation. Properly

**FIGURE 2.4** Cultivation of oyster mushrooms in various plastic containers: (A) Food-grade container, (B) Repurposed paint bucket, and (C) Chocolate storage box.

hydrated (75% moisture) and pasteurized at 60 °C, these paper-based substrates demonstrated enhanced pinhead initiation, fruiting, and yield, particularly in combinations like waste paper and rice husk (4:1 ratio), as demonstrated by Baysal et al. (2003).

Coffee waste, both husk and parchment, presents another eco-friendly option. When composted, it becomes even more effective. Among four *Pleurotus* species tested, *P. ostreatus* not only exhibited the shortest time to pinhead formation but also the highest BE on composted coffee husk. This is attributed to the species' secretion of pectinase enzymes that degrade complex plant fibers, releasing nutrients for mycelial absorption (Dissasa, 2022).

Studies have explored the potential of incorporating *Ginkgo biloba* leaf powder into cultivation substrates for *P. eryngii* as a substitute for conventional components such as sawdust and sugarcane bagasse. Results indicated that the inclusion of 21% leaf powder in the substrate mixture led to a noticeable improvement in mushroom yield, highlighting its suitability as a beneficial additive for enhanced production (Li et al., 2024).

In recent studies, alfalfa pulp has emerged as a highly efficient substrate, demonstrating BE exceeding 166% and outperforming traditional wheat straw (Table 2.3), while also shortening the cultivation cycle from 31 to just 24 days. Depending on the scale and infrastructure, cultivation setups may range from PE bags to sterilized PP boxes or micropore-sealed plastic containers. Notably, *P. ostreatus* grown on alfalfa pulp in small PP boxes under tightly controlled environmental conditions exhibited rapid substrate colonization and earlier fruiting initiation. Key parameters such as maintaining 24 °C during mycelial colonization, reducing to 18 °C for fruiting, coupled with 85%–90% RH and appropriate light exposure, were critical to maximizing productivity (Zhou et al., 2022).

Innovative methods have also explored the potential of anaerobically digested distillery effluents and spent mushroom substrate (SMS). When supplemented with wheat bran or soybean flour, even SMS (a by-product once seen as waste) can support additional cultivation cycles, significantly contributing to the circular economy (Economou et al., 2020). Moreover, additions like manganese and ammonium ions to sunflower seed hull substrates further boost mushroom output (Curvetto et al., 2002). As research continues to identify more efficient and low-cost substrates, it becomes increasingly evident that sustainability and BE can go hand in hand. Whether leveraging agricultural residues, industrial byproducts, or urban organic waste, oyster mushroom cultivation provides a remarkable opportunity to turn potential waste into nutritional wealth with significant ecological and economic benefits.

#### 2.2.3.2 Optimal Temperature for Primordia and Fruit Body Development

Under natural conditions, oyster mushrooms typically initiate fruiting when temperature range between 20 and 25 °C (Mahari et al., 2020). However, in controlled cultivation, cooler temperature is required for primordia initiation and maturation. Studies report that 15–18 °C is ideal for primordia formation, while 17–22 °C supports optimal fruiting and harvesting (Zhang et al., 2002; Baysal et al., 2003). For certain species like *P. tuoliensis*, even lower temperatures (12–14 °C) are necessary to induce fruit body formation (Zou et al., 2020). *Pleurotus ostreatus* shows robust performance during cooler months, reflecting its preference for moderate winter conditions (Sher et al., 2010). A recent investigation by Hu et al. (2023) found that 22 °C was optimal for *P. ostreatus* cultivation in bag systems, yielding a mycelial extension rate of 0.56 cm/d and BE nearing 99.5%. When ambient temperature exceeds the tolerable limits for a given strain, microclimate control strategies such as frequent misting and enhanced nighttime ventilation become critical to sustaining productive conditions (Oei and Nieuwenhuijzen, 2005).

#### 2.2.3.3 Optimal Moisture for Primordia and Fruit Body Development

Moisture plays a pivotal role in the physiological processes of *Pleurotus* cultivation, particularly in regulating primordia initiation and fruiting body development. Inadequate

**TABLE 2.3**
**Biological Efficiency of Various *Pleurotus* Species Cultivated on Different Substrates Across Countries**

| *Pleurotus* sp | Country | Substrate | BE (%) | Reference |
|---|---|---|---|---|
| *P. citrinopileatus* | Ethiopia | Non-composted coffee parchment | 20.86 | Dissasa, 2022 |
| | | Composted coffee parchment | 35.86 | |
| *P. eryngii* | India | Wheat straw | 76.7 | Deora et al., 2021 |
| | | Paddy straw | 88.4 | |
| | | Maize straw | 75 | |
| | | Sorghum straw | 47.5 | |
| | | Wheat straw + 5% Rice bran | 72.1 | |
| | | Wheat straw + 5% Wheat bran | 53.5 | |
| | Ethiopia | Non-composted coffee parchment | 14.79 | Dissasa, 2022 |
| | | Composted coffee parchment | 23.45 | |
| *P. floridanus* | China | Peach sawdust (210 kg), corncob (90 kg), and broken egg mixture (16.80 kg) | 100.19% | Wei et al., 2024 |
| *P. ostreatus* | Ethiopia | Non-composted coffee parchment | 39.13 | Dissasa, 2022 |
| | | Composted coffee parchment | 55.27 | |
| | Ethiopia | Waste paper (50%) + cornstalk (25%) + wheat bran (25%) | 646.4 | Tesfay et al., 2020 |
| | Brazil | Sugarcane bagasse (39.09 kg) + brizantha grass (48.51 kg) + wheat bran (8.9 kg) | 123.95 | Vieira and de Andrade, 2016 |
| | | Sugarcane bagasse (19.55 kg) + sugarcane straw (48.48 kg) + wheat bran (8.9 kg) | 103.7 | |
| | Greece | Spent mushroom substrate of *P. ostretaus* supplemented with wheat bran, soybean flour, and calcium carbonate | 185.38 | Economou et al., 2020 |
| | Denmark | Alfalfa pulp | 166.30 | Zhou et al., 2022 |
| | | Alfaalfa pulp:wheat straw = 1:1 | 101.85 | |
| | | Wheat straw | 22.72 | |
| | Thailand | Rubber sawdust + 9.00% rice bran + 0.03% $MgSO_4$ + 0.03% $CaCO_3$ + diluted lemon basil straw extract | 109.67 | Chanprapai et al., 2022 |
| | Saudi Arabia | Date palm: wheat straw = 25:75 | 250 | Alananbeh et al., 2014 |
| | Turkey | Sawdust:hazelnut leaves = 8:2 | 73.8 | Yildiz et al., 2002 |
| | | Sawdust:hazelnut leaves = 1:1 | 102.1 | |
| | | Wheat straw | 79.4 | |
| | | Wheat straw:waste paper = 1:1 | 121.2 | |
| | | Wheat straw:hazelnut leaves:waste paper = 3:5:2 | 91.7 | |
| | Argentina | Sunflower seed hull substrate with 100 ppm Mn(II) and 500 ppm $NH^{4+}$ | 112.2 | Curvetto et al., 2002 |
| *P. ostreatus* var. *florida* | Brazil | 2% $Ca(OH)_2$ treated *Brachiaria* straw | 37.63 | Iossi et al., 2018 |
| *P. pulmonarius* | Greece | Spent mushroom substrate of *P. ostretaus* supplemented with wheat bran, soybean flour, and calcium carbonate | 207.86 | Economou et al., 2020 |
| *P. sajor-caju* | USA | Rice straw (ground or chopped) | 75.5–131 | Zhang et al., 2002 |
| | | Wheat straw | 97 | |
| | | Mixture of wheat straw and rice straw | 101.2–108.3 | |
| | India | Rice straw:biogas residual slurry manure = 1:1 | 120 | Banik and Nandi, 2004 |
| *P. sapidus* | Ethiopia | Non-composted coffee parchment | 32.74 | Dissasa, 2022 |
| | | Composted coffee parchment | 47.18 | |
| | Brazil | 2% $Ca(OH)_2$ treated *Brachiaria* straw | 159.75 | Iossi et al., 2018 |
| *P. tuoliensis* | China | Cottonseed hull:corncob:wheat bran:maize powder:lime:gypsum = 65:15:10:5:4:1 | 48 | Zou et al., 2020 |
| | | Cottonseed hull:corncob:wheat bran:maize powder:lime:gypsum = 55:25:10:5:4:1 | 56.2 | |
| | | Cottonseed hull:corncob:wheat bran:maize powder:lime:gypsum = 45:35:10:5:4:1 | 54.4 | |
| | | Cottonseed hull:corncob:wheat bran:maize powder:lime:gypsum = 35:45:10:5:4:1 | 46.3 | |
| | | Cottonseed hull:corncob:wheat bran:maize powder:lime:gypsum = 25:55:10:5:4:1 | 46.5 | |

moisture can lead to desiccation of primordia, resulting in malformed or underdeveloped mushrooms, whereas excessive moisture often encourages abnormal fruiting patterns, including smaller and misshapen fruit bodies (Hoa et al., 2015; Silva et al., 2024). Generally, the ideal substrate moisture content falls within the range of 50% to 75%, though optimal levels vary by species and substrate composition (Chang and Miles, 2004). For successful primordia initiation and mature fruit body formation, RH should typically be maintained between 80%–95% and 85%–90%, respectively (Zhang et al., 2002; Baysal et al., 2003). Species-specific requirements have also been observed. For instance, *P. eryngii* demonstrates optimal mycelial growth at 65%–68% substrate moisture (Moonmoon et al., 2010; Ryu et al., 2015). For other *Pleurotus* species such as *P. albidus*, *P. cystidiosus*, *P. djamor*, and *P. ostreatus*, a substrate moisture content of approximately 74% has been deemed suitable for optimal productivity (Lechner and Albertó, 2011). Beyond quantity, the quality of moisture is equally critical. Spraying with contaminated water containing heavy metals like mercury, lead, or copper not only alters mushroom taste but also poses potential health risks (Ihugba et al., 2018; Ab Rhaman et al., 2021). Therefore, careful moisture management and the use of clean water sources are essential for producing high-quality, safe, and sustainable oyster mushrooms (Mahari et al., 2020).

#### 2.2.3.4 Optimal Luminosity for Primordia and Fruit Body Development

Although *Pleurotus* spp. are heterotrophic organisms incapable of photosynthesis, light significantly influences their preharvest growth and development (Bellettini et al., 2019). Light exposure affects the formation and maturation of fruiting bodies by regulating metabolic pathways and developmental processes such as stipe elongation, cap expansion, and spore dispersal (El Kolaly et al., 2020; Kim et al., 2020). In wild conditions, when optimal conditions that are suitable for fruiting, such as appropriate light intensity (e.g., 200–640 lux, 8–12 h/d), is available, oyster mushrooms undergo the reproductive phase (Ahmed et al., 2013; Mejía and Albertó, 2013; Mahari et al., 2020). However, the relationship between yield and light intensity may vary, with some studies reporting nonsignificant correlations (Kibar et al., 2008). Insufficient light or darkness typically results in pale, elongated, and malformed mushrooms, whereas excessive light can cause undesirable traits such as paleness and deformities (Marino et al., 2003; Urben, 2004). Kues and Liu (2000) described that blue and ultraviolet (UV) light wavelengths are most effective in initiating fruiting and maturation. In the absence of light, mushrooms may form abnormal stipe-like structures without caps (Oei and Nieuwenhuijzen, 2005). Innovative technologies like pulsed UV light enhance vitamin $D_2$ synthesis by converting ergosterol in mushrooms, with species-specific efficiency observed among *Pleurotus* spp. (Koyyalamudi et al., 2009, 2011; Chen et al., 2015).

### 2.2.4 Common Challenges in *Pleurotus* Cultivation

Despite the relative ease of cultivating *Pleurotus* species compared to other mushrooms, several recurring technical issues can hinder successful production, particularly during the preparation of pure cultures and spawn. One frequent problem is the improper preparation of agar medium. If the agar concentration is too low, the medium fails to solidify adequately, resulting in a soft texture unsuitable for optimal mycelial growth. This can be avoided by maintaining the recommended agar concentration during media formulation (Sharma and Kumar, 2011). Additionally, if the medium is allowed to cool too much before pouring, it may form an uneven, lumpy surface. This is best addressed by pouring the medium while it remains hot and uniformly fluid (El-Sayed et al., 2024). Surface contamination appearing within 48–72 hours post-sterilization often signals lapses in sterilization or aseptic technique. To curb such issues, meticulous sterilization and strict aseptic handling during media transfer are imperative (Sahu et al., 2014; Sofi et al., 2014). Contaminants on cotton plugs during incubation can result from already infected cultures, poor airflow filtration, or overcrowded incubation rooms which raise spore loads in the environment (Bandura et al., 2022). Preventative steps include using freshly prepared cultures, regularly maintaining laminar flow hoods, and spacing culture vessels adequately during incubation (Muswati et al., 2021). A lack of visible mycelial regrowth after tissue transfer (indicative of non-viable mycelium) is usually traced to aged or contaminated cultures, unsuitable media pH, or accidental thermal damage to the inoculum. These issues can be mitigated by using young, actively growing cultures, selecting media with appropriate pH, and allowing inoculating instruments to cool sufficiently before use (Telang et al., 2010; Raman et al., 2020; Prasad et al., 2024). Another notable challenge is slow, fluffy, or sparse mycelial growth, which often indicates genetic degeneration of the fungal strain. In such cases, reverting to preserved stock cultures or sourcing vigorous strains from verified repositories can restore performance (Thakur and Rathod, 2021).

Grain contamination after sterilization typically arises from infected raw materials or insufficient sterilization. Employing clean grains and extending sterilization cycles helps resolve such concerns (Stanley et al., 2010; Kumar et al., 2020). Improper grain moisture is another critical factor. Excessively wet or overly dry grains lead to weak or uneven mycelial colonization and poor substrate penetration. Adjusting moisture levels to optimal standards before autoclaving enhances spawn viability (Borah et al., 2019; Singh et al., 2022). Patchy colonization across the substrate bed may suggest bacterial interference or the use of low-vigor fungal strains. Rigorous sterilization and robust strain selection are essential to avoid such issues (Pal et al., 2008). Contamination during inoculation, particularly on grains or plugs, generally arises from poor aseptic handling or contaminated tools. Maintaining strict cleanliness, using sterile instruments, and following sterile technique protocols remain fundamental to avoid contamination (Sahu et al., 2014; Hoa and Wang, 2015). In some cases, sluggish mycelial progression is tied to nonoptimal substrates, substandard incubation conditions, or deteriorated fungal cultures. Ensuring appropriate substrate choice, maintaining incubation temperature within the recommended range, and using healthy cultures can significantly improve mycelial growth and spawn quality (Oei and Nieuwenhuijzen, 2005; Muswati et al., 2021; Prasad et al., 2024).

Another major challenge lies in selecting and processing substrates that are nutritionally balanced and free from contaminants. Poorly processed substrates may harbor competing microorganisms or insufficient nutrients. For example, substrates with high lignin and low nitrogen contents, like untreated sawdust, often lead to reduced mycelial colonization and BE (Akcay et al., 2023). Moreover, improper pasteurization or sterilization can leave behind residual pathogens that compromise mushroom yield (Ghimire et al., 2021). Inconsistent bed size, compaction, and inoculum distribution are common errors during bed preparation. Overcompaction of substrates can reduce aeration, impairing mycelial spread, whereas poor spawn distribution may lead to patchy colonization (Borah et al., 2019). Additionally, external biotic and abiotic factors can profoundly affect mushroom yield. Environmental variables including humidity, temperature, oxygen availability, and $CO_2$ concentration further influence the growth, development, and productivity of fruiting bodies (Patil et al., 2024). Mites and green molds are frequently observed during fruiting. These pests may not only damage fruiting bodies but also vector secondary infections (Stanley et al., 2010). Timing is critical during harvesting. Overmature mushrooms become spongy and lose market value, whereas immature harvesting reduces biomass yield. Manual harvesting must be done with care to avoid damaging adjacent pins (Zhang et al., 2002).

## 2.3 CONCLUSION AND FUTURE DIRECTIONS

*Pleurotus* cultivation is more than just mushroom farming—it is a smart, sustainable solution to some of today's biggest challenges: food insecurity, agro-waste pollution, and rural unemployment. With the right approach, it transforms low-cost materials into high-value nutrition and income, making it a perfect fit for circular bioeconomy models. The journey begins with robust mycelial growth. PDA remains the gold standard medium, while glucose and dextrose provide fast energy for expansion. Nitrogen sources like ammonium chloride and sulfate fuel stronger growth, especially under optimal conditions of 25–28 °C and pH 5.5–6.5. Spawn quality makes or breaks success. Grain spawn, especially from wheat or sorghum, is widely preferred for its strong, fast colonization. Sawdust spawn offers a sustainable option using local wood waste, while liquid spawn, though highly efficient, requires more technical input. At the heart of it all is the substrate. Treated wheat or rice straw, rich in lignocellulose and ideal C:N ratios, creates the perfect base for fruiting. Supplementing with bran, poultry litter, leaf powder, or even innovative inputs like broken eggs and waste paper can dramatically enhance yields and nutritional value. Emerging substrates like coffee husk or SMS showcase the power of waste-to-wealth innovations. The future depends on empowering communities through affordable, eco-friendly cultivation systems that foster women- and youth-led enterprises, align with multiple Sustainable Development Goals, and integrate mushroom farming into climate-smart agriculture. *Pleurotus* cultivation thus embodies resilience, sustainability, and a vision of hope for a better tomorrow.

## REFERENCES


Ab Rhaman, S. M. S., L. Naher, and S. Siddiquee. 2021. Mushroom quality related with various substrates' bioaccumulation and translocation of heavy metals. *Journal of Fungi* 8(1): 42. https://doi.org/10.3390/jof8010042

Abdullah, N., R. Ismail, N. M. K. Johari, and M. S. M. Annuar. 2013. Production of liquid spawn of an edible grey oyster mushroom, *Pleurotus pulmonarius* (Fr.) Quél by submerged fermentation and sporophore yield on rubber wood sawdust. *Scientia Horticulturae* 161: 65–69. https://doi.org/10.1016/j.scienta.2014.05.028

Ahmed, M., N. Abdullah, K. U. Ahmed, and M. H. M. Bhuyan. 2013. Yield and nutritional composition of oyster mushroom strains newly introduced in Bangladesh. *Pesquisa Agropecuária Brasileira* 48: 197–202. https://doi.org/10.1590/S0100-204X2013000200010

Akcay, C., F. Ceylan, and R. Arslan. 2023. Production of oyster mushroom (*Pleurotus ostreatus*) from some waste lignocellulosic materials and FTIR characterization of structural changes. *Scientific Reports* 13: 12897. https://doi.org/10.1038/s41598-023-40200-x

Alananbeh, K. M., N. A. Bouqellah, and N. S. Al Kaff. 2014. Cultivation of oyster mushroom *Pleurotus ostreatus* on date-palm leaves mixed with other agro-wastes in Saudi Arabia. *Saudi Journal of Biological Sciences* 21(6): 616–625. https://doi.org/10.1016/j.sjbs.2014.08.001

Ban, G. H., B. K. Kim, S. R. Kim, M. S. Rhee, and S. A. Kim. 2022. Bacterial microbiota profiling of oyster mushrooms (*Pleurotus ostreatus*) based on cultivation methods and distribution channels using high–throughput sequencing. *International Journal of Food Microbiology* 382: 109917. https://doi.org/10.1016/j.ijfoodmicro.2022.109917

Bandura, I., S. Makohon, T. S. Y. Z. Oleksandr, I. Ivanova, O. Khareba, V. Khareba, and S. Chausov. 2022. Effect of different grain spawn materials on *Pleurotus ostreatus* (Jacq.) P. Kumm. mushroom cultivation under unregulated and regulated fruiting conditions. *Acta agriculturae Slovenica* 118(1): 1–13. https://doi.org/10.14720/aas.2022.118.1.1862

Banik, S., and R. Nandi. 2004. Effect of supplementation of rice straw with biogas residual slurry manure on the yield, protein and mineral contents of oyster mushroom. *Industrial Crops and Products* 20: 311–319. https://doi.org/10.1016/j.indcrop.2003.11.003

Baysal, E., H. Peker, M. K. Yalinkiliç, and A. Temiz. 2003. Cultivation of oyster mushroom on waste paper with some added supplementary materials. *Bioresource Technology* 89(1): 95–97. https://doi.org/10.1016/s0960-8524(03)00028-2

Bellettini, M. B., F. A. Fiorda, H. A. Maieves, G. L. Teixeira, S. Ávila, P. S. Hornung, and R. H. Ribani. 2019. Factors affecting mushroom *Pleurotus* spp. *Saudi Journal of Biological Sciences* 26(4): 633–646. https://doi.org/10.1016/j.sjbs.2016.12.005

Bhatia, J. N., and A. N. Yadav. 2024. Characterization and yield performance of spawn prepared from *Hypsizygus ulmarius* (Bull.) Redhead and some *Pleurotus* species (Agaricomycetes). *Biocatalysis and Agricultural Biotechnology* 56: 103047. https://doi.org/10.1016/j.bcab.2024.103047

Bhatti, M. I., M. M. Jiskani, K. H. Wagan, M. A. Pathan, and M. R. Magsi. 2007. Growth, development and yield of oyster mushroom, *Pleurotus ostreatus* (Jacq. Ex. Fr.) Kummer as affected by different spawn rates. *Pakistan Journal of Botany* 39(7): 2685–2692.

Borah, T. R., A. R. Singh, P. Paul, H. Talang, B. Kumar, and S. Hazarika. 2019. Spawn production and mushroom cultivation technology. *ICAR Research Complex for NEH Region* 46.

Carrasco-Cabrera, C. P., T. L. Bell, and M. A. Kertesz. 2019. Caffeine metabolism during cultivation of oyster mushroom (*Pleurotus ostreatus*) with spent coffee grounds. *Applied Microbiology and Biotechnology* 103(14): 5831–5841. https://doi.org/10.1007/s00253-019-09883-z

Chang, S.-T., and P. G. Miles. 2004. *Mushrooms: Cultivation, Nutritional Value, Medicinal Effect and Environmental Impact.* CRC Press Inc. https://doi.org/10.1201/9780203492086

Chanprapai, P., T. Wichai, S. Sooksai, et al. 2022. Aqueous extracts of lemon basil straw as chemical stimulator for gray oyster mushroom cultivation. *Foods* 11(9): 1370. https://doi.org/10.3390/foods11091370

Chauhan, S., and D. C. Pant. 1988. Effect of spawn substrate and storage conditions on sporophore production in *Pleurotus sajor-caju. Indian Journal of Mycology and Plant Pathology* 18(3): 231–234.

Chen, S. Y., S. J. Huang, M. C. Cheng, Y. K. Chen, S. C. Yang, and J. L. Mau. 2015. Enhancement of vitamin D2 content in *Pleurotus* mushrooms using pulsed light. *Journal of Food Processing and Preservation* 39(6): 2027–2034. https://doi.org/10.1111/jfpp.12443

Choi, I. Y., G. T. Joung, J. Ryu, J. S. Choi, and Y. G. Choi. 2003. Physiological characteristics of green mold (*Trichoderma* spp.) isolated from oyster mushroom (*Pleurotus* spp.). *Mycobiology* 31(3): 139–144. https://doi.org/10.4489/MYCO.2003.31.3.139

Chuku, S. B., E. O. Nwachukwu, and H. O. Stanley. 2015. Effects of selected culture media on mycellial growth of oyster mushroom (*Pleurotus ostreatus*). *African Journal of Biotechnology* 14: 1471–1474.

Chukwu, S. C., C. A. Ibeji, C. Ogbu, et al. 2022. Primordial initiation, yield and yield component traits of two genotypes of oyster mushroom (*Pleurotus* spp.) as affected by various rates of lime. *Scientific Reports* 12(1): 19054. https://doi.org/10.1038/s41598-022-16833-9

Curvetto, N. R., D. Figlas, R. Devalis, and S. Delmastro. 2002. Growth and productivity of different *Pleurotus ostreatus* strains on sunflower seed hulls supplemented with N–$NH^{4+}$ and/or Mn(II). *Bioresource Technology* 84(2): 171–176. https://doi.org/10.1016/s0960-8524(02)00013-5

Das, N., and M. Mukherjee. 2007. Cultivation of *Pleurotus ostreatus* on weed plants. *Bioresource Technology* 98(14): 2723–2726. https://doi.org/10.1016/j.biortech.2006.09.061

De Mastro, F., A. Traversa, F. Matarrese, C. Cocozza, and G. Brunetti. 2023. Influence of growing substrate preparation on the biological efficiency of *Pleurotus ostreatus. Horticulturae* 9: 439. https://doi.org/10.3390/horticulturae9040439

Deora, A., S. S. Sharma, P. Kumari, V. Dahima, S. Kumar, and M. Rohith. 2021. Cultivation of Kabul Dhingri (*Pleurotus eryngii*) mushroom by standardizing protocols in subtropical zones of world. *Scientific Reports* 11(1): 14692. https://doi.org/10.1038/s41598-021-94038-2

Dissasa, G. 2022. Cultivation of different oyster mushroom (*Pleurotus* species) on coffee waste and determination of their relative biological efficiency and pectinase enzyme production, Ethiopia. *International Journal of Microbiology* 2022: 5219939. https://doi.org/10.1155/2022/5219939

Economou, C. N., A. N. Philippoussis, and P. A. Diamantopoulou. 2020. Spent mushroom substrate for a second cultivation cycle of *Pleurotus* mushrooms and dephenolization of agro–industrial wastewaters. *FEMS Microbiology Letters* 367(8): fnaa060. https://doi.org/10.1093/femsle/fnaa060

El Kolaly, W., W. Ma, M. Li, and M. Darwesh. 2020. The investigation of energy production and mushroom yield in greenhouse production based on mono photovoltaic cells effect. *Renewable Energy* 159: 506–518. https://doi.org/10.1016/j.renene.2020.05.144

El-Sayed, O., E. A. Bardisi, H. G. Zyada, and M. H. Arisha. 2024. Using different grain substrates for spawn production and maximization of oyster mushroom (*Pleurotus florida*) production. *Zagazig Journal of Agricultural Research* 51(2): 151–156. https://doi.org/10.21608/ZJAR.2024.353637

Gapiński, M., W. Woźniak, M. Ziombra, and J. Murawska. 2007. Oyster mycelium on the liquid medium. *Acta Mycologica* 42(1): 125–128.

Ghimire, A., K. R. Pandey, Y. R. Joshi, and S. Subedi. 2021. Major fungal contaminants of mushrooms and their management. *International Journal of Applied Sciences and Biotechnology* 9(2): 80–93.

Girmay, Z., W. Gorems, G. Birhanu, and S. Zewdie. 2016. Growth and yield performance of *Pleurotus ostreatus* (Jacq. Fr.) Kumm (oyster mushroom) on different substrates. *AMB Express* 6(1): 87. https://doi.org/10.1186/s13568-016-0265-1

Hernández, D., J. E. Sánchez, and K. Yamasaki. 2003. A simple procedure for preparing substrate for *Pleurotus ostreatus* cultivation. *Bioresource Technology* 90(2): 145–150. https://doi.org/10.1016/s0960-8524(03)00118-4

Hoa, H. T., and C. L. Wang. 2015. The effects of temperature and nutritional conditions on mycelium growth of two oyster mushrooms (*Pleurotus ostreatus* and *Pleurotus cystidiosus*). *Mycobiology* 43(1): 14–23. https://doi.org/10.5941/MYCO.2015.43.1.14

Hu, Y., F. Xue, Y. Chen, et al. 2023. Effects and mechanism of the mycelial culture temperature on the growth and development of *Pleurotus ostreatus* (Jacq.) P. Kumm. *Horticulturae* 9(1): 95. https://doi.org/10.3390/horticulturae9010095

Ihugba, U. A., C. O. Nwoko, F. R. Tony-Njoku, A. A. Ojiaku, and L. Izunobi. 2018. Heavy metal determination and health risk assessment of oyster mushroom *Pleurotus tuberregium* (Fr.) Singer, collected from selected markets in Imo State. Nigeria. *American Journal of Environmental Protection* 6(1): 22–27. https://doi.org/10.12691/env-6-1-4

Iossi, M. R., J. D. V. Cobos, F. J. G. Alegria, and D. C. Zied. 2018. *Pleurotus* spp. cultivation on *Brachiaria* sp. straw treatment with alkaline water: oyster mushroom and substrate treatment. *Brazilian Journal of Microbiology* 49(Suppl 1): 64–67. https://doi.org/10.1016/j.bjm.2018.06.003

Irshad, A., A. Tahir, S. Sharif, et al. 2023. Determination of nutritional and biochemical composition of selected *Pleurotus* spp. *BioMed Research International* 2023(1): 8150909. https://doi.org/10.1155/2023/8150909

Kerketta, V., C. S. Shukla, H. K. Singh, and A. Kerketta. 2019. Effect of grain substrates on spawn development and their impact on yield and yield attributing characters of *Pleurotus* spp. *Journal of Pharmacognosy and Phytochemistry* 8(6): 722–725.

Kibar, B., and A. Peksen. 2008. Modelling the effects of temperature and light intensity on the development and yield of different *Pleurotus* species. *Agricultura Tropica Et Subtropica* 41(2): 68–73.

Kim, H. S., J. M. Lohmar, M. Busman, et al. 2020. Identification and distribution of gene clusters required for synthesis of sphingolipid metabolism inhibitors in diverse species of the filamentous fungus *Fusarium*. *BMC Genomics* 21: 510. https://doi.org/10.1186/s12864-020-06896-1

Koul, B., M. Yakoob, and M. P. Shah. 2022. Agricultural waste management strategies for environmental sustainability. *Environmental Research* 206: 112285. https://doi.org/10.1016/j.envres.2021.112285

Koyyalamudi, S. R., S. C. Jeong, G. Pang, A. Teal, and T. Biggs. 2011. Concentration of vitamin D2 in white button mushrooms (*Agaricus bisporus*) exposed to pulsed UV light. *Journal of Food Composition and Analysis* 24(7): 976–979. https://doi.org/10.1016/j.jfca.2011.02.007

Koyyalamudi, S. R., S. C. Jeong, C. H. Song, K. Y. Cho, and G. Pang. 2009. Vitamin D2 formation and bioavailability from *Agaricus bisporus* button mushrooms treated with ultraviolet irradiation. *Journal of Agricultural and Food Chemistry* 57(8): 3351–3355. https://doi.org/10.1021/jf803908q

Kumar, S., G. Singh, R. Singh, et al. 2020. Effect of different cereals flour additives on sporophores production of Oyster mushroom (*Pleurotus sapidus* and *Pleurotus flabellatus*). *International Journal of Chemical Studies* 8(1): 698–701. https://doi.org/10.22271/chemi.2020.v8.i1j.8348

Lechner, B. E., and E. Albertó. 2011. Search for new naturally occurring strains of *Pleurotus* to improve yields. *Pleurotus albidus* as a novel proposed species for mushroom production. *Revista Iberoamericana de Micología* 28(4): 148–154. https://doi.org/10.1016/j.riam.2010.12.001

Lenka, K. C., B. Padhan, N. Pradhan, T. Mantry, R. Sahu, and S. Venkatlaxmi. 2022. The effect of growth conditions on mycelial run of oyster mushrooms spp. (*Pleurotus* spp.): implication for agricultural practices. *Bhartiya Krishi Anusandhan Patrika* 37(2): 137–143.

Li, H., P. Liu, Z. Li, et al. 2024. Valorization of *Ginkgo biloba* Leaf powder as a substrate in king oyster mushroom (*Pleurotus eryngii*) cultivation. *Life (Basel)* 14(5): 639. https://doi.org/10.3390/life14050639

Lima, U. A., E. Aquarone, and W. Borzani. 1975. *Biotecnologia: Tecnologia das fermentacoes*. Editora Universidade de Sao Paulo, Sao Paulo.

Liu, H., J. Long, K. Zhang, et al. 2024. Agricultural biomass/waste–based materials could be a potential adsorption–type remediation contributor to environmental pollution induced by pesticides–a critical review. *Science of the Total Environment* 946: 174180. https://doi.org/10.1016/j.scitotenv.2024.174180

Liu, S. R., W. R. Zhang, and Y. B. Kuang. 2018. Production of stalk spawn of an edible mushroom (*Pleurotus ostreatus*) in liquid culture as a suitable substitute for stick spawn in mushroom cultivation. *Scientia Horticulturae* 240: 572–577. https://doi.org/10.1016/j.scienta.2018.06.068

Mahadevan, K., and K. Shanmugasundaram. 2018. Comparative effect of different culture media on mycelial growth performance of *Pleurotus sapidus*. *Journal of Pharmacognosy and Phytochemistry* 7(4): 874–878.

Mahari, W. A. W., W. Peng, W. L. Nam, et al. 2020. A review on valorization of oyster mushroom and waste generated in the mushroom cultivation industry. *Journal of Hazardous Materials* 400: 123156. https://doi.org/10.1016/j.jhazmat.2020.123156

Marino, R. H., A. F. D. Eira, E. E. Kuramae, and E. C. Queiroz. 2003. Morphomolecular characterization of *Pleurotus ostreatus* (Jacq. Fr.) kummer strains in relation to luminosity and temperature of frutification. *Scientia Agricola* 60: 531–535. https://doi.org/10.1590/S0103-90162003000300018

Maseko, K. H., T. Regnier, O. C. Wokadala, P. Bartels, and B. Meiring. 2024. Effect of culture media on the yield and protein content of *Pleurotus ostreatus* (Jacq.) Kumm mycelia. *International Journal of Food Science* 2024: 5562732. https://doi.org/10.1155/ijfo/5562732

Mejía, S. J., and E. Albertó. 2013. Heat treatment of wheat straw by immersion in hot water decreases mushroom yield in *Pleurotus ostreatus*. *Revista iberoamericana de micologia* 30(2): 125–129. https://doi.org/10.1016/j.riam.2012.11.004

Moonmoon, M., M. N. Uddin, S. Ahmed, N. J. Shelly, and M. A. Khan. 2010. Cultivation of different strains of king oyster mushroom (*Pleurotus eryngii*) on saw dust and rice straw in Bangladesh. *Saudi Journal of Biological Sciences* 17(4): 341–345.

Mshandete, A. M., and J. R. Mgonja. 2009. Submerged liquid fermentation of some Tanzanian basidiomycetes for the production of mycelial biomass, exopolysaccharides and mycelium protein using wastes peels media. *Journal of Agricultural and Biological Science* 4: 1–13.

Muswati, C., K. Simango, L. Tapfumaneyi, M. Mutetwa, and W. Ngezimana. 2021. The effects of different substrate combinations on growth and yield of oyster mushroom (*Pleurotus ostreatus*). *International Journal of Agronomy* 2021: Article ID 9962285. https://doi.org/10.1155/2021/9962285

Neelam, S., S. Chennupati, and S. Singh. 2013. Comparative studies on growth parameters and physio-chemical analysis of *Pleurotus ostreatus* and *Pleurotus florida*. *Asian Journal of Plant Science and Research* 3(1): 163–169.

Nguyen, T. M., and S. L. Ranamukhaarachchi. 2020. Effect of different culture media, grain sources and alternate substrates on the mycelial growth of *Pleurotus eryngii* and *Pleurotus ostreatus*. *Pakistan Journal of Biological Sciences* 23(3): 223–230. https://doi.org/10.3923/pjbs.2020.223.230

Obodai, M., J. Cleland-Okine, and K. A. Vowotor. 2003. Comparative study on the growth and yield of *Pleurotus ostreatus* mushroom on different lignocellulosic by-product. *Journal of Industrial Microbiology and Biotechnology* 30: 146–149. https://doi.org/10.1007/s10295-002-0021-1

Oei, P. 2003. *Mushroom Cultivation, Appropriate Technology for Mushroom Growers*. Backhugs Publishers, Leiden. The Netherlands.

Oei, P., and B. V. Nieuwenhuijzen. 2005. *Small-Scale Mushroom Cultivation: Oyster, Shiitake and Wood Ear Mushrooms*. Agromisa Foundation and CTA, Wageningen.

Oh, S. J., P. G. Shin, H. Y. Weon, K. H. Lee, and G. H. Chon. 2003. Effect of fermented sawdust on *Pleurotus* spawn. *Mycobiology* 31(1): 46–49.

Pal, D. P., C. S. Shukla, A. K. Deo, and M. K. Tripathi. 2008. Effects of grain substrate on spawn run, pinhead and yield of wild *Pleurotus* sp. and *Pleurotus eous*. *Plant Archives* 8(1): 401–402.

Pandey, N., A. Kerketta, B. Sahu, and H. K. Singh. 2017. Screening of suitable grains substrates for spawn development, growth and yield of *Pleurotus flabellatus*. *Research Journal of Agricultural Sciences* 8(2): 536–541.

Pant, D., U. G. Reddy, and A. Adholeya. 2006. Cultivation of oyster mushroom on wheat straw and bagasse substrate amended with distillery effluent. *World Journal of Microbiology and Biotechnology* 22: 267–275.

Patil, R., A. N. M. Ramli, A. S. Xuan, N. Z. Xin, N. I. W. Azelee, and P. Bhuyar. 2024. Unlocking the growth potential: harnessing the power of synbiotics to enhance cultivation of *Pleurotus* spp. *Journal of Zhejiang University Science B* 25(4): 293–306. https://doi.org/10.1631/jzus.B2300383

Philippoussis, A., G. Zervakis, and P. Diamantopoulou. 2001. Bioconversion of agricultural lignocellulosic wastes through the cultivation of the edible mushrooms *Agrocybe aegerita*, *Volvariella volvacea* and *Pleurotus* spp. *World Journal of Microbiology and Biotechnology* 17: 191–200.

Pradhan, P., J. De, and K. Acharya. 2024. Cultivation strategies of *Ganoderma* or Reishi mushroom. In: *Ganoderma: Cultivation, Chemistry and Medicinal Applications, Volume 1*, ed. K. Acharya and S. Khatua, 19–34. CRC Press. https://doi.org/10.1201/9781003354789-2

Prasad, R., I. Singh, and A. Milan. 2024. Production of spawn and storage techniques. In: *Scientific Inputs in Modern Indian Agriculture*, ed. S. Siddiqui, 249–259. Weser Books.

Rajarathnam, S., and Z. Bano. 1987. *Pleurotus* mushrooms. Part I A. Morphology, life cycle, taxonomy, breeding, and cultivation. *Critical Reviews in Food Science and Nutrition* 26(2): 157–223. https://doi.org/10.1080/10408398709527465

Raman, J., K. Y. Jang, Y. L. Oh, et al. 2020. Cultivation and nutritional value of prominent *Pleurotus* spp.: an overview. *Mycobiology* 49(1): 1–14. https://doi.org/10.1080/12298093.2020.1835142

Rawte, H., and R. Diwan. 2011. Growth response of *Pleurotus* spp. on different basal media and different pH levels. *Journal of Ecobiotechnology* 3(4): 10–12.

Royse, D. J., T. W. Rhodes, S. Ohga, and J. E. Sánchez. 2004. Yield, mushroom size and time to production of *Pleurotus cornucopiae* (oyster mushroom) grown on switch grass substrate spawned and supplemented at various rates. *Bioresource Technology* 91: 85–91. https://doi.org/10.1016/S0960-8524(03)00151-2

Ruiz-Rodríguez, A., C. Soler-Rivas, I. Polonia, and H. J. Wicher. 2010. Effect of olive mill waste (OMW) supplementation to oyster mushrooms substrates on the cultivation parameters and fruiting bodies quality. *International Biodeterioration and Biodegradation* 64: 638–645.

Ryu, J. S., M. K. Kim, C. H. Im, and P. G. Shin. 2015. Development of cultivation media for extending the shelf-life and improving yield of king oyster mushrooms (*Pleurotus eryngii*). *Scientia Horticulturae* 193: 121–126. https://doi.org/10.1016/j.scienta.2015.07.005

Sahu, S. K., D. P. Singh, R. Patel, G. K. Awadhiya, and I. G. K. Vishwavidhyalaya. 2014. Screening of suitable grains substrates for Spawn development, growth and yield of *Pleurotus eous*. *American International Journal of Research in Formal, Applied and Natural Sciences* 5(1): 86–89.

Sánchez, C. 2010. Cultivation of *Pleurotus ostreatus* and other edible mushrooms. *Applied Microbiology and Biotechnology* 85(5): 1321–1337. https://doi.org/10.1007/s00253-009-2343-7

Sardar, H., M. A. Ali, C. M. Ayyub, and R. Ahmed. 2015. Effects of different culture media, temperature and pH levels on the growth of wild and exotic *Pleurotus* species. *Pakistan Journal of Phytopathology* 27(2): 139–145.

Sariyer, T., Ç. Kaya, M. A. Gündoğdu, et al. 2023. Determination of the effects of different tree logs on yield and some quality properties in oyster mushroom [*Pleurotus ostreatus* (Jacq.) P. Kumm.]. *International Journal of Food, Agriculture and Animal Sciences* 3(2): 23–33.

Sarker, N. C., M. M. Hossain, N. Sultana, I. H. Mian, A. J. M. S. Karim, and S. M. R. Amin. 2007. Effect of different levels of pH on the growth and yield of *Pleurotus ostreatus* (Jacquin ex. Fr.) Kummer. *Bangladesh Journal of Mushroom* 1(1): 57–62.

Sharma, B. B. 2003. Effect of different substrates (grains/straws) on spawn growth and yield of pink oyster mushroom *Pleurotus djamor* [Fr.] Boedijn. *Journal of Mycology and Plant Pathology* 33: 265–268.

Sharma, V. P., and S. Kumar. 2011. *Spawn Production Technology*. Mushrooms Cultivation, Marketing and Consumption, Directorate of Mushroom Research (ICAR), Chambaghat, Solan 35–42.

Sher, H., M. Al-Yemeni, A. H. A. Bahkali, and H. Sher. 2010. Effect of environmental factors on the yield of selected mushroom species growing in two different agro ecological zones of Pakistan. *Saudi Journal of Biological Sciences* 17(4): 321–326. https://doi.org/10.1016/j.sjbs.2010.06.004

Silva, M., A. C. Ramos, F. J. Lidon, F. H. Reboredo, and E. M. Gonçalves. 2024. Pre- and postharvest strategies for *Pleurotus ostreatus* mushroom in a circular economy approach. *Foods* 13(10): 1464. https://doi.org/10.3390/foods13101464

Singh, J., R. K. Chugh, N. Singh, S. K. Mehta, and N. K. Yadav. 2022. Evaluation of different substrates for spawn preparation of monkey head mushroom. *Plant Disease Research* 37(2): 173–177.

Sofi, B., M. Ahmad, and M. Khan. 2014. Effect of different grains and alternate substrates on oyster mushroom (*Pleurotus ostreatus*) production. *African Journal of Microbiology Research* 8(14): 1474–1479. https://doi.org/10.5897/AJMR2014.6697

Stanley, H. O., and G. D. Awi-Waadu. 2010. Effect of substrates of spawn production on mycelial growth of oyster mushroom species. *Agriculture and Biology Journal of North America* 1(5): 817–820.

Subedi, S., N. Kunwar, K. R. Pandey, and Y. R. Joshi. 2023. Performance of oyster mushroom (*Pleurotus ostreatus*) on paddy straw, water hyacinth and their combinations. *Heliyon* 9(8): e19051. https://doi.org/10.1016/j.heliyon.2023.e19051

Suharban, M., and M. C. Nair. 1994. Physiological studies on *Pleurotus* spp. *Mushroom Research* 3(1): 99.

Telang, S. M., S. S. Patil, and M. M. V. Baig. 2010. Comparative study on yield and nutritional aspect of *Pleurotus eous* mushroom cultivated on different substrate. *Food Science Research Journal* 1: 60–63.

Tesfay, T., T. Godifey, R. Mesfin, and G. Kalayu. 2020. Evaluation of waste paper for cultivation of oyster mushroom (*Pleurotus ostreatus*) with some added supplementary materials. *AMB Express* 10(1): 15. https://doi.org/10.1186/s13568-020-0945-8

Thakur, G. M., and M. G. Rathod. 2021. Spawn preparation techniques in mushroom cultivation. *Research Insights of Life Science Students* 3: 712–714.

Tinoco, R., M. A. Pickard, and R. Vazquez-Duhalt. 2001. Kinetic differences of purified laccases from six *Pleurotus ostreatus* strains. *Letters in Applied Microbiology* 32(5): 331–335. https://doi.org/10.1046/j.1472-765X.2001.00913.x

Ufitinema, B., M. Hakizimana, and J. D. D. Niyomugabo. 2023. Mycelial growth of oyster mushroom in spawn made using different cereal grains, sawdust and vermiculite as substrate. *Mushroom Research* 32(1): 35–40. https://doi.org/10.36036/MR.32.1.2023.136300

Urben, A. F. 2004. *Produção de cogumelos por meio de tecnologia chinesa modificada Embrapa Recursos Genéticos e Biotecnologia*. Brasília.

Velioglu, Z., and R. O. Urek. 2015. Optimization of cultural conditions for biosurfactant production by *Pleurotus djamor* in solid state fermentation. *Journal of Bioscience and Bioengineering* 120(5): 526–531. https://doi.org/10.1016/j.jbiosc.2015.03.007

Vieira, F. R., and M. C. N. de Andrade. 2016. Optimization of substrate preparation for oyster mushroom (*Pleurotus ostreatus*) cultivation by studying different raw materials and substrate preparation conditions (composting: phases I and II). *World Journal of Microbiology and Biotechnology* 32: 190. https://doi.org/10.1007/s11274-016-2152-y

Waghmare, C. S. 2016. Development of protocol for liquid spawn of Oyster Mushroom. Doctoral dissertation, Mahatma Phule Krishi Vidyapeeth (Pune).

Wei, J., Y. X. Wang, T. K. Guan, et al. 2024. Bacterial communities during composting cultivation of oyster mushroom *Pleurotus floridanus* using broken eggs as the nitrogen source and study of agronomic and nutritional properties. *Frontiers in Microbiology* 14: 1274032. https://doi.org/10.3389/fmicb.2023.1274032

Werghemmi, W., S. Abou Fayssal, H. Mazouz, H. Hajjaj, and L. Hajji. 2022. Olive and green tea leaves extract in *Pleurotus ostreatus* var. *florida* culture media: effect on mycelial linear growth rate, diameter and growth induction index. *IOP Conference Series: Earth and Environmental Science* 1090(1): 012020. https://doi.org/10.1088/1755-1315/1090/1/012020

Xiao, X., Y. Li, X. Li, et al. 2024. Identification, nutrient composition, and evaluation of a wild *Pleurotus citrinopileatus* strain (X21156) from Tibet for antioxidant and cytotoxic activities. *Horticulturae* 10(4): 377. https://doi.org/10.3390/horticulturae10040377

Yang, W., F. Guo, and Z. Wan. 2013. Yield and size of oyster mushroom grown on rice/wheat straw basal substrate supplemented with cotton seed hull. *Saudi Journal of Biological Sciences* 20(4): 333–338. https://doi.org/10.1016/j.sjbs.2013.02.006

Yang, Y. R., Y. X. Guo, Q. Y. Wang, et al. 2022. Impacts of composting duration on physicochemical properties and microbial communities during short-term composting for the substrate for oyster mushrooms. *Science of the Total Environment* 847: 157673. https://doi.org/10.1016/j.scitotenv.2022.157673

Yildiz, S., U. C. Yildiz, E. D. Gezer, and A. Temiz. 2002. Some lignocellulosic wastes used as raw material in cultivation of the *Pleurotus ostreatus* culture mushroom. *Process Biochemistry* 38: 301–306. https://doi.org/10.1016/S0032-9592(02)00040-7

Zhang, R., X. Li, and J. G. Fadel. 2002. Oyster mushroom cultivation with rice and wheat straw. *Bioresource Technology* 82(3): 277–284. https://doi.org/10.1016/s0960-8524(01)00188-2

Zhang, R. Y., D. D. Hu, X. T. Ma, et al. 2014. Adapting stick spawn reduced the spawn running time and improved mushroom yield and biological efficiency of *Pleurotus eryngii*. *Scientia Horticulturae* 175: 156–159. https://doi.org/10.1016/j.scienta.2014.05.028

Zhang, W. R., S. R. Liu, Y. B. Kuang, and S. Z. Zheng. 2019. Development of a novel spawn (block spawn) of an edible mushroom, *Pleurotus ostreatus*, in liquid culture and its cultivation evaluation. *Mycobiology* 47(1): 97–104. https://doi.org/10.1080/12298093.2018

Zharare, G. E., S. M. Kabanda, and J. Z. Poku. 2010. Effects of temperature and hydrogen peroxide on mycelial growth of eight *Pleurotus* strains. *Scientia Horticulturae* 125(2): 95–102. https://doi.org/10.1016/j.scienta.2010.03.006

Zhou, F., M. Hansen, T. J. Hobley, and P. R. Jensen. 2022. Valorization of green biomass: alfalfa pulp as a substrate for oyster mushroom cultivation. *Foods* 11(16): 2519. https://doi.org/10.3390/foods11162519

Zou, Y., F. Du, Q. Hu, X. Yuan, D. Dai, and M. Zhu. 2020. Integration of *Pleurotus tuoliensis* cultivation and biogas production for utilization of lignocellulosic biomass as well as its benefit evaluation. *Bioresource Technology* 317: 124042. https://doi.org/10.1016/j.biortech.2020.124042

# 3 A Comprehensive Overview of the Nutritional Value in Oyster Mushrooms

*Charalampia Amerikanou, Stamatia-Angeliki Kleftaki, and Andriana C. Kaliora**
Department of Nutrition and Dietetics, School of Health Sciences & Education, Harokopio University of Athens, Athens, Greece

## 3.1 INTRODUCTION

The *Pleurotus* genus includes species that produce mushrooms with a characteristic morphology (commonly known as "oyster mushrooms"). These have a global distribution and primarily grow as saprotrophs on a wide range of substrates (mainly dead wood and plant residues). Mushrooms of this genus have gills and a distinctive "pleurotoid" shape, with an eccentric or lateral stipe, or even a rudimentary one. Together with species of the *Hohenbuehelia* genus, they are unique in their ability to attack and feed on live nematodes. Currently, 39 species of *Pleurotus* have been identified worldwide, with 22 found in India. Among these, commonly cultivated species include *Pleurotus ostreatus*, *Pleurotus pulmonarius*, *Pleurotus eryngii*, *Pleurotus citrinopileatus*, *Pleurotus djamor*, *Pleurotus cornucopiae*, *Pleurotus tuber-regium*, *Pleurotus florida*, *Pleurotus salmoneostramineus*, *Pleurotus eous*, *Pleurotus flabellatus*, *Pleurotus porrigens*, *Pleurotus sajor-caju*, and *Pleurotus cystidiosus*. Species of the *Agaricus* genus (mainly *A. bisporus*) make up one-third of total production, they are followed by species of the *Pleurotus* genus (primarily *P. ostreatus* and secondarily *P. eryngii* and *P. pulmonarius*) (FAOSTAT, 2023). Certain oyster mushroom species are well known for their high nutritional content, offering a good source of protein, vitamins, minerals, and antioxidants, which contribute to various health benefits. Commonly known as the king oyster mushroom, *P. eryngii* is valued for its unique morphology, adaptability to various substrates, and significant culinary and nutraceutical properties. The name *P. eryngii* encompasses various varieties differentiated primarily by their host plants (Zervakis et al., 2014). This species is distributed from Europe (mainly Mediterranean countries) to North Africa and the Middle East and has also been recorded in other Asian countries. Its hosts belong to the Apiaceae family, including plants of the *Eryngium* genus (where *P. eryngii var. eryngii* develops), *Ferula communis* (*P. eryngii var. ferulae*), *Elaeoselinum asclepium* and *E. asclepium subsp. meioides* (*P. eryngii var. elaeoselini*), and *Thapsia garganica* (*P. eryngii var. thapsiae*). The species produces large fruiting bodies with a convex to funnel-shaped cap in shades of pale beige to dark gray or brown, often with cracks. The cap margin remains curled downward until near full maturity. The decurrent gills extend to the stipe, which is central or eccentric. The spores are cylindrical to elliptical, smooth, and measure 10–12.5 × 4.5–5.5 μm (Zervakis et al., 2014).

*Pleurotus* species can be commercially grown on various substrates, such as paddy straw, wheat straw, and maize stalks, without requiring additional nitrogen supplements (Singh et al., 2021). Cultivation represents a remarkable example of integrated management and utilization of agricultural, forestry, and agri-industrial byproducts through controlled solid fermentation that converts these byproducts into edible biomass. Key stages include substrate preparation (hydration, mixing, and often composting), pasteurization or sterilization, inoculation, incubation (complete colonization of the substrate by fungal mycelium), and mushroom growth and harvesting. Maintaining suitable environmental conditions is essential for reducing cultivation time, maximizing yield, and ensuring high-quality products. *Pleurotus* mushrooms have been successfully cultivated on olive and wine byproducts, which are rich in lignocellulosic material, and thus are excellent substrates for the cultivation of mushrooms with distinct characteristics and added value (Koutrotsios et al., 2018; Tagkouli et al., 2020, 2021).

## 3.2 NUTRITIONAL CONTENT

Oyster mushrooms are highly nutritious, low in calories and fat, typically containing only 0.4 to 0.5 g of fat/100 g. Conversely, they are rich in protein, fiber, vitamins, and minerals (Table 3.1). Specifically, *Pleurotus* species contain a variety of vitamins, including niacin, riboflavin, and vitamin D, as well as essential minerals such as potassium, phosphorus, and iron. In addition, oyster mushrooms are abundant in bioactive compounds, such as dietary fibers, essential amino acids, sterols, terpenoids, statins, β-glucans, and antioxidant microconstituents like ergothioneine. These compounds are associated with a wide range of health benefits, including immunomodulatory, hypoglycemic, hypocholesterolemic, antimicrobial, anti-inflammatory, and

* Corresponding author: akaliora@hua.gr

DOI: 10.1201/9781003527572-3

**TABLE 3.1**
**Nutritional Composition of Oyster Mushrooms**

| | g/100 g | | | | | | |
|---|---|---|---|---|---|---|---|
| **Species** | **Carbohydrate** | **Crude Fiber** | **Protein** | **Lipid** | **Ash** | **Moisture** | **Reference** |
| *P. citrinopileatus* | 53.10 ± 0.36 | 6.85 | 18.20 ± 2.56 | 7.90 ± 0.44 | 9.10 ± 0.58 | 3.90 ± 0.29 | Acosta-Urdapilleta et al., 2020 |
| *P. djamor* | 61.1 ± 0.00 | 4.07 | 17.00 ± 0.90 | 6.70 ± 0.13 | 8.90 ± 1.17 | 2.10 ± 0.21 | |
| *P. eryngii* | 59.69 | - | 25.36 | 5.18 | - | - | Sardar et al., 2017 |
| *P. florida* | 40.33 ± 0.10 | 12.56 ± 0.10 | 27.81 ± 0.10 | 2.49 ± 0.05 | 7.60 ± 0.20 | 9.21 ± 0.30 | Priyadarshni et al., 2022 |
| *P. floridanus* | 55.83 ± 1.00 | 18.09 ± 0.16 | 11.10 ± 1.03 | 8.73 ± 0.04 | 6.25 ± 0.17 | - | Chai et al., 2021 |
| *P. pulmonarius* | 60.63 ±4.21 | 13.70 ±0.95 | 15.09 ±3.93 | 4.77 ±0.77 | 5.81 ± 0.84 | - | |
| *P. ostreatus* | 43.42±0.01 | 23.63±0.01 | 17.06±0.17 | 1.21±0.02 | 8.22±0.04 | 6.46±0.04 | Effiong et al., 2024c |
| *P. sajor-caju* | 82.47 | 54.12 | 23.51 | 1.88 | 2.1 | - | Irshad et al., 2023 |
| *P. sapidus* | 72.52 | 8.48 | 16.07 | 0.64 | 5.87 | - | |
| *P. tuber—regium* | 42.17 ± 0.02 | 17.49 ± 0.01 | 18.29 ± 0.00 | 6.81 ± 0.02 | 0.88 ± 0.01 | 13.36 ± 0.02 | Afolabi et al., 2024 |

antihypertensive activities. Furthermore, *Pleurotus* mushrooms have demonstrated antithrombotic, antiatherogenic, antitumor, and prebiotic properties, all of which contribute to overall health and may help prevent chronic diseases.

The moisture content in dried oyster mushrooms, ranging from 2.1% to 13.36% of dry weight (DW), reflects their residual water. Lower moisture levels correlate with reduced water activity, which helps inhibit microbial growth and enhances food preservation (Chen et al., 2012). Dried oyster mushrooms typically have lower moisture levels than oats and barley, making them less prone to spoilage, though their moisture content is similar to that of other edible mushrooms (Lai et al., 2024) The stalk has the highest moisture content at 6.33 g/100 g; while the cap contains the lowest moisture content at 3.48 g/100 g (Oluwafemi et al., 2016).

In terms of ash content, which represents the residue left after incinerating the mushrooms at high temperatures (500–600 °C for 24 hours), dried oyster mushrooms generally have higher ash content compared to oats and barley. However, their ash content is similar to that of other edible mushrooms. Ash content also varies between different parts of the mushroom, ranging from 5.30 g to 8.24 g/100 g (Ho et al., 2020).

### 3.2.1 Carbohydrates

Oyster mushrooms contain various forms of carbohydrates, including starches, pentoses, hexoses, disaccharides, amino sugars, sugar alcohols, and sugar acids. Among the species, *P. sajor-caju* demonstrates the highest carbohydrate concentration at 82.47% (Irshad et al., 2023). In contrast, *P. florida* has the lowest carbohydrate content at 40.33%, indicating its suitability for lower-calorie diets (Priyadarshni et al., 2022). Other species have moderate carbohydrate ranges, such as *P. pulmonarius* (60.63%), *Pleurotus floridanus* (55.83%), as well as P. *citrinopileatus* (53.10%) (Acosta-Urdapilleta et al., 2020; Chai et al., 2021). According to another study by Singh et al. (2021) in seven *Pleurotus* species, *P. eryngii* showed the lowest carbohydrate content, while *Pleurotus membranaceus* had the highest. It is notable that oyster mushrooms have lower total carbohydrates, but higher crude fiber compared to oats and barley. However, white button mushrooms have similar levels of both carbohydrates and fiber to oyster mushrooms (Lai et al., 2024). The content of polysaccharides and dietary fiber in oyster mushrooms varies between species, with polysaccharides ranging from 36 to 60 g/100 g dry fruit bodies and dietary fiber from 10 to 31 g/100 g DW (Chatterjee et al., 2021). The polysaccharides extracted from *P. eryngii* include glucose, galactose, glucuronic acid, ribose, rhamnose, and mannose (El-Zaher et al., 2022). *Pleurotus ostreatus*, a valuable source of both carbohydrates and dietary fiber, contains approximately 6–8 g of carbohydrates and 3–4 gm of dietary fibers per 100 g of fresh weight (Deepalakshmi et al., 2014; Medihi et al., 2024). Mostly, carbohydrates are present as polysaccharides, such as chitin with several components including monosaccharides and their derivatives. Soluble sugars are present in the form of oligosaccharides. Additionally, it contains glycogen with a range of glycosidic linkages, including branched (1→3) and (1→6) β-glucans, as well as linear (1→3) α-glucans (Lesa et al., 2022).

As for the different parts of the mushroom, the crude fiber content varies, with the stalk having the highest at 7.53 g/100 g, and the cap having the lowest at 3.14 g/100 g (Ho et al., 2020). Pink oyster mushroom (*P. djamor*) contains 14.60% fiber, whereas *P. eryngii* has a significantly higher fiber content (28.29%) and *P. citrinopileatus* also shows a high fiber level (20.78%). Other *Pleurotus* species such as *P. florida, P. ferula, P. sajor-caju, P. pulmonarius, P. ostreatus*, and *P. tuber-regium* show fiber content ranging from 10.21% to 29.9% (Raman et al., 2020). According to Irshad et al. (2023) *P. sajor-caju* stands out with the highest crude fiber content at 54.12%, making it an exceptional source for dietary fiber among the analyzed species.

In recent research analyzing *P. ostreatus*, erythrose, glucose, xylose, fructose, and galactose were identified as the primary monosaccharides, with glucose being the most abundant. Additionally, nine disaccharides were detected, with sucrose as the most prominent, followed by maltose, chitobiose, trehalose, and cellobiose. These findings suggest that *P. ostreatus* may serve as a natural sweetener, binder, and food additive, and it could also act as a substrate for the gut microbiome to support the production of short-chain fatty acids (Effiong et al., 2024c).

It is also notable that variations in carbohydrate content arise depending on the substrate composition. *Pleurotus ostreatus* grown on pure sawdust has a lower carbohydrate content compared to those cultivated on sawdust mixed with other materials, regardless of the country of origin. However, the variation in fiber content, despite the similarity in substrate types across studies, indicates that the type of substrate likely has minimal impact on the fiber content of *P. ostreatus* (Effiong et al., 2024c).

### 3.2.2 Proteins

Oyster mushrooms are famous for their high crude protein content, typically ranging from 12% to 35%, depending on the species and environmental factors like cultivation conditions, offering all essential amino acids in amounts that vary by species and growth stage (Lesa et al., 2022). In *P. sajor-caju* and *P. florida*, protein content varies based on the cultivation substrate, typically ranging between 24% and 26% (Singh et al., 2021). The protein content of oyster mushrooms is influenced by the specific strain of *P. ostreatus* and the environmental factors in the growing medium. Research has shown that the protein content can vary between 17 g and 42 g/100 g of dried fruit bodies, depending on these factors. Protein levels in dried fruit bodies range from 17 g to 42 g/100 g, as reported in various studies on *P. ostreatus* mushrooms (Lesa et al., 2022). In addition, the different parts of the oyster mushroom play distinct roles in its nutritional composition. The cap contains 34.19 g of protein/100 g, while the stalk contains 20.96 g of protein/100 g. The combined protein content of the cap and stalk is 30.48 g/100 g (Ho et al., 2020).

The protein content of *Pleurotus* species ranges from 15.09 g to 27 g/100 g DW, which is higher than that of oats at 14.4 g/100 g DW and barley at 11 g/100 g. However, white button mushrooms have a higher protein content than oyster mushrooms, with 39.84 g/100 g DW (Chatterjee et al., 2021; Lai et al., 2024). They could also serve as a viable alternative to red and processed meats, which are associated with health risks such as an increased incidence of breast and colorectal cancers, as well as other diseases related to oxidative stress (González et al., 2020).

Additionally, mushroom proteins are rich in amino acids such as glutamic acid, L-aspartic acid, arginine, and threonine. Some species also contain leucine and lysine (two essential amino acids typically limited in plant-based proteins) as well as two rare amino acids, γ-aminobutyric acid (GABA) and ornithine (Rathore et al., 2017; Zhou et al., 2023). According to Effiong et al. (2024c), lysine is the most abundant essential amino acid in *P. ostreatus*, followed by leucine, phenylalanine, methionine, and threonine. Among the non-essential amino acids, aspartic acid is most abundant, followed by glutamic acid and proline. Protein extracts from *P. ostreatus*, along with nebrodeolysin, a notable protein isolated from *Pleurotus nebrodensis*, offer significant health benefits. Finally, proteins such as eryngin from *P. eryngii* and pleurostrin from *P. ostreatus* demonstrate antifungal and antibacterial properties. *Pleurotus citrinopileatus* is also a rich source of ergothioneine, a water-soluble amino acid with strong antioxidant properties. Various enzymes, such as ribonuclease from *P. djamor* and laccase from *P. ostreatus*, further contribute to these health-promoting effects (Golak-Siwulska et al., 2018).

### 3.2.3 Lipids

Oyster mushrooms display a wide range of crude lipid content from 0.64 to 8.73 g/100 g, with most varieties having lower lipid content than oats at 8.4 g/100 g DW, except for *P. floridanus*, which matches oats at 8.73 g/100 g DW. In comparison, white button mushrooms, with lipid levels between 1.48 and 1.73 g/100 g DW, as well as barley at 2.69 g/100 g DW, have considerably lower lipid levels than most oyster mushroom species, except for *P. florida* at 2.49 g/100 g DW, *P. ostreatus* at 1.6 g/100 g DW, *P. sapidus* at 0.64 g/100 g DW, as well as *Pleurotus geesteranus* at 1.2 g/100 g DW (Lai et al., 2024).

The crude fat content in *Pleurotus* species includes various lipid components, such as free fatty acids, mono-, di-, and triglycerides, sterols, sterol esters, and phospholipids. There is variation in fat content among *Pleurotus* species, likely due to the type of agro-waste used during cultivation, which may influence the mushrooms' nutritional profile. Despite these differences, it is reported that there is a range of 0.64%–8.73% crude fat in *Pleurotus* species, with the highest recorded in *P. floridanus* at 8.73% and the lowest in *P. sapidus* at 0.64%. The average fat content was observed in *P. citrinopileatus* (7.90%) and *P. djamor* (6.70%) (Acosta-Urdapilleta et al., 2020; Chai et al., 2021; Irshad et al., 2023). In *P. sajor-caju* and *P. florida*, the crude fat content is also low, generally between 1.88% and 1.98% (Singh et al., 2021).

Factors such as growth conditions, oxygen, nutrition, and temperature significantly influence the lipid content of mushrooms. The lipid content across different *Pleurotus* species generally ranges from 0.2 g to 8 g/100 g of dried fruit bodies. *Pleurotus ostreatus* contains oleic acid (363 μg/g dried mushroom), linoleic acid (533 μg/g dried mushroom), linolenic acid (11.6 μg/g dried mushroom), and arachidonic acid (10.8 μg/g dried mushroom). In *P. florida*, the major fatty acids are linoleic and palmitic acids, while *P. sajor-caju* contains pentadecanoic acid.

### 3.2.4 Vitamins and Minerals

Oyster mushrooms generally contain higher levels of vitamins A, B2, B3, B5, and C compared to oats and barley.

**TABLE 3.2**
**Vitamin Content in Oyster Mushrooms**

| | Vitamins | | | | | | | | | | |
|---|---|---|---|---|---|---|---|---|---|---|---|
| **Species** | **C** | **E** | **β-carotene** | **A** | **B1** | **B2** | **B3** | **B4** | **B5** | **B6** | **Reference** |
| *P. ostreatus* | 16.46 mg/100 g | 21.50 mg/100 g | 0.06 mg/100 g | 2.93 IU/100 g | 0.7 mg/kg | 92.97 mg/kg | 0.79 mg/kg | 0.005 mg/kg | 0.68 mg/kg | 0.68 mg/kg | Effiong et al., 2024c |
| *P. pulmonarius* | 92 g/100 g | - | - | - | 60.8 g/100 g | 20.6 g/100 g | 40.8 g/100 g | - | - | - | Okwulehie and Nosike, 2015 |

When compared to white button mushrooms, oyster mushrooms have lower levels of vitamin B3 but higher levels of vitamins A, B5, and D (Lai et al., 2024). *Pleurotus eryngii* and *P. ostreatus* exhibit higher levels of niacin compared to other mushrooms (Zhou et al., 2023). Also, *P. ostreatus* has low levels of vitamins A and C, but it is an excellent source of vitamin E, a powerful antioxidant. Vitamin B2 is the most abundant B vitamin found in *P. ostreatus*. While the contents of vitamins B1, B3, B5, and B6 are below their Recommended Daily Allowance (RDA), the levels of vitamins B2, B7, B9, and B12 exceed their respective RDAs (Effiong et al., 2024c; Zhou et al., 2023) (Table 3.2).

Agricultural practices can influence the bioavailability of these important vitamins, as well as the methods used for mineral analysis. In *P. eous*, thiamine and riboflavin contents were 2.23 and 8.97 mg/100 g, respectively, while *P. florida* exhibited 73.3 mg/100 g of niacin and 144 mg/100 g of ascorbic acid. However, when *P. pulmo* was cultivated on indigenous fruit tree barks, it displayed lower levels of B-complex vitamins (Raman et al., 2020).

As for the minerals, the copper content in *Pleurotus* mushrooms is relatively high compared with other mushroom species (Singh et al., 2021). Iron content in *Pleurotus* species varies between 5.5 and 13.4 mg/kg, with *P. eryngii* having the highest iron concentration at 0.062%. The potassium and sodium levels in *P. djamor*, *P. salmoneostramineus*, *P. ostreatus*, and *P. ferulae* range from 1537 to 2218 mg/kg for potassium and 67 to 686 mg/kg for sodium. Additionally, selenium, a trace element found in *Pleurotus* mushrooms, acts as a cofactor in antioxidants, with bioavailability ranging from 0.011 to 0.512 mg/100 g in *P. ostreatus*, *P. ferulae*, *P. salmoneostramineus*, and *P. eryngii* (Gąsecka et al., 2016; Raman et al., 2020). *Pleurotus ostreatus*, specifically, was found to contain substantial levels of potassium (K), magnesium (Mg), iron (Fe), zinc (Zn), and sodium (Na), with lower levels of calcium (Ca). The concentrations of heavy metals such as lead (Pb), nickel (Ni), cadmium (Cd), and chromium (Cr) were very low (Zhou et al., 2023; Effiong et al., 2024c) (Table 3.3). It is notable that the higher ash content observed in oyster mushrooms, compared with oats and barley, correlates with elevated levels of minerals such as potassium, phosphorus, calcium, sodium, and iron, though their zinc content is similar and their magnesium content is comparatively lower (Lai et al., 2024).

## 3.3 MYCOCHEMICALS OF OYSTER MUSHROOMS

### 3.3.1 Phenolic Compounds

Phenolic compounds are secondary metabolites of plants but are also present in abundance in mushrooms. The most common phenolic compounds in mushrooms are phenolic acids, flavonoids, tannins, and other polyphenols (Zhou et al., 2023) and in *Pleurotus* species are mainly phenolics and flavonoids, such as quercetin, catechin, and chrysin (Devi et al., 2024; Zhou et al., 2023). Extracts obtained using dichloromethane (DCM) from the mycelia of *P. djamor* and *P. florida* contain bioactive compounds, including flavonoids, anthraquinones, and terpenoids (Illuri et al., 2022). Jaworska et al. (2015) found that the fruiting bodies of *P. ostreatus* contain 708 mg of total phenolics/100 g dry matter (DM), including 170 mg of flavonoids/100 g DM. The primary phenolic acids in oyster mushrooms are ferulic and *p*-coumaric acids, with concentrations of 30.00 and 10.54 μg/g DM in *P. ostreatus*, and 29.00 and 13.49 μg/g DM in *P. eryngii*, respectively (Gąsecka et al., 2016).

Phenolic acids and flavonoids in edible mushroom species exhibit several bioactivities, such as anti-inflammatory, antioxidant, anticancer, antidiabetic, antiosteoporotic, and antimicrobial activities (Abdelshafy et al., 2022). Generally, oyster mushrooms (except *P. tuber-regium* and *P. djamor*) have higher total phenolic content, ranging from 21.23 to 91.29 mg gallic acid equivalent (GAE)/100 g, compared to white button mushrooms, which have 4.41 mg GAE/100 g. It is also comparable to oats (36.07–101.56 mg GAE/100 g), though it is lower than barley (127.57 mg GAE/100 g) (Lai et al., 2024). Additional studies suggest that *P. ostreatus* cultivated on diverse substrates displays variability in its antioxidant content (Mihai et al., 2022).

Oyster mushrooms contain significant levels of phenolic acids, divided into two subclasses: hydroxybenzoic acid and hydroxycinnamic acid. The hydroxybenzoic acid group includes compounds like gallic acid, while the hydroxycinnamic acid group contains cinnamic acid, caffeic acid, ferulic acid, and others. A study by Selli et al. (2021) found cinnamic acid to be the dominant phenolic compound in oyster mushrooms, with concentrations ranging from 3.47 to 13.10 mg/kg. Other phenolic acids, such as *p*-hydroxybenzoic acid and protocatechuic acid, have also been detected in

**TABLE 3.3**
**Mineral Composition of Oyster Mushrooms**

| | Mineral Content (mg/kg) | | | | | | | |
|---|---|---|---|---|---|---|---|---|
| **Species** | **Ca** | **Mg** | **K** | **P** | **Na** | **Zn** | **Fe** | **Reference** |
| *P. citrinopileatus* | 1.69 | 15.06 | 286.43 | 83.51 | – | 1.05 | 0.93 | Rodrigues et al., 2015 |
| *P. djamor* | 61.90 | 11.10 | 170 | 136.5 | 6.90 | 4.74 | 6.90 | Krakowska et al., 2020; Bermúdez et al., 2023 |
| *P. eryngii* | 2267.4 | 978.7 | 133.3 | 18.1 | – | 4.17 | 8.57 | Rodrigues et al., 2015 |
| *P. florida* | 4.64 | 11.01 | 153.10 | 52.36 | 3.29 | 0.59 | 1.21 | Priyadarshni et al., 2022 |
| *P. floridanus* | 100 | 15.11 | 52.12 | | 18.62 | 2.44 | 8.48 | Chai et al., 2021; Dávila Giraldo et al., 2023 |
| *P. ostreatus* | 0.08 | 7.00 | 12.25 | – | 1.93 | 2.73 | 9.66 | Effiong et al., 2024c |
| *P. pulmonarius* | 11.66 | 3.53 | 177.45 | 29.43 | 6.48 | 0.63 | 3.082 | Chai et al., 2021 |
| *P. tuber-regium* | 1.93 | 1.63 | 92.76 | 8.03 | 11.60 | 0.34 | 0.07 | Audrey et al., 2021 |

various mushroom species, including oyster mushrooms (Zhou et al., 2023).

Oyster mushrooms contain tannins, which are secondary plant metabolites classified into hydrolyzable and condensed tannins. Hydrolyzable tannins are made up of gallic or ellagic acid esterified to glucose, while condensed tannins, also known as procyanidins, consist of flavan-3-ols or flavan-3,4-diols. A study by Akindahunsi and Oyetayo (2006) found that the total tannin content in *P. tuber-regium* was 310 µg catechin equivalent per gram of DW. However, research on tannins in mushrooms, including oyster mushrooms, is still limited, and further studies are required to better understand the qualitative and quantitative aspects of tannin content in different mushroom species (Zhou et al., 2023).

### 3.3.2 Other Bioactive Compounds

β-glucans are among the most significant mycochemicals in oyster mushrooms. These are complex polysaccharides that have immune-modulating properties, helping to enhance the body's defense mechanisms (Chamutpong et al., 2019). Sari et al. (2017) reported that the total glucan content ranges from 18.260 g/100 g DM in *P. citrinopileatus* to 25.636 g/100 g DM in *P. ostreatus*. The β-glucan concentration varies, with *P. eryngii* containing 15.321 g/100 g DM and *P. ostreatus* reaching 24.230 g/100 g DM. β-glucans are also present in the mycelium of oyster mushrooms, with *P. ostreatus* fruiting bodies containing a total β-glucan concentration of 9 g/100 g DM. The total concentration of α-glucans in the fruiting bodies of different oyster mushroom species varies from 2.0% DM in *P. eryngii* to 4.0% DM in *P. citrinopileatus*. Pleuran, a well-known bioactive glucan, has also been identified in several other species of the *Pleurotus* genus, including *P. eryngii*, *P. florida*, *P. pulmonarius*, *P. tuber-regium*, *P. sajor-caju*, and *P. ostreatoroseus* (Baeva et al., 2019).

In a study by Koutrotsios et al. (2018), researchers found that β-glucans made up the majority of total glucans, comprising 80%–91% in *P. ostreatus* and *P. eryngii*, as well as 70%–77% in *P. nebrodensis*, depending on the growth substrate used. *P. eryngii* had the highest β-glucan content among all species analyzed, with values reaching 32.76%–42.43% of DW—among the highest levels reported for both cultivated and wild edible mushrooms. While α-glucan content remained largely unchanged across various substrate types, the β-glucan content showed a substantial increase of about 8%–23%, depending on the mushroom species, when a combination of grape marc compost and wheat straw substrate was used (Koutrotsios et al., 2018).

In the mycelium, β-glucan levels range from 2.5 g/100 g DM in *P. pulmonarius* to 4.6 g/100 g DM in *P. ostreatus* (Nitschke et al., 2011). Additionally, it has been found that β-glucan concentrations in the stems of *P. ostreatus* fruiting bodies are between 32.5% and 50% DM, which is higher than in the caps (27.4% to 39.2% DM). For *P. eryngii*, the β-glucan content is 39.1% DM in the stem and 20.4% DM in the cap. In general, the concentration of β-glucans ranges from 0.21 to 0.53 g/100 g DW across different *Pleurotus* species (Chatterjee et al., 2021).

It has been observed that β-glucans in oats and barley primarily consist of linear polymer chains with mixed beta (1→3) and beta (1→4) glycosidic bonds. In contrast, β-glucans in oyster mushrooms tend to have lower molecular weight, shorter linear chains of β (1→3) glycosidic bonds, and branching linked through β (1→6) glycosidic bonds (Du et al., 2019). Oyster mushrooms, with the exception of *P. tuber-regium*, generally possess significantly higher β-glucan content, ranging from 15.32 to 47.62 g/100 g DW. This is substantially greater than the β-glucan levels found in oats (7.6 g/100 g DW) and barley (4.5 to 8.1 g/100 g DW) (Belobrajdic et al., 2024).

It has also been shown that agricultural byproducts can be utilized to produce valuable compounds, such as exopolysaccharides (EPS) with potential health benefits. In one study, *P. pulmonarius* was grown on various agro-waste substrates—including plantain peels, pineapple peels, mango peels, groundnut shells, coconut coir, and walnut husks—demonstrating the potential of these waste materials to support EPS production (Ogidi et al., 2020).

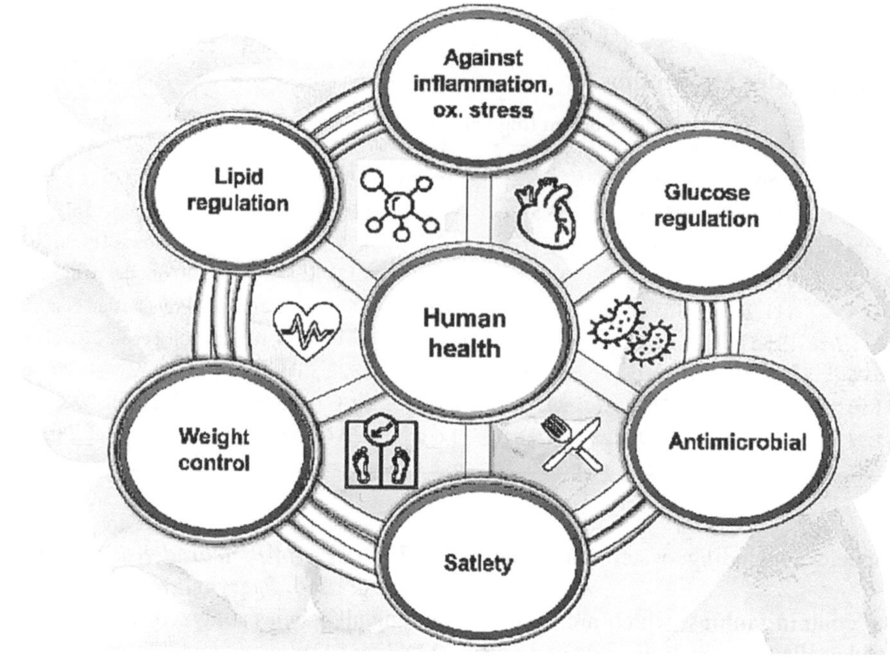


**FIGURE 3.1** The beneficial effects of Oyster mushrooms in human health.

Terpenoids isolated from *P. cornucopiae* mycelium, ergosterol derivatives from *P. cystidiosus*, as well as fatty acid esters from *P. eous* extracts, are considered beneficial components. Sterols, a class of terpenoids, were found in *P. ostreatus* in notable amounts. Koutrotsios et al. (2017) identified sterol content ranging from 1.9% to 87.4%, including key metabolites such as ergosta-7-enol at 12.7%, ergosta-5,7-dienol at 7.6%, and ergosta-7,22-dienol at 6%. Another study reported ergosterol levels at 20 mg/100 g DW in *P. ostreatus (*Souilem et al., 2017). Additionally, Tagkouli et al. (2020) used gas chromatography–mass spectrometry to detect various terpenes in raw *P. ostreatus*, highlighting the diverse terpenoid profile in this species.

Oyster mushrooms also contain lovastatin, a natural statin that regulates cholesterol metabolism. The average lovastatin content in oyster mushrooms ranges from 0.7% to 2.8% DM, with concentrations varying among species from 101 mg/kg dry mass in *P. cystidiosus* to 216 mg/kg DM in *P. ostreatus* (Chen et al., 2012). *Pleurotus species, particularly P. pulmonarius and P. ostreatus, have been identified as potential producers of lovastatin analogs, compounds that could serve as natural sources of lovastatin. According to Lam et al. (2022), these mushrooms possess the capability to synthesize lovastatin-related compounds, which are known for their cholesterol-lowering properties.*

In oyster mushrooms, tocopherols, a form of vitamin E with strong antioxidant properties, have been identified. Ergosterol, a sterol found in fungi but not plants, is common in mushrooms like oyster mushrooms and contributes to their biological activity (Zhou et al., 2023).

Ergothioneine, a sulfur-containing amino acid, plays a significant role in protecting cells from oxidative stress and inflammation. It was found in high concentrations in species of the *Pleurotus* genus (Chen et al., 2012). According to Chen et al. (2012), among the fruiting bodies, *P. citrinopileatus*, *P. ostreatus*, *P. ostreatus*, and *P. salmoneostramineus* had the highest ergothioneine content, with concentrations of 2850.7, 1829.4, 1458.4, and 1245.0 mg/kg, respectively. In contrast, among the mycelia, *P. eryngii* contained the highest ergothioneine level at 1514.6 mg/kg.

## 3.4 MECHANISMS OF ACTION OF OYSTER MUSHROOMS BIOACTIVE COMPOUNDS AND HEALTH BENEFITS

The core pathways via which bioactive components of Oyster mushrooms exhibit their numerous activities have been revealed through several *in vitro* and *in vivo* studies. Also, human studies offer substantial contributions to the understanding of the medical value of Oyster mushrooms. Figure 3.1 depicts the most representative beneficial activities in humans.

### 3.4.1 *In Vitro* and Animal Studies

#### 3.4.1.1 Antimicrobial Activities

*Pleurotus* mushrooms possess several antimicrobial activities and may be considered important alternative antimicrobial sources in food systems, as well as very strong candidates for new antibiotics development (Dash et al., 2024). Mycelial extracts of *P. djamor* and *P. florida* cultures have shown antimicrobial effects against bacterial and fungal pathogens, such as *Escherichia coli*, *Vibrio cholerae*, *Staphylococcus aureus*, and *Aspergillus niger*, with flavanol glycosides, isoflavones, and flavanones being the main phenolic compounds identified in the mycelial extracts after separation with thin-layer chromatography (Illuri et al., 2022).

Similarly, mycelial extracts and filtrates of several *Pleurotus* species, *P. ostreatus, P. cornucopiae* and *P. salmoneostramineus* were examined for their antibacterial and antifungal activities. *Pleurotus salmoneostramineus* filtrate showed the best inhibitory activity against *Pseudomonas aeruginosa* and *Candida parapsilosis. Pleurotus cornucopiae* mycelia showed effect against *Enterococcus faecalis* and *Candida parapsilosis*, and *P. salmoneostramineus* against *E. coli* (Owaid et al., 2015).

In another study, the methanolic extract of the edible pink oyster mushroom *P. eous* showed remarkable growth inhibition activity against *S. aureus*. In the same study, the authors employed scanning electron microscopy to examine morphological changes in bacterial cells and observed membrane rupture induced by the extract. This damage was attributed to the extract's molecules adhering to the negatively charged bacterial surface, leading to cytoplasmic shrinkage and detachment of the cell membrane. In the same study, the antibacterial activity of *P. eous* was tested *in vivo* in BALB/c mice infected with *Staphylococcus aureus*. Treatment with methanolic extracts of *P. eous* caused a reduction of bacterial load in liver, spleen, and peritoneal fluid and reduced white blood cells, C-reactive protein, interleukin (IL)-1β, and IL-6 levels in blood, along with a significant improvement in the histological findings in liver and spleen (Dash et al., 2024).

A polar extract of *P. ostreatus*, with total phenolic and flavonoid content of 6.94 mg/g and 0.15 mg/g, respectively, exhibited antibacterial activity against *C. albicans, S. aureus, Micrococcus luteus*, and *E. coli* and antifungal activity against *Fusarium oxysporum, Fusarium solani*, and *Rhizoctonia solani*. Gas chromatography–mass spectrometry (GC-MS) detected 15 bioactive compounds in the extract, with ethyl iso-allocholate, a steroid derivative, being the most prominent one (62.5%) and known for its antimicrobial activities (Malathi et al., 2016; Hamad et al., 2022).

The antibacterial activity of *P. ostreatus* against the Gram-positive bacteria *S. aureus* was also shown in its ethanolic extract, along with effects on Gram-negative bacteria, such as *Pseudomonas aeruginosa, Proteus vulgaris*, and *Proteus mirabilis*. GC-MS analysis identified 32 compounds in the ethanolic extract, with linoleic acid ethyl ester (32.78%) and hexadecanoic acid ethyl ester (21.42%) accounting for more than 50% of the peak areas (Mishra et al., 2022).

### 3.4.1.2 Anti-inflammatory and Antioxidant Activities

Oyster mushrooms possess several antioxidant and anti-inflammatory activities due to their high content of bioactive compounds. Recently, a study compared the phytochemical content and antioxidant capacity of five different *P. ostreatus* extracts and concluded that chloroform is the most efficient solvent, as its extracts show the highest flavonoid content and the best hydroxyl radical activity (Effiong et al., 2024b).

Also, *P. ostreatus* inhibited nitric oxide (NO) and tumor necrosis factor alpha (TNF-α) in RAW264.7 murine macrophages that had been activated with lipopolysaccharide (LPS) and interferon-γ (IFN-γ) (Gunawardena et al., 2014). A chloroform extract of *P. flabellatus* showed anti-inflammatory cyclooxygenase (COX)-2 activity, whereas a methanolic extract of *P. ostreatus* showed nuclear factor (NF)-κB inhibition activity and high radical scavenging activity. Interestingly, the first one had the highest content of ergosterol, ergothioneine, and mannitol, and the second one had the highest content of β-glucans (Stastny et al., 2022).

Nevertheless, the above mycochemicals have been extensively studied for their antioxidant and immunoregulatory activities in several *in vitro* and *in vivo* studies. Ergosterol has been suggested to downregulate COX-2 through the attenuation of TNF-α expression (Kuo et al., 2010), whereas β-glucans trigger macrophage activation through Toll-like receptors which regulate several signaling pathways such as NF-κB (Bergandi et al., 2021).

Furthermore, β-glucans from oyster mushrooms are known for their antioxidant properties as well. The antioxidant capacity of the enriched polysaccharide fraction from *P. eryngii*, which is rich in β-glucans, has been tested with various assays. This fraction is capable of scavenging 2,2-diphenyl-1-picrylhydrazyl (DPPH), superoxide, hydroxyl, and nitric oxide radicals in a very effective way. Also, it has a high total reducing power, which increases rapidly at higher concentrations (Petraglia et al., 2023). This bioactivity of the polysaccharide fraction may be related to hydrogen and hydroxyl groups and may use a similar mechanism to the one used by phenolic compounds (Kishk et al., 2007; Petraglia et al., 2023).

*Pleurotus eryngii* was administered in dextran sulfate sodium (DSS) mice in the form of powder and suppressed colitis, as expressed by the prevention of colon shortening and damage to the epithelial wall. Also, it downregulated the production of several inflammatory cytokines and increased short-chain fatty acids and the diversity of microbiota in the colon. The doses of *P. eryngii* that were administered in this study can be considered equivalent to 5–15 g of dry mushroom/day in humans, which can be easily achieved (Hu et al., 2019).

The methanolic extract of *P. pulmonarius*, also known as Indian oyster or lung oyster mushroom, has also been studied for its anti-inflammatory activities *in vitro* and *in vivo*. LPS-stimulated RAW264.7 cells exhibited lower concentrations of NO in their media after being treated with *P. pulmonarius* methanolic extracts of different concentrations, and the effect was manifested in a concentration-dependent manner. In an acute inflammation rat model with carrageenan-induced paw edema, treatment with the same extracts inhibited paw edema development, suggesting an action on histamine and prostaglandin release (Nguyen et al., 2016).

The same inflammation animal model was used by Jayasuriya and his colleagues in 2020, where they tested freeze-dried powdered and acetone extracts of *P. ostreatus* in both healthy and diabetic Wistar rats. The acetone extract showed higher paw edema inhibition activity in healthy rats and was even higher in diabetic rats, highlighting a potential role of *P. ostreatus* in the management of inflammation in diabetes. Regarding the potential mechanism of action, the same study has shown antihistamine activity, inhibition of cell migration, and NO production (Jayasuriya et al., 2020).

In order to elucidate the role of β-glucans in immunoregulatory activities of oyster mushrooms, aqueous extracts from different *Pleurotus* species were prepared and tested for their bioactivities after the chemical characterization of the fractions. The purified fractions, enriched in β-glucans and poor in low molecular weight compounds, had significant immunomodulatory capacity when tested in THP-1 macrophages and reduced IL-8 production after LPS stimulation (Pérez-Bassart et al., 2023).

Recently, it has been shown that β-glucans conformation is responsible for their biological activity due to modulating actions on the membrane organization and dynamics of Dectin-1A, a pattern recognition receptor of innate immunity (Anaya et al., 2023). The above findings do not imply that polyphenols and other antioxidant substances do not possess anti-inflammatory activities, but they support the hypothesis that their presence may affect structure–function relationships and therefore the immunostimulatory activity of β-glucans (Ferreira et al., 2015).

Although phenolic compounds possess anti-inflammatory activities with the ability to downregulate the production of proinflammatory cytokines, and despite the fact that total phenolic content has been correlated with anti-inflammatory effects in some mushroom extracts, this cannot be confirmed in all studies. This may be attributed to the fact that bioactivity can be influenced by the diversity of structures of different phenolic compounds existing in mushrooms. Also, potential synergisms between different bioactive compounds in the extracts may affect the activity and explain the observed differences (Moro et al., 2011).

Phenolic compounds are also known for their antioxidant activities, with a prominent one being the iron chelation capacity. Therefore, iron-overloaded mice were administered methanolic extracts of *P. porrigens* for 20 days, and plasma iron, as well as liver-deposited iron, was measured and compared to that of control mice. Treated animals exhibited significantly lower plasma iron levels, as well as fewer necrotic hepatocytes, fibrous tissues, and pseudo-lobules in the liver compared to the control group. High-performance liquid chromatography revealed rutin and *p*-coumaric acid as the main flavonoid and phenolic acids in the methanolic extract, respectively, which are well known for their chelating activity (Khalili et al., 2015).

Not only β-glucans and mycochemicals contribute to the anti-inflammatory and antioxidant activities of oyster mushrooms, but other bioactive compounds may also play a significant role. Yamashita and his colleagues used the lipophilic extract of *P. citrinopileatus*. DSS mice, models of chronic colon inflammation, were treated with ethanolic extract of the mushroom, which was rich in sphingolipids, sterylglycoside derivatives, and free and acyl sterols. This treatment decreased the damage to the chorionic crypts of large intestinal villi, as well as the levels of five inflammatory cytokines and chemokines compared to the control group (Yamashita et al., 2021).

González and his colleagues (2021) showed that the protein concentrate of *P. ostreatus* exhibited lower 2,2′-Azino-bis(3-ethylbenzothiazoline-6-sulfonic acid) (ABTS) and similar DPPH scavenging activities. ABTS activity is attributed to phenolic compounds and flavonoids, while DPPH activity is based on of high-pigmented and hydrophilic antioxidants (Floegel et al., 2011) suggesting that protein concentrate may be low in antioxidant compounds such as flavonoids and phenolic compounds, but hydrophilic and high-pigmented antioxidants (i.e., ascorbic acid, carotenoids) may be unaffected (González et al., 2021).

#### 3.4.1.3 Health Benefits Against Metabolic Syndrome Components

Several studies have shown the cardioprotective, antiobesity, antihypertensive, antidiabetic, and lipid-lowering properties of Oyster mushrooms, mainly via *in vitro* and *in vivo* studies, whereas human studies are limited (Lai et al., 2024). 3T3-L1 cells are usually used for the study of adipocyte differentiation. These cells, when treated with *P. eryngii* extracts, displayed reduced adipocyte separation/lipid collection due to downregulation of transcription factors, such as peroxisome proliferator-activated receptor γ (PPARγ), and proteins that regulate lipid transport, such as fatty acid-binding protein (FABP4) and fatty acid transport protein (FATP1) (Bindu and Das, 2019).

The antiobesity and hypolipidemic effects of *P. eryngii* have also been shown in mice fed with a high-fat diet. Soluble polysaccharides extracted from the mushroom were administered to C57BL/6J high-fat diet mice, and the effects on adiposity-related parameters were measured compared to controls. The group receiving the extracts exhibited lower body weight gain and blood lipids, and higher expression of low-density lipoprotein (LDL) receptor gene in liver and G-protein-coupled receptor 43 (GPR43) in adipocytes than the control group, indicating a potential role in the prevention and management of obesity and metabolic syndrome (Nakahara et al., 2020).

*Pleurotus ostreatus* methanolic extracts exhibit strong inhibition of α-glucosidase activity *in vitro* (Nnemolisa et al., 2024) and its polysaccharides improve insulin resistance by activating glycogen synthase kinase-3 phosphorylation and glucose transporter-4 (GLUT4) translocation in diabetic rats (Zhang et al., 2016). Similarly, *P. sajor-caju* aqueous extract improved blood sugar and oral glucose tolerance test compared to a control group, with comparable results to those of metformin administration (Ng et al., 2015).

The antihypertensive effect of Oyster mushrooms was shown for *P. pulmonarius*, with its mycelial water extract possessing stronger angiotensin-converting enzyme (ACE) inhibitory activity than six other non-Oyster mushrooms (Ibadallah et al., 2015). A similar activity on ACE was described for *P. ostreatus* aqueous extracts, which inhibited strongly not only ACE but also arginase and acetylcholinesterase, enzymes known for their implication in the pathogenesis of hypertension (Agunloye, 2021). The same group also demonstrated an effect of *P. ostreatus* on enzymes associated with the development of hypertension in both normal and diabetic rats treated with the mushroom, with ACE and arginase activities being significantly reduced compared to untreated diabetic rats. In this study, an antidiabetic activity was also revealed, with α-amylase and α-glucosidase being significantly reduced as well (Agunloye and Oboh, 2022).

#### 3.4.1.4 Other Activities

Oyster mushrooms contain several bioactive compounds with known antiproliferative and antitumor activities, such as β-glucan, ergosterol, linoleic acid, and others. The proposed mechanisms of antineoplasmatic actions include inhibition of pathways that drive angiogenesis, tumor growth, cell migration, invasion, and anti-apoptosis, as well as induction of cell cycle arrest and DNA fragmentation in cancerous cells (Mishra et al., 2021).

Three pentacyclic triterpenoid compounds isolated from *P. eryngii* ethanolic extracts exhibited inhibitory activity against MCF-7 cell proliferation, a breast cancer cell line. One of the three compounds showed the highest antiproliferative activity, confirming that an additional –OH group may enhance the cytotoxic effect (Xue et al., 2015).

*Pleurotus sajor-caju n*-hexane extract exerted significant cytotoxicity against HCT116wt colonic cancer cells, which was associated with apoptosis and cell cycle-arrest pathways. More specifically, Caspase-3, Bax, and p53 were upregulated, whereas Bcl-2 was downregulated in the treated cells, suggesting that apoptosis may be induced in a p53-dependent way. Additionally, the fact that cell arrest was also evident at the G2/M phase and no cytotoxicity was observed in HTC116–p21 cells supports a potential disruption of the p21/p53 cell cycle regulation pathway by *P. sajor-caju* compounds (Finimundy et al., 2018).

The incorporation of *P. eryngii* powder into the diet of C26 tumor-bearing mice, experimental model of cancer cachexia, increased their survival and was accompanied by changes in the levels of heat shock proteins that regulate apoptosis. Also, inflammatory cytokines, such as IL-6 and IL1, were downregulated, and autophagy-related protein 7 (ATG7), an enzyme essential for autophagy, was upregulated (Barone et al., 2021).

A polysaccharide extracted from *P. citrinopileatus* containing arabinose, galactose, glucose, xylose, mannose, and glucuronic acid was intragastrically administered to H22 tumor-bearing mice, inhibiting tumor growth and inducing tumor cell apoptosis *in vivo* in a dose-dependent manner (Wang et al., 2021).

*In silico* analysis of compounds identified in *P. ostreatus* extracts through GC-MS and HPLC revealed several molecules with the potential to interact with breast cancer receptors, including $ER^+$, $PR^+$, and $HER2^+$. Notably, phenolic and flavonoid compounds demonstrated the highest binding affinities to these receptors, such as the progesterone receptor and epidermal growth factor receptor, when compared to standard controls, highlighting their promising potential for development as anticancer agents (Effiong et al., 2024a).

A potential beneficial role of Oyster mushrooms in bone metabolic diseases has also emerged. *Pleurotus sajor-caju* glucan oligosaccharide's effect on osteoclast differentiation was tested on preosteoclastic RAW264.7 cells, resulting in the inhibition of the production of mature osteoclast multinucleated cells and the downregulation of the expression of osteoclast modulators (Rattajak et al., 2024).

Two polysaccharide-peptides were extracted from *Pleurotus* and showed hepatoprotective and gut microbiota regulatory effects on a hepatoma cell model, HepG2 cells. The results showed that both peptides reduced extracellular alanine transaminase (ALT) and aspartate transaminase, suggesting that they could inhibit the release of extracellular transaminase. Also, the two peptides were fermented in human fecal samples and were responsible for changes to the relative abundances of several bacterial taxa, suggesting a regulatory role in gut microbiota (Huang et al., 2022).

Wistar rats with hyperoxaluria that were treated with *P. ostreatus* aqueous extracts showed improved kidney architecture and reduced urine and kidney oxalate compared to the control group. The above changes were paralleled by a decrease in the expression of kidney NF-κB p65 and NF-κB p50 and the apoptotic mediators Bax and Bak of the Bcl-2 family, highlighting a nephroprotective effect of the extract that may be facilitated by NF-κB activation and apoptotic pathways (Ahmed et al., 2020).

### 3.4.2 Clinical Trials

Data on human trials using Oyster mushrooms are scarce. In 2022, researchers conducted a randomized controlled trial with a *P. eryngii* snack in adults with central obesity and metabolic abnormalities. One hundred adults were randomized to either the intervention group that received the snack for 3 months or to a control group that received only standard nutritional counseling. Participants consuming the snack showed reduced body weight, fat, glucose, and LDL levels compared to the control group that did not receive the snack. All the above suggested an antiobesity, antidiabetic, and lipid-lowering potential of snacks made with Oyster mushrooms with known cardioprotective activities. An anti-inflammatory and antioxidant potential of the snack was also evident, with a significant decrease in serum IL-6 and oxidized LDL observed only in the intervention group, with the mean changes being significantly different between the two groups, suggesting that the incorporation of a snack made with Oyster mushrooms in the diet can have a positive effect on inflammation and oxidative stress associated with metabolic abnormalities (Kleftaki et al., 2022a). The same group conducted an acute, randomized, crossover trial with a meal with *P. eryngii* versus a control isocaloric and macronutrient-matched meal and showed that the test meal resulted in better postprandial glucose and ghrelin response, and in lower desire to eat and higher fullness. The above were conducted in 19 adults with central obesity and metabolic abnormalities and pointed toward not only a glucose but also an appetite regulation effect (Kleftaki et al., 2022b).

A potential anti-inflammatory action of β-glucans from *P. ostreatus* was also shown when incorporated in a cream and tested in an open study in 105 patients with atopic dermatitis, a chronic inflammatory skin disease. The application of the cream led to a reduction in the duration and intensity of flares, as well as in eczema area and severity index scores. It was well tolerated, with no reported clinical adverse events (Jesenak et al., 2016).

Another postprandial study administered three different meals in a double-blind randomized controlled crossover trial in 22 subjects with impaired glucose tolerance. The participants consumed either a meal enriched with 20 g of *P. ostreatus* or a non-enriched meal and exhibited lower non-esterified free fatty acids and glucagon-like peptide-1 levels compared to control, with hunger sensation being lower as well (Dicks et al., 2022).

The hypoglycemic effect of *P. ostreatus* and *P. cystidiosus* was also tested in both healthy individuals and Type 2 diabetic patients who received powdered samples of the above for 2 weeks. A significant reduction in fasting and postprandial serum glucose was observed in healthy volunteers, whereas postprandial serum glucose decreased and insulin increased in diabetic patients (Jayasuriya et al., 2014).

Another study conducted in a Mexican population involved 30 participants who followed either a healthy traditional diet alone or the same diet supplemented with specific portions of *P. ostreatus*. The results showed a significant reduction in glucose levels, cholesterol, and visceral fat among women who consumed the mushroom-enriched diet—effects that were not observed in the control group (González-Bonilla et al., 2022).

## 3.5 APPLICATIONS OF OYSTER MUSHROOMS IN FOOD PRODUCT DEVELOPMENT

Mushrooms exhibit several biotechnological applications in the food industry and medicine and can be considered ideal candidates for the development of high-value-added foods, as well as dietary supplements and myconutraceuticals, due to high nutritional and medical properties and low water and energy requirements for cultivation (Llanaj et al., 2023). Oyster mushrooms are considered economically and environmentally friendly due to their ability to degrade organic lignocellulosic materials; therefore, the use of agricultural byproducts as their cultivation substrates contributes to reduction of wastes and production of enriched foods (Aditya et al., 2024). Biotechnological applications of Oyster mushrooms, such as bioconversion and production of nanoparticles or bioactive compounds, can be used in food packaging, food processing, food security, and pharmaceutical biotechnology (Llanaj et al., 2023). Apart from their nutritional and medicinal value, Oyster mushrooms can also be used as a natural adjuvant due to their delicate flavor and texture (Devi et al., 2024).

Oyster mushroom flours and powders have been extensively used for food fortification. For example, the addition of mushroom flour in a mixture of wheat and cassava flour increased protein and energy content of the bread, a fortification that can have significant nutritional value in countries that substitute wheat with cassava due to high prices of imported wheat flour (Irakiza et al., 2021). Pasta fortified with 10% Oyster mushroom flour showed increased protein, ash, and fat, and decreased carbohydrates and energy levels, with sensory evaluation indicating no significant differences in most characteristics, except for darker color (Nordiana et al., 2019). Similarly, mushroom powder enrichment of noodles showed higher levels of protein, fiber, and minerals, such as iron, calcium, and potassium, and were organoleptically acceptable (Parvin et al., 2020). Recently, in Kenya, Oyster mushrooms were used to create composite flours with sorghum and pearl millet in a proportion of 0% to 50%, with the protein content increasing from 11.15% to 19.74% and fiber from 2.68% to 10.66%. An improvement in minerals and vitamins was also observed, suggesting a potential role of this fortification in micronutrient deficiencies and protein-energy malnutrition that are observed in the country (Ndunge Charles et al., 2024).

Bioactive compounds and extracts derived from Oyster mushrooms are also used in food fortification. Crude water-soluble polysaccharides from *P. ostreatus* were added to cow milk yogurt production before fermentation and improved their antioxidant potential as measured by ferric-reducing antioxidant power and ABTS assays (Radzki et al., 2023). Another study in yogurts fortified with fresh Pacific oysters (*Crassostrea gigas*) hydrolysates showed that digestates of low molecular weight possessed antioxidant and anti-inflammatory activity in Caco-2 and HT-29 cell lines by suppression of NF-κB and proinflammatory cytokines production (Liu et al., 2022). β-glucans extracted from *P. ostreatus* were used for the fortification of ovine milk cheese and increased the moisture content compared to a non-fortified cheese, leading to better scores in flavor characteristics in sensory analysis (Kondyli et al., 2022).

## 3.6 CONCLUSION

*Pleurotus* mushrooms have gained attention in recent years due to their significant nutritional value, as they are rich in proteins, fiber, vitamins, and minerals. Additionally, their unique mycochemical profile confers various health benefits. *In vitro*, animal, and human studies reveal their antioxidant, anti-inflammatory, antimicrobial, cardioprotective, and chemopreventive properties, highlighting their potential as a promising resource for the development of dietary supplements and myconutraceuticals.

## REFERENCES


Abdelshafy, A. M., T. Belwal, Z. Liang, et al. 2022. A comprehensive review on phenolic compounds from edible mushrooms: occurrence, biological activity, application and future prospective. *Critical Reviews in Food Science and Nutrition* 62(22): 6204–6224. https://doi.org/10.1080/10408398.2021.1898335

Acosta-Urdapilleta, L., E. Villegas, A. Estrada-Torres, M. Téllez-Téllez, and G. Diaz-Godínez. 2020. Antioxidant activity and proximal chemical composition of fruiting bodies of mushroom, *Pleurotus* spp. produced on wheat straw. *Journal of Environmental Biology* 41: 1075–1081. https://doi.org/10.22438/jeb/41/5/MRN-1307

Aditya, N., R. S. Jarial, K. Jarial, and J. N. Bhatia. 2024. Comprehensive review on oyster mushroom species (Agaricomycetes): morphology, nutrition, cultivation and future aspects. *Heliyon* 10(5): e26539. https://doi.org/10.1016/j.heliyon.2024.e26539

Afolabi, F., B. Babaniyi, M. Obagunwa, F. Olu, J. Ajagun, O. Julius, and S. Fasiku. 2024. Ethnomycology of *Pleurotus tuber-regium* and its use in food, medicine and bioremediation. *Botanica* 30: 183–192. https://doi.org/10.35513/Botlit.2024.4.7

Agunloye, O. M. 2021. Effect of aqueous extracts of *Pleurotus* ostreatus and *Lentinus subnudus* on activity of adenosine deaminase, arginase, cholinergic enzyme, and angiotensin-1-converting enzyme. *Journal of Food Biochemistry* 45(3): e13490. https://doi.org/10.1111/jfbc.13490

Agunloye, O. M., and G. Oboh. 2022. Blood glucose lowering and effect of oyster (*Pleurotus* ostreatus)- and shiitake (*Lentinus subnudus*)-supplemented diet on key enzymes linked diabetes and hypertension in streptozotocin-induced diabetic in rats. *Food Frontiers* 3: 161–171. https://doi.org/10.1002/fft2.111

Ahmed, O. M., H. Ebaid, E. S. El-Nahass, M. Ragab, and I. M. Alhazza. 2020. Nephroprotective effect of *Pleurotus* ostreatus and *Agaricus bisporus* extracts and carvedilol on ethylene glycol-induced urolithiasis: roles of NF-κB, p53, Bcl-2, Bax and Bak. *Biomolecules* 10(9): 1317. https://doi.org/10.3390/biom10091317

Akindahunsi, A. A., and F. L. Oyetayo. 2006. Nutrient and antinutrient distribution of edible mushroom, Pleurotus tuber-regium (fries) singer. *LWT—Food Science and Technology* 39(5): 548–553. https://doi.org/10.1016/j.lwt.2005.04.005.

Anaya, E. U., A. E. Amin, M. J. Wester, M. E. Danielson, K. S. Michel, and A. K. Neumann. 2023. Dectin-1 multimerization and signaling depends on fungal β-glucan structure and exposure. *Biophysical Journal* 122(18): 3749–3767. https://doi.org/10.1016/j.bpj.2023.07.021

Audrey, T. T. K., E. N. H. Calixte, N. André-Ledoux, N. F. Nico, and M. F. Paul. 2021. Proximate and minerals composition of 12 wild mushrooms from the Noun division, West region in Cameroon. *Cameroon Journal of Biological Biochemistry and Sciences* 29: 100–107.

Baeva, E., R. Bleha, E. Lavrova, et al. 2019. Polysaccharides from basidiocarps of cultivating mushroom *Pleurotus ostreatus*: isolation and structural characterization. *Molecules* 24: Article 2740. https://doi.org/10.3390/molecules24152740

Barone, R., C. Caruso Bavisotto, F. Rappa, et al. 2021. JNK pathway and heat shock response mediate the survival of C26 colon carcinoma bearing mice fed with the mushroom *Pleurotus eryngii* var. *eryngii* without affecting tumor growth or cachexia. *Food and Function* 12(7): 3083–3095. https://doi.org/10.1039/d0fo03171b

Belobrajdic, D. P., H. Brook, G. James-Martin, et al. 2024. β-Glucan content and in vitro bile-acid binding capacity of Agaricus bisporus and Pleurotus spp. *Food & Function* 15(19): 9880–9887. https://doi.org/10.1039/d4fo02416h

Bergandi, L., G. Apprato, and F. Silvagno. 2021. Vitamin D and Beta-glucans synergically stimulate human macrophage activity. *International Journal of Molecular Sciences* 22(9): 4869. https://doi.org/10.3390/ijms22094869

Bermúdez, R., E. Rangel-Vargas, J. M. Lorenzo, J. A. Rodríguez, P. E. S. Munekata, A. Teixeira, M. Pateiro, L. Romero, and E. M. Santos. 2023. Effect of partial meat replacement by *Hibiscus sabdariffa* by-product and *Pleurotus djamor* powder on the quality of beef patties. *Foods* 12: 391. https://doi.org/10.3390/foods12020391

Bindu, J., and A. Das. 2019. An edible fungi *Pleurotus* ostreatus inhibits adipogenesis via suppressing expression of PPAR γ and C/EBP α in 3T3-L1 cells: *in vitro* validation of gene knock out of RNAs in PPAR γ using CRISPR spcas9. *Biomedicine and Pharmacotherapy* 116: 109030. https://doi.org/10.1016/j.biopha.2019.109030

Chai, Y., U. Krishnan, V. Sabaratnam, and J. Tan. 2021. Assessment of coffee waste in formulation of substrate for oyster mushrooms *Pleurotus pulmonarius* and *Pleurotus floridanus*. *Future Foods* 4: 100075. https://doi.org/10.1016/j.fufo.2021.100075

Chamutpong, S., T. Pummarin, C. Noysang, et al. 2019. Nutritional properties, antioxidant and anti-acetylcholinesterase activities of *Pleurotus ostreatus*. *Applied Mechanics and Materials* 891: 14–20. https://doi.org/10.4028/www.scientific.net/AMM.891.14

Chatterjee, D., D. Halder, and S. Das. 2021. Varieties of mushrooms and their nutraceutical importance: a systematic review. *Journal of Clinical and Diagnostic Research* 15(3): OE01–OE06.

Chen, S.-Y., K.-J. Ho, Y.-J. Hsieh, L.-T. Wang, and J.-L. Mau. 2012. Contents of lovastatin, γ-aminobutyric acid, and ergothioneine in mushroom fruiting bodies and mycelia. *LWT—Food Science and Technology* 47(2): 274–278. https://doi.org/10.1016/j.lwt.2012.01.019

Dash, P., B. Kar, M. Gochhi, et al. 2024. Antimicrobial properties of the edible pink oyster mushroom, *Pleurotus* eous: *in-vivo* and *in-vitro* studies. *Microbial Pathogenesis* 196: 106915. https://doi.org/10.1016/j.micpath.2024.106915

Dávila Giraldo, L. R., C. C. Pérez Jaramillo, J. J. Méndez Arteaga, and W. Murillo-Arango. 2023. Nutritional value and antioxidant, antimicrobial, and cytotoxic activity of wild macrofungi. *Microorganisms* 11: Article 1158. https://doi.org/10.3390/microorganisms11051158

Deepalakshmi, K., and S. Mirunalini. 2014. *Pleurotus ostreatus*: an oyster mushroom with nutritional and medicinal properties. *Journal of Biochemical Technology* 5(3): 718–726.

Devi, P. V., J. Islam, P. Narzary, D. Sharma, and F. Sultana. 2024. Bioactive compounds, nutraceutical values and its application in food product development of oyster mushroom. *Journal of Future Foods* 4(4): 335–342. https://doi.org/10.1016/j.jfutfo.2023.11.005

Dicks, L., L. Jakobs, M. Sari, et al. 2022. Fortifying a meal with oyster mushroom powder beneficially affects postprandial glucagon-like peptide-1, non-esterified free fatty acids and hunger sensation in adults with impaired glucose tolerance: a double-blind randomized controlled crossover trial. *European Journal of Nutrition* 61(2): 687–701. https://doi.org/10.1007/s00394-021-02674-1

Du, B., M. Meenu, H. Liu, and B. Xu. 2019. A concise review on the molecular structure and function relationship of β-Glucan. *International Journal of Molecular Sciences* 20(16): 4032. https://doi.org/10.3390/ijms20164032

Effiong, M. E., M. Bella-Omunagbe, I. S. Afolabi, and S. N. Chinedu. 2024a. *In silico* evaluation of potential breast cancer receptor antagonists from GC-MS and HPLC identified compounds in *Pleurotus* ostreatus extracts. *RSC Advances* 14(33): 23744–23771. https://doi.org/10.1039/d4ra03832k

Effiong, M. E., C. P. Umeokwochi, I. S. Afolabi, and N. C. Chinedu. 2024b. Comparative antioxidant activity and phytochemical content of five extracts of *Pleurotus* ostreatus (oyster mushroom). *Scientific Reports* 14: 3794. https://doi.org/10.1038/s41598-024-54201-x

Effiong, M. E., C. P. Umeokwochi, I. S. Afolabi, and S. N. Chinedu. 2024c. Assessing the nutritional quality of *Pleurotus ostreatus* (oyster mushroom). *Frontiers in Nutrition* 10: Article 1279208. https://doi.org/10.3389/fnut.2023.1279208

El-Zaher, A., E. Hasan, E. M. Tousson, et al. 2022. Production of endo polysaccharides from cultivated *Pleurotus eryngii* fruiting bodies. *Delta Journal of Science* 44: 135–144. https://doi.org/10.21608/djs.2022.127995.1020

FAOSTAT. 2023. *Crops. Livestock Products*. Food and Agricultural Organization of the United Nations, Statistics Division (FAOSTAT).

Ferreira, S. S., C. P. Passos, P. Madureira, M. Vilanova, and M. A. Coimbra. 2015. Structure-function relationships of immunostimulatory polysaccharides: a review. *Carbohydrate Polymers* 132: 378–396. https://doi.org/10.1016/j.carbpol.2015.05.079

Finimundy, T. C., R. M. V. Abreu, N. Bonetto, et al. 2018. Apoptosis induction by *Pleurotus sajor-caju* (Fr.) Singer extracts on colorectal cancer cell lines. *Food and Chemical Toxicology* 112: 383–392. https://doi.org/10.1016/j.fct.2018.01.015

Floegel, A., D.-O. Kim, S.-J. Chung, S. I. Koo, and O. K. Chun. 2011. Comparison of ABTS/DPPH assays to measure antioxidant capacity in popular antioxidant-rich US foods. *Journal of Food Composition and Analysis* 24(7): 1043–1048. https://doi.org/10.1016/j.jfca.2011.01.008

Gąsecka, M., M. Mleczek, M. Siwulski M, et al. 2016. Phenolic composition and antioxidant properties of *Pleurotus ostreatus* and *Pleurotus eryngii* enriched with selenium and zinc. *European Food Research and Technology* 242(5): 723–732.

Golak-Siwulska, I., A. Kałużewicz, T. Spiżewski, M. Siwulski, and K. Sobieralski. 2018. Bioactive compounds and medicinal properties of Oyster mushrooms (*Pleurotus sp.*). *Folia Horticulturae* 30. https://doi.org/10.2478/fhort-2018-0012

González, A., C. Nobre, L. S. Simões, et al. 2021. Evaluation of functional and nutritional potential of a protein concentrate from *Pleurotus* ostreatus mushroom. *Food Chemistry* 346: 128884. https://doi.org/10.1016/j.foodchem.2020.128884

González-Bonilla, A., M. E. Meneses, A. Pérez-Herrera, D. Armengol-Álvarez, and D. Martínez-Carrera. 2022. Dietary Supplementation with Oyster culinary-medicinal mushroom, *Pleurotus ostreatus* (Agaricomycetes), reduces visceral fat and hyperlipidemia in inhabitants of a rural community in Mexico. *International Journal of Medicinal Mushrooms* 24(9): 49–61. https://doi.org/10.1615/IntJMedMushrooms.2022044837

Gunawardena, D., L. Bennett, K. Shanmugam, et al. 2014. Anti-inflammatory effects of five commercially available mushroom species determined in lipopolysaccharide and interferon-γ activated murine macrophages. *Food Chemistry* 148: 92–96. https://doi.org/10.1016/j.foodchem.2013.10.015

Hamad, D., H. El-Sayed, W. Ahmed, H. Sonbol, and M. A. H. Ramadan. 2022. GC-MS analysis of potentially volatile compounds of *Pleurotus ostreatus* polar extract: in vitro antimicrobial, cytotoxic, immunomodulatory, and antioxidant activities. *Frontiers in Microbiology* 13: 834525. https://doi.org/10.3389/fmicb.2022.834525

Ho, L. H., N. Asyikeen Zulkifli, and T. C. Tan. 2020. *Edible Mushroom: Nutritional Properties, Potential Nutraceutical Values, and Its Utilisation in Food Product Development*. IntechOpen. http://dx.doi.org/10.5772/intechopen.91827

Hu, Q., B. Yuan, X. Wu, et al. 2019. Dietary intake of *Pleurotus eryngii* ameliorated dextran-sodium-sulfate-induced colitis in mice. *Molecular Nutrition and Food Research* 63(17): e1801265. https://doi.org/10.1002/mnfr.201801265

Huang, Y., Y. Gao, X. Pi, S. Zhao, and W. Liu. 2022. *In vitro* hepatoprotective and human gut microbiota modulation of polysaccharide-peptides in *Pleurotus citrinopileatus*. *Frontiers in Cellular and Infection Microbiology* 12: 892049. https://doi.org/10.3389/fcimb.2022.892049

Ibadallah, B., N. Abdullah, and A. Shuib. 2015. Identification of angiotensin-converting enzyme inhibitory proteins from mycelium of *Pleurotus pulmonarius* (Oyster mushroom). *Planta Medica* 81(02): 123–129. https://doi.org/10.1055/s-0034-1383409

Illuri, R., E. M, K. M, et al. 2022. Bio-prospective potential of *Pleurotus* djamor and *Pleurotus* florida mycelial extracts towards Gram positive and Gram negative microbial pathogens causing infectious disease. *Journal of Infection and Public Health* 15(2): 297–306. https://doi.org/10.1016/j.jiph.2021.10.012

Irakiza, P. N., G. B. Chuma, T. Z. Lyoba, et al. 2021. Fortification with mushroom flour (*Pleurotus* ostreatus (Jacq.) P. Kumm) and substitution of wheat flour by cassava flour in breadmaking: nutritional and technical implications in eastern DR Congo. *Agriculture and Food Security* 10: 28. https://doi.org/10.1186/s40066-021-00301-0

Irshad, A., A. Tahir, S. Sharif, et al. 2023. Determination of nutritional and biochemical composition of selected *Pleurotus* spp. *BioMed Research International* 2023: Article 8150909. https://doi.org/10.1155/2023/8150909

Jaworska, G., K. Pogoń, E. Bernaś, and A. Duda-Chodak. 2015. Nutraceuticals and antioxidant activity of prepared for consumption commercial mushrooms *Agaricus bisporus* and *Pleurotus ostreatus*. *Journal of Food Quality* 38(2): 111–122.

Jayasuriya, W. J. A. B. N., S. M. Handunnetti, C. A. Wanigatunge, G. H. Fernando, D. T. U. Abeytunga, and T. S. Suresh. 2020. Anti-inflammatory activity of *Pleurotus* ostreatus, a culinary medicinal mushroom, in wistar rats. *Evidence-Based Complementary and Alternative Medicine* 2020: 6845383. https://doi.org/10.1155/2020/6845383

Jayasuriya, W. J. A. B. N., C. A. Wanigatunge, G. H. Fernando, D. T. U. Abeytunga, and T. S. Suresh. 2014. Hypoglycaemic activity of culinary *Pleurotus ostreatus* and *P. cystidiosus* mushrooms in healthy volunteers and type 2 diabetic patients on diet control and the possible mechanisms of action. *Phytotherapy Research* 29(2): 303–309. https://doi.org/10.1002/ptr.5255

Jesenak, M., S. Urbancek, J. Majtan, P. Banovcin, and J. Hercogova. 2016. β-Glucan-based cream (containing pleuran isolated from *Pleurotus* ostreatus) in supportive treatment of mild-to-moderate atopic dermatitis. *The Journal of Dermatological Treatment* 27(4): 351–354. https://doi.org/10.3109/09546634.2015.1117565

Khalili, M., M. A. Ebrahimzadeh, and M. Kosaryan. 2015. *In vivo* iron-chelating activity and phenolic profiles of the angel's wings mushroom, *Pleurotus porrigens* (higher Basidiomycetes). *International Journal of Medicinal Mushrooms*

17(9): 847–856. https://doi.org/10.1615/IntJMedMushrooms.v19.i2.20

Kishk, Y. F. M., M. A. Hanan, and H. M. A. Al-Sayed. 2007. Free-radical scavenging and antioxidative activities of some polysaccharides in emulsions. *LWT—Food Science and Technology* 40(2): 270–277. https://doi.org/10.1016/j.lwt.2005.11.004

Kleftaki, S. A., C. Amerikanou, A. Gioxari, et al. 2022a. A randomized controlled trial on *Pleurotus eryngii* mushrooms with antioxidant compounds and vitamin D2 in managing metabolic disorders. *Antioxidants (Basel)* 11(11): 2113. https://doi.org/10.3390/antiox11112113

Kleftaki, S. A., S. Simati, C. Amerikanou, et al. 2022b. *Pleurotus eryngii* improves postprandial glycaemia, hunger and fullness perception, and enhances ghrelin suppression in people with metabolically unhealthy obesity. *Pharmacological Research* 175: 105979. https://doi.org/10.1016/j.phrs.2021.105979

Kondyli, E., E. C. Pappa, D. Arapoglou, M. Metafa, C. Eliopoulos, and C. Israilides. 2022. Effect of fortification with mushroom polysaccharide β-glucan on the quality of ovine soft spreadable cheese. *Foods (Basel)* 11(3): 417. https://doi.org/10.3390/foods11030417

Koutrotsios, G., N. Kalogeropoulos, A. C. Kaliora, and G. I. Zervakis. 2018. Toward an increased functionality in oyster (*Pleurotus*) mushrooms produced on grape marc or olive mill wastes serving as sources of bioactive compounds. *Journal of Agricultural and Food Chemistry* 66(24): 5971–5983. https://doi.org/10.1021/acs.jafc.8b01532

Koutrotsios, G., N. Kalogeropoulos, P. Stathopoulos, et al. 2017. Bioactivecompounds and antioxidant activity exhibit high intraspecific variability in *Pleurotus ostreatus* mushrooms and correlate well with cultivation performance parameters. *World Journal of Microbiology and Biotechnology* 33: 1–14. https://doi.org/10.1007/s11274-017-2262-1

Krakowska, A., P. Zięba, A. Włodarczyk, et al. 2020. Selected edible medicinal mushrooms from *Pleurotus* genus as an answer for human civilization diseases. *Food Chemistry* 327: Article 127084. https://doi.org/10.1016/j.foodchem.2020.127084

Kuo, C.-F., C.-H. Hsieh, and W.-Y. Lin. 2010. Proteomic response of LAP-activated RAW264.7 macrophages to the anti-inflammatory property of fungal ergosterol. *Food Chemistry* 126(1): 207–212. https://doi.org/10.1016/j.foodchem.2010.10.101

Lai, M. D., K. C. Ong, B. Arumugam, and U. R. Kuppusamy. 2024. Nutritional composition, efficacy and mechanisms of oyster mushrooms (*Pleurotus* spp.) in preventing metabolic syndrome: insights into perspectives and challenges. *Food Bioscience* 61: 104768. https://doi.org/10.1016/j.fbio.2024.104768

Lam, V., N. Bui, T. Le, H. Dang, V. Nguyen, Q. H. B. Thuy, T. Nguyen, N. Pham, and M. Dinh. 2022. Identification of lovastatin analogs-producing *Pleurotus cultivars* in southern Vietnam. *International Journal of Agricultural Technology* 18(1): 231–240.

Lesa, K. N., M. U. Khandaker, F. M. R. Iqbal, et al. 2022. Nutritional value, medicinal importance, and health-promoting effects of dietary mushroom (*Pleurotus ostreatus*). *Journal of Food Quality* Article ID 2454180. https://doi.org/10.1155/2022/2454180

Liu, L., S. Jiang, W. Xie, J. Xu, Y. Zhao, and M. Zeng. 2022. Fortification of yogurt with oyster hydrolysate and evaluation of its *in vitro* digestive characteristics and anti-inflammatory activity. *Food Bioscience* 47: 101472. https://doi.org/10.1016/j.fbio.2021.101472.

Llanaj, X., G. Törős, P. Hajdú, et al. 2023. Biotechnological applications of mushrooms under the water-energy-food nexus: crucial aspects and prospects from farm to pharmacy. *Foods (Basel)* 12(14): 2671. https://doi.org/10.3390/foods12142671

Malathi, K., A. Anbarasu, and R. Sudha. 2016. Ethyl iso-allocholate from a medicinal rice karungkavuni inhibits dihydropteroate synthase in *Escherichia coli*: a molecular docking and dynamics study. *Indian Journal of Pharmaceutical Sciences* 78. https://doi.org/10.4172/pharmaceutical-sciences.1000184

Medihi, N. I., Z. A. Haiyee, R. Patmawati, S. Sukor, and S. Raseetha. 2024. Exploring the functional properties and nutritional values of colored oyster mushrooms species (*Pleurotus, Agaricomycetes*): a review. *International Journal of Medicinal Mushrooms* 26(6): 25–38. doi:10.1615/IntJMedMushrooms.2024053563

Mihai, R. A., E. J. Melo Heras, L. I. Florescu, et al. 2022. The edible gray oyster fungi *Pleurotus ostreatus* (Jacq. ex Fr.) P. Kumm a potent waste consumer, a biofriendly species with antioxidant activity depending on the growth substrate. *Journal of Fungi* 8(274). https://doi.org/10.3390/jof8030274

Mishra, V., S. Tomar, P. Yadav, and M. P. Singh. 2021. Promising anticancer activity of polysaccharides and other macromolecules derived from oyster mushroom (*Pleurotus* sp.): an updated review. *International Journal of Biological Macromolecules* 182: 1628–1637. https://doi.org/10.1016/j.ijbiomac.2021.05.102

Mishra, V., S. Tomar, P. Yadav, S. Vishwakarma, and M. P. Singh. 2022. Elemental analysis, phytochemical screening and evaluation of antioxidant, antibacterial and anticancer activity of *Pleurotus ostreatus* through *in vitro* and *in silico* approaches. *Metabolites* 12(9): 821. https://doi.org/10.3390/metabo12090821

Moro, C., I. Palacios, M. Lozano, et al. 2011. Anti-inflammatory activity of methanolic extracts from edible mushrooms in LPS activated RAW264.7 macrophages. *Food Chemistry* 130(2): 350–355. https://doi.org/10.1016/j.foodchem.2011.07.049

Nakahara, D., C. Nan, K. Mori, et al. 2020. Effect of mushroom polysaccharides from *Pleurotus eryngii* on obesity and gut microbiota in mice fed a high-fat diet. *European Journal of Nutrition* 59(7): 3231–3244. https://doi.org/10.1007/s00394-019-02162-7

Ndunge Charles, A., M. Mburu, D. Njoroge, et al. 2024. Chemical composition and consumer acceptability of oyster mushroom and sorghum-pearl millet based composite flours. *Discover Food* 4: 152. https://doi.org/10.1007/s44187-024-00219-z

Ng, S. H., M. S. Mohd Zain, F. Zakaria, W. R. Wan Ishak, and W. A. Wan Ahmad. 2015. Hypoglycemic and antidiabetic effect of *Pleurotus sajor-caju* aqueous extract in normal and streptozotocin-induced diabetic rats. *BioMed Research International* 2015: 214918. https://doi.org/10.1155/2015/214918

Nguyen, T. K., K. H. Im, J. Choi, P. G. Shin, and T. S. Lee. 2016. Evaluation of antioxidant, anti-cholinesterase, and anti-inflammatory effects of culinary mushroom *Pleurotus pulmonarius*. *Mycobiology* 44(4): 291–301. https://doi.org/10.5941/MYCO.2016.44.4.291

Nitschke, J., H. J. Altenbach, T. Malolepszy, and H. Mölleken. 2011. A new method for the quantification of chitin and chitosan in edible mushrooms. *Carbohydrate Research* 346(11): 1307–1310. https://doi.org/10.1016/j.carres.2011.03.040

Nnemolisa, S. C., C. C. Chukwurah, S. C. Edeh, et al. 2024. Antidiabetic and antioxidant potentials of *Pleurotus* ostreatus-derived compounds: an *in vitro* and *in silico* approach. *Food Chemistry Advances* 4: 100639. https://doi.org/10.1016/j.focha.2024.100639

Nordiana, A. B., W. W. Rosli, and W. A. Nizam. 2019. The effect of oyster mushroom (*Pleurotus sajor-caju*) flour incorporation on the physicochemical quality and sensorial acceptability of pasta. *International Food Research Journal* 26: 1249–1257.

Ogidi, C. O., A. M. Ubaru, T. Ladi-Lawal, et al. 2020. Bioactivity assessment of exopolysaccharides produced by *Pleurotus pulmonarius* in submerged culture with different agro-waste residue. *Heliyon* 6: e05685. https://doi.org/10.1016/j.heliyon.2020.e05685

Okwulehie, I. C., and E. N. Nosike. 2015. Phytochemicals and vitamin compositions of *Pleurotus pulmonarius* cultivatedon barks of some indigenous fruit trees supplemented with agro-wastes. *Asian Journal of Plant Sciences* 5(2): 1–7.

Oluwafemi, G., and K. Araoye, Fagbemi, et al. 2016. Chemical composition, functional properties and protein fractionation of edible oyster mushroom (pleurotus ostreatus). *Annals. Food Science and Technology* 17.

Owaid, M., N., S. Salem, Al-Saeedi, and I. A. A. Al-Assaffii. 2015. Antimicrobial activity of mycelia of Oyster mushroom species (*Pleurotus* spp.) and their liquid filtrates (*in vitro*). *Journal of Medical and Bioengineering* 4(5): 376–380. https://doi.org/10.12720/jomb.4.5.376–380

Parvin, R., T. Farzana, S. Mohajan, H. Rahman, S. S. Rahman. 2020. Quality improvement of noodles with mushroom fortified and its comparison with local branded noodles. *NFS Journal* 20: 37–42. https://doi.org/10.1016/j.nfs.2020.07.002

Pérez-Bassart, Z., C. Bäuerl, M. J. Fabra, A. Martínez-Abad, M. C. Collado, and A. López-Rubio. 2023. Composition, structural properties and immunomodulatory activity of several aqueous *Pleurotus* β-glucan-rich extracts. *International Journal of Biological Macromolecules* 253(Pt 6): 127255. https://doi.org/10.1016/j.ijbiomac.2023.127255

Petraglia, T., T. Latronico, A. Fanigliulo, A. Crescenzi, G. M. Liuzzi, R. Rossano. 2023. Antioxidant activity of polysaccharides from the edible mushroom *Pleurotus eryngii*. *Molecules* 28: 2176. https://doi.org/10.3390/molecules28052176

Priyadarshni, K. C., R. Krishnamoorthi, C. Mumtha, and P. U. Mahalingam. 2022. Biochemical analysis of cultivated mushroom, *Pleurotus florida* and synthesis of silver nanoparticles for enhanced antimicrobial effects on clinically important human pathogens. *Inorganic Chemistry Communications* 142: 109673. https://doi.org/10.1016/j.inoche.2022.109673

Radzki, W., K. Skrzypczak, B. Sołowiej, E. Jabłońska-Ryś, and W. Gustaw. 2023. Properties of yogurts enriched with crude polysaccharides extracted from *Pleurotus ostreatus* cultivated mushroom. *Foods (Basel)* 12(21): 4033. https://doi.org/10.3390/foods12214033

Raman, J., K. Y. Jang, Y. L. Oh, M. Oh, J. H. Im, H. Lakshmanan, and V. Sabaratnam. 2020. Cultivation and nutritional value of prominent *Pleurotus spp.*: an overview. *Mycobiology* 49(1): 1–14. doi: 10.1080/12298093.2020.1835142

Rathore, H., S. Prasad, and S. Sharma. 2017. Mushroom nutraceuticals for improved nutrition and better human health: a review. *PharmaNutrition* 5(2): 35. https://doi.org/10.1016/j.phanu.2017.02.001

Rattajak, P., A. Aroonkesorn, C. Smythe, R. Wititsuwannakul and T. Pitakpornpreecha. 2024. *Pleurotus sajor-caju* (Fr.) Singer β-1,3-glucanoligosaccharide (Ps-GOS) suppresses RANKL-induced osteoclast differentiation and function in pre-osteoclastic RAW 264.7 cells by inhibiting the RANK/NFκB/cFOS/NFATc1 signalling pathway. *Molecules (Basel)* 29(9): 2113. https://doi.org/10.3390/molecules29092113

Rodrigues, D. M., A. C. Freitas, T. A. Rocha Santos, et al. 2015. Chemical composition and nutritive value of *Pleurotus citrinopileatus* var *cornucopiae, P. eryngii, P. salmoneo stramineus, Pholiota nameko*, and *Hericium erinaceus*. *Journal of Food Science and Technology* 52: 6927–6939.

Sardar, H., M. A. Ali, M. A. Anjum, et al. 2017. Agro-industrial residues influence mineral elements accumulation and nutritional composition of king oyster mushroom (Pleurotus eryngii). *Scientia Horticulturae* 225: 327–334. https://doi.org/10.1016/j.scienta.2017.07.010

Sari, M., A. Prange, Jan I. Lelley, and R. Hambitzer. 2017. Screening of beta-glucan contents in commercially cultivated and wild growing mushrooms. *Food Chemistry* 216: 45–51. https://doi.org/10.1016/j.foodchem.2016.08.010

Selli, S., G. Guclu, O. Sevindik, and H. Kelebek. 2021. Variations in the key aroma and phenolic compounds of champignon (*Agaricus bisporus*) and oyster (*Pleurotus ostreatus*) mushrooms after two cooking treatments as elucidated by GC-MS-O and LC-DAD-ESI-MS/MS. *Food Chemistry* 30: 354. https://doi.org/10.1016/j.foodchem.2021.129576

Singh, A., S. Singh, U. Pradesh, et al. 2021. Nutritional and health importance of fresh and dehydrated oyster mushroom (*Pleurotus florida*). *Journal of Current Research in Food Science* 2: 10–14.

Souilem, F., Â. Fernandes, R. C. Calhelha, et al. 2017. Wild mushrooms and their mycelia as sources of bioactive compounds: Antioxidant, anti-inflammatory, and cytotoxic properties. *Food Chemistry* 230: 40–48. https://doi.org/10.1016/j.foodchem.2017.03.026

Stastny, J., P. Marsik, J. Tauchen, et al. 2022. Antioxidant and anti-inflammatory activity of five medicinal mushrooms of the genus *Pleurotus*. *Antioxidants (Basel)* 11(8): 1569. https://doi.org/10.3390/antiox11081569

Tagkouli, D., G. Bekiaris, S. Pantazi, et al. 2021. Volatile profiling of *Pleurotus eryngii* and *Pleurotus ostreatus* cultivated on agricultural and agro-industrial by-products. *Foods* 10(6): 1287. https://doi.org/10.3390/foods10061287

Tagkouli, D., A. Kaliora, G. Bekiaris, et al. 2020. Free amino acids in three *Pleurotus* species cultivated on agricultural and agro-industrial by-products. *Molecules* 25(17): 4015. https://doi.org/10.3390/molecules25174015

Wang, Q., L. L. Niu, H. P. Liu, Y. R. Wu, M. Y. Li, and Q. Jia. 2021. Structural characterization of a novel polysaccharide from *Pleurotus citrinopileatus* and its antitumor activity on H22 tumor-bearing mice. *International Journal of Biological Macromolecules* 168: 251–260. https://doi.org/10.1016/j.ijbiomac.2020.12.053

Xue, Z., J. Li, A. Cheng, et al. 2015. Structure identification of triterpene from the mushroom *Pleurotus eryngii* with inhibitory effects against breast cancer. *Plant Foods for Human Nutrition* 70(3): 291–296. https://doi.org/10.1007/s11130-015-0492-7

Yamashita, S., K. Akada, S. Matsumoto, and M. Mikio Kinoshita. 2021. Effects of dietary ethanol extract from fruiting bodies of golden oyster mushroom (*Pleurotus citrinopileatus*) on chronic colon inflammation in mice treated with dextran sulfate sodium salt. *Mushroom Science and Biotechnology* 28(1): 7–14. https://doi.org/10.24465/msb.28.1_7

Zervakis, G. I., S. Ntougias, M. L. Gargano, et al. 2014. A reappraisal of the *Pleurotus eryngii* complex: new species and taxonomic combinations based on the application of a polyphasic approach, and an identification key to *Pleurotus* taxa associated with Apiaceae plants. *Fungal Biology 118*(9): 814–834. https://doi.org/10.1016/j.funbio.2014.06.002

Zhang, Y., T. Hu, H. Zhou, Y. Zhang, G. Jin, and Y. Yang. 2016. Antidiabetic effect of polysaccharides from *Pleurotus* ostreatus in streptozotocin-induced diabetic rats. *International Journal of Biological Macromolecules* 83: 126–132. https://doi.org/10.1016/j.ijbiomac.2015.11.045

Zhou, Y., M. Chu, F. Ahmadi, et al. 2023. A comprehensive review on phytochemical profiling in mushrooms: occurrence, biological activities, applications and future prospective. *Food Reviews International* 40(3): 924–951. https://doi.org/10.1080/87559129.2023.2202738

# 4 Biofortification of *Pleurotus* spp. and Their Medicinal Effects

*Madhumita Karmakar*[1], *Somanjana Khatua*[2], *and* **Krishnendu Acharya*[1]
[1]Molecular and Applied Mycology and Plant Pathology Laboratory, Centre of Advanced Study, Department of Botany, University of Calcutta, Kolkata, West Bengal, India
[2]Molecular and Applied Mycology Laboratory, Department of Botany, Faculty of Science, University of Allahabad, Prayagraj, Uttar Pradesh, India

## 4.1 INTRODUCTION

The human body requires a complex array of nutrients, at least 22 essential elements, for optimal physiological growth and metabolic functioning (Słyszyk et al., 2024). However, micronutrient malnutrition remains a pervasive global challenge, with an estimated two billion people lacking adequate intake of critical vitamins and minerals (Kong et al., 2022). This deficiency, often referred to as "hidden hunger," is largely attributed to the poor bioavailability of nutrients in agricultural soils, which results in diminished nutrient concentrations in edible plant parts and disrupts the nutrient flow across the food chain (Słyszyk et al., 2024). Deficiencies in essential macronutrients and micronutrients, including proteins, vitamins, amino acids, iron, zinc, and calcium, not only impair human health but also lead to long-term developmental and cognitive issues, particularly in vulnerable populations such as children, pregnant women, and the elderly (Yadava et al., 2018). According to the World Health Organization (2023), iron-deficiency anemia affects a substantial proportion of the global population, including 40% of children under five, 37% of pregnant women, and nearly one-third of women of reproductive age. Furthermore, the COVID-19 pandemic has significantly exacerbated nutritional vulnerabilities, exposing the need for enhanced dietary sources rich in immunity-supporting nutrients like vitamins D and C, zinc, and calcium. Inadequate intake of these micronutrients is closely associated with impaired immunity, developmental delays, stunted growth, and an increased risk of neonatal and maternal mortality (Black et al., 2013).

Addressing micronutrient malnutrition demands multifaceted interventions. One of the most promising approaches is biofortification, the process of increasing the nutrient content of food crops through agronomic practices, conventional plant breeding, or modern biotechnological methods. First proposed in the early 1990s as a low-cost and sustainable alternative to supplementation and food fortification, biofortification initially focused on staple food crops commonly consumed in low- and middle-income countries (Li et al., 2024). Over the years, the scope of biofortification has expanded to include a wide array of food sources, reflecting its potential to improve nutritional health on a global scale. Golden Rice, enriched with provitamin A, remains a well-known example of genetically biofortified food (Monika et al., 2023). However, with increasing limitations in cultivable land and the challenges posed by climate change, alternative sources of nutrition that do not depend on traditional farmland are gaining attention (Khatua and Acharya, 2022).

In this context, edible mushrooms, particularly those of the genus *Pleurotus*, offer a unique and underexplored avenue for biofortification. Commonly referred to as oyster mushrooms, *Pleurotus* species are among the most widely cultivated mushrooms globally, prized for their high yield, low production cost, adaptability to a broad range of lignocellulosic agricultural wastes, and rich nutritional profile (Rosado et al., 2002; Obodai et al., 2003; Mane et al., 2007; Galappaththi et al., 2021). In addition to providing quality proteins, dietary fiber, and essential vitamins and minerals, they exhibit a wide spectrum of bioactivities, including immunomodulatory, anticancer, antioxidant, and anti-aging properties (Mitra et al., 2013; Patra et al., 2013; Zięba et al., 2020). Importantly, *Pleurotus* mushrooms possess an exceptional ability to absorb, accumulate, and metabolize mineral elements from their substrates, particularly through their highly active mycelial networks (Muñoz et al., 2006; Rabinovich et al., 2007; da Silva et al., 2012; Rzymski et al., 2016). This intrinsic trait positions them as ideal candidates for mineral biofortification. Emerging research indicates that mineral-enriched *Pleurotus* mushrooms not only exhibit higher concentrations of essential nutrients like iron, selenium, zinc, and calcium but also display enhanced therapeutic benefits (Fasoranti et al., 2019; Oyetayo et al., 2024).

Given the growing interest in functional foods and the urgent need for nutrition-sensitive interventions, biofortified *Pleurotus* mushrooms present a promising strategy to enhance public health and nutritional security. However, despite scattered efforts and promising outcomes, a comprehensive synthesis of methodologies, outcomes, and nutritional implications in mushroom biofortification is lacking. This chapter thus seeks to consolidate current knowledge on biofortification of different oyster mushrooms and explore their potential as a sustainable and accessible solution for global health challenges.

* Corresponding author: krish_paper@yahoo.com


DOI: 10.1201/9781003527572-4

## 4.2 REQUIRED AMOUNTS OF CRUCIAL MICRONUTRIENTS FOR HUMAN

To lead healthy and productive lives, humans require adequate intake of approximately 40 essential nutrients. Among these, several mineral elements, including sodium, potassium, calcium, magnesium, phosphorus, chlorine, and sulfur, are required in modest quantities to support various physiological functions. Micronutrients, a specific subset of essential nutrients needed in trace amounts, include iron, zinc, copper, manganese, iodine, selenium, molybdenum, cobalt, nickel, and vitamin A (Prashanth et al., 2015; Khatua and Acharya, 2022). Despite being required in small quantities, these micronutrients are vital for proper physical and cognitive development (White and Broadley, 2005). Many of them function as enzyme cofactors, playing crucial roles in regulating metabolic pathways and maintaining homeostasis (Welch and Graham, 2004). The Recommended Daily Allowance (RDA) refers to the average daily intake level sufficient to meet the nutrient requirements of nearly all healthy individuals. The RDA values and physiological roles of key micronutrients are summarized in Table 4.1.

**TABLE 4.1**
**RDA and Physiological Roles of Essential Micronutrients in Humans**

| Essential Minerals | RDA (for Adults, >19 years) | | Importance | References |
|---|---|---|---|---|
| | Men | Women | | |
| Iron | 19 mg/d | 29 mg/d | Essential for hemoproteins like hemoglobin and myoglobin; acts as a cofactor for enzymes; non-heme iron proteins play key roles in DNA synthesis and repair, cell growth, gene regulation, and drug metabolism | Pantopoulos et al., 2012; Yiannikourides and Latunde-Dada, 2019; Budzyńska et al., 2022; Abbaspour et al., 2014 |
| Copper | 1.7 mg/d | 1.7 mg/d | Supports carbohydrate, amino acid, and cholesterol metabolism; aids bone formation and blood sugar regulation; acts as a cofactor for superoxide dismutase; essential for hemoglobin production, catecholamine biosynthesis, and cross-linking of keratin, collagen, and elastin in hair | Silva et al., 2019; Avram et al., 2021; Binesh and Venkatachalam, 2024 |
| Manganese | 2.3–4 mg/d | 1.8–4 mg/d | Important for normal growth, immune function, wound healing, DNA synthesis, cell division; needed for fetal development, growth and survival; regulate skeletal system development, energy metabolism, nervous system, function of reproductive hormone | Santamaria, 2008; Silva et al., 2019; Studer et al., 2022 |
| Chromium | 50 µg/d | 50 µg/d | Supports connective tissue formation, normal immune function, neural development, blood health, and antioxidant defense; lowers total cholesterol, LDL-cholesterol, triglycerides, and non-esterified fatty acids, while increasing HDL-cholesterol and β-oxidation rate | Lewicki et al., 2014; Silva et al., 2019; Islam et al., 2023 |
| Selenium | 40–55 µg/d | 40–55 µg/d | Possesses antioxidant and anticancer properties, regulates the immune and reproductive systems (including thyroid function), and is a key component of selenoproteins essential for immune response, muscle activity, central nervous system, circulatory system, and cancer prevention | Jovanović et al., 2013; Ruseva et al., 2013; Flis et al., 2015; Kieliszek and Błażejak, 2016; Minich, 2022; Genchi et al., 2023 |
| Iodine | 140–150 µg/d | 140–150 µg/d | Essential for development of bones and teeth, and a vital component of thyroid hormones | Zimmermann, 2011; Nicol et al., 2024 |
| Zinc | 11–17 mg/d | 8–13.2 mg/d | Plays a key role in thyroid hormone synthesis, oxidative stress defense, blood pressure regulation, wound healing, hair growth, and immune function. It supports night vision and senses of taste and smell, acts as a cofactor for numerous enzymes and transcription factors, and exhibits antioxidant, anti-inflammatory, anticancer, antidiabetic, and anti-aging properties, aiding recovery from atherosclerosis, cancer, Alzheimer's, and autoimmune diseases | Chasapis et al., 2020; Sharma et al., 2021; Maywald and Rink, 2022 |
| Fluoride | 4 mg/d | 3 mg/d | Essential for preventing early childhood caries and required for the synthesis of hemoglobin, myoglobin, and certain hormones | Sham et al., 2021; Adeghe, 2024 |

*HDL, high density lipoprotein; LDL, low density lipoprotein*

## 4.3 BIOFORTIFICATION: A PROMISING TECHNIQUE TO IMPROVE NUTRITION

For decades, agricultural research in developing countries has prioritized boosting the yield of staple crops. However, this focus has often neglected the nutritional quality of food, leading to widespread "hidden hunger," a chronic lack of essential vitamins and minerals (Aggarwal et al., 2020). The modern agricultural agenda now increasingly emphasizes producing nutrient-rich crops rather than just increasing calorie output, particularly in low- and middle-income nations (Khush et al., 2012).

Biofortification is an innovative and sustainable approach aimed at enhancing the nutritional quality of food crops to combat malnutrition and micronutrient deficiencies (Neill et al., 2021; El-Ramady et al., 2021). Unlike post-harvest fortification, biofortification enriches food with essential nutrients during the plant's growth phase, ensuring the nutrients are available in their natural form (Yadava et al., 2018; Kumar et al., 2019; Malik and Maqbool, 2020). This is typically achieved by modifying agronomic practices or employing selective breeding techniques (Neill et al., 2023), distinguishing it from conventional fortification, which adds nutrients during food processing (Odiketa et al., 2020). Biofortification is also economically viable, as it typically requires a one-time investment with long-term benefits (Nestel et al., 2006; Pfeiffer and Mc Clafferty, 2007). Although the concept emerged during the Green Revolution (1966–1985), the term "biofortification" was officially coined by Steve Beebe in 2001. The first biofortified crop was released in 2013 following national approval, with nutritional efficacy trials initiated soon after. Since then, over 140 biofortified varieties across 10 crops enriched with iron, zinc, and provitamin A have been released in more than 30 countries, benefiting rural populations with limited access to diverse and nutritious food (Aggarwal et al., 2020).

Due to their high capacity for bioaccumulation, mushrooms, particularly their mycelial networks, serve as effective candidates for biofortification (Rzymski et al., 2016). Trace elements essential for human health can be incorporated into mushrooms through substrate enrichment, allowing them to function as nutrient-dense functional foods (Zięba et al., 2020). It is crucial, however, to maintain balance, as some elements such as lead, mercury, cadmium, and lithium can become toxic beyond trace levels. Controlled biofortification ensures that only beneficial quantities of essential elements are assimilated into the mushroom tissues. For instance, mushrooms biofortified with selenium and lithium have shown potential in supporting therapies for HIV, cancer, cardiovascular disorders, neurodegenerative diseases, and aging-related conditions. Additionally, biofortified mushrooms may contribute to nutritional security while offering a pathway to economic empowerment, especially in resource-constrained rural communities (Pattanayak and Das, 2022).

## 4.4 BIOFORTIFICATION STRATEGIES FOR *PLEUROTUS* SP.

*Pleurotus* species, commonly known as oyster mushrooms, are among the most widely cultivated mushrooms globally, accounting for approximately 25% of total mushroom production (Raman et al., 2020). These mushrooms are highly valued not only for their nutritional properties being rich in proteins, vitamins, minerals, amino acids, and dietary fibers but also for their therapeutic potential, including antioxidant, immunomodulatory, anticancer, antibacterial, and anti-aging effects (Mitra et al., 2013; Patra et al., 2013; Zięba et al., 2020). A remarkable characteristic of *Pleurotus* spp. is their capacity to absorb and bioaccumulate essential minerals from their substrate, making them ideal candidates for biofortification (Madaan et al., 2024). Recent years have seen a surge in studies investigating the mineral biofortification of *Pleurotus* species to elevate their value as functional foods (Figure 4.1). For instance, Ogidi and Oyebode (2023) successfully biofortified *Pleurotus pulmonarius* with copper, lithium, and zinc by enriching the cultivation substrate with specific mineral salts. In their study, spawn was prepared using sterilized parboiled sorghum grains inoculated with pure mycelial cultures. Once colonized, the mycelium was transferred to a substrate composed of dried, sterilized spent mushroom substrate (SMS), supplemented with 100 mg/kg of $CuSO_4$, $ZnSO_4$, and $LiNO_3$. Eight different treatments were established, including single and combined mineral applications. Following inoculation, the substrates were incubated at 26 °C in darkness until the formation of primordia, after which the temperature was reduced to 19 °C and humidity maintained at 91%. Mature fruiting bodies were harvested and evaluated based on their mineral enrichment profiles.

In a similar vein, Zięba et al. (2020) conducted a study on biofortification of *Pleurotus eryngii* using selenium and zinc, targeting both the mycelial biomass and fruiting bodies. For mycelial biofortification, cultures were grown in modified Oddoux liquid medium enriched with sodium selenite (50 mg/L), zinc sulfate (87.2 mg/L), or zinc hydroaspartate (100 mg/L), both sources containing the same concentration of $Zn^{2+}$ ions (20 mg/L). Flasks containing 250 mL of liquid culture were incubated at 25 °C under a 16-hour light/8-hour dark photoperiod. After four weeks, the mycelial biomass was harvested, triple-washed with distilled water, frozen, and lyophilized for drying. For fruitbody enrichment, the cultivation substrate consisted of a 1:1 mixture of beech sawdust and powdered wheat straw. The substrate was supplemented with the same concentrations of selenium and zinc as in the liquid culture and inoculated with wheat grain spawn of *P. eryngii*. Colonization was carried out in darkness at 24 °C for four weeks. Once fully colonized, the jars were moved to fruiting conditions: 18 °C temperature and a 12-hour light cycle. Mature fruiting bodies were harvested and analyzed for their mineral content.

Another comprehensive study by Budzyńska et al. (2022) investigated iron biofortification in *P. ostreatus*, *P. eryngii*, and *Pleurotus nameko* using three iron salts, such

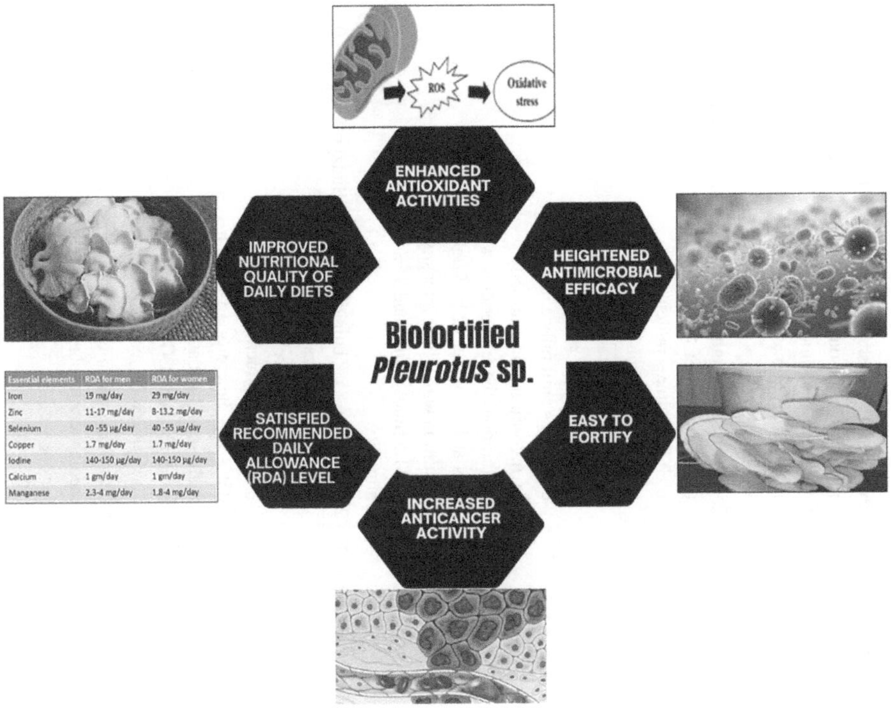


**FIGURE 4.1** Potential benefits of biofortified oyster mushrooms.

as $FeCl_3$, $FeSO_4$, and FeHBED, at concentrations of 5, 10, and 50 mM. For *P. eryngii* and *P. nameko*, a substrate of alder and beech sawdust (1:1) enriched with gypsum, saccharose, cornmeal, soy meal, and wheat bran was used, whereas *P. ostreatus* was cultivated on wheat straw. These substrates were moistened with the iron solutions, sterilized, inoculated, and incubated under optimized temperature and humidity conditions. Fruiting was induced under a controlled environment, and mineral content in the fruiting bodies was measured post-harvest.

The substrate-based approach was further expanded by Ajewole et al. (2012), who incorporated agro-wastes such as cassava peels, orange pulp, and maize straw into sawdust-based substrates for *P. ostreatus* cultivation in different ratios (1:1, 1:2, 1:3). After sterilization and inoculation, bags were incubated in the dark at 25–30 °C for two weeks. Complementary to this, Oyetayo (2023) emphasized the importance of mineral-enriched substrates in the biofortification process, reinforcing substrate optimization as a key strategy for enhancing the nutritional value of cultivated mushrooms.

This aligns with another finding by Ogidi et al. (2016) who assessed the mycelial growth and biomass production of various *Pleurotus* strains, namely *P. ostreatus*, *P. djamor*, *P. djamor* var. *roseus*, *P. pulmonarius*, and *P. cornucopiae*, in potato dextrose agar media supplemented with different concentrations of $FeSO_4$. Colony diameter and biomass were measured to determine iron uptake at the mycelial stage. For fruiting, the fungus was cultivated on iron-enriched *Urochloa decumbens* substrate, and fruiting conditions were carefully regulated to ensure successful yield and mineral incorporation.

Further advancing the field, Poursaeid et al. (2015) focused on liquid-based zinc biofortification of *Pleurotus florida*. Seed cultures were grown in glucose–yeast extract-based medium and inoculated into basal liquid media enriched with zinc sulfate (0–300 mg/L). Cultures were incubated at 25 °C in darkness, then exposed to LED light at 20 °C with controlled humidity for fruiting, highlighting the role of both nutrient medium and environmental cues in achieving effective biofortification.

Odiketa et al. (2020) enriched *Pleurotus floridanus* with calcium using organic supplements such as Mycroliquid and Mycrowhitener. Two substrate variants, one with wheat bran and one without, were fortified with calcium carbonate, sterilized, and inoculated. Following four weeks of incubation at 28 °C, calcium supplements were added at various concentrations. This method proved successful in producing calcium-fortified fruiting bodies, indicating that post inoculation mineral application could be an effective method for enhancing the elemental profile of mushrooms. All in all, these studies underscore the versatility of *Pleurotus* species for biofortification using a wide range of mineral sources, substrates, and cultivation conditions, as represented in Table 4.2.

## 4.5 EFFECTS OF IRON FORTIFICATION IN *PLEUROTUS* SP.

Iron fortification in mushrooms has emerged as a promising strategy to address iron deficiency, and numerous studies have explored its efficacy across different mushroom species and fortification techniques. Budzyńska et al. (2022)

**TABLE 4.2**
**Summary of Biofortification Studies in *Pleurotus* Species.**

| Mushroom | Substrate | Sterilization of Substrate | Inoculation of Substrate | Minerals Used for Fortification | Required Environment for Mycelial Growth and Fruit-body Development | References |
|---|---|---|---|---|---|---|
| *P. pulmonarius* | Dried SMS of *Pleurotus* sp. | Autoclaved at 121°C for 30 minutes | Sorghum spawn | Copper sulfate, zinc sulfate and lithium nitrate (100 mg/kg of substrate) | 26°C until the mushroom primordia came out and then 19°C and 91% humidity were maintained for maturation of fruit body | Ogidi and Oyebode, 2023 |
| *P. eryngii* | Modified liquid Oddoux medium for mycelial biomass and mixture of beech sawdust and powdered wheat straw (1:1 v/v) for fruit body production | Sterilized at 121 °C for 90 minutes | Wheat grain | Sodium selenite (50 mg/L), zinc sulfate (87.2 mg/L), and zinc hydro-aspartate (100 mg/L) | 25°C under 16 h light condition (light intensity of 11.5 μmol $s^{-1}$ $m^{-2}$) for cultures in liquid medium. For fruit-body formation, substrate-containing jars were kept at 24°C in dark conditions until fully colonized by mycelium. For fruit-body maturation, the temperature was kept at 18°C under 12 h of light condition with a light intensity of 11.5 μmol $s^{-1}$ $m^{-2}$ and 95% humidity. | Zięba et al., 2020 |
| *P. ostreatus*, *P. eryngii* and *P. nameko* | Wheat straw for *P. ostreatus*. For other two species, alder and beech sawdust (1:1) mixed with 1% gypsum, 1% saccharose, 5% cornmeal, 5% soy meal and 20% wheat bran were taken as substrate. | Autoclaved for one hr at 121 °C | Wheat grain spawn of *P. ostreatus* at the ratio of 3:100 of spawn: wet weight of substrate and for *P. eryngii* and *P. nameko,* 10 g of wheat grain spawn of was added in 450 g substrate. | 5 mM, 10 mM and 50 mM solution of three different iron salts, i.e., $FeCl_3{\cdot}6H_2O$, $FeSO_4.7H_2O$ and FeHBED were added in different sets of substrates. | 25°C and 80–85% humidity for complete colonization of substrate by mycelium. For fruit-body production, 85%–90% humidity, 15°C, 12 h fluorescent light at 500 lx intensity/d and $CO_2$ level below 1000 ppm were maintained. | Budzyńska et al., 2022 |
| Mycelial growth of *Pleurotus ostreatus* (PLO6), *Pleurotus ostreatus* (P93), *P. djamor* (PLO13), *P. djamor* var. *roseus* (Psfarm), *P. pulmonarius* (Pindo) and *P. cornucopiae* (Plocor). | Potato dextrose agar media | Autoclaved | Fungal isolates were cultured on normal potato dextrose agar (PDA) media and kept at 25 °C for one week. Then one disc of 8 mm from was transferred in iron-enriched PDA media. | $FeSO_4.7H_2O$ at 0, 50, and 100 mg/L were added to PDA media | Cultures in iron-enriched PDA media were incubated at 25 °C for one week. | Ogidi et al., 2016 |

| | | | | | | |
|---|---|---|---|---|---|---|
| Fruit-body production of *Pleurotus ostreatus* (PLO6), *Pleurotus ostreatus* (P93), *P. djamor* (PLO13), *P. djamor* var. *roseus* (Psfarm), *P. pulmonarius* (Pindo) and *P. cornucopiae* (Plocor). | Milled *Urochloa decumbens* | Soaked in 2% $Ca(OH)_2$ lime water solution for 24 h | Substrates were inoculated by culture of these fungal isolates of normal PDA media. | $FeSO_4.7H_2O$ at 0, 500, and 1000 mg/kg was added to the substrate. | 90% humidity was required at 25 °C | Ogidi et al., 2016 |
| *P. florida* | Basal liquid media containing beet molasses (50 g/L), soybean flour (5 g/L), $KH_2PO_4$ (1 g/L) and $MgSO_4.7H_2O$ (1 g/L) | Sterilized by autoclaving at 121 °C for 20 minutes | Seed culture of *P. florida* [5% (v/v) seed culture for 100 mL media] was used as inoculum. | Zinc sulfate solution at the concentrations of 0, 50,100, 150, 200, and 300 mg/L were added in basal liquid media | Containers were incubated in the dark at 25 °C for 15 days, then exposed to 500 lx LED light at 20 °C, with 90%–95% humidity for pinhead initiation and 80%–90% during fruiting body development | Poursaeid et al., 2015 |
| *Pleurotus floridanus* | Two types of substrates—one is sawdust, wheat bran, calcium carbonate and water at the ratio 12:1:0.13:5. Another one is sawdust, calcium carbonate and water were mixed at 12:0.13:5 | Sterilization of substrate was done at 100 °C for 4 h in a fabricated sterilization drum. | For production of spawn, guinea corn (*Sorghum bicolor*) was used. | Two supplements of calcium —Organic Mycroliquid and Organic Mycrowhitener. 20 mL solution of these calcium supplements at the ratio (Ca:water) 1:1, 1:5, 1:10, and 2:1 were applied to the respective substrate bags | 28 °C as the optimum temperature for fruit-body development | Odiketa et al., 2020 |

conducted a comprehensive investigation on three *Pleurotus* species, as described earlier. Their findings revealed that the 50 mM concentration led to the highest iron accumulation, with $FeCl_3$ showing superior enrichment in *P. eryngii* and *P. nameko*, increasing iron content by 145% and 185%, respectively. Interestingly, *P. ostreatus* responded better to FeHBED, exhibiting a 108% rise in iron levels. Beyond mineral enrichment, biofortification also enhanced functional compounds such as phenolic acids; for instance, 4-hydroxybenzoic acid in *P. ostreatus* increased from 75.3 μg/g in the control to 108 μg/g with $FeCl_3$ supplementation.

Complementing this, Condé et al. (2017) reported that *P. ostreatus* cultivated on sugarcane bagasse supplemented with 1 mg/kg iron exhibited higher levels of iron, calcium, and protein. Similarly, Oyetayo et al. (2021) fortified *P. pulmonarius* using Fe-enriched rice bran and maize stalk, resulting in a remarkable iron content of 417.6 mg/100 g. Importantly, there was no significant reduction in biological efficiency among iron-fortified, zinc-fortified, and control treatments. Their study further observed that iron and zinc enrichment increased total phenolics, flavonoids, and antioxidant activity, with combined mineral treatment exhibiting 92.2% 2,2-diphenyl-1-picrylhydrazyl (DPPH) radical scavenging activity, a strong indication of enhanced bioactive potential.

Scheid et al. (2020) adopted a liquid culture approach using sugarcane molasses medium to biofortify *P. eryngii* and *P. ostreatus*, and reported a notable enhancement in both biomass and iron accumulation, with levels surpassing those typically found in green vegetables. In another noteworthy study, Umeo et al. (2020) observed that the *P. ostreatus* strain U2–9 accumulated the highest iron concentration when grown on malt extract agar enriched with 50 mg/L iron. However, this came at the cost of a 64% decrease in biomass, pointing to a trade-off between mineral accumulation and fungal growth.

Beyond nutritional enhancement, iron fortification was also shown to improve therapeutic potential. Ariyo et al. (2023) demonstrated that iron-enriched *P. ostreatus* displayed increased levels of moisture, ash, protein, and phytochemicals, along with significantly enhanced antibacterial activity. The iron content rose from 0.34 mg/g in controls to 1.01 mg/g post-fortification, and the fortified mushrooms exhibited greater inhibitory effects against *Escherichia coli*, suggesting improved antimicrobial potential.

However, not all outcomes of iron biofortification have been consistently positive. Vieira et al. (2013) observed that *P. ostreatus* cultivated on coffee husk with minimal iron supplementation exhibited lower iron content and antioxidant activity compared to controls. Although there was a modest rise in phenolic content and inhibition of β-carotene/linoleic acid oxidation, overall nutritional enhancement remained limited. Furthermore, Ogidi et al. (2016) highlighted a critical challenge, reporting that high iron concentrations in the substrate negatively impacted mycelial growth and biomass yield. This inhibition may stem from iron toxicity or poor solubility and bioavailability under aerobic conditions, where ferric iron tends to form insoluble hydroxides. In addition, the permeability of the fungal cell wall may further influence the efficiency of mineral uptake.

Despite these variations and limitations, iron-fortified mushrooms continue to be a promising and sustainable non-animal source of bioavailable iron. Their dual role as a nutritious and therapeutic food makes them an important candidate in combating iron deficiency, particularly in regions where conventional iron sources are inaccessible or culturally restricted. Future research aimed at optimizing iron sources, concentrations, and delivery methods can further enhance the nutritional and functional potential of mushrooms in biofortification programs.

## 4.6 EFFECTS OF ZINC FORTIFICATION IN *PLEUROTUS* SP.

To address the widespread issue of zinc deficiency, numerous studies have turned their focus toward enhancing zinc accumulation in edible mushrooms, particularly oyster mushrooms. Among these, *Pleurotus* species have shown considerable promise due to their inherent ability to bioaccumulate minerals from their growing substrates. Umeo et al. (2020) cultivated several fungal strains, including *P. ostreatus*, *P. eryngii*, *Schizophyllum commune*, and *Lentinula edodes* on malt extract agar supplemented with 7.5 mg/L zinc. Notably, *P. ostreatus* exhibited the highest mycelial zinc content, reaching 440.4 mg/kg, underscoring its potential for biofortification.

Moving beyond laboratory media to more applicable food systems, Luvitaa et al. (2023) demonstrated a practical intervention by integrating zinc-fortified *P. ostreatus* powder into finger millet porridge, a staple in many zinc-deficient regions. The mushrooms were cultivated on rice straw soaked in zinc sulfate solutions ranging from 50 to 600 mg/kg. Zinc accumulation in fruiting bodies increased proportionally with substrate concentration, peaking at 11.65 mg/100 g at the highest supplementation level. However, the best yield was obtained at 100 mg/kg $ZnSO_4$, suggesting that while zinc accumulation benefits from higher concentrations, excessive amounts can hinder fungal growth.

Indeed, factors such as fungal strain, growth stage, and tissue type play significant roles in mineral uptake efficiency, as emphasized by Kalač and Svoboda (2000), Yilmaz et al. (2003), and Keleş et al. (2017). Zięba et al. (2020) further supported this view by demonstrating that both zinc sulfate and zinc hydroaspartate significantly improved the dry weight yields of *P. eryngii*, with respective outputs of 12.28 g and 13.97 g/100 g of dry substrate, suggesting that the chemical form of zinc is also crucial in determining fortification success.

However, zinc biofortification is not without trade-offs. Vieira et al. (2013) observed that *P. ostreatus* grown on zinc-enriched substrates exhibited a slight reduction in total phenolic content and decreased antioxidant activity. Interestingly, the fortified group exhibited greater β-carotene/linoleic acid inhibition than the control group, suggesting that fortification may influence bioactive compound activity in complex and potentially divergent ways.

In contrast, Oyetayo et al. (2021) presented more encouraging results. They found that *P. pulmonarius* grown on zinc-supplemented rice bran and maize stalk not only accumulated 349.5 mg/100 g zinc but also exhibited elevated flavonoid content. The fortified mushrooms showed exceptional free radical scavenging potential, with 96.3% DPPH and 96.8% nitric oxide scavenging activities, reinforcing their functional food appeal.

Adding a multidimensional layer, Ogidi and Oyebode (2023) explored combined mineral supplementation using copper, lithium, and zinc in SMS. The highest antioxidant activity was observed in mushrooms cultivated on a combination of Cu, Li, and Zn, where DPPH scavenging activity reached 81%, significantly surpassing the 66.7% recorded in the control group. Additionally, the zinc-fortified mushrooms displayed enhanced hydroxyl radical scavenging and iron chelation capacities. They even fostered the growth of beneficial probiotic bacteria such as *Lactobacillus acidophilus* and *Lactobacillus plantarum*, opening new avenues for synbiotic mushroom-based products.

In further support, Gąsecka et al. (2016) reported increased levels of phenolics, flavonoids, and ascorbic acid, along with improved antioxidant activity in *P. ostreatus* and *P. eryngii* enriched with zinc nitrate hexahydrate and selenium salts. These findings collectively highlight how zinc supplementation can positively modulate not just mineral content but also antioxidant and nutritional profiles.

Recently, a study by Oyetayo et al. (2024) underscored the synergistic potential of co-fortification strategies. Their work demonstrated that the combined application of zinc and iron slightly enhanced the antimicrobial efficacy of *P. pulmonarius* extracts against pathogens such as *Candida* and *Staphylococcus* species, further showcasing the versatile biofunctional enhancements achievable through mineral biofortification.

## 4.7 EFFECTS OF SELENIUM FORTIFICATION IN *PLEUROTUS* SP.

Cultivated mushrooms are typically low in selenium, generally containing less than 1 to 8.5 μg/g; while their wild counterparts often boast significantly higher levels, sometimes ranging from 12 to as much as 200 μg/g (Falandysz, 2008; Costa-Silva, 2011; Kora, 2020). Despite their nutritional appeal, the commercial cultivation of these selenium-rich wild mushrooms remains difficult due to their complex growth requirements and specific ecological dependencies. As a result, a more feasible and scalable strategy to enhance selenium content in edible mushrooms lies in the biofortification of cultivated species through the use of selenium-enriched substrates, a technique that effectively harnesses the natural bioaccumulation potential of mushrooms (Pattanayak and Das, 2022).

The effectiveness of such selenium enrichment, however, is closely tied to the chemical form of selenium used. Savić et al. (2009) highlighted that sodium selenite proved to be a more efficient selenium source compared to other inorganic salts. Further advancing this approach, Savić et al. (2012) demonstrated that enriching *P. ostreatus* and *P. cornucopiae* with selenized yeast resulted in remarkably high selenium accumulation.

In contrast, da Silva et al. (2013) reported noticeable changes in colony morphology, growth rate, and pigmentation in *P. ostreatus* and *P. eryngii* cultivated on selenium-supplemented media. Similarly, Nunes et al. (2012) observed that selenium concentrations exceeding 12.8 mg/kg in the substrate led to reduced mycelial expansion and distinct alterations in fruit-body architecture, including enlarged stipes and smaller caps. These morphological changes suggest that while biofortification can elevate nutritional value, it must be balanced carefully to avoid compromising mushroom quality and productivity.

In a separate study, Milovanović et al. (2014) examined how *P. ostreatus* mycelium absorbs selenium from media enriched with varying concentrations of sodium selenite. They found that lower concentrations (5–20 mg/L) supported optimal growth, with maximum selenium accumulation (938.9 μg/g) recorded at 20 mg/L. However, levels above 500 mg/L significantly hampered mycelial development.

Fasoranti et al. (2018) assessed the phytochemical makeup of selenium-fortified *Pleurotus* species. Selenium was not detected in the control mushrooms but was present in fortified *P. ostreatus* and *P. pulmonarius* at 51.79 μg/g and 59.29 μg/g, respectively. These fortified mushrooms exhibited antifungal properties against pathogens such as *Aspergillus flavus*, *Candida albicans*, and *Candida parapsilosis*, which were absent in the control group.

In a follow-up study, Fasoranti et al. (2019) evaluated the nutritional and antioxidant profiles of selenium-enriched mushrooms. With 50 mg/kg sodium selenite supplementation, both *P. ostreatus* and *P. pulmonarius* exhibited improved crude fiber content and enhanced levels of essential and non-essential amino acids. Additionally, selenium fortification significantly boosted antioxidant activities, including DPPH and hydroxyl radical scavenging, as well as ferric-reducing power.

Poniedziałek et al. (2017) explored how selenium alone and in combination with zinc and copper affected *P. ostreatus* and *P. eryngii*. At mineral concentrations of 0.1–1.2 mM, they observed a decline in biomass at higher levels (0.6–1.2 mM). Nonetheless, the greatest reductions in reactive oxygen species and lipid peroxidation occurred at 1.2 mM Se and Se+Zn, highlighting their potential for oxidative stress mitigation.

Yan and Chang (2012) investigated the bioactivity of selenium- and zinc-enriched *P. ostreatus* (SZM), which contained 39.8 mg Se and 201.2 mg Zn/kg. Mice supplemented with SZM in their diet showed increased levels of antioxidant enzymes, namely glutathione peroxidase and superoxide dismutase, relative to both control and inorganic Se-Zn (ISZ) groups, correlating with promising antitumor effects in lung cancer models.

Velez et al. (2019) compared the impacts of sodium selenite and selenate on *P. djamor*. At 12.5 mg/L, selenite enhanced β-glucan content, mycelial growth, protein levels,

**TABLE 4.3**
**Comparison of Selenium Accumulation in Different *Pleurotus* spp. Cultivated on Substrates with Varying Selenium Concentrations**

| *Pleurotus* Species | Amended Se Form | Concentration of Se in Substrate | Amount of Se Accumulation in Fruit Body | References |
|---|---|---|---|---|
| *P. cornucopiae* | Selenium-enriched yeast | 25 μg/g | 42.8 μg/g | Savić et al., 2012 |
| *P. djamor* | Sodium selenite | 25.6 μg/g | 76 μg/g | de Oliveira and Naozuka, 2019 |
| | Wheat straw hyperaccumulated with Se | 24 μg/g | 145.4 μg/g | Bhatia et al., 2013b |
| *P. eryngii* | Sodium selenite | 5–10 μg/g | 4.6–9.3 μg/g | Rodriguez et al., 2009 |
| *P. florida* | Wheat straw hyperaccumulated with Se | 27.9 μg/g | 141 μg/g | Bhatia et al., 2013a |
| | Wheat straw hyperaccumulated with Se | 24 μg/g | 110 μg/g | Bhatia et al., 2014 |
| *P. ostreatus* | Sodium selenite | 25.4 μg/g | 261 μg/g | da Silva et al., 2010 |
| | Sodium selenite | 6.4 μg/g | 19 μg/g | de Oliveira and Naozuka, 2019 |
| | Sodium selenate | 100 μg/g | 154.7–169.3 μg/g | Savić et al., 2009 |
| | Se enriched yeast | 25 μg/g | 42.4–50.2 μg/g | Savić et al., 2012 |
| | Wheat straw hyperaccumulated with Se | 24 μg/g | 44.3 μg/g | Bhatia et al., 2013b |
| *P. pulmonarius* | Sodium selenite | 56.2 μg/g | 23.1 μg/g | Milovanovic et al., 2019 |
| *P. sajor-kaju* | Wheat straw hyperaccumulated with Se | 27 μg/g | 43.5 μg/g | Bhatia et al., 2014 |

and antioxidant activity, whereas selenate diminished these benefits. Notably, selenate reduced β-glucan by 50% and ergosterol by 30%, suggesting a less favorable profile than selenite for functional enhancement. Together, these findings emphasize the beneficial role of selenium, particularly in the selenite form, in boosting the nutritional and therapeutic properties of oyster mushrooms, as represented in Table 4.3.

## 4.8 FORTIFICATION OF *PLEUROTUS* SP. WITH OTHER MINERALS

Odiketa et al. (2020) evaluated the influence of different calcium sources on the yield of *P. floridanus* cultivated on substrates with and without wheat bran. They observed that organic calcium supplements such as Organic Mycroliquid and Organic Mycrowhitener enhanced mushroom yield when applied at specific calcium-to-water ratios, with effects varying based on substrate composition. Similarly, Julian et al. (2018) fortified *P. ostreatus* with calcium using agricultural lime, eggshell powder, and oyster shell powder on rice straw. Optimal yields were achieved with agricultural lime (8%–10%) and eggshell powder (6%), while oyster shell powder had negligible impact.

Lithium enrichment has also been explored in oyster mushrooms. Using coffee husk as the base substrate, *P. ostreatus* and *P. djamor* were cultivated with increasing concentrations of lithium chloride and lithium carbonate (de Souza et al., 2022). Lithium accumulation in fruiting bodies increased proportionally with supplementation. However, elevated lithium chloride levels negatively affected yield in *P. ostreatus*, whereas *P. djamor* maintained stable biomass and biological efficiency. Lithium carbonate, on the other hand, reduced growth in both species, a trend also noted by Mleczek et al. (2017), who reported a more pronounced toxic effect from lithium carbonate compared to lithium acetate. Fortification with lithium also influenced the biochemical profile of mushrooms. Vieira et al. (2013) found that lithium-treated *P. ostreatus* exhibited decreased antioxidant activity and total phenolic content, as indicated by higher half-maximal inhibitory concentration ($IC_{50}$) values in DPPH assays. Nonetheless, these mushrooms showed a modest increase in β-carotene/linoleic acid oxidation inhibition, suggesting selective enhancement of certain antioxidant mechanisms. Interestingly, lithium fortification appeared to improve the antimicrobial potential of mushrooms. Ogidi and Oyebode (2023) reported that lithium-enriched *P. pulmonarius* exhibited broader antimicrobial activity, with inhibition zones extending to pathogens such as *Proteus mirabilis* and *Candida tropicalis*, against which non-fortified samples showed no activity.

The same researchers also investigated copper enrichment in *P. pulmonarius* by adding $CuSO_4$ (100 mg/kg) to the spent mushroom substrate. Copper accumulation reached 98.09 mg/100 g dry weight in fortified mushrooms, compared to 41.04 mg/100 g in the controls. While phytochemical profiles (phenols, flavonoids, alkaloids, saponins, terpenoids) remained largely unchanged, copper-fortified mushrooms demonstrated enhanced antimicrobial activity.

## 4.9 CONCLUSION AND FUTURE PERSPECTIVES

The abundant agricultural residues and favorable climate create excellent conditions for cultivating a variety of edible mushrooms. Fortifying mushrooms with essential minerals offers a promising way to combat malnutrition while providing sustainable income opportunities for landless farmers, unemployed youth, rural women, and self-help groups. These nutrient-enriched mushrooms serve not only as functional foods and dietary supplements but also possess antioxidant and anticancer properties. However, mineral concentrations need to be carefully optimized, as excessive levels can inhibit mushroom growth.

Future research should focus on determining safe and effective doses of fortified mushrooms for different health conditions, as well as studying mineral bioavailability after cooking and preservation. Monitoring consumption is crucial to avoid toxicity from exceeding recommended dietary allowances. Additionally, health education is vital for raising awareness about the benefits and risks of fortified mushrooms. More studies are also needed on how fortified mushrooms interact with medicines to ensure their safe use. With continued research and education, fortified mushroom cultivation can play a key role in improving nutrition, health, and rural livelihoods.

## REFERENCES


Abbaspour, N., R. Hurrell, and R. Kelishadi. 2014. Review on iron and its importance for human health. *Journal of Research in Medical Sciences: The Official Journal of Isfahan University of Medical Sciences* 19(2): 164.

Adeghe, E. P. 2024. The multifaceted role of fluoride in preventing early childhood caries: a comprehensive review. *International Journal of Life Science Research Updates* 2(1): 009–017. https://doi.org/10.53430/ijlsru.2024.2.1.0022

Aggarwal, N., P. Upadhyay, and S. B. Tigadi. 2020. Biofortification to improve nutrition: a review. *International Journal of Current Microbiology and Applied Sciences* 9: 763–779. https://doi.org/10.20546/ijcmas.2020.901.083

Ajewole, K., M. A. Abibu, and A. A. T. Taleat. 2012. Substrate fortification and proximate analysis of cultivated mushroom (*Pleurotus ostreatus*). *Natural Products: An Indian Journal* 8(4): 152–156.

Ariyo, O. O., O. P. Odeleye, O. O. Ajuwon, and M. D. Adeniyi. 2023. Evaluation of antibacterial activities of iron-fortified *Pleurotus ostreatus* on selected enteric bacteria. *FUOYE Journal of Pure and Applied Sciences* 8(2): 45–56. https://doi.org/10.55518/fjpas.RHFI4778

Avram, O. R., G. Caragea, and C. A. Varzaru. 2021. Copper and its role in the human body—the importance of establishing copper concentrations in the body. *Romanian Journal* 124(2): 254.

Bhatia, P., F. Aureli, M. D'Amato, et al. 2013a. Selenium bioaccessibility and speciation in biofortified *Pleurotus* mushrooms grown on selenium-rich agricultural residues. *Food Chemistry* 140(1–2): 225–230. https://doi.org/10.1016/j.foodchem.2013.02.054

Bhatia, P., R. Prakash, and N. T. Prakash. 2013b. Selenium uptake by edible oyster mushrooms (*Pleurotus* sp.) from selenium-hyperaccumulated wheat straw. *Journal of Nutritional Science and Vitaminology* 59(1): 69–72. https://doi.org/10.3177/jnsv.59.69

Bhatia, P., R. Prakash, and N. T. Prakash. 2014. Enhanced antioxidant properties as a function of selenium uptake by edible mushrooms cultivated on selenium-accumulated waste postharvest wheat and paddy residues. *International Journal of Recycling of Organic Waste in Agriculture* 3: 127–132. https://doi.org/10.1007/s40093-014-0074-y

Binesh, A., and K. Venkatachalam. 2024. Copper in human health and disease: a comprehensive review. *Journal of Biochemical and Molecular Toxicology* 38(11): e70052. https://doi.org/10.1002/jbt.70052

Black, R. E., C. G. Victora, S. P. Walker, et al. 2013. Maternal and child undernutrition and overweight in low-income and middle-income countries. *Lancet* 382(9890): 427–451. https://doi.org/10.1016/S0140-6736(13)60937-X

Budzyńska, S., M. Siwulski, M. Gąsecka, et al. 2022. Biofortification of three cultivated mushroom species with three iron salts-potential for a new iron-rich superfood. *Molecules* 27(7): 2328. https://doi.org/10.3390/molecules27072328

Chasapis, C. T., P. S. A. Ntoupa, C. A. Spiliopoulou, and M. E. Stefanidou. 2020. Recent aspects of the effects of zinc on human health. *Archives of Toxicology* 94: 1443–1460. https://doi.org/10.1007/s00204-020-02702-9

Condé, V. F., J. E. Z. de Oliveira, and D. M. F. de Oliveira. 2017. *Pleurotus ostreatus* (Jacq.) P. Kumm. flour (Hiratake mushroom) enriched in iron. *Ciência e Natura* 39(1): 1–6.

Costa-Silva, F., G. Marques, C. C. Matos, A. I. Barros, and F. M. Nunes. 2011. Selenium contents of Portuguese commercial and wild edible mushrooms. *Food Chemistry* 126(1): 91–96. https://doi.org/10.1016/j.foodchem.2010.10.082

da Silva, M. C., J. Naozuka, J. M. R. da Luz, et al. 2012. Enrichment of *Pleurotus ostreatus* mushrooms with selenium in coffee husks. *Food Chemistry* 131(2): 558–563. https://doi.org/10.1016/j.foodchem.2011.09.023

da Silva, M. C., J. Naozuka, P. V. Oliveira, et al. 2010. *In vivo* bioavailability of selenium in enriched *Pleurotus ostreatus* mushrooms. *Metallomics* 2(2): 162–166. https://doi.org/10.1039/b915780h

da Silva, M. D. C. S., M. D. Nunes, J. M. R. da Luz, and M. C. M. Kasuya. 2013. Mycelial growth of *Pleurotus* spp. in Se-enriched culture media. *Advances in Microbiology* 2013. https://doi.org/10.4236/aim.2013.38A003

de Oliveira, A. P., and J. Naozuka. 2019. Preliminary results on the feasibility of producing selenium-enriched pink (*Pleurotus djamor*) and white (*Pleurotus ostreatus*) oyster mushrooms: bioaccumulation, bioaccessibility, and Se-proteins distribution. *Microchemical Journal* 145: 1143–1150. https://doi.org/10.1016/j.microc.2018.12.046

de Souza Lopes, L., M. de Casssia Silva, A. de Oliveira Faustino, L. L. de Oliveira, and M. C. M. Kasuya. 2022. Bioaccessibility, oxidizing activity and co-accumulation of minerals in Li-enriched mushrooms. *LWT* 155: 112989. https://doi.org/10.1016/j.lwt.2021.112989

El-Ramady, H., N. Abdalla, H. Elbasiouny, et al. 2021. Nanobiofortification of different crops to immune against COVID-19: a review. *Ecotoxicology and Environmental Safety* 222: 112500. https://doi.org/10.1016/j.ecoenv.2021.112500

Falandysz, J. 2008. Selenium in edible mushrooms. *Journal of Environmental Science and Health Part C* 26(3): 256–299. https://doi.org/10.1080/10590500802350086

Fasoranti, O. F., C. O. Ogidi, and V. O. Oyetayo. 2018. Phytochemical constituents and antimicrobial evaluation of ethanolic extracts from *Pleurotus* spp. cultivated on substrate fortified with selenium. *Microbial Biosystems* 3(2): 29–39. https://doi.org/10.21608/mb.2018.26274

Fasoranti, O. F., C. O. Ogidi, and V. O. Oyetayo. 2019. Nutrient contents and antioxidant properties of *Pleurotus* spp. cultivated on substrate fortified with selenium. *Current Research in Environmental & Applied Mycology* 9(1): 66–76. https://doi.org/10.5943/cream/9/1/7

Flis, A., P. Suchocki, M. A. Krolikowska, et al. 2015. Selenite triglycerides—redox-active agents. *Pharmacological Reports* 67: 1–8. https://doi.org/10.1016/j.pharep.2014.07.017

Galappaththi, M. C. A., L. Dauner, S. Madawala, and S. C. Karunarathna. 2021. Nutritional and medicinal benefits of oyster (*Pleurotus*) mushrooms: a review. *Fungal Biotec* 1(2): 65–87.

Gąsecka, M., M. Mleczek, M. Siwulski, and P. Niedzielski. 2016. Phenolic composition and antioxidant properties of *Pleurotus ostreatus* and *Pleurotus eryngii* enriched with selenium and zinc. *European Food Research and Technology* 242: 723–732. https://doi.org/10.1007/s00217-015-2580-1

Genchi, G., G. Lauria, A. Catalano, M. S. Sinicropi, and A. Carocci. 2023. Biological activity of selenium and its impact on human health. *International Journal of Molecular Sciences* 24(3): 2633. https://doi.org/10.3390/ijms24032633

Islam, M. R., S. Akash, M. H. Jony, et al. 2023. Exploring the potential function of trace elements in human health: a therapeutic perspective. *Molecular and Cellular Biochemistry* 478(10): 2141–2171. https://doi.org/10.1007/s11010-022-04638-3

Jovanović, I. B., M. Veličković, D. Vuković, S. Milanović, O. Valčić, and D. Gvozdić. 2013. Effects of different amounts of supplemental selenium and vitamin E on the incidence of retained placenta, selenium, malondialdehyde, and thyronines status in cows treated with prostaglandin F2α for the induction of parturition. *Journal of Veterinary Medicine* 2013(1): 867453. https://doi.org/10.1155/2013/867453

Julian, A. V., M. R. Umagat, and R. G. Reyes. 2018. Mineral composition and yield of *Pleurotus ostreatus* on rice straw-based substrate enriched with natural calcium sources. *International Journal of Biology, Pharmacy and Allied Sciences* 7(8): 1489–1499. https://doi.org/10.31032/IJBPAS/2018/7.8.4501

Kalač, P., and L. Svoboda. 2000. A review of trace element concentrations in edible mushrooms. *Food Chemistry* 69(3): 273–281. https://doi.org/10.1016/S0308-8146(99)00264-2

Keles, A., H. Genccelep, and K. Demirel. 2017. Elemental composition of naturally growing wild edible mushroom. *Journal of Natural Product and Plant Resources* 7: 37–44.

Khatua S., and K. Acharya. 2022. Mushroom ingestion for mineral supplementation. In: *Wild Mushrooms: Characteristics, Nutrition, and Processing*, ed. S. J. Dhull, A. Bains, P. Chawla, P. K. Sadh, 135–167. CRC Press.

Khush, G. S., S. Lee, J. I. Cho, and J. S. Jeon. 2012. Biofortification of crops for reducing malnutrition. *Plant Biotechnology Reports* 6: 195–202. https://doi.org/10.1007/s11816-012-0216-5

Kieliszek, M., and S. Błażejak. 2016. Current knowledge on the importance of selenium in food for living organisms: a review. *Molecules* 21(5): 609. https://doi.org/10.3390/molecules21050609

Kong, D., S. A. Khan, H. Wu, Y. Liu, and H. Q. Ling. 2022. Biofortification of iron and zinc in rice and wheat. *Journal of Integrative Plant Biology* 64(6): 1157–1167. https://doi.org/10.1111/jipb.13262

Kora, A. J. 2020. Nutritional and antioxidant significance of selenium-enriched mushrooms. *Bulletin of the National Research Centre* 44: 19. https://doi.org/10.1186/s42269-020-00289-w

Kumar, S., A. Palve, C. Joshi, and R. K. Srivastava. 2019. Crop biofortification for iron (Fe), zinc (Zn) and vitamin A with transgenic approaches. *Heliyon* 5(6). https://doi.org/10.1016/j.heliyon.2019.e01914

Lewicki, S., R. Zdanowski, M. Krzyzowska, et al. 2014. The role of Chromium III in the organism and its possible use in diabetes and obesity treatment. *Annals of Agricultural and Environmental Medicine* 21(2): 331–335. https://doi.org/10.5604/1232-1966.1108599

Li, J., C. Martin, and A. Fernie. 2024. Biofortification's contribution to mitigating micronutrient deficiencies. *Nature Food* 5(1): 19–27. https://doi.org/10.1038/s43016-023-00905-8

Luvitaa, K. S., M. A. Wambui, M. Fredrick, and O. D. Otieno. 2023. Zinc bioaccessibility in finger millet porridge blended with zinc-dense mushroom. *Heliyon* 9(8). https://doi.org/10.1016/j.heliyon.2023.e18901

Madaan, K., S. Sharma, and A. Kalia. 2024. Effect of selenium and zinc biofortification on the biochemical parameters of *Pleurotus* spp. under submerged and solid-state fermentation. *Journal of Trace Elements in Medicine and Biology* 82: 127365. https://doi.org/10.1016/j.jtemb.2023.127365

Malik, K. A., and A. Maqbool. 2020. Transgenic crops for biofortification. *Frontiers in Sustainable Food Systems* 4: 571402. https://doi.org/10.3389/fsufs.2020.571402

Mane, V. P., S. S. Patil, A. A. Syed, and M. M. V. Baig. 2007. Bioconversion of low quality lignocellulosic agricultural waste into edible protein by *Pleurotus sajor-caju* (Fr.) Singer. *Journal of Zhejiang University Science B* 8: 745–751.https://doi.org/10.1631/jzus.2007.B0745

Maywald, M., and L. Rink. 2022. Zinc in human health and infectious diseases. *Biomolecules* 12(12): 1748. https://doi.org/10.3390/biom12121748

Milovanović, I., I. Brčeski, M. Stajić, A. Korać, J. Vukojević, and A. Knežević. 2014. Potential of *Pleurotus ostreatus* mycelium for selenium absorption. *The Scientific World Journal* 2014(1): 681834. https://doi.org/10.1155/2014/681834

Milovanovic, I., B. Lajin, S. Braeuer, O. Steiner, F. Lisa, and W. Goessler. 2019. Simultaneous selenium and sulfur speciation analysis in cultivated *Pleurotus pulmonarius* mushroom. *Food Chemistry* 279: 231–236. https://doi.org/10.1016/j.foodchem.2018.12.009

Minich, W. B. 2022. Selenium metabolism and biosynthesis of selenoproteins in the human body. *Biochemistry (Moscow)* 87(1): S168–S177. https://doi.org/10.1134/S0006297922140139

Mitra, P., S. Khatua, and K. Acharya. 2013. Free radical scavenging and NOS activation properties of water soluble crude polysaccharide from *Pleurotus ostreatus*. *Asian Journal of Pharmaceutical and Clinical Research* 6(3): 67–70.

Mleczek, M., M. Siwulski, P. Rzymski, et al. 2017. Cultivation of mushrooms for production of food biofortified with lithium. *European Food Research and Technology* 243: 1097–1104. https://doi.org/10.1007/s00217-016-2823-9

Monika, G., M. K. S. R. Kim, P. S. Kumar, K. V. Gayathri, G. Rangasamy, and A. Saravanan. 2023. Biofortification: a long-term solution to improve global health—a review. *Chemosphere* 314: 137713. https://doi.org/10.1016/j.chemosphere.2022.137713

Muñoz S., A. H., K. Kubachka, K. Wrobel, et al. 2006. Se-enriched mycelia of *Pleurotus ostreatus*: distribution of selenium in cell walls and cell membranes/cytosol. *Journal of Agricultural and Food Chemistry* 54(9): 3440–3444. https://doi.org/10.1021/jf052973u

Neill, H. R., C. I. Gill, E. J. McDonald, W. C. McRoberts, and L. K. Pourshahidi. 2021. Vitamin D biofortification of pork may offer a food-based strategy to increase vitamin D intakes in the UK population. *Frontiers in Nutrition* 8: 777364. https://doi.org/10.3389/fnut.2021.777364

Neill, H. R., C. I. Gill, E. J. McDonald, W. C. McRoberts, and L. K. Pourshahidi. 2023. The future is bright: biofortification of common foods can improve vitamin D status. *Critical Reviews in Food Science and Nutrition* 63(4): 505–521. https://doi.org/10.1080/10408398.2021.1950609

Nestel, P., H. E. Bouis, J. V. Meenakshi, and W. Pfeiffer. 2006. Biofortification of staple food crops. *The Journal of Nutrition* 136(4): 1064–1067. https://doi.org/10.1093/jn/136.4.1064

Nicol, K., A. Nugent, J. V. Woodside, K. H. Hart, and S. C. Bath. 2024. Iodine and plant-based diets: a narrative review and calculation of iodine content. *British Journal of Nutrition* 131(2): 265–275. https://doi.org/10.1017/S0007114523001873

Nunes, R. G. F. L., J. M. R. D. Luz, R. D. B. Freitas, A. Higuchi, M. C. M. Kasuya, and M. C. D. Vanetti. 2012. Selenium bioaccumulation in shiitake mushrooms: a nutritional alternative source of this element. *Journal of Food Science* 77(9): 983–986. https://doi.org/10.1111/j.1750-3841.2012.02837.x

Obodai, M., J. Cleland-Okine, and K. A. Vowotor. 2003. Comparative study on the growth and yield of *Pleurotus ostreatus* mushroom on different lignocellulosic by-products. *Journal of Industrial Microbiology and Biotechnology* 30(3): 146–149. https://doi.org/10.1007/s10295-002-0021-1

Odiketa, J. K., S. Whitehall, and O. M. Adedokun. 2020. Biofortification of mushroom (*Pleurotus floridanus*) using calcium-based supplements. *Journal of Mushroom* 18(4): 287–291. http://doi.org/10.14480/JM.2020.18.4.287

Ogidi, C. O., M. D. Nunes, V. O. Oyetayo, B. J. Akinyele, and M. C. M. Kasuya. 2016. Mycelial growth, biomass production and iron uptake by mushrooms of *Pleurotus* species cultivated on *Urochloa decumbens* (Stapf) R.D. Webster. *Journal of Food Research* 5(3): 13–19. https://doi.org/10.5539/jfr.v5n3p13

Ogidi, C. O., and K. O. Oyebode. 2023. Assessment of nutrient contents and bio-functional activities of edible fungus bio-fortified with copper, lithium and zinc. *World Journal of Microbiology and Biotechnology* 39(2): 56. https://doi.org/10.1007/s11274-022-03500-4

Oyetayo, V. O. 2023. Mineral element enrichment of mushrooms for the production of more effective functional foods. *Asian Journal of Biological Sciences* 16(1): 18–29. https://doi.org/10.17311/ajbs.2023.18.29

Oyetayo, V. O., C. O. Ogidi, S. O. Bayode, and F. F. Enikanselu. 2021. Evaluation of biological efficiency, nutrient contents and antioxidant activity of *Pleurotus pulmonarius* enriched with Zinc and Iron. *Indian Phytopathology* 74: 901–910. https://doi.org/10.1007/s42360-021-00410-7

Oyetayo, V. O., C. O. Ogidi, F. F. Enikanselu, A. E. Fadugba, and S. O. Bayode. 2024. Anticandidal and Antistaphylococcal potentials of extracts of *Pleurotus pulmonarius* fortified with Zinc and Iron. *Vegetos* 37(2): 695–700. https://doi.org/10.1007/s42535-024-00852-0

Pantopoulos, K., S. K. Porwal, A. Tartakoff, and L. Devireddy. 2012. Mechanisms of mammalian iron homeostasis. *Biochemistry* 51(29): 5705–5724. https://doi.org/10.1021/bi300752r

Patra, S., P. Patra, K. K. Maity, et al. 2013. A heteroglycan from the mycelia of *Pleurotus ostreatus*: structure determination and study of antioxidant properties. *Carbohydrate Research* 368: 16–21. https://doi.org/10.1016/j.carres.2012.12.003

Pattanayak, S., and S. Das. 2022. Value addition and nutritional enrichment of wonder food mushroom. *Journal of Mycopathological Research* 60(4): 507–514. https://doi.org/10.57023/JMycR.60.4.2022.507

Pfeiffer, W. H., and B. Mc Clafferty. 2007. HarvestPlus: breeding crops for better nutrition. *Crop Science* 47: 88–105. https://doi.org/10.2135/cropsci2007.09.0020IPBS

Poniedziałek, B., M. Mleczek, P. Niedzielski, et al. 2017. Bio-enriched *Pleurotus* mushrooms for deficiency control and improved antioxidative protection of human platelets? *European Food Research and Technology* 243: 2187–2198. https://doi.org/10.1007/s00217-017-2921-3

Poursaeid, N., A. Azadbakht, and G. R. Balali. 2015. Improvement of zinc bioaccumulation and biomass yield in the mycelia and fruiting bodies of *Pleurotus florida* cultured on liquid media. *Applied Biochemistry and Biotechnology* 175: 3387–3396. https://doi.org/10.1007/s12010-015-1510-9

Prashanth, L., K. K. Kattapagari, R. T. Chitturi, V. R. Baddam, and L. K. Prasad. 2015. A review on role of essential trace elements in health and disease. *Journal of NTR University of Health Sciences* 4(2): 75–85. https://doi.org/10.4103/2277-8632.158577

Rabinovich, M., D. Figlas, S. Delmastro, and N. Curvetto. 2007. Copper-and zinc-enriched mycelium of *Agaricus blazei* Murrill: bioaccumulation and bioavailability. *Journal of Medicinal Food* 10(1): 175–183. https://doi.org/10.1089/jmf.2005.064

Raman, J., K. Y. Jang, Y. L. Oh, et al. 2020. Cultivation and nutritional value of prominent *Pleurotus* spp.: an overview. *Mycobiology* 49(1): 1–14. https://doi.org/10.1080/12298093.2020.1835142

Rodriguez Estrada, A. E., H. J. Lee, R. B. Beelman, M. D. M. Jimenez-Gasco, and D. J. Royse. 2009. Enhancement of the antioxidants ergothioneine and selenium in *Pleurotus eryngii* var. *eryngii* basidiomata through cultural practices. *World Journal of Microbiology and Biotechnology* 25: 1597–1607. https://doi.org/10.1007/s11274-009-0049-8

Rosado, F. R., E. R. Carbonero, C. Kemmelmeier, C. A. Tischer, P. A. Gorin, and M. Iacomini. 2002. A partially 3-O-methylated (1→4)-linked α-D-galactan and α-D-mannan from *Pleurotus ostreatoroseus* Sing. *FEMS Microbiology Letters* 212(2): 261–265. https://doi.org/10.1111/j.1574-6968.2002.tb11276.x

Ruseva, B., I. Himcheva, and D. Nankova. 2013. Importance of selenoproteins for the function of the thyroid gland. *Medicine* 3(1): 60–64.

Rzymski, P., M. Mleczek, P. Niedzielski, M. Siwulski, and M. Gąsecka. 2016. Potential of cultivated *Ganoderma lucidum* mushrooms for the production of supplements enriched with essential elements. *Journal of Food Science* 81(3): C587–C592. https://doi.org/10.1111/1750-3841.13212

Santamaria, A. B. 2008. Manganese exposure, essentiality and toxicity. *Indian Journal of Medical Research* 128(4): 484–500.

Savić, M., I. Andjelković, D. Duvnjak, D. Matijasević, A. Avramović, and M. Nikšić. 2012. The fungistatic activity of organic selenium and its application to the production of cultivated mushrooms *Agaricus bisporus* and *Pleurotus* spp. *Archives of Biological Sciences* 64(4): 1455–1463. https://doi.org/10.2298/ABS1204455S

Savić, M. D., J. Petrović, A. Klaus, et al. 2009. Growth and fruit body formation of *Pleurotus ostreatus* on media supplemented with inorganic selenium. *Zbornik Matice Srpske za Prirodne Nauke* (116): 209–215. https://doi.org/10.2298/ZMSPN0916209S

Scheid, S. S., M. G. I. Faria, L. G. Velasquez, et al. 2020. Iron biofortification and availability in the mycelial biomass of edible and medicinal basidiomycetes cultivated in sugarcane molasses. *Scientific Reports* 10(1): 12875. https://doi.org/10.1038/s41598-020-69699-0

Sham, S. M., N. F. Jalil, and F. N. Athirah. 2021. Human health risk and benefit from fluoride intake: a systematic review. https://doi.org/10.26911/ICPHepidemiology.FP.08.2021.03

Sharma, P., P. K. Reddy, and B. Kumar. 2021. Trace element zinc, a nature's gift to fight unprecedented global pandemic COVID–19. *Biological Trace Element Research* 199: 3213–3221. https://doi.org/10.1007/s12011-020-02462-8

Silva, C. S., C. Moutinho, A. Ferreira da Vinha, and C. Matos. 2019. Trace minerals in human health: iron, zinc, copper, manganese and fluorine. *International Journal of Science and Research Methodology* 13(3): 57–80.

Słyszyk, K., M. Siwulski, A. Wiater, M. Tomczyk, and A. Waśko. 2024. Biofortification of mushrooms: a promising approach. *Molecules* 29(19): 4740. https://doi.org/10.3390/molecules29194740

Studer, J. M., W. P. Schweer, N. K. Gabler, and J. W. Ross. 2022. Functions of manganese in reproduction. *Animal Reproduction Science* 238: 106924. https://doi.org/10.1016/j.anireprosci.2022.106924

Umeo, S. H., M. G. I. Faria, D. C. Dragunski, J. S. Valle, N. B. Colauto, and G. A. Linde. 2020. Iron or zinc bioaccumulated in mycelial biomass of edible basidiomycetes. *Anais da Academia Brasileira de Ciências* 92(suppl 2). https://doi.org/10.1590/0001-3765202020191350

Velez, M. E. V., J. M. R. da Luz, M. D. C. S. da Silva, et al. 2019. Production of bioactive compounds by the mycelial growth of *Pleurotus djamor* in whey powder enriched with selenium. *LWT* 114: 108376. https://doi.org/10.1016/j.lwt.2019.108376

Vieira, P. A. F., D. C. Gontijo, B. C. Vieira, et al. 2013. Antioxidant activities, total phenolics and metal contents in *Pleurotus ostreatus* mushrooms enriched with iron, zinc or lithium. *LWT—Food Science and Technology* 54(2): 421–425. https://doi.org/10.1016/j.lwt.2013.06.016

Welch, R. M., and R. D. Graham. 2004. Breeding for micronutrients in staple food crops from a human nutrition perspective. *Journal of Experimental Botany* 55: 353–364. https://doi.org/10.1093/jxb/erh064

White, P. J., and M. R. Broadley. 2005. Biofortifying crops with essential mineral elements. *Trends in Plant Science* 10(12): 586–593.

World Health Organization (WHO). 2023. *Anaemia*. Retrieved from https://www.who.int/news-room/fact-sheets/detail/anaemia

Yadava, D. K., F. Hossain, and T. Mohapatra. 2018. Nutritional security through crop biofortification in India: status & future prospects. *The Indian Journal of Medical Research* 148(5): https://doi.org/621. 10.4103/ijmr.IJMR_1893_18

Yan, H., and H. Chang. 2012. Antioxidant and antitumor activities of selenium- and zinc-enriched oyster mushroom in mice. *Biological Trace Element Research* 150(1): 236–241. https://doi.org/10.1007/s12011-012-9454-1

Yiannikourides, A., and G. O. Latunde-Dada. 2019. A short review of iron metabolism and pathophysiology of iron disorders. *Medicines* 6(3): 85. https://doi.org/10.3390/medicines6030085

Yilmaz, F., M. Işıloğlu, and M. Merdivan. 2003. Heavy metal levels in some macrofungi. *Turkish Journal of Botany* 27(1): 45–56.

Zięba, P., K. Kała, A. Włodarczyk, A. Szewczyk, E. Kunicki, A. Sękara, and B. Muszyńska. 2020. Selenium and zinc biofortification of *Pleurotus eryngii* mycelium and fruiting bodies as a tool for controlling their biological activity. *Molecules* 25(4): 889. https://doi.org/10.3390/molecules25040889

Zimmermann, M. B. 2011. The role of iodine in human growth and development. *Seminars in Cell and Developmental Biology*, 22(6): 645–652. Academic Press. https://doi.org/10.1016/j.semcdb.2011.07.009

# 5 Current Insights into Bioactive Primary and Secondary Metabolites of Oyster Mushroom

*Sudeshna Nandi* and Sudatta Maity*
Department of Botany, Vedanta College, Kolkata, West Bengal, India

## 5.1 INTRODUCTION

Since the dawn of humanity, people have lived as hunter-gatherers for the majority of their existence, relying entirely on biodiversity for food, medicine, and overall sustenance (Łysakowska et al., 2023). During gathering forays, fruiting bodies of macromycetes, or large fungi, were collected, and these fruiting bodies, i.e., the visible part above the substrate, have been commonly referred to as mushrooms or toadstools, which are the edible elements of some filamentous fungi (Feeney et al., 2014). Mushrooms, considered one of the curiosities of nature, have been utilized in daily food as well as in folk medicine throughout the world since ancient times (Govorushko et al., 2019). Possessing exceptional flavor, aroma, and nutritional content, they are designated as "functional foods" due to their potential to increase overall health and confer nutritional advantages (Fogarasi et al., 2024). Presently, mushrooms are encountering a resurgence in popularity within the modern human diet, as they are emerging as a source of nourishment as well as impeccably merging into a burgeoning market category known as nutra-mycoceuticals (Kozarski et al., 2023). Within the scientific community, mushrooms are hailed as nature's prolific sources of pharmaceutical compounds (Cateni et al., 2021; Kozarski et al., 2023), and this has increased the worldwide urge for mushrooms to grow around 15%/annum (Grujic et al., 2015; Kamarudzaman et al., 2015).

However, in addition to their medicinal benefits and nutritional constituents, mushrooms are also observed to contain various bioactive elements, including both primary and secondary metabolites, like polysaccharides, peptides, proteoglycans, terpenes, lectins, and primarily phenolic compounds (Yadav and Negi, 2021; Abdelshafy et al., 2022; Hamza et al., 2023). As of now, around 2.2–3.8 million species of fungi in the world have been identified, of which 150,000 species have been described and characterized throughout the globe (Beulah and Margret, 2013; Hyde, 2022). Among these, 3000 species have been reported as edible and over 200 species of wild mushrooms are found to have medicinal importance (Choudhary et al., 2015; Meenu and Xu, 2019; Hyde, 2022), and approximately 1400 are identified to be toxic (Mowsurni and Chowdhury, 2010). Among all types of mushrooms, more than 200 categories belong to oyster mushrooms (Zieba et al., 2021). The study reported that there are several mushroom species, that is, *Agaricus bisporus*, *Pleurotus* species, *Lentinus edodes*, and *Volvariella* species, which are the most acceptable varieties among the cultivated mushrooms considering their health benefit and biochemical properties (Lesa et al., 2022). *Pleurotus* sp. is commonly referred to as "oyster mushrooms" as they produce shell or oyster-shaped basidiocarps (Khan et al., 2008; Raman et al., 2020). They may be of different colors like white, cream, gray, yellow, pink, or light brown (Raman et al., 2020) with basidiocarp stalked or sessile, above or underground and even epiphytic but rarely parasitic (Patel et al., 2012). *Pleurotus*, comprising about 40 species, enjoys worldwide distribution from temperate to tropical regions, growing saprophytically at a temperature range of 12–32 °C (Zadrazil, 1978; Golak-Siwulska et al., 2018). Commercially, button mushroom ranks first, followed by oyster mushrooms, which occupy the second largest commercially produced mushroom globally (Aditya et al., 2024), especially in Europe, Africa (Sánchez, 2010; Royse et al., 2017), and Asian countries (mainly India, South Korea, China, Taiwan, Japan, Thailand, and Vietnam) which are the main producers and consumers of *Pleurotus* with approximately 99% of the total volume (Royse et al., 2017). To date, around 26 species, including *Pleurotus eryngii*, *Pleurotus citrinopileatus*, *Pleurotus flabellatus*, *Pleurotus ostreatus*, *Pleurotus djamor* var. *roseus*, and *Pleurotus florida*, have been reported to be cultivated utilizing various types of lignocellulosic uncomposted agro wastes (Bano and Srivastava, 1962; Suguimoto et al., 2001; Mandeel et al., 2005; Islam et al., 2009; Bernardi et al., 2013; Bumanlag et al., 2018; Golak-Siwulska et al., 2018; Kumla et al., 2020).

The species of *Pleurotus* is acclaimed for its comestible use and serves as a valuable dietary component (Yingyue et al., 2014). Plentiful in carbohydrates, proteins, fibers, lipids, amino acids, vitamins and minerals, these mushrooms boast a plethora of nutritional benefits (Effiong et al., 2024; Aditya et al., 2024). Beyond their nutritive value, the species of *Pleurotus* accommodate an array of bioactive compounds like polysaccharides, fatty acids, peptides, oligosaccharides, glycoproteins, terpenoids, lectins and nucleosides, which determine a wide range of tantalizing

* Corresponding author: sudeshna.nandi8@gmail.com

DOI: 10.1201/9781003527572-5

aromas and distinctive flavor characteristics in their fruiting body. They have also been employed in traditional medicine to battle against various ailments (Fernandes et al., 2015; Correa et al., 2016; Sharma et al., 2021; Sreedharan et al., 2025). Furthermore, the species of *Pleurotus* have been found to possess abundant health-promoting properties due to the presence of various bioactive compounds such as lectins with immunomodulatory, antiproliferative, and antitumor activities; phenolic compounds with antioxidant activities; and polysaccharides (polysaccharide-peptides and polysaccharide-proteins) with immunoenhancing and anticancer activities (Khan and Tania, 2012; Valverde et al., 2015). This review provides an up-to-date and comprehensive overview of the primary and secondary metabolites present in *Pleurotus* species, aiming to support future researchers in uncovering the biological activities of these bioactive compounds. It also seeks to inspire further exploration of untapped metabolites from this mushroom, with the potential to develop next-generation functional foods or life-saving therapeutic agents.

## 5.2 ACTIVE METABOLITES IN MUSHROOM

Fungi are considered nature's most skilled chemists because the compounds they synthesize range from simple, ubiquitous molecules, often referred to as primary metabolites, to complex, unusual molecules, which are called secondary metabolites (Bhambri et al., 2022). Biologically active compounds, often called bioactive compounds, are defined as "essential and nonessential compounds" that occur in nature and are part of the food chain. Primary metabolites in mushrooms include proteins, lipids, carbohydrates, vitamins, and sugars, which are essential for the growth and development of mushrooms (Venditti et al., 2016). On the other hand, mushrooms have distinct metabolic mechanisms to fabricate an assemblage of secondary metabolites, owing to diverse functionalities and bioactivity, during long-term evolution, and simultaneously furnish a copious and engrossing resource for new drug discovery (Xiao and Zhong, 2016). Mushroom's secondary metabolites are chemical compounds produced by fungi that are not directly involved in the organism's growth, development, or reproduction but often play vital roles in their survival and interaction with the environment (Elhamouly et al., 2022). Secondary metabolites found in mushrooms include phenolic compounds (Xiaokang et al., 2020), polysaccharides and ß-glucan (Sakdasri et al., 2022; Figure 5.1A), triterpenoids (Zhao et al., 2023), and ergosterols (Sun et al., 2022). A vast chemodiversity of the compounds has been reported in various mushrooms, and important among them have been discussed below.

### 5.2.1 Primary Metabolites

#### 5.2.1.1 Polysaccharides

Edible mushroom polysaccharides are complex carbohydrates mostly found in the cell walls of mushrooms to maintain their structural integrity (Friedman, 2016). Biologically active polysaccharides are expansive among higher basidiomycetous mushrooms, and the majority have distinctive structures in different species (Chun et al., 2021). Mushrooms depict a noteworthy source of polysaccharides like α-glucans, β-glucans, heteroglycans, and proteoglycan (Kozarski et al., 2014), as well as chitin, mannans, galactans, and xylans. Among glucans, those with high molecular weight seem to be more effective than those with low molecular weight (Friedman, 2016). The monosaccharides, galactose, glucose, trehalose, rhamnose, xylose, mannitol, mannose, arabinose, fucose, fructose, and so on, portray the larger part of the mushroom polysaccharides (Valverde et al., 2015; Wang et al., 2017). These polysaccharides are representatives of the major bioactive compound group, engaging in vital roles like nutritional and possible health-promoting aspects of mushrooms. Mushroom polysaccharides consist of numerous different sugar units that are linked together and depict a diversity of chemical configurations (Bolton et al., 2019). The characteristics of each mushroom species' polysaccharide profile are determined by the distinct arrangement of the monosaccharides inside the polysaccharide chains (Villares et al., 2012). In edible mushrooms, the concentration of polysaccharides differs enormously depending on factors like growing conditions, type of species, and harvesting time. Few mushrooms are chiefly high in the amount of polysaccharides, and their concentrations are customarily given as a percentage of dry weight. The polysaccharide content in various mushroom species ranges from 5% to 60% or more of their dry weight (Vetter, 2023). Lately, polysaccharides, the most common and potent compounds derived from mushrooms, have demonstrated a fair amount of potential human-beneficial activities for their extensive biological properties such as anti-inflammatory, antioxidative, and antitumor effects (Selvamani et al., 2018; Pant et al., 2020) and immunomodulatory properties (Pant et al., 2020; Venturella et al., 2021). The usual functioning mode of action of numerous mushroom polysaccharides requires the activation of natural killer cells, monocytes, neutrophils, dendritic cells, cytotoxic cytokines (interferons, interleukins, and colony-stimulating factors), and macrophages (Wasser, 2011). Hence, edible mushroom polysaccharides are the chief components of these macrofungi, adding to their nutritional value as well as prospective health betterment. The varied composition and complex structures of these polysaccharides extend directions for further investigation into their bioactive properties and applications.

#### 5.2.1.2 Protein and Peptides

In mushrooms, proteins and peptides are prime bioactive nutraceuticals having several wellness benefits such as increased absorption and digestion of exogenous nutritional ingredients, certain enzyme inhibition or retardation activities, and immune function modulation to assist the host in guarding against the invasion of pathogens (Valverde et al., 2015). Studies report that proteins and peptides such as ribonucleases, ribosome-inactivating proteins, lectins, laccases,

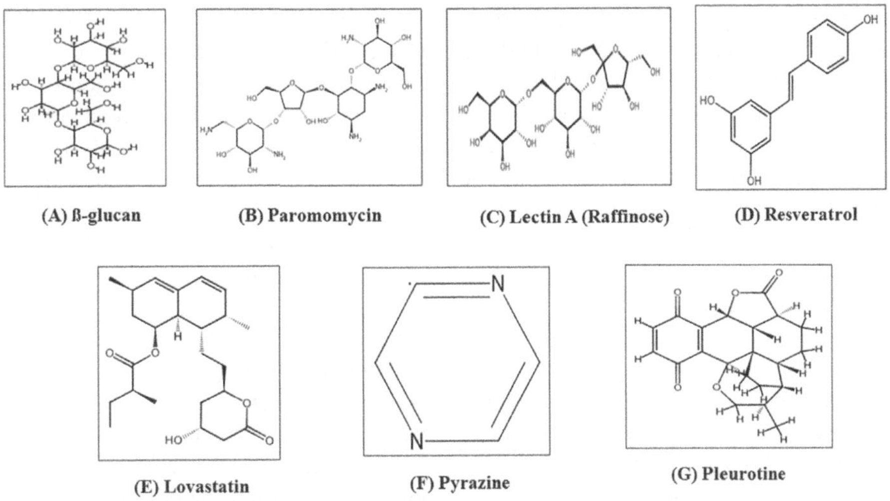


**FIGURE 5.1** Major primary and secondary metabolites derived from *Pleurotus* sp. (Figure obtained from www.chemspider.com).

and immunomodulatory proteins derived from mushrooms have remedial potential (Xu et al., 2011). Lectins are glycol or nonimmune proteins that bind with the carbohydrates of cell surfaces and possess vital activities like antibacterial, antiviral, antifungal, antitumor, and immunomodulatory properties (Singh et al., 2014), whereas fungal immunomodulatory proteins are utilized as an adjunct for the medication of tumors because of their essential role in inhibiting the invasion and metastasis of tumor cells (Lin et al., 2010). Studies demonstrate that ribosome-inactivating proteins are capable of preventing the proliferation of pathogenic fungi and exhibit HIV-1 reverse transcriptase activity (Puri et al., 2012). Usually, edible mushrooms possess between 19% and 39% protein by dry weight while being a part of a complex network of fungal cells (Reddy, 2016).

### 5.2.2 Secondary Metabolites

#### 5.2.2.1 Terpenes

Terpenes, the volatile unsaturated hydrocarbon clusters, are regarded as a significant variety of naturally occurring bioactive metabolites produced by a variety of macrofungi (Duru and Çayan, 2015). Diterpenoids, triterpenoids, and sesquiterpenoids are the model representatives of terpenes with fascinating biological activities, which have been identified or isolated from diverse genera of mushrooms (Trepa et al., 2024). Among these compounds, triterpenoids are widely distributed in edible mushrooms and have attracted considerable attention due to their notable health benefits and diverse bioactivities. Although typically present in small amounts, certain mushroom species have been reported to accumulate substantial levels of these compounds (Kumar et al., 2021). Sesquiterpenoids of unique structure are also characterized by several basidiomycetes (Masunaga et al., 2023). Accumulated reports demonstrated that terpenes possess potent antioxidant, antitumor, anti-inflammatory, antiviral, cytotoxicity, insecticidal, and nematicidal activity (Wu et al., 2013; Song et al., 2013; Duru and Çayan, 2015; Dudekula et al., 2020). Therefore, while the concentration and presence of terpenes in edible mushrooms vary, their presence and health benefits drive captivating research on mushroom bioactive compounds.

#### 5.2.2.2 Phenolic Compounds

Phenolic compounds are one of the most studied secondary metabolites found in edible mushrooms (Sun et al., 2024). The phenolic compounds are aromatic hydroxylated compounds with one or more hydroxyl groups as well as aromatic rings, mainly comprised of phenolic acids, oxidized polyphenols, lignans, tannins, hydroxybenzoic acids, stilbenes, flavonoids, and hydroxycinnamic acids (D'Archivio et al., 2010). Cinnamic, gallic, ferulic, chlorogenic, salicylic, caffeic acid, *p*-hydroxybenzoic, syringic, and *p*-coumaric acids are the chief phenolic groups playing an indispensable role in the biological activity of mushrooms (Nowacka et al., 2015). The contents of polyphenol vary notably between the species of mushroom. The amount of free and bound phenolic compounds noted in several edible mushrooms ranges from several milligrams to over a hundred milligrams/gram of dry weight. The highest total polyphenol concentration exceeded 100 mg/100 g fresh mass (FM), ranging from 109.67 to 159.33 mg/100 g FM, which corresponds to approximately 650 to 1675 mg/100 g dry matter (DM) (Witkowska et al., 2011). Further, depending on the species, phenolic contents ranged from 1 to 6 mg/g of dried mushroom, whereas flavonoid values ranged from 0.9 to 3.0 mg/g of dried matter (Palacios et al., 2011). As per the scientific and literary texts, most mushroom species can be utilized to produce extracts that contain flavonoid compounds (Bahadori et al., 2019). Phenolic compounds derived from mushrooms furnish defense against several degenerative disorders like cardiovascular diseases, brain dysfunction, cancer, and aging (Finimundy et al., 2013). They also contribute to mushroom's unique antioxidant, antimicrobial, anti-inflammatory, antiviral, antithrombotic, antiallergenic, and vasodilator effects (Balasundram et al., 2006; Ferreira et al., 2009; Heleno et al., 2012; Tan et al., 2022). Overall, the concentrations of polyphenol and their chemical structures derived from edible mushrooms furnish a perception of their prospective bioactivity. The diverse array of phenolic acids and flavonoids, each with distinct structural characteristics, contributes significantly to the complex roles of mushrooms in both gourmet cuisine and promising medicinal applications.

#### 5.2.2.3 Sterols (Ergosterol)

Ergosterol is one of the primary sterols found in fungi belonging to basidiomycetes (Guillamon et al., 2010). Ergosterols are a subgroup of steroids that are structurally similar to cholesterol in animals, and their chemical structure includes a steroid nucleus with a side chain. It is known that ergosterol is a kind of sterol, which is further metabolized into vitamin D2 after ultraviolet (UV) irradiation (Tong et al., 2023). Ergosterols are crucial components of fungal cell membranes, which nourish membrane fluidity as well as integrity (Moura et al., 2022). Ergosterol levels have been observed to range between 2290 and 6200 μg $g^{-1}$ in edible mushrooms, with *A. bisporus* being the richest, followed by *P. ostreatus* and *L. edodes* having the lowest concentration (Sun et al., 2022). Ergosterols available in edible mushrooms hold importance due to their role as a precursor to vitamin D and their budding health-related properties (Rangsinth et al., 2023). Ergosterol is detected as tocopherols and considered an effective as well as novel natural antioxidant (Guillamon et al., 2010) and has major biological activities for protection against microbial and cardiovascular diseases (Kalac, 2013) and degenerative malfunctions (Heleno et al., 2012; Jaworska et al., 2015).

## 5.3 BIOACTIVE METABOLITES DERIVED FROM *Pleurotus* sp.

### 5.3.1 Primary Metabolites

High levels of α-glucan and ß-glucan, alongside substantial amounts of protein, have been obtained from fruiting bodies of *P. citrinopileatus* (Figure 5.1A). The glucan content of *P. citrinopileatus* mushrooms ranged from 27.2% to 30.3% dry cell weight (DCW), with β-glucans constituting over 70% DCW of the total glucans (i.e., 20.5%–22.7% DCW) (Figure 5.1A). These polysaccharides demonstrated effective scavenging capabilities against free radicals such as 2,2′-azino-bis(3-ethylbenzothiazoline-6-sulfonic acid) (ABTS), 2,2-diphenyl-1-picrylhydrazyl (DPPH), and hydroxyl, with half-maximal effective concentration ($EC_{50}$) values of 0.06 mg $mL^{-1}$, 1.21 mg $mL^{-1}$, and 3.62 mg $mL^{-1}$, respectively. Additionally, the cytotoxicity of polysaccharides toward hepatocellular carcinoma cells (HepG2) was more significant ($IC_{50}$: 1.69 mg $mL^{-1}$) compared to that of triple-negative breast cancer cells (MDA-MB-468) ($IC_{50}$: 1.76 mg $mL^{-1}$) (Koutrotsios et al., 2022). In another article, Huang et al. (2022) reported that conjugated polysaccharide-peptides have been derived from *P. citrinopileatus* fruiting bodies through hot-water extraction methods. Gut bacteria metabolize these conjugated polysaccharides and generate short-chain fatty acids (SCFAs), thus exhibiting hepatoprotective activity (Huang et al., 2022). *Pleurotus citrinopileatus* is a valuable medicinal fungus from which a novel neutral polysaccharide (CMP) has been isolated from its mycelia. This polysaccharide has a molecular weight of $3.22 \times 10^6$ Da, highlighting its potential for therapeutic applications. The CMP consists of D-mannose, D-glucose, and D-galactose in a molar ratio of 0.33:1.03:0.21. The CMP demonstrates effective inhibition of α-glucosidase, characterized as mixed-type inhibition. The CMP can substantially enhance glucose uptake and glycogen production to modulate insulin resistance in HepG2 insulin-resistant cells (HepG2-IR). This research article presents an alternate source of polysaccharides from *P. citrinopileatus* mycelia and develops a novel material for hypoglycemic meals (Hao et al., 2020).

A polysaccharide has been extracted from *Pleurotus abalonus* and *Pleurotus geesteranus* fruiting bodies which were cultivated on asparagus (*Asparagus officinalis* L.) straw. The ethanolic extracts of both the mushrooms exhibited scavenging effect on 1,1-diphenyl-2-picrylhydrazyl radicals and reducing power. Studies revealed that the antioxidant properties of *P. abalones* was superior to that of *P. geesteranus*. The supplementation of sucrose + $MgSO_4$ to the asparagus substrate was observed to be favorable for the enhancement of antioxidant activity of both the mushrooms (Wang et al., 2012). An acidic selenium-enriched heteropolysaccharide has been derived from *Pleurotus abalonus* fruiting bodies through a hot-water extraction method. This heteropolysaccharide is predominantly composed of D-glucose (87.85%), D-galactose (4.41%), D-mannose (3.42%), L-rhamnose (1.89%), D-glucuronic acid (1.35%), and D-ribose (1.08%), which indicates its superoxide anion radical scavenging activity, leading to the development of an antitumor compound (Ren et al., 2015). Lectin, a carbohydrate-binding protein known for its diverse biological activities including antibacterial, antifungal, and anticancer effects has been reported to be derived from *P. flabellatus* (Figure 5.1C). The *P. flabellatus* lectin (PFL-L) was isolated utilizing DEAE-cellulose anion-exchange chromatography, which was succeeded by gel-filtration chromatography. PFL-L exhibited the most significant radical scavenging activity in the DPPH experiment at a concentration of 100 μg $mL^{-1}$ (Murugesan et al., 2021; Table 5.1). The exopolysaccharides (EPS) extracted from *P. pulmonarius* in a submerged culture supplemented with groundnut shell (20.0 g $L^{-1}$) demonstrated biofunctional characteristics, including antibacterial, antioxidant, and prebiotic actions. The functional groups in EPS, including hydroxyl (-OH), methyl (-CH3), ketone (-RCOH), and carbonyl (-C=O), corroborated its phytochemical properties (Ogidi et al., 2020). A recent study demonstrated a comparative experiment analyzing metabolite production in *P. djamor* mushrooms cultivated on agave bagasse against a traditional barley straw substrate. Five treatments were suggested based on the following ratios of agave bagasse to barley straw: T1 (1:0), T2 (3:1), T3 (1:1), T4 (1:3), and T5 (0:1), where T1 and T5 exhibited the highest quantity of primary metabolites, comprising hydroxycitric acid, 2-deoxy-D-galactose, D-mannose, paromomycin, palmitic acid, pyrrole, mannitol, and D- and L-arabinose. These metabolites demonstrated a wide range of therapeutic potential, including the antibiotic paromomycin (Figure 5.1B), the anticancer metabolite pyrrole, and the tumor growth inhibitor 2-deoxy-D-galactose (Cruz-Moreno et al., 2023).

The maximum polysaccharide yield of *P. eryngii* fruiting bodies was attained when cultivated on wheat straw, rice straw, sugarcane bagasse, and wood. The resultant polysaccharides obtained from the fruiting body were characterized using Fourier transform infrared spectroscopy (FTIR) spectrum analysis, revealing the presence of OH, CH, $CH_3$, $CH_2$, SH, C=O, C-C, and C-S functional groups. The $^1$H and $^{13}$C nuclear magnetic resonance (NMR) spectra indicated that the polysaccharides exhibited both α and β configurations, with signals corresponding to C-1, C-2, C-3, C-4, and C–5. The obtained polysaccharides demonstrated antioxidant activity (El-Zaher et al., 2022). On the other hand, He et al. (2016) reported successfully utilizing the wasted substrate of *P. eryngii* to yield a high percentage (12.18%) of crude polysaccharide (referred to as CPS), from which a polysaccharide-protein complex (identified as Released polysaccharide [RPS]) was further purified using a hot alkali extraction protocol. *In vitro* pharmacological investigations indicated that both CPS and RPS demonstrated significant antioxidant activity in a dose-dependent manner, utilizing several *in vitro* models such as ABTS, superoxide, hydroxyl, and DPPH (He et al., 2016). Furthermore, sulfated polysaccharides were also extracted from *P. eryngii*, and they exhibited antioxidant and antibacterial properties during pharmacological analysis (Li and Shah, 2014). Intracellular polysaccharide (IPS) and its two fractions, IPS-1 and IPS-2, were extracted and purified via DEAE-52 cellulose chromatography from *P. eryngii* SI-04 mycelia, and their hepatoprotective effects were examined. The findings indicated that IPS-2 exhibited enhanced hepatoprotective benefits by elevating blood enzyme activities and bilirubin levels, reducing serum albumin and triglyceride levels, enhancing hepatic antioxidant status, and thus mitigating hepatic structural damage (Zhang et al., 2016). The hepatoprotective and antioxidant properties of acidic- and alkali-extractable mycelial zinc polysaccharides (AcMZPS, AlMZPS) derived from *P. eryngii* var. *tuoliensis* were examined in hyperlipidemic mice caused by a high-fat–high-cholesterol emulsion. The *in vivo* experiments indicated that both AcMZPS and AlMZPS exhibited potential hepatoprotective effects by significantly reducing the levels of low-density lipoprotein cholesterol, very-low-density lipoprotein cholesterol, total cholesterol, triglycerides, alanine aminotransferase, aspartate aminotransferase, alkaline phosphatase, malondialdehyde, and lipid peroxidation—while markedly increasing high-density lipoprotein cholesterol, superoxide dismutase, glutathione peroxidase, and catalase in serum lipid/liver homogenate, respectively (Xu et al., 2017).

Similarly, a novel polysaccharide, pleuran, has been derived from the fruiting body of *P. ostreatus*. An alkali-insoluble β-D-glucan with a branched structure features a backbone composed of (1→3)-linked β-D-glucopyranosyl residues, where every fourth residue is substituted at the O-6 position by a single D-glucopyranosyl side chain. The polysaccharide comprised a minor fraction (7%) of (1→6)- and (1→4)-linked internal residues. Pleuran exhibits effectiveness in enhancing the survival of mice vulnerable to bacterial infections (Karácsonyi and Kuniak, 1994). In another article, Rizkyana et al. (2022) reported that the maximum quantity of ß-glucan and the sulfated polysaccharides (PS) produced from the fruiting body of *P. ostreatus* through the subcritical water extraction method. These polysaccharides showed superior anticoagulant action and improved digestive and immune systems (Rizkyana et al., 2022).

### 5.3.2 Secondary Metabolites

Eleven phytochemicals, comprising essential oil, triterpenes, anthraquinones, tannins, flavonoids, phenols, fatty acids, alkaloids, steroids, sugars, and coumarins, were identified from the mycelia of *P. cystidiosus* WS218–2. Ethanolic extract of the strain showed significant radical scavenging activity at 79% because of its high phenolic content of 95.67 mg of gallic acid equivalents per gram of material. Therefore, the mycelia of *P. cystidiosus* could serve as a significant source of natural antioxidants and bioactive compounds (Garcia et al., 2020). *Pleurotus citrinopileatus* is recognized for its antioxidant qualities due to the presence of α- and γ-tocopherols, ascorbic acid, lovastatin (Figure 5.1E), and phenolic compounds. Seven phenolic compounds were found in *P. citrinopileatus*; among them, four phenolic acids, namely *p*-OH benzoic, *p*-OH phenylacetic, vanillic, and cinnamic acids, were quantified at values between 0.09% and 0.87% DCW. Moreover, resveratrol is a stilbene synthesized by various plants and fungi; in *P. citrinopileatus* samples, the content of resveratrol varied from 4.0% to 6.9% DCW (Koutrotsios et al., 2022; Figure 5.1D).

Different ratios of agave bagasse to barley straw were used for the cultivation of *P. djamor*, where pyrrole, pyrazine 1,4, dizone, and hexahydro 3-(2-methylpropyl) were accumulated in the species at the ratios of 3:1 and 1:1 agave bagasse to barley straw (Figure 5.1F). All these compounds show a significant amount of anticancer activity. The total phenolic and flavonoid concentrations, along with the scavenging activities (DPPH), ferric thiocyanate, and thiobarbituric acid, were assessed in methanol and water extracts from primordia, basidiomata, and mycelium. Both basidiomata and mycelial methanol extracts demonstrated considerable antioxidant activity. Along with that, a novel secondary metabolite, quinoline, was also extracted from

**TABLE 5.1**
**List of Primary and Secondary Metabolites Derived from *Pleurotus* sp.**

| Species | Source | Type of Metabolite | Bioactive Metabolites | Extraction Technique | Biological Activity | References |
|---|---|---|---|---|---|---|
| *P. abalonus* | Fruiting body | Primary | Polysaccharides | Ethanol extract | Antioxidant | Wang et al., 2012 |
| *P. abalonus* | Fruiting body | Primary | Selenium-enriched polysaccharides | Hot water extract | Super oxide anion | Ren et al., 2015 |
| *P. citrinopileatus* | Mycelia | Primary | Polysaccharides | Aqueous extract | Enhance glucose uptake and glycogen production | Hao et al., 2020 |
| | Fruiting body | Primary | α-glucan<br>ß-glucan | Methanolic extract | Antioxidant activity | Koutrotsios et al., 2022 |
| | | Primary | Polysaccharides-peptides | Hot-water extract | Exhibits hepatoprotectve and antioxidant activity | Huang et al., 2022 |
| | Fruiting body | Secondary | Tri-terpenic acids, lovastatin, α- and γ-tocopherols, ergosterol, *p*-OH benzoic, *p*-OH phenylacetic, vanillic and cinnamic acids, resveratrol, and ergothioneine | Methanolic extract | Antioxidant activity | Koutrotsios et al., 2022 |
| *P. cystidiosus* WS218–2 | Mycelia | Secondary | Triterpenes, anthraquinones, tannins, flavonoids, phenols, alkaloids, steroids, and coumarins | Ethanol extract | Antioxidant | Garcia et al., 2020 |
| *P. djamor* | Fruiting body | Primary | Paromomycin | Methanol extract | Prevent parasitic infections | Cruz-Moreno et al., 2023 |
| | | Secondary | Pyrrole<br>Pyrazine 1,4, dizone,<br>Hexahydro 3- (2- methyl propyl) | Methanol extract | Anticancer activity | |
| | | | Quinoline | N.D. | Antiseptic, amebicide and antibiotic | |
| | Mycelia | | Secondary | Flavonoids, anthraquinone and terpenoid | Dichlorometh-ane extracts | antibacterial activity and antimicrobial activity |
| Illuri et al., 2022 | *P. erynjii* | Fruiting body | Primary | Polysaccharide | Ethanol extract | Antioxidant activity |
| El-Zaher et al., 2022 | | Spent mushroom substrate | | Released polysaccharide (RPS) | Alkali extract | Hydroxyl radical |
| He et al., 2016 | | Fruiting body | | Sulfated polysaccharide | Aqueous extract | Superoxide |
| Li and Shah, 2014 | *P. eryngii* | Fruiting body | Secondary | 2,5-dihydroxybenzoi-c acid, 4-hydroxybenzoic, ferulic, *p*-coumaric, protocatechuic, trans-cinnamic, and vanillic acids | Methanol extract | Antioxidant activity |
| Gąsecka et al., 2016 | *P. eryngii* SI-04 | Mycelia | Primary | Intracellular polysaccharide (IPS) | Acidic extract | Antioxidant assay |
| Zhang et al., 2016 | *P. eryngii* var. tuoliensis | Mycelia | Primary | Zinc polysaccharide | Acidic and alkali extract | DPPH radical scavenging assay |
| Xu et al., 2017 | *P. florida* | Mycelial extracts | Secondary | Flavonoids, anthraquinones and terpenoid | Dicholometh-ane extract | Antibacterial and antimicrobial activity |
| Illuri et al., 2022 | | Fruiting bodies | Secondary | Alkaloid,<br>Anthraquinone | Methanol extract | Anti-cancerous and antioxidant activity |

| Species | Source | Type of Metabolite | | Bioactive Metabolites | Extraction Technique | Biological Activity | References |
|---|---|---|---|---|---|---|---|
| Ganesan et al., 2023 | | | | | Flavonoid | Choloform extract | |
| | | | | | Glycosides | | |
| | | | | | Phenol | | |
| | | | | | Saponin | | |
| | | | | Alkaloid, tannin | Hexane extract | Tanin, terpenoid | |
| | | | | Aqueous extract | *P. flabellatus* | Fruiting body | |
| | Primary | Lectin | Acid extract | Increase hemagglutinating activity, potent antibacterial activity against gram-negative bacteria | Murugesan et al., 2021 | *P. ostreatus* | Fruiting body |
| Primary | Pleuran | Alkali extract | Antidiabetic and lipid-lowering effects. Immunomodulatory. | Karácsonyi and Kuniak, 1994 | Mycelium | | Pleurotin |
| | Ethyl acetate extract | N.D. | Al-Temimay et al., 2015 | Fruiting body | | | Primary |
| Sulfated polysaccharides | Subcritical water extract | Anticoagulant | Rizkyana et al., 2022 | | | | |
| | ß-glucan | Improvement of digestive and immune systems, boots metabolic activity | | Fruiting body | | | Secondary |
| Methanol extract | 4-hydroxybenzoic, ferulic, p-coumaric, protocatechuic, trans-cinnamic, vanillic acid and ferulic acid | Antioxidant activity | Gąsecka et al., 2016 | *P. ostreatus* | Fruiting bodies | | Secondary |
| Methanol extract | Phenolic, flavonoid, condensed tannin | Radical scavenging activity | Yılmaz et al., 2017 | | Mycelium | | Secondary |
| Aqueous extract | Flavonoids, polyketones, terpenoids | N.D. | Hu et al., 2022 | *P. ostreatus* SY10 | Fruiting body | | Secondary |
| N.D. | Pleurotusin A | Antibacterial acid | Ge et al., 2022 | *P. eryngii* SY302 | | | |
| | Pleurotusin B | antioxidant acid | | *P. pulmonaris* | Fruiting body | | Primary |
| Ethanol extract | Exopolysaccharides | Antimicrobial, antioxidant and prebiotic activities | Ogidi et al., 2020 | | | | Secondary |
| Ethanol extract | Phenolic compounds, hydroxybenzoic acids | Anti-inflammatory bioactivities | Amirullah et al., 2021 | *P. tuberregium* | Fruiting body | | Secondary |
| Ethanol extract | Triterpene | Hypolipidemic activity | Wang et al., 2022 | | | | |

*N.D. Not determined*

the fruiting body of *P. djamor*, which demonstrated a significant amount of antiseptic, amebicide, and antibiotic activities (Cruz-Moreno et al., 2023). Moreover, mycelial extracts of *P. djamor* and *P. florida* demonstrated pronounced antibacterial activity. Dichloromethane mycelial extracts were acquired using Soxhlet extraction, yielding positive results in several phytochemical analyses. The identification of metabolites in thin-layer chromatography (TLC) utilizing spray reagents represents one of the earliest known examples of flavonoids, anthraquinones, and terpenoid chemicals in *P. djamor* and *P. florida*. *Pleurotus djamor* and *P. florida* exhibited dose-dependent antiproliferative effects on MCF7 cells, demonstrating inhibitory rates of 55.7% and 64.5%, respectively, at a concentration of 125 μg $mL^{-1}$ (Illuri et al., 2022).

Total triterpenes were isolated from *Pleurotus tuber-regium* which exhibited a potent hypolipidemic activity. *In vivo* trials revealed that zebrafish could safely tolerate these mushroom-derived triterpenes at concentrations up to 500 μg/mL. Further correlation analysis indicated that the compounds significantly decreased lipid accumulation in zebrafish subjected to a high-fat diet, with the lipid-lowering activity showing a clear dose-dependent relationship (Wang et al., 2022).

Different phytochemical substances, including alkaloids, anthraquinones, coumarins, flavonoids, phenols, saponins, steroids, tannins, and terpenoids, utilizing various solvent extracts such as water, methanol, chloroform, and hexane, were investigated from *P. florida* by Ganesan et al. (2023). Chemical tests indicated a significant presence of saponins and terpenoids in the aqueous extract. Additional chemicals include alkaloids, anthraquinones, flavonoids, and glycosides. Phenol and tannin were also present in moderate quantities.

The methanolic extract exhibited a pronounced presence of the alkaloid compound, and a moderate presence of anthraquinone, steroids, and tannin was also observed. Other compounds, such as coumarin and tannin, were notably absent in this chloroform solvent extract. However, the hexane extract exhibited a significant presence of saponin and terpenoid compounds, with alkaloids and tannin in moderate quantities. The comprehensive findings of the initial analysis of bioactive chemical compounds demonstrate significant antioxidant activities (Ganesan et al., 2023).

The impacts of traditional (Soxhlet) and ultrasound-assisted (UAE) extractions on the composition, antioxidative properties, and anti-inflammatory bioactivities of gray oyster mushroom, *P. pulmonarius*, were examined by Amirullah et al. (2021). Liquid chromatography–tandem mass spectrometry (LCMS/MS) detected many phenolic compounds in the Soxhlet and UAE extracts; however, hydroxybenzoic acids were exclusively identified in the UAE extracts. The UAE extract demonstrated exceptional anti-inflammatory efficacy by markedly inhibiting nitric oxide generation in lipopolysaccharide-stimulated murine macrophage cells and by inhibiting nuclear factor (NF)-κB (Amirullah et al., 2021).

Pleurotin, a significant secondary metabolite, was isolated from *P. ostreatus* using TLC and high-performance liquid chromatography (HPLC) (Figure 5.1G). Pleurotin was purified using two methods: chloroform and ethyl acetate. The results indicated that ethyl acetate was more effective in pleurotin production, yielding 14.6 μg $mL^{-1}$ compared to 9.8 μg $mL^{-1}$ using chloroform (Al-Temimay et al., 2015). Environmental conditions significantly influence the accumulation of microbial secondary metabolites. The effects of 24-h heat stress and 0.05 mmol $L^{-1}$ salicylic acid therapy on the secondary metabolism of *P. ostreatus* by metabolome, transcriptome, and gene differential expression analyses were examined, where metabolome and transcriptome analysis indicated that salicylic acid markedly enhanced the accumulation of antibiotics and polyketones in mycelia, whereas heat stress elevated the levels of flavonoids, polyketones, terpenoids, and polysaccharides (Hu et al., 2022). A novel isovaleryl-monoterpene, pleurotusin A, and a new cyclopentenone derivative, pleurotusin B, along with five related known terpenoids, were derived from the coculture broth of two edible fungi, *P. ostreatus* SY10 and *P. eryngii* SY302. The absolute configurations of these novel secondary metabolites were determined through an extensive study of density functional theory calculations of NMR and optical rotation measurements. The antibacterial efficacy of these compounds, except the unstable compound pleurotusin B, was assessed against two species of human-pathogenic fungus, *Candida albicans* and *Cryptococcus neoformans*, as well as three species of human-pathogenic bacteria, *Staphylococcus aureus*, *Escherichia coli*, and *Shigella* sp. Pleurotusin A exhibited modest efficacy against *S. aureus*, with a minimum inhibitory concentration (MIC) value of 90.3 μM (Ge et al., 2022).

Furthermore, the antioxidant properties of five high-yielding terpenoids were evaluated using DPPH, revealing that the compound exhibited moderate activity with an $EC_{50}$ value of 573 μM (Ge et al., 2022). The potential applications of various agro-industrial wastes, including peanut residues, potato farm by-products, walnut, and orange tree sawdust, in the cultivation of *P. ostreatus* were investigated, where the total phenolic, flavonoid, and condensed tannin content, as well as the antioxidant properties of the methanolic extracts from these mushrooms, were analyzed. The maximum overall phenolic content was observed in mushrooms cultivated on walnut sawdust, whereas the mushroom extract cultivated on potatoes exhibited the highest levels of condensed tannin and the highest total flavonoid content and free radical scavenging activity against DPPH (Yılmaz et al., 2017: Table 5.1). Six phenolic acids, including 4-hydroxybenzoic, ferulic, *p*-coumaric, protocatechuic, trans-cinnamic, and vanillic acids, were identified in the extract of *P. ostreatus*, where naringenin was also quantified. Similarly, all of them were obtained from *P. eryngii*, and as an added advantage, 2,5-dihydroxybenzoic acid was also obtained. Ferulic acid was the predominant component found in both *Pleurotus* species. Chromatogram analysis indicated that mushrooms contain several phenolic compounds. The addition of substrates with micronutrients enhanced their antioxidant capabilities (Gąsecka et al., 2016).

## 5.4 CONCLUSION AND OUTLOOK

Mushrooms are prized fungi that occur as an indispensable element of the ecosystem and are universally considered customary for their nutritional and therapeutic values. They are a nutritive, frugally, and biotechnologically valued group of organisms because of their significance in dietary, medication, biocontrol, cosmeceutical, and various other industries. This literature review summarized that species of *Pleurotus* are an affluent source of a high number of biologically active (primary as well as secondary) metabolites, with a focus on their utilization in various applications and health benefits. Lately, wild edible and medicinal species of *Pleurotus* have been a topic of interest since the bioactive compounds like polysaccharides, terpenoids, protein and peptides, sterols, and phenols derived from them were observed to possess several biological activities like antioxidant, antitumor, antihyperglycemic, antimicrobial, and anti-inflammatory activities. To date, a combination of around 14 primary and secondary metabolites have been reported from different species of *Pleurotus* (Figure 5.2). It is also worth mentioning that there are still many indigenous species of *Pleurotus* whose metabolites are undivulged; therefore, this imparts to researchers a vast prospect of finding or discovering many other new, interesting, or unknown metabolites from the world's untapped regions, where mycological studies are still lagging. In the coming years, genomics, proteomics, and metabolomics might play an essential role in improving the active metabolic component of mushroom research. However, international and interdisciplinary collaborations can be encouraged and reinforced in the future to amplify further discoveries of beneficial mushroom metabolites—with the knowledge of understanding their mode of action—to establish stable therapeutic applications.

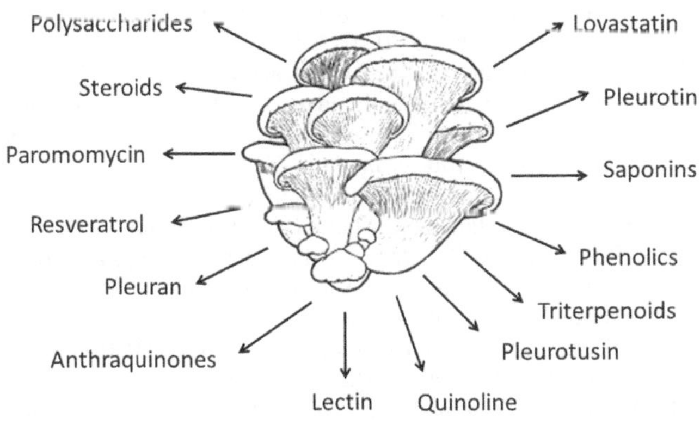


**FIGURE 5.2** Types of bioactive metabolites derived from *Pleurotus* sp.

## REFERENCES

Abdelshafy, A. M., T. Belwal, Z. Liang, et al. 2022. A comprehensive review on phenolic compounds from edible mushrooms: occurrence, biological activity, application and future prospective. *Critical Reviews in Food Science and Nutrition* 62(22): 6204–6224. https://doi.org/10.1080/10408398.2021.1898335

Aditya, N., R. S. Jarial, K. Jarial, and J. N. Bhatia. 2024. Comprehensive review on Oyster mushroom species (Agaricomycetes): morphology, nutrition, cultivation and future aspects. *Heliyon* 10(5): 26539. https://doi.org/10.1016/j.heliyon.2024.e26539

Al-Temimay, I. A., M. H. AL-Jibouri, A. A. Hassan, et al. 2015. Test the cytotoxicityof Pleurotin extracted from an edible mushroom *Pleurotus osteratus* against three human carcinoma cellline. *Iraqi Journal of Science* 56: 2773–2781.

Amirullah, N. A., N. Z. Abidin, N. Abdullah, and S. Manickam. 2021. Application of ultrasound towards improving the composition of phenolic compounds and enhancing *in vitro* bioactivities of *Pleurotus pulmonarius* (Fr.) Quél extracts. *Biocatalysis and Agricultural Biotechnology* 31: 101881. https://doi.org/10.1016/j.bcab.2020.101881

Bahadori, M. B., C. Sarikurkcu, O. U. Yalcin, M. Cengiz, and H. Gungor. 2019. Metal concentration, phenolics profiling, and antioxidant activity of two wild edible *Melanoleuca* mushrooms (*M. cognata* and *M. stridula*). *Microchemical Journal* 150: 104172. https://doi.org/10.1016/j.microc.2019.104172

Balasundram, N., K. Sundram, and S. Samman. 2006. Phenolic compounds in plants and agri-industrial by-products: antioxidant activity, occurrence and potential uses. *Food Chem*istry 99: 191–203. https://doi.org/10.1016/j.foodchem.2005.07.042

Bano, Z., and H. C. Srivastava. 1962. Studies on cultivation of *Pleurotus* sp. on paddy straw. *Food Science* 11: 363–365.

Bernardi, E., E. Minotto, and J. S. do Nascimento. 2013. Evaluation of growth and production of *Pleurotus* sp. in sterilized substrates. *Arquivos do Instituto Biológico* 80(3): 318–324. https://doi.org/10.1590/S1808-16572013000300009

Beulah, H., A. A. Margret, and J. Nelson. 2013. Marvelous medicinal mushrooms. *International Journal of Pharmacy and Biological Sciences* 3: 611–615.

Bhambri, A., M. Srivastava, V. G. Mahale, S. Mahale, and S. K. Karn. 2022. Mushrooms as potential sources of active metabolites and medicines. *Frontiers in Microbiology* 13: 837266. https://doi.org/10.3389/fmicb.2022.837266

Bolton, J. L., T. L. Dunlap, A. Hajirahimkhan, et al. 2019. The multiple biological targets of hops and bioactive compounds. *Chemical Research in Toxicology* 32(2): 222–233. https://doi.org/10.1021/acs.chemrestox.8b00345

Bumanlag, C. P. B., S. P. Kalaw, R. M. R. Dulay, and R. G. Reyes, 2018. Optimum conditions for mycelia growth and basidiocarp production of *Pleurotus djamor* on cornbased media. *International Journal of Biology, Pharmacy and Allied Sciences* 7(4): 558–575.

Cateni, F., M. L. Gargano, G. Procida, G. Venturella, F. Cirlincione, and V. Ferraro. 2021. Mycochemicals in wild and cultivated mushrooms: nutrition and health. *Phytochemistry Reviews* 21(2): 339–383. https://doi.org/10.1007/s11101-021-09748-2

Choudhary, M., R. Devi, A. Datta, A. Kumar, and H. S. Jat. 2015. Diversity of wild edible mushrooms in Indian subcontinent and its neighboring countries. *Recent Advances in Biology and Medicine* 1: 69–76. https://doi.org/10.18639/RABM.2015.01.200317

Chun, S., J. Gopal, and M. Muthu. 2021. Antioxidant activity of mushroom extracts/polysaccharides-their antiviral properties and plausible anti COVID-19 properties. *Antioxidants* 10(12): 1899. https://doi.org/10.3390/antiox10121899

Correa, R. C. G., T. Brugnari, A. Bracht, R. M. Peralta, and I. C. Ferreira. 2016. Biotechnological, nutritional and therapeutic uses of *Pleurotus* spp. (Oyster mushroom) related with its chemical composition: a review on the past decade findings. *Trends in Food Science and Technology* 50: 103–117. https://doi.org/10.1016/j.tifs.2016.01.012

Cruz-Moreno, B. A., A. A. F. Pérez, J. F. García-Trejo, S. A. Pérez-García, and C. Gutiérrez-Antonio. 2023. Identification of secondary metabolites of interest in *Pleurotus djamor* using Agave tequilana Bagasse. *Molecules* 28(2): 557. https://doi.org/10.3390/molecules28020557

D'Archivio, M., C. Filesi, R. Vari, B. Scazzocchio, and Masella, R. 2010. Bioavailability of the polyphenols: status and controversies. *International Journal of Molecular Sciences* 11: 1321–1342. https://doi.org/10.3390/ijms11041321

Dudekula, U. T., K. Doriya, and S. K. Devarai. 2020. A critical review on submerged production of mushroom and their bioactive metabolites. *3 Biotech* 10: 1–12. https://doi.org/10.1007/s13205-020-02333-y

Duru, M. E., and G. T. Çayan. 2015. Biologically active terpenoids from mushroom origin: a review. *Records of Natural Products* 9: 456.

Effiong, M. E., C. P. Umeokwochi, I. S. Afolabi, and S. N. Chinedu. 2024. Assessing the nutritional quality of *Pleurotus ostreatus* (Oyster mushroom). *Frontiers in Nutrition* 10: 1279208. https://doi.org/10.3389/fnut.2023.1279208

Elhamouly, N. A., O. A. Hewedy, and A. Zaitoon. 2022. The hidden power of secondary metabolites in plant-fungi interactions and sustainable phytoremediation. *Frontiers in Plant Science* 13: 1044896. https://doi.org/10.3389/fpls.2022.1044896

El-Zaher, A., E. Hasan, E. M. Tousson, A. A. Mostafa, and E. M. El-Gaar. 2022. Production of endo Polysaccharides from cultivated *Pleurotus eryngii* fruiting bodies. *Delta Journal of Science* 44(1): 135–144. https://doi.org/10.21608/djs.2022.127995.1020

Feeney, M. J., A. M. Miller, and P. Roupas. 2014. Mushrooms biologically distinct and nutritionally unique: exploring a "Third Food Kingdom". *Nutrition today* 49: 301–307. https://doi.org/10.1097/NT.0000000000000063

Fernandes, Â., L. Barros, A. Martins, P. Herbert, and I. C. Ferreira. 2015. Nutritional characterization of *Pleurotus Ostreatus* (Jacq. ex Fr.) P. Kumm. produced using paper scraps as substrate. *Food Chemistry* 169: 396–400. https://doi.org/10.1016/j.foodchem.2014.08.027

Ferreira, I. C. F. R., L. Barros, and R. M. V. Abreu. 2009. Antioxidants in wild mushrooms. *Current Medicinal Chemistry* 16: 1543–1560. https://doi.org/10.2174/092986709787909587

Finimundy, T. C., G. Gambato, R. Fontana, et al. 2013. Aqueous extracts of *Lentinula edodes* and *Pleurotus sajor-caju* exhibit high antioxidant capability and promising *in vitro* antitumor activity. *Nutrition Research* 33(1): 76–84. https://doi.org/10.1016/j.nutres.2012.11.005

Fogarasi, M., S. A. Nemeş, A. Farcaş, et al. 2024. Bioactive secondary metabolites in mushrooms: a focus on polyphenols, their health benefits and applications. *Food Bioscience* 62: 105166. https://doi.org/10.1016/j.fbio.2024.105166

Friedman, M. 2016. Mushroom polysaccharides: chemistry and antiobesity, antidiabetes, anticancer, and antibiotic properties in cells, rodents, and humans. *Foods* 5(4): 80. https://doi.org/10.3390/foods5040080

Ganesan, G., G. Selvam, A. Varatharasu, and P. Annamalai. 2023. Physico-chemical and qualitative, quantitative chemical compounds analysis of *Pleurotus florida* in Thanjavur, Tamilnadu, India. *International Journal of Life Science and Pharma Research* 13(2): L8–L17. https://doi.org/10.22376/ijlpr.2023.13.2.L8-L17

Garcia, K., C. J. Garcia, R. Bustillos, et al. 2020. Mycelial biomass, antioxidant, and myco-actives of mycelia of abalone mushroom *Pleurotus cystidiosus* in liquid culture. *Journal of Applied Biology & Biotechnology* 8(2): 94–97. https://doi.org/10.7324/JABB.2020.80215

Gąsecka, M., M. Mleczek, M. Siwulski, and P. Niedzielski. 2016. Phenolic composition and antioxidant properties of *Pleurotus ostreatus* and *Pleurotus eryngii* enriched with selenium and zinc. *European Food Research and Technology* 242: 723–732. https://doi.org/10.1007/s00217-015-2580-1

Ge, X., Y. Wang, C. Sun, et al. 2022. Secondary metabolites produced by coculture of *Pleurotus ostreatus* SY10 and *Pleurotus eryngii* SY302. *Chemistry and Biodiversity* 19(4): 202100832. https://doi.org/10.1002/cbdv.202100832

Golak-Siwulska, I., A. Kaluzewicz, T. Spizewski, M. Siwulski, and K. Sobieralski. 2018. Bioactive compounds and medicinal properties of Oyster mushrooms (*Pleurotus* sp). *Folia Horticulturae* 30(2): 191–201. https://doi.org/10.2478/fhort-2018-0012

Govorushko, S., R. Rezaee, J. Dumanov, and A. Tsatsakis. 2019. Poisoning associated with the use of mushrooms: a review of the global pattern and main characteristics. *Food and Chemical Toxicology* 128: 267–279. https://doi.org/10.1016/j.fct.2019.04.016

Grujic, M., B. Dojnov, I. Potocnik, B. Duduk, and Vujcic, Z. 2015. Spent mushroom compost as substrate for the production of industrially important hydrolytic enzymes by fungi *Trichoderma* sp. and *Aspergillus niger* in solid state fermentation. *International Biodeterioration and Biodegradation* 104: 290–298. https://doi.org/10.1016/j.ibiod.2015.04.029

Guillamon, E., A. Garcia-Lafuente, M. Lozano, et al. 2010. Edible mushrooms: role in the prevention of cardiovascular diseases. *Fitoterapia* 81(7): 715–723. https://doi.org/10.1016/j.fitote.2010.06.005

Hamza, A., S. Ghanekar, and D. Santhosh Kumar. 2023. Current trends in health promoting potential and biomaterial applications of edible mushrooms for human wellness. *Food Bioscience* 51: 102290. https://doi.org/10.1016/j.fbio.2022.102290

Hao, Y., H. Sun, X. Zhang, L. Wu, and Z. Zhu. 2020. A novel polysaccharide from *Pleurotus citrinopileatus* mycelia: structural characterization, hypoglycemic activity and mechanism. *Food Bioscience* 37: 100735. https://doi.org/10.1016/j.fbio.2020.100735

He, P., F. Li, L. Huang, D. Xue, W. Liu, and C. Xu. 2016. Chemical characterization and antioxidant activity of polysaccharide extract from spent mushroom substrate of *Pleurotus eryngii*. *Journal of the Taiwan Institute of Chemical Engineers* 69: 48–53. https://doi.org/10.1016/j.jtice.2016.10.017

Heleno, S. A., L. Barros, A. Martins, M. J. Queiroz, C. Santos-Buelga, and I. C. Ferreira. 2012. Phenolic, polysaccharidic and lipidic fractions of mushrooms from North-eastern Portugal: chemical compounds with antioxidant properties. *Journal of Agricultural and Food Chemistry* 60(18): 4634–4640. https://doi.org/10.1021/jf300739m

Hu, Y., Q. Chai, Y. Wang, et al. 2022. Effects of heat stress and exogenous salicylic acid on secondary metabolites biosynthesis in *Pleurotus ostreatus* (Jacq.) P. Kumm. *Life* 12(6): 915. https://doi.org/10.3390/life12060915

Huang, Y., Y. Gao, X. Pi, S. Zhao, and W. Liu. 2022. *In vitro* hepatoprotective and human gut microbiota modulation of polysaccharide-peptides in *Pleurotus citrinopileatus*. *Frontiers in Cellular and Infection Microbiology* 12: 892049. https://doi.org/10.3389/fcimb.2022.892049

Hyde, K. D. 2022. The numbers of fungi. *Fungal Diversity* 114: 1. https://doi.org/10.1007/s13225-022-00507-y

Illuri, R., M. Eyini, M. Kumar, et al. 2022. Bio-prospective potential of *Pleurotus djamor* and *Pleurotus florida* mycelial extracts towards Gram positive and Gram negative microbial pathogens causing infectious disease. *Journal of Infection and Public Health* 15(2): 297–306. https://doi.org/10.1016/j.jiph.2021.10.012

Islam, M. Z., M. H. Rahman, and F. Hafiz. 2009. Cultivation of Oyster mushroom (*Pleurotus flabellatus*) on different substrates. *International Journal of Sustainable Crop Production* 4(1): 45–48.

Jaworska, G., K. Pogon, E. Bernas, and Duda-Chodak, A. 2015. Nutraceuticals and antioxidant activity of prepared for consumption commercial mushrooms *Agaricus bisporus* and *Pleurotus ostreatus*. *Journal of Food Quality* 38: 111–122. https://doi.org/10.1111/jfq.12132

Kalac, P. 2013. A review of chemical composition and nutritional value of wild growing and cultivated mushrooms. *Journal of the Science of Food and Agriculture* 93: 209–218. https://doi.org/10.1002/jsfa.5960

Kamarudzaman, A. N., T. C. Chay, A. Amir, and S. A. Talib. 2015. Biosorption of Mn (II) ions from aqueous solution by *Pleurotus* spent mushroom compost in a fixed-bed column. *Procedia-Social and Behavioral Sciences* 195: 2709–2716. https://doi.org/10.1016/j.sbspro.2015.06.379

Karácsonyi, S., and L. Kuniak. 1994. Polysaccharides of *Pleurotus ostreatus*: isolation and structure of pleuran, an alkali-insoluble β-D-glucan. *Carbohydrate Polymers* 24(2): 107–111. https://doi.org/10.1016/0144-8617(94)90019-1

Khan, M. A., S. M. R. Amin, M. N. Uddin, M. Tania, and N. Alam. 2008. Comparative study of the nutritional composition of Oyster mushrooms cultivated in Bangladesh. *Bangladesh Journal of Mushroom*. 2(1): 9–14.

Khan, M. A., and M. Tania. 2012. Nutritional and medicinal importance of *Pleurotus* mushrooms: an overview. *Food Reviews International* 28(3): 313–329. https://doi.org/10.1080/87559129.2011.637267

Koutrotsios, G., D. Tagkouli, G. Bekiaris, et al. 2022. Enhancing the nutritional and functional properties of *Pleurotus citrinopileatus* mushrooms through the exploitation of winery and olive mill wastes. *Food Chemistry* 370: 131022. https://doi.org/10.1016/j.foodchem.2021.131022

Kozarski, M., A. Klaus, M. Niksic, L. J. L. D. Van Griensven, M. Vrvic, and Jakovljevic, D. 2014. Polysaccharides of higher fungi: biological role, structure and anti-oxidative activity. *Hemijska Industrija* 68(3): 305–320. https://doi.org/10.2298/HEMIND121114056K

Kozarski, M., A. Klaus, L. van Griensven, et al. 2023. Mushroom β-glucan and polyphenol formulations as natural immunity boosters and balancers: nature of the application. *Food Science and Human Wellness* 12(2): 378–396. https://doi.org/10.1016/j.fshw.2022.07.040

Kumar, K., R. Mehra, R. P. F. Guine, et al. 2021. Edible mushrooms: a comprehensive review on bioactive compounds with health benefits and processing aspects. *Foods* 10(12): 2996. https://doi.org/10.3390/foods10122996

Kumla, J., N. Suwannarach, K. Sujarit, et al. 2020. Cultivation of mushrooms and their lignocellulolytic enzyme production through the utilization of agro-industrial waste. *Molecules* 25(12): 2811. https://doi.org/10.3390/molecules25122811

Lesa, K. N., M. U. Khandaker, F. M. R. Iqbal, et al. 2022. Nutritional value, medicinal importance, and health-promoting effects of dietary mushroom (*Pleurotus ostreatus*). *Journal of Food Quality* 2022(2454180): 1–9. https://doi.org/10.1155/2022/2454180

Li, S., and N. P. Shah. 2014. Antioxidant and antibacterial activities of sulphated polysaccharides from *Pleurotus eryngii* and *Streptococcus thermophilus* ASCC 1275. *Food Chemistry* 165: 262–270. https://doi.org/10.1016/j.foodchem.2014.05.110

Lin, C. H., G. T. Sheu, Y. W. Lin, et al. 2010. A new immunomodulatory protein from *Ganoderma microsporum* inhibits epidermal growth factor mediated migration and invasion in A549 lung cancer cells. *Process Biochem*istry 45: 1537–1542. https://doi.org/10.1016/j.procbio.2010.06.006

Łysakowska, P., A. Sobota, and A. Wirkijowska. 2023. Medicinal mushrooms: their bioactive components, nutritional value and application in functional food production-a review. *Molecules* 28(14): 5393. https://doi.org/10.3390/molecules28145393

Mandeel, Q. A., A. A. Al-Laith, and S. A. Mohamed. 2005. Cultivation of Oyster mushrooms (*Pleurotus* spp.) on various lignocellulosic wastes. *World Journal of Microbiology and Biotechnology* 21(4): 601–607. https://doi.org/10.1007/s11274-004-3494-4

Masunaga, N., T. Kitaoka, and H. Ichinose. 2023. Biocatalyst collection and heterologous expression of sesquiterpene synthases from basidiomycetous fungi: discovery of a novel sesquiterpene hydrocarbon. *Microbial Biotechnology* 16(3): 632–644. https://doi.org/10.1111/1751-7915.14204

Meenu, M., and B. Xu. 2019. Application of vibrational spectroscopy for classification, authentication and quality analysis of mushroom: a concise review. *Food Chemistry* 289: 545–557. https://doi.org/10.1016/j.foodchem.2019.03.091

Moura, B., H. Monteiro, T. M. Mata, M. Iten, and A. A. Martins. 2022. Environmental life cycle assessment of early-stage development of ergosterol extraction from mushroom bio-residues. *Journal Clean Production* 355: 131623. https://doi.org/10.1016/j.jclepro.2022.131623

Mowsurni, F., and M. Chowdhury. 2010. Oyster mushroom: biochemical and medicinal prospects. *Bangladesh Journal of Medical Biochemistry* 3(2): 23–28. https://doi.org/10.3329/bjmb.v3i1.13804

Murugesan, A. K., and K. S. Gunasagaran. 2021. Purification and characterization of a synergistic bioactive lectin from *Pleurotus flabellatus* (PFL-L) with potent antibacterial and in-vitro radical scavenging activity. *Analytical Biochemistry* 635: 114450. https://doi.org/10.1016/j.ab.2021.114450

Nowacka, N., R. Nowak, M. Drozd, M. Olech, R. Los, and A. Malm. 2015. Antibacterial, antiradical potential and phenolic compounds of thirty-one polish mushrooms. *PloS one* 10(10): 0140355. https://doi.org/10.1371/journal.pone.0140355

Ogidi, C. O., A. M. Ubaru, T. Ladi-Lawal, O. A. Thonda, O. M. Aladejana, and Malomo, O. 2020. Bioactivity assessment of exopolysaccharides produced by *Pleurotus pulmonarius* in submerged culture with different agro-waste residues. *Heliyon* 6(12): e05685. https://doi.org/10.1016/j.heliyon.2020.e05685

Palacios, I., M. Lozano, C. Moro, et al. 2011. Antioxidant properties of phenolic compounds occurring in edible mushrooms. *Food Chemistry* 128(3): 674–678. https://doi.org/10.1016/j.foodchem.2011.03.085

Pant, C., S. K. Karn, and R. K. Singh. 2020. Herbs as immunomodulators and increase the immunity against viral infection. *International Journal of Science and Innovative Research* 8: 1–5. https://doi.org/10.3390/molecules28248045

Patel, Y., R. Naraian, and V. K. Singh. 2012. Medicinal properties of *Pleurotus* Species (Oyster Mushroom): a review. *World Journal of Fungal and Plant Biology* 3(1): 1–12.

Puri, M., I. Kaur, M. A. Perugini, and R. C. Gupta. 2012. Ribosome-inactivating proteins: current status and biomedical applications. *Drug Discovery Today* 17: 774–783. https://doi.org/10.1016/j.drudis.2012.03.007

Raman, J., K. Y. Jang, Y. L. Oh, et al. 2020. Cultivation and nutritional value of prominent *Pleurotus* spp.: an overview. *Mycobiology*. 49(1): 1–14. https://doi.org/10.1080/12298093.2020.1835142

Raman, J., H. Lakshmanan, K. Y. Jang, M. Oh, Y. L. Oh, and J. H. Im. 2020. Nutritional composition and antioxidant activity of pink oyster mushrooms (*Pleurotus djamor* var. *roseus*) grown on a paddy straw substrate. *Journal of Mushroom* 18(3): 189–200. https://doi.org/10.14480/JM.2020.18.3.189

Rangsinth, P., R. Sharika, N. Pattarachotanant, et al. 2023. Potential beneficial effects and pharmacological properties of ergosterol, a common bioactive compound in edible mushrooms. *Foods* 12(13): 2529. https://doi.org/10.3390/foods12132529

Reddy, S. M. 2016. Diversity and applications of mushrooms. In: *Plant Biology and Biotechnology*, ed. B. Bahadur, M. Venkat Rajam, I. Sahijram, K. V. Krishnamurthy, 231–261. Springer, Heidelberg. https://doi.org/10.1007/978-81-322-2286-6_9

Ren, D., Y. Jiao, X. Yang, et al. 2015. Antioxidant and antitumor effects of polysaccharides from the fungus *Pleurotus abalonus*. *Chemico-Biological Interactions* 237: 166–174.

Rizkyana, A. D., T. C. Ho, V. C. Roy, et al. 2022. Sulfation and characterization of polysaccharides from Oyster mushroom (*Pleurotus ostreatus*) extracted using subcritical water. *The Journal of Supercritical Fluids* 179: 105412. https://doi.org/10.1016/j.supflu.2021.105412

Royse, D. J., J. Baars, and Q. Tan. 2017. Current overview of mushroom production in the world. In: *Edible and Medicinal Mushrooms: Technology and Applications*, ed. C. Z. Diego, A. Pardo-Giménez, Wiley-Blackwell, USA. 5(13): 5–13. https://doi.org/10.1002/9781119149446.ch2

Sakdasri, W., P. Arnutpongchai, S. Phonsavat, E. Bumrungthaichaichan, and R. Sawangkeaw. 2022. Pressurized hot water extraction of crude polysaccharides, β-glucan, and phenolic compounds from dried gray oyster mushroom. *LWT* 168: 113895. https://doi.org/10.1016/j.lwt.2022.113895

Sánchez, C. 2010. Cultivation of *Pleurotus Ostreatus* and other edible mushrooms. *Applied Microbiology and Biotechnology* 85(5): 1321–1337. https://doi.org/10.1007/s00253-009-2343-7

Selvamani, S., H. A. El-Enshasy, D. J. Dailin, et al. 2018. Antioxidant compounds of the edible mushroom *Pleurotus ostreatus*. *International Journal of Biotechnology for Wellness Industries* 7: 1–14. https://doi.org/10.6000/1927-3037.2018.07.01

Sharma, A., A. Sharma, and A. Tripathi. 2021. Biological activities of *Pleurotus* spp. polysaccharides: a review. *Journal of Food Biochemistry* 45(6): 13748. https://doi.org/10.1111/jfbc.13748

Singh, S. S., H. Wang, Y. S. Chan, et al. 2014. Lectins from edible mushrooms. *Molecules* 20(1): 446–469. https://doi.org/10.3390/molecules20010446

Song, F. Q., Y. Liu, X. S. Kong, W. Chang, and G. Song. 2013. Progress on understanding the anticancer mechanisms of medicinal mushroom: *Inonotus obliquus*. *Asian Pacific Journal of Cancer Prevention* 14(3): 1571–1578. https://doi.org/10.7314/APJCP.2013.14.3.1571

Sreedharan, P. L., M. Kishorkumar, E. G. Neumann, and S. S. Kurup. 2025. The emerging role of oyster mushrooms as a functional food for complementary cancer therapy. *Foods* 14: 128. https://doi.org/10.3390/foods14010128

Suguimoto, H. H., A. M. Barbosa, R. F. Dekker, and R. J. Castro-Gomez. 2001. Veratryl alcohol stimulates fruiting body formation in the oyster mushroom, *Pleurotus ostreatus*. *EMS Microbiology Letters* 194(2): 235–238. https://doi.org/10.1111/j.1574-6968.2001.tb09475.x

Sun, X., Y. Shi, D. Shi, Y. Tu, and L. Liu. 2024. Biological activities of secondary metabolites from the edible-medicinal macrofungi. *Journal of Fungi* 10(2): 144. https://doi.org/10.3390/jof10020144

Sun, Y., F. K. Nzekoue, S. Vittori, G. Sagratini, and G. Caprioli. 2022. Conversion of ergosterol into vitamin D2 and other photoisomers in *Agaricus bisporus* mushrooms under UV-C irradiation. *Food Bioscience* 50: 102143. https://doi.org/10.1016/j.fbio.2022.102143

Tan, Y., N. K. Zeng, and B. Xu. 2022. Chemical profiles and health-promoting effects of porcini mushroom (*Boletus edulis*): a narrative review. *Food Chemistry* 390: 133199. https://doi.org/10.1016/j.foodchem.2022.133199

Tong, Z., G. Chu, C. Wan, et al. 2023. Multiple metabolites derived from mushrooms and their beneficial effect on Alzheimer's diseases. *Nutrients* 15(12): 2758. https://doi.org/10.3390/nu15122758

Trepa, M., K. Sułkowska-Ziaja, K. Kała, and B. Muszyńska. 2024. Therapeutic potential of fungal terpenes and terpenoids: application in skin diseases. *Molecules* 29(5): 1183. https://doi.org/10.3390/molecules29051183

Valverde, M. E., T. Hernandez-Perez, and O. Paredes-Lopez. 2015. Edible mushrooms: improving human health and promoting quality life. *International Journal of Microbiology* 2015:376387. https://doi.org/10.1155/2015/376387

Venditti, A., C. Frezza, F. Sciubba, M. Serafini, and A. Bianco. 2016. Primary and secondary metabolites of an European edible mushroom and its nutraceutical value: *Suillus bellinii* (Inzenga) Kuntze. *Natural Product Research* 31(16): 1910–1919. https://doi.org/10.1080/14786419.2016.1267731

Venturella, G., V. Ferraro, F. Cirlincione, and M. L. Gargano. 2021. Medicinal mushrooms: bioactive compounds, use, and clinical trials. *International Journal of Molecular Sciences* 22(2): 634. https://doi.org/10.3390/ijms22020634

Vetter, J. 2023. The mushroom glucans: molecules of high biological and medicinal importance. *Foods* 12(5): 1009. https://doi.org/10.3390/foods12051009

Villares, A., L. Mateo-Vivaracho, and E. Guillamon. 2012. Structural features and healthy properties of polysaccharides occurring in mushrooms. *Agriculture* 2: 452–471. https://doi.org/10.3390/agriculture2040452

Wang, C., Y. Liu, Y. Lan, and J. Yuan. 2022. Extraction of a triterpene solution and evaluation of the hypolipidemic efficacy of the *Pleurotus tuber-regium* (Fr.) *Sing Sclerotium*. *Foods* 11(18): 2881. https://doi.org/10.3390/foods11182881

Wang, Q., H. Li, T. T. Chen, and J. R. Han. 2012. Yield, polysaccharides content and antioxidant properties of *Pleurotus abalonus* and *Pleurotus geesteranus* produced on asparagus straw as substrate. *Scientia Horticulturae* 134: 222–226. https://doi.org/10.1016/j.scienta.2011.11.005

Wang, Q., F. Wang, Z. Xu, and Z. Ding. 2017. Bioactive mushroom polysaccharides: a review on monosaccharide composition, biosynthesis and regulation. *Molecules* 22(6): 955. https://doi.org/10.3390/molecules22060955

Wasser, S. P. 2011. Current findings, future trends and unsolved problems in studies of medicinal mushrooms. *Applied Microbiology and Biotechnology* 89: 1323–1332. https://doi.org/10.1007/s00253-010-3067-4

Witkowska, A. M., M. E. Zujko, and I. Mironczuk-Chodakowska. 2011. Comparative study of wild edible mushrooms as sources of antioxidants. *International Journal of Medicinal Mushrooms* 13(4): 335–341. https://doi.org/10.1615/IntJMedMushr.v13.i4.30

Wu, G. S., J. J. Guo, J. L. Bao, et al. 2013. Anticancer properties of triterpenoids isolated from *Ganoderma lucidum* - a review. *Expert Opinion on Investiational Drugs* 22(8): 981–992. https://doi.org/10.1517/13543784.2013.805202

Xiao, H., and J. J. Zhong. 2016. Production of useful terpenoids by higher-fungus cell factory and synthetic biology approaches. *Trends Biotechnology* 34(3): 242–255. https://doi.org/10.1016/j.tibtech.2015.12.007

Xiaokang, W., J. G. Lyng, N. P. Brunton, et al. 2020. Monitoring the effect of different microwave extraction parameters on the recovery of polyphenols from shiitake mushrooms: comparison with hot-water and organic-solvent extractions. *Biotechnology Reports* 27: e00504. https://doi.org/10.1016/j.btre.2020.e00504

Xu, N., Z. Ren, J. Zhang, et al. 2017. Antioxidant and antihyperlipidemic effects of mycelia zinc polysaccharides by *Pleurotus eryngii* var. *tuoliensis*. *International Journal of Biological Macromolecules* 95: 204–214.

Xu, X., H. Yan, J. Chen, and X. Zhang. 2011. Bioactive proteins from mushrooms. *Biotechnology Advances* 29: 667–674. https://doi.org/10.1016/j.biotechadv.2011.05.003

Yadav, D., and P. S. Negi. 2021. Bioactive components of mushrooms: processing effects and health benefits. *Food Research International* 148: 110599. https://doi.org/10.1016/j.foodres.2021.110599

Yılmaz, A., S. Yıldız, C. Kılıç, and Z. Can. 2017. Total phenolics, flavonoids, tannin contents and antioxidant properties of *Pleurotus ostreatus* cultivated on different wastes and sawdust. *International Journal of Secondary Metabolite* 4(1): 1–9. https://doi.org/10.21448/ijsm.252052

Yingyue, S., G. Min, J. Qunil, et al. 2014. Effects of cold stimulation on primordial initiation and yield of *Pleurotus pulmonarius*. *Scientia Horticulture* 167: 100–106. https://doi.org/10.1016/j.scienta.2013.12.021

Zadrazil, F. 1978. Cultivation of *Pleurotus*. In: *The Biology and Cultivation of Edible Mushrooms*, ed. S. T. Chang, and W. A. Hayes, 521–557. Academic Press, USA. https://doi.org/10.1016/B978-0-12-168050-3.50031-1

Zhang, C., S. Li, J. Zhang, et al. 2016. Antioxidant and hepatoprotective activities of intracellular polysaccharide from *Pleurotus eryngii* SI–04. *International Journal of Bioogical Macromolecules* 91: 568–577. https://doi.org/10.1016/j.ijbiomac.2016.05.104

Zhao, Z. Z., B. Y. Ji, Z. Z. Wang, et al. 2023. Lanostane triterpenoids with antiproliferative and anti-inflammatory activities from medicinal mushroom *Ganoderma lingzhi*. *Phytochemistry* 213: 113791. https://doi.org/10.1016/j.phytochem.2023.113791

Zieba, P., A. Sękara, K. Sułkowska-Ziaja, and B. Muszyńska. 2021. Culinary and medicinal mushrooms: insight into growing technologies. *Acta Mycologica* 55(2): 1–19. https://doi.org/10.5586/am.5526

# 6 Exploring the Antioxidant Activity of Oyster Mushrooms
## *An Updated Vision*

*Maura Téllez-Téllez**
Biological Research Center, Autonomous University of Morelos State, Cuernavaca, Morelos, Mexico

## 6.1 INTRODUCTION

Great importance has been given to oxidative stress because reactive species damage cellular macromolecules that can trigger changes at the cellular level, causing chronic inflammation, DNA damage, and alterations in cell signaling pathways, which can be the cause of various diseases. It has been associated with the incidence and/or progression of certain diseases such as diabetes, metabolic disorders, atherosclerosis, cardiovascular diseases, cancer, and so on (Krishnamurthy et al., 2024). Cells constantly exposed to oxidative stress can present genetic mutations, promoting angiogenesis and facilitating metastasis, which contributes to the development and promotion of cancer (Andreadi et al., 2022). Reactive species are oxidizing molecules; some contain oxygen or oxygen-like electronegative elements produced inside the cell during cellular metabolism or under pathological conditions. Endogenous free radicals are formed during cellular metabolism, the respiratory process, in β-oxidation of fatty acids, cytochrome P450-mediated reactions, immune defense, inflammation, excessive exercise, ischemia, infection, and so on. Exogenous free radicals are obtained due to exposure to solvents, radiation, metals, or habits such as smoking and alcohol consumption (Valko et al., 2007; Nandi et al., 2019). Currently, edible and medicinal mushrooms have been seen as an alternative to prevent cellular damage, and one of the most cultivated mushrooms worldwide is the *Pleurotus* genus, which consists of several species belonging to the Basidiomycota phylum and is distributed in tropical and subtropical regions.

There is a discrepancy in the number of species. Hilber (1982) and Singer (1986) report 30 to 40 species; Hawksworth et al. (1995) indicated about 50 valid species for the genus, and Kirk et al. (2008) suggest 20 species. The genus *Pleurotus* (Fr.) P. Kumm. is a cosmopolitan group that grows on living or dead deciduous plants (mainly trees) and rarely on conifers. Some species of the *Pleurotus* can remove lignin without significantly damaging cellulose (Cohen et al., 2002). *Pleurotus* are grown on agricultural residues using simple and low-cost production techniques (Jegadeesh et al., 2018). Worldwide, about 998 million tons of agricultural waste, including rice, wheat, and other cereal straws, are produced annually. Mushrooms of the genus *Pleurotus* can grow in almost all agro-industrial waste due to their high capacity to degrade lignocellulosic material, making it an alternative to recycle waste and, at the same time, obtain food of gastronomic importance with medicinal properties (Philippoussis, 2009; Kakon et al., 2012). Cultivating fungi produces approximately 50% carbon dioxide, 20% water, 10% fungi, and 20% residual compost (Stamets, 1993). Most species are suitable for cultivation and can be proposed to farmers as an alternative to unprofitable crops today (Venturella et al., 2015).

The genus is characterized by its high nutritional value and several biotechnological applications. Their medicinal effect has been indicated, since they present antitumor, antibiotic, hypocholesterolemic, and immunomodulatory activity, which is why they are considered nutraceuticals. These mushrooms present several molecules with biological activity that frequently act as antioxidants, including polysaccharides, tocopherols, carotenoids, ergosterol, lycopene, ergothioneine, ascorbic acid, minerals, and phenolic compounds (phenolic acids and flavonoids), and so on. Antioxidants neutralize free radicals by transferring electrons, breaking autooxidation reactions, chelating metals, inhibiting hydroperoxides, and protecting the cell membrane, structural proteins, membrane lipids, and nucleic acids (Chaudhary et al., 2023). Therefore, the genus *Pleurotus* is an essential alternative for obtaining natural antioxidants; based on the above, this chapter shows a bibliographical review of the importance and antioxidant capacity of different species of *Pleurotus*. Therefore, the genus *Pleurotus* is an essential alternative to obtaining natural antioxidants; based on the above, this chapter shows a bibliographical review of the importance and antioxidant capacity of different species of *Pleurotus*.

## 6.2 OXIDATIVE STRESS

Sies (1985) defined oxidative stress as an alteration between the pro-oxidant–antioxidant balance in favor of the former, a definition later updated as an imbalance between oxidant (reactive species) and antioxidant in favor of the oxidants, which leads to an interruption of signaling and redox control

* Corresponding author: maura.tellez@uaem.mx



DOI: 10.1201/9781003527572-6

and/or molecular damage (Sies and Jones, 2007). When the intensity is average within the cell, it is called eustress (oxidative stress at a physiological level); the loss of balance between oxidants and antioxidants is called distress (toxic oxidative load that damages molecules). Cells exposed to eustress during metabolism are subject to continuous exposure to oxidizing compounds (endogenous), and if the effect of exogenous oxidants (external factors) is added, cell damage increases. When intracellular defense mechanisms can control reactive species, there is no molecular damage; however, when they are present in excess (distress), they can directly or indirectly damage molecules (nucleic acids, oxidize proteins, and cause lipid peroxidation). Reactive species are molecules or molecular fragments from which one or more electrons are missing, which gives them a considerable degree of reactivity (Sies, 2018).

Some authors have indicated that it is crucial to differentiate between free radicals (molecules with one or more electrons that disappear in their outer shell) and reactive oxygen species, since they are used interchangeably. However, in some cases, this is incorrect; for example, $^{\bullet}O_2^-$, $H_2O_2$ and $^{\bullet}HO$ are all reactive oxygen species, but only $^{\bullet}O_2^-$ and $^{\bullet}HO$ are free radicals, whereas $H_2O_2$ is not. Weidinger and Kozlov (2015) reported that harmful effects of oxidants do not appear if only one primary reactive species (superoxide radical, nitric oxide) is present in a biological system, even at high concentrations. However, it is observed that before highly reactive secondary species (hydroxyl radical, peroxynitrite) with harmful effects are formed, these species result from the reaction of a primary species or a transition metal. The toxicity of secondary species is due to their unstable and uncontrolled molecules, which can cause irreversible damage to cellular biomolecules (Chaudhary et al., 2023).

In contrast, primary reactive oxygen or nitrogen species are enzymatically controlled (superoxide dismutase, catalase), and the reaction with biomolecules is reversible. Therefore, these reactive species are essential to physiology/pathophysiology (Droge, 2002). Reactive oxygen species (ROS), together with reactive nitrogen species (RNS) (Table 6.1), play a dual role as both harmful and beneficial species (Valko et al., 2006).

It has been reported that ROS produced at a given time during the cell cycle and in adequate physiological concentrations can significantly participate in the homeostasis of the organism, cellular functions (energy production, regulation of cell growth, synthesis of biological components, intracellular signaling), chemotaxis, during gene expression, in defense against infections, proliferation, apoptosis, and in cellular ageing (Droge, 2002; Halliwell, 2006; Reverberi et al., 2008). In mammals, it has been seen that appropriate levels of ROS stimulate antioxidant reactions and regulation of cell proliferation (Bokoch and Knaus, 2003), increase capillary permeability, which is necessary for a normal inflammatory process, and direct cells toward the apoptotic pathway, which allows the elimination of cells that have been seriously damaged and could represent a risk for the organism (cancer cells). By maintaining homeostasis between these processes, the functioning of the organism and the modulation of the immune system are ensured; however, it has been shown that ROS and RNS that are produced in excess by the immune system can damage the host cells (Schwab et al., 2014; Jakubczyk et al., 2020).

Overproduction of ROS, decreased antioxidant defenses, and alteration of enzymatic pathways in humans with diabetes mellitus can contribute to endothelial, vascular, and neurovascular dysfunction due to the reactivity of these compounds. Such reactivity can cause deleterious effects, producing cellular damage by damaging proteins (aggregation, denaturation), lipids (lipoperoxidation), carbohydrates, and nucleotides (changes in the DNA structure that generate mutations or cytotoxic effects), which can result in cell death and aging. Therefore, oxidative stress has an essential role in the incidence of certain diseases: degenerative (Parkinson's, Alzheimer's), cancer, cardiovascular, liver, diabetes mellitus, kidney failure, deterioration of the immune system, circulatory system (atherosclerosis), in the aging process, among others (Halliwell et al., 1992; Halliwell, 1994; Jakubczyk et al., 2020).

## 6.3 CELLULAR DEFENSE SYSTEM

The cell has antioxidant defense strategies; it produces endogenous antioxidants that delay or prevent oxidation, and antioxidant compounds effectively donate their electrons to ROS and neutralize possible cell damage (Birben et al., 2012). When ROS accumulate, endogenous antioxidant defense is insufficient; the cell performs oxidative modification in the cell membrane or intracellular molecules to balance the antioxidant defense mechanisms of ROS, which is favored by the consumption of exogenous antioxidants. The defense system (Table 6.2) works by blocking the production of free radicals, eliminating oxidants, converting toxic free radicals into less harmful substances, blocking the production of toxic metabolites or inflammatory mediators, blocking the chain propagation of secondary oxidants, repairing damaged molecules, or potentiating the endogenous antioxidant defense system (Aguilar et al., 2016; Adwas et al., 2019).

When free radicals cause damage to molecules, antioxidants come into play that reconstitute, recognize, and eliminate oxidized or damaged proteins, DNA, and lipids to prevent their accumulation. Among these antioxidants are DNA repair enzymes (polymerases, glycosylases, and nucleases), proteolytic enzymes (proteinases, proteases, and peptidases), and finally, the free radicals formed generate signals that induce the formation and transport of antioxidants (Ighodaro and Akinloye, 2018). Exogenous antioxidants help control free radicals by blocking propagation reactions, by exchanging one or more protons with a free radical or peroxide radical, inhibiting the formation of free radicals by forming complexes with the metal (chelating agent), and decreasing the concentration of reactive oxygen or oxygen species (oxygen scavenging agent) (Bazinet and

**TABLE 6.1**
**Reactive Oxygen and Nitrogen Species**

| Reactive Species | | Characteristics |
|---|---|---|
| Oxygen | Singlet oxygen ($O_2^{\cdot}$) | In its molecular state, it has no restriction on electron transfer, making it highly reactive. It is produced by phagocytosis, light induction, autoxidation, and enzymatic reactions in different cellular organelles. It damages DNA and proteins through the oxidation of certain essential groups of amino acids (tryptophan, methionine, histidine, and cysteine). It also initiates lipoperoxidation, generating radicals such as alkoxy ($RO^{\cdot}$) and peroxyl ($ROO^{\cdot}$) radicals. |
| | Superoxide anion radical ($O_2^{-\cdot}$) | Intermediate in the reduction of $O_2$ to $H_2O_2$. Good reducer and poor oxidant. Generates ROS (OH• and $H_2O_2$). It is generated in the mitochondrial respiratory chain, and endoplasmic reticulum; damages lipids, proteins, nucleic acids. |
| | Hydroxyl radical (OH•) | It is highly reactive and has a short half-life. The damage it causes is because cells do not have an antioxidant enzyme system against this radical. It is formed by several mechanisms: when the superoxide radical reacts with hydrogen peroxide, the hydroxyl radical is produced through the Haber-Weiss reaction and in the electron transport chain. It damages DNA, proteins, lipids and carbohydrates. |
| | Peroxyl radical (ROO•) | It is formed by radical reactions with cellular components such as lipids and nucleobases, when the other oxygen carries a hydrogen (hydroperoxide, R–O–O–H). |
| | Alkoxyl radical (RO•) | In lipid nature structures they participate in lipid peroxidation reactions, which occur in the presence of oxygen and by the addition of radicals to double bonds or hydrogen abstraction. |
| Nitrogen | Nitric oxide = nitrogen monoxide (NO•) | It is an apolar molecule, which allows it to concentrate and diffuse through biological membranes. It often functions as an intercellular messenger, which can rapidly diffuse from one cell type to another. It is synthesized in endothelial cells by the NADPH-dependent oxidation of the guanidinium nitrogen of the amino acid L-arginine. |
| | Peroxynitrite (ONOO-) | Reaction between nitric oxide (NO·) and superoxide anion ($O_2$-), NADPH oxidase and xanthine oxidase. It acts on amino acids (cysteine, methionine, tryptophan, among others), metal ions, proteins and cofactors (tetrahydrobiopterin, cofactor of nitric oxide synthases and tyrosine hydroxylase), among others. |

*Source:* Jomova et al., 2023.

Doyen, 2017). Antioxidants act by preventing the initiation of reactive species reactions, blocking the propagation of free radical reactions, and repairing the modifications (damage) caused by radicals in the body (Toyokuni, 2016). Therefore, supplementation with exogenous antioxidants can be a preventive, therapeutic alternative to reduce disease risk. Mushrooms are a natural resource with high added value (antioxidant) due to their nutrient composition and the presence of molecules that are essential sources of nutrients for human consumption (Tan et al., 2024).

## 6.4 IMPORTANCE OF *PLEUROTUS* GENUS AS FUNCTIONAL FOOD

The species that is mainly used for cultivation for food purposes throughout the world is *Pleurotus ostreatus* (oyster mushroom), but other species are also cultivated, such as *P. pulmonarius* (phoenix oyster), *P. sajor-caju* (grey abalone oyster), *P. eryngii* (king oyster), *P. citrinopileatus* (golden oyster), *P. djamor* (pink oyster), *P. nebrodensis* (white ferula mushroom), *P. tuber-regium* (king tuber oyster), and some species are rarely cultivated, for example, *P. calyptratus*, *P. cystidiosus* (abalone mushroom), and *P. dryinus*, where the last two species exhibit an anamorphic stage (asexual sporulation) in the life cycle that is not typical for most members of the genus *Pleurotus* (Shnyreva et al., 2017). Mushrooms of the *Pleurotus* genus are among the most widely cultivated worldwide, accounting for approximately 19% of global production. Other major genera include *Lentinula* (22%), *Auricularia* (18%), *Agaricus* (15%), and *Flammulina* (11%). In 2013, the global mushroom market was valued at approximately USD 63 billion. This market comprised cultivated edible mushrooms (54%), medicinal mushrooms (38%), and wild-harvested mushrooms (8%) (Royse et al., 2017). Similarly, Hamza et al. (2024) indicate that global mushroom production increased to 43 million metric tons between 2018 and 2019, representing one of the fastest-growing branches of horticulture. The global mushroom market was valued at $45.3 US dollars in 2020 and is expected to grow at an annual growth rate of 7% between 2022 and 2027 (Niego et al., 2021).

Food production is one of the activities that damage the environment. In the face of climate change, it is essential to change certain habits, using food alternatives that support environmental sustainability and meet human needs to improve human health (Willett et al., 2019; van Dam et al., 2024). Among the proteins considered in the world diet, animal and vegetable sources are included, but mushroom protein is not mentioned (Feeney et al., 2014; Herforth et al., 2019). On average, the protein content of *Pleurotus* is between 20% and 25%, with a good distribution of essential and non-essential amino acids (Shukla and Jaitly, 2011). Mushroom proteins have high thermal and pH stability (Erjavec et al., 2012), and the digestibility of the protein of the *Pleurotus* species is comparable to that of meat. Lavelli et al. (2018) reported that the digestibility of mushroom

**TABLE 6.2**
**The Function of the Main Endogenous Molecules of the Antioxidant Defense System**

| Molecules Type | Function | References |
|---|---|---|
| **Enzymatic** | | |
| Superoxide dismutase | Constitutes the first and one of the main links of the defense process against free radicals. Inactivates the superoxide ion by transforming it into hydrogen peroxide. Regulates oxidative stress, lipid metabolism, inflammation, and oxidation in cells. Can prevent lipid peroxidation, the oxidation of low-density lipoprotein in macrophages, lipid droplets' formation, and the adhesion of inflammatory cells into endothelial monolayers | Islam et al., 2021 |
| Catalase | Breakdown of hydrogen peroxide into oxygen and water. Important role in the acquisition of tolerance to oxidative stress in the adaptive response of cells. Important enzyme implicated in mutagenesis and inflammation conditions as well as during the suppression of apoptosis | Nandi et al., 2019 |
| Glutathione peroxidase | Protects cells and tissues from oxidation toxicity and oxidative damage. Is one of the most essential antioxidative defense mechanisms. Protects lipids from peroxidation at intracellular distance. Protection against peroxidation of the membrane in cases where a membrane-bound antioxidant is insufficient | Maiorino et al., 2011; Kabel, 2014 |
| Glutathione reductase | Catalyzes the reduction of oxidized glutathione to reduced glutathione. Deficiency of this enzyme is characterized by hemolysis due to increased sensitivity of erythrocyte membranes to $H_2O_2$ | |
| Glutathione-S-transferases | Represent a major group of detoxifying enzymes. Involved in the cellular removal of cytotoxic and genotoxic compounds and in the protection of tissues against oxidative damage | |
| **Non-enzymatic** | | |
| Glutathione | Has antioxidant properties since the thiol group in its cysteine is a reducing agent and can be reversibly oxidized and reduced. Interrelated with a metabolic pathway and its intracellular modulation could have an impact on the entire cell. Biological antioxidant that acts as a quencher of oxidative molecule, thereby playing a significant role in the inhibition of melanogenesis | Sastre et al., 2005 |
| Coenzyme Q | Participation as electron carrier in the mitochondrial respiratory chain. Endogenously synthesized, lipid-soluble antioxidant. Regulation of mitochondrial permeability transition pores. Regulation of the physicochemical properties of membranes. Improvement of endothelial dysfunction (probably by increasing nitrogen oxide) | Turunen et al., 2004 |
| Ferritin | Ferritin has been characterized as a cytosolic protein, but it is also present in other cell compartments, such as the nucleus, mitochondria, and lysosomes, and a small part of ferritin is in the serum. Iron sequestration from invading pathogens, and oxidative stress protection. Mitochondrial ferritin can protect cells against ROS and thus against ferroptosis. Nuclear ferritin can protect DNA by sequestering iron and preventing uncontrolled Fenton chemistry, or by directly binding to DNA | Plays et al., 2021 |
| Uric acid | Is the predominant antioxidant molecule in plasma and necessary for induction of immune responses. Uric acid is a strong ROS and peroxynitrite scavenger and antioxidant. Role in healing via initiating the inflammatory process that is necessary for tissue repair, scavenging oxygen free radicals, and mobilizing progenitor endothelial cells | Plays et al., 2021 |
| Bilirubin | Affecting various cell regulatory systems, binding to cell membranes and as well as to nuclear receptors. Protect lipids from oxidation. Major contributor to the total antioxidant capacity in blood plasma. Prevents endothelial cell injury at multiple levels by both inhibiting the release of superoxide from neutrophils, and by directly scavenging superoxide | Adin, 2021 |

proteins ranges between 60% and 70%. Mushroom proteins are inexpensive and high-quality and can be a viable alternative to counteract malnutrition (Yadav and Negi, 2021). Edible mushrooms have been reported to possess bioactive peptides, which are inactive within the protein but can be released during enzymatic proteolysis (gastrointestinal digestion) by food processing (cooking, fermentation, ripening), causing them to have functional properties (González et al., 2021). Several studies have shown that mushroom protein concentrates, hydrolysates, and peptides have angiotensin-converting enzyme inhibitory, antimicrobial, anticancer, antiviral, gut microbiota modulation, and antioxidant properties (Yadav and Negi, 2021). Liao et al. (2020) reported that a protein hydrolysate from *P. geesteranus* presented a higher content of hydrophobic, negatively charged, and aromatic amino acids, which contributed to its antioxidant properties. It inhibited 2,2-diphenyl-1-picrylhydrazyl (DPPH) radical by 91.62%, 2,2'-azino-bis(3-ethylbenzothiazoline-6-sulfonic acid (ABTS) by 90.53%, and 82.16% of ferric reducing antioxidant power and metal ion chelating activity, and also showed cytoprotective effects on $H_2O_2$-induced oxidative stress in PC12 cells by decreasing intracellular ROS accumulation and stimulating antioxidant enzymatic activities. Therefore, the hydrolysate can be used as an ingredient with antioxidant properties and protective effects against oxidative stress.

Mushrooms have a desirable nutritional composition due to low-fat levels, low caloric value, and a high proportion of polyunsaturated fatty acids, which makes them suitable for consumption by cardiac patients, as well as for diabetic patients due to the absence of starch, low glycemic index, and high amount of mannitol (Reis et al., 2012; Sharma et al., 2024). Similarly, mushrooms represent a source of nutrients; therefore, they are considered functional foods, as they contain bioactive compounds that promote health and are used as prophylactics for numerous human diseases. This has been related in some cases to their effect on the intestinal microbiota, since they can act as direct prebiotics, indirect probiotics, or symbiotics. Probiotics have clinical or health care functions in the prevention or control of intestinal, respiratory, and urogenital infections, allergic reactions, inflammatory bowel disease, irritable bowel syndrome, and other aspects (You et al., 2022). Sharma et al. (2024) indicate that polyphenolic compounds (catechin, gallic acid, vanillic acid, and protocatechuic acid) can selectively promote the growth of probiotics and inhibit the growth of pathogenic bacteria. Even mushrooms can accumulate trace elements and heavy metals such as Zn, Cu, Mn, Fe, Se, Cd, Pb, and so on, which are cofactors of enzymes and have antioxidant activity (Rašeta et al., 2022). Various selenium compounds such as seleno-methionine, selenocysteine, Se-methyl selenocysteine, selenite, and seleno-polysaccharides in mushrooms reduce the cytotoxic effect of ROS (Milovanović et al., 2015). In addition, they present potassium and phosphorus that help patients suffering from musculoskeletal problems, and they have low sodium concentrations, which is essential in hypertension (Chang and Wasser, 2012).

Furthermore, the population with a vegan diet has increased, so they often lack vitamin B12, which is responsible for maintaining the myelin surrounding nerve cells, the mental capacity, formation of red blood cells, and breakdown of fats and amino acids to produce energy. Some mushrooms contain bioavailable and stable vitamin B12 and D2 (Fulgoni and Agarwal, 2021). The low plasma level of vitamin D2 is associated with the development of many metabolic diseases such as diabetes, hypertension, inflammatory and autoimmune diseases, cancer, immunity disorders, and infectious diseases (Nowak et al., 2022). Bengu et al. (2019) quantified 66.86 µg/g of vitamin A, 654.16 µg/g of vitamin E, and 334.36 µg/g of vitamin C in *P. ostreatus*. This reinforces the consideration of mushrooms as a rich source of bioactive nutrients, including vitamins and minerals. They are also one of the best dietary sources of sulfur-containing amino acids with antioxidant activity, such as ergothioneine, glutathione, and ergosterol (Kalaras et al., 2017; Halliwell et al., 2018). Work has been extensively worked with fruit bodies, but van Dam et al. (2024) indicate that several secondary metabolites with health benefits were detected in the fungal mycelium, so it is also an alternative for developing natural and biotherapeutic functional dietary supplements, as they can modulate the immune system and act as anti-inflammatory agents through their high antioxidant content (Phillips et al., 2022; Matsugo et al., 2022). It has been reported that mushrooms have anti-inflammatory effects indirectly by inhibiting free radicals and preventing oxidative damage, since, by presenting antioxidant activity, they protect cells against oxidative damage (Bell et al., 2022).

## 6.5 ANTIOXIDANT ACTIVITY OF *PLEUROTUS* SPP.

Effiong et al. (2024) reported that the suggested mechanisms of action of antioxidants include inhibition of the expression or activity of free radical-generating enzymes, enhancement of the activity or expression of antioxidant enzymes, or direct reaction with free radicals to terminate their chain reactions. In this regard, the molecule of *Pleurotus* spp. has been attributed to the ability to scavenge and/or reduce free radicals through hydrogen donation, chelation of transition metal ion catalysts, inhibition of lipid peroxidation, and so on (Jayakumar et al., 2011; Figure 6.1). Heleno et al. (2015) classified phenolic compounds into four groups: phenolic acids, phenolic diterpenes, flavonoids and volatile oils. It has been stated that antioxidant properties in mushrooms were mainly in the form of phenolic acids and flavonoids (Arbaayah and Umi, 2013; Kalaras et al., 2017). Table 6.3 shows phenolic compounds found in some mushrooms of the genus *Pleurotus*. The presence of phenolic compounds reduces protein oxidation through a radical scavenging and metal chelation mechanism (Estévez et al., 2008). Most phenolic compounds are soluble in organic solvents and slightly soluble in water (Walter and Purcell, 1979). The solubility of polyphenols is high in methanol, followed by ethanol and the lowest in water (Lapornik et al., 2005).

Among the main phenolic compounds of mushrooms, including the genus *Pleurotus*, are phenolic acids that are classified as: (1) hydroxybenzoic acid, which is derived from benzoic acid and consists of complex structures linked to lignin; hydrolysable tannins; and some of the plant sugars or organic acids, such as 4-hydroxybenzoic acid, 2,5-dihydroxybenzoic acid, protocatechuic acid, gallic acid, veratric acid, syringic acid, and vanillic acid; and (2) hydrocinnamic acids, which are derived from cinnamic acid and are compounds that bind to the cell wall structure, such as cellulose, lignin, and proteins—some also bind to organic acids such as tartaric and quinic acids (example: ρ-coumaric, caffeic, ferulic, and *t*-cinnamic acids) (El-Enshasy et al., 2018). It has been mentioned that the accumulation of secondary metabolites such as flavonoids and phenolic compounds is a response of fungi and plants to biotic and abiotic stress and a protection strategy against ultraviolet light, insects, and viral and bacterial infections. Dong et al. (2012) reported that when the fruit bodies were exposed to nitrogen dioxide, the antioxidant activity, phenol, and flavonoid content

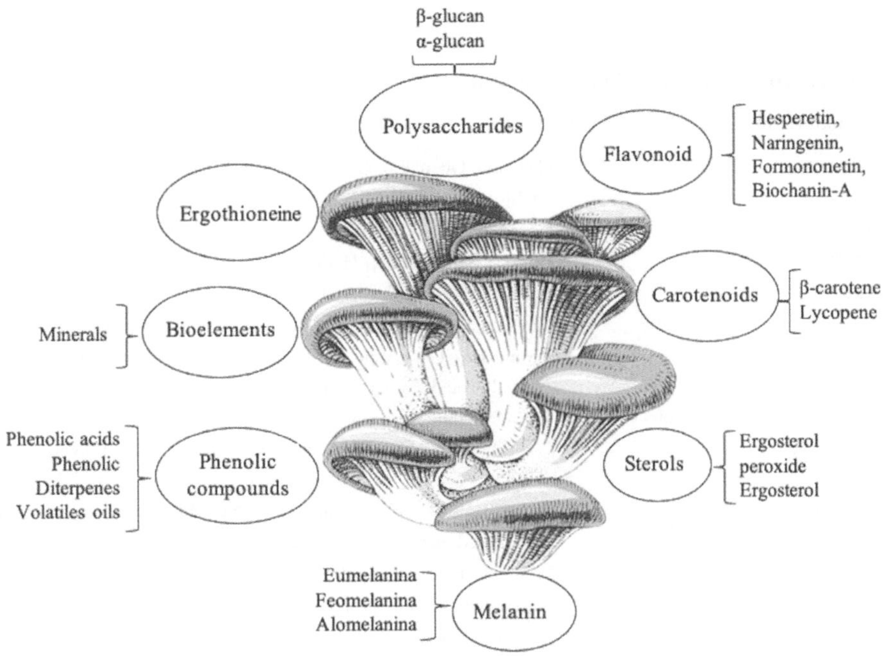


**FIGURE 6.1** Some compounds with antioxidant activity of the genus *Pleurotus*.

were significantly increased, and the activity of phenylalanine ammonia-lyase and chalcone synthase was stimulated. The composition and biological properties of mushrooms vary within and between species, as does the maturity of the fruiting body (Roupas et al., 2012).

Ergosterol (5,7,22-ergostrythren-3β-ol) is a naturally synthesized compound necessary to maintain the structure and function of the cell membranes of fungi. It is known as a precursor of vitamin D—ultraviolet light with wavelengths in the range of 280 to 320 nm causes photoirradiation of ergosterol, resulting in the production of different products (Nowak et al., 2022). Ergosterol peroxide (5α,8α-epidioxy-22E-ergosta-6,22-dien-3β-ol) is produced by converting ergosterol into its epidioxid and is considered a product of the enzymatic oxidation of ergosterol, $H_2O_2$ dependent (Krzyczkowski et al., 2009), It has immunosuppressive, anti-inflammatory, antitumor, and antioxidant properties (Suwannarach et al., 2020). Ergosterol-containing *P. ostreatus* extracts showed lipid peroxidation inhibitory activity, so the freeze-dried mushroom powder itself may be helpful as a dietary supplement for its antioxidant activity (Dissanayake et al., 2009). In addition, exposure to UV-B radiation (2.00 W/cm$^2$ for 60 minutes) induces the conversion of ergosterol to vitamin $D_2$. In the case of *P. florida*, this treatment also enhances the levels of total phenolic compounds and flavonoids. These findings support the potential development of fortified mushroom-based products and dietary supplements (Kaur et al., 2023).

Ergothioneine is a betaine thio-histidine amino acid that has antioxidant activity. It is synthesized by a variety of organisms, especially fungi and actinobacteria. The consumption of mushrooms seems to be a great alternative to prevent oxidative stress in a wide variety of systems. Ergothioneine is transported in the blood through a specific transporter (OTCN1), and ergothioneine levels seem to decrease during neurodegenerative and cardiovascular disorders (Borodina et al., 2020). Stastny et al. (2022) indicated that antioxidant activity might be related to ergothioneine; it has been proposed that this compound may be a nutritional biomarker for mushroom consumption, although mushrooms usually contain different amounts that vary depending on physiological or environmental conditions (Wang et al., 2018). It has been reported that mushrooms have the biosynthetic machinery necessary for ergothioneine biosynthesis (Stastny et al., 2022). Many studies confirmed the presence of ergothioneine in edible mushrooms (approximately 5.8 to 1812.4 mg/kg dry weight), including *P. eryngii*, which reported 541.7 mg/kg dry weight (Liang et al., 2013). In *P. citrinopileatus* and *P. ostreatus*, 3.94 and 1.21 mg/g of dry weight were found, respectively (Kalaras et al., 2017). Dubost et al. (2006) found ergothioneine in *P. ostreatus* and *P. eryngii* (2010 and 1720 mg/kg dry weight, respectively). It is essential to mention that the consumption of ergothioneine by mammals at levels much higher than those found in food has not presented any adverse effects, so it has been declared safe by relevant committees such as the European Food Standards Agency, as well as being found to lack detectable mutagenicity or genotoxicity in these tests (Borodina et al., 2020).

Fungi produce several pigments, including carotenoids, melanin, polyketides, and polyketide/azaphyllone derivatives. Polyketides are compounds with two or more carbonyl functional groups connected to a single carbon atom. They are of variable colors and are represented by

**TABLE 6.3**
**Phenolic Compounds Identified in Some *Pleurotus* spp.**

| Species | Mushroom Tissue | Extraction Solvent | Compounds | References |
|---|---|---|---|---|
| *Pleurotus citrinopileatus* | Fruiting body | 80% acetone | Catechin, Quercetin, Gallic acid, Vanillic acid, Ferulic acid, Chlorogenic acid | Yin et al., 2019 |
| | Fruiting body | Aqueous | Catechin, Quercetin, Gallic acid, Ferulic acid, Chlorogenic acid | |
| | Fruiting body | Methanol | Protocatechuic acid, ρ-Hydroxybenzoic acid, Vanillic acid | Krakowska et al., 2020 |
| | Mycelium | Methanol | Protocatechuic acid, ρ-Hydroxybenzoic acid, Vanillic acid | |
| *Pleurotus djamor* | Fruiting body | Methanol | Tannic acid, Gallic acid, Protocatechuic acid, Gentisic acid | Puttaraju et al., 2006 |
| | Fruiting body | Water | Tannic acid, Gallic acid, Protocatechuic acid, Gentisic acid, Vanillic acid, Cinnamic acid | |
| | Fruiting body | Methanol | Protocatechuic acid, Gallic acid, Gentisic acid, Vanillic acid, Syringic acid, Cinnamic acid, ρ-Coumaric acid, Ferulic acid, Tannic acid | Ferreira et al., 2009 |
| | Fruiting body | Methanol | Protocatechuic acid, ρ-Hydroxybenzoic acid | Krakowska et al., 2020 |
| | Mycelium | Methanol | ρ-Hydroxybenzoic acid | |
| *Pleurotus eryngii* | Fruiting body | Methanol | Protocatechuic acid, Gallic acid | Ferreira et al., 2009 |
| | Fruiting body | Methanol | Protocatechuic acid, ρ-Hydroxybenzoic acid, ρ-Coumaric acid, Cinnamic acid | Reis et al., 2012 |
| | Fruiting body<br>Fruiting body with added Zn | Methanol 80% | 2,5-Dihydroxybenzoic acid, 4-Hydroxybenzoic acid, Ferulic acid, ρ-Coumaric acid, Protocatechuic acid, t-Cinnamic acid, Vanillic acid, Naringenin | Gąsecka et al., 2016 |
| | Mycelium | Methanol | Gallic acid, ρ-Hydroxybenzoic acid, Cinnamic acid | Reis et al., 2012 |
| | Fruiting body/Mycelium | Methanol | Protocatechuic acid, ρ-Hydroxybenzoic acid. | Krakowska et al., 2020 |
| *Pleurotus ostreatus* | Fruiting body | Acetonitrile and hydrochloric acid | Gallic acid, homogentisic acid, protocatechuic acid, chlorogenic acid, naringin, myricetin | Kim et al., 2008 |
| | Fruiting body | Methanol | Protocatechuic acid, Gallic acid, Homogentisic acid, Rutin, Myricetin, Chrysin, Naringin | Ferreira et al., 2009 |
| | Fruiting body | Ethanol | Rutin, Chrysin | Jayakumar et al., 2009 |
| | Fruiting body | Methanol | ρ-Coumaric acid, Ferulic acid, Gallic acid, Gentisic acid, ρ-Hydroxybenzoic acid, Homogentisic acid, Myricetin, Protocatechuic acid | Palacios et al., 2011 |
| | Fruiting body | Methanol | Protocatechuic acid, ρ-Hydroxybenzoic acid, ρ-Coumaric acid, Cinnamic acid | Reis et al., 2012 |
| | Fruiting body | Methanol | Caffeic acid, Gallic acid, ρ-Hydroxybenzoic acid, Myricetin | Woldegiorgis et al., 2014 |
| | Fruiting body/with added Zn | Methanol 80% | 4-Hydroxybenzoic acid, Ferulic acid, ρ-Coumaric acid, Protocatechuic acid, t-Cinnamic acid, Vanillic acid, Naringenin | Gąsecka et al., 2016 |
| | Fruiting body | Ethanol 50% | Gallic acid, Protocatechuic acid, Chrysin | Vargas-Sánchez et al., 2018 |
| | Fruiting body | Methanol | Protocatechuic acid | Krakowska et al., 2020 |
| | Mycelium | | ρ-Hydroxybenzoic acid, Cinnamic acid | Reis et al., 2012 |
| | Mycelium | Methanol | Protocatechuic acid | Krakowska et al., 2020 |
| *Pleurotus pulmonarius* | Fruiting body | Methanol 80% | Gallic acid, Homogentisic acid, procatechuic acid, (+)-Catechin, Chlorogenic acid, Vanillin, Naringin, Myricetin, Resveratrol, Quercetin | Nguyen et al., 2016 |
| | Fruiting body | Water | Gallic acid, Caffeic acid, Vanillic acid, ρ-Coumaric acid | Contato et al., 2020 |
| | Mycelium | Water | Gallic acid, Caffeic acid, Ferulic acid | |
| | Mycelium | Methanol | Protocatechuic acid, ρ-Hydroxybenzoic acid | Krakowska et al., 2020 |
| *Pleurotus sajor-caju* | Fruiting body | Methanol | Protocatechuic acid, Gallic acid, Gentisic acid, Cinnamic acid, Ferulic acid, Tannic acid | Ferreira et al., 2009 |
| | Fruiting body | Methanol | Tannic acid, Gallic acid, Protocatechuic acid, Gentisic acid, Vanillic acid | Puttaraju et al., 2006 |
| | Fruiting body | Water | Tannic acid, Gallic acid, Protocatechuic acid, Gentisic acid, Vanillic acid, Ferulic acid, cinnamic acid | |

anthraquinone, hydroxyanthraquinone, naphthoquinone, and azaphyllone structures (Mapari et al., 2010). Melanin is a group of high-weight molecules formed by complex heterogeneous polymers of phenolic and/or indole monomers. It is considered a dermo-cosmetic component due to its high UV absorption and antioxidant activity. All three types of melanin (eumelanin, pheomelanin, and allomelanin) are present in *Pleurotus*, usually existing as a mixture (Toledo et al., 2017; Smith and Casadevall, 2019). Colored *Pleurotus* species have immense potential for use as natural dyes, in addition to nutritive and antioxidant-rich compounds (Medihi et al., 2024).

Flavonoids are polyphenols of diverse structures that can be found as aglycones or glycosides. The antioxidant action of flavonoids depends on the arrangement of functional groups in the nuclear structure. Flavonoids have been recognized to have antioxidant properties in preventing lipid peroxidation, scavenging free radicals, and chelating ferrous ions (Kumar and Pandey, 2013). Arbaayah and Umi (2013) reported high flavonoid content of ethanol extracts of various *Pleurotus* species. Jeena et al. (2016) indicated the antioxidant effect of flavonoids by finding a correlation between the methanolic extract of the mature fruiting body of *P. ostreatus* and antioxidant activity. González-Palma et al. (2016) did not find flavonoids at any growth stage, including the mycelial and reproductive phases of *P. ostreatus*; these studies analyzed the metabolite by spectrophotometric assays. It is essential to mention that there is a controversy about the ability of the fungal cell to produce flavonoids (Gil-Ramírez et al., 2016). Alam et al. (2011a) performed HPLC-DAD analysis on *P. nebrodensis* fruiting bodies, reporting flavonoids (hesperetin, naringenin, formononetin, and biochanin-A) but did not provide information on the culture medium, so it could be due to the absorption of flavonoids or the stimulation of the production of these compounds caused by interactions with other organisms. Therefore, to clarify this controversy, it has been recommended that detailed information on the composition of the culture media and the culture conditions should be included in all flavonoid studies to confirm the ability of mushrooms to synthesize these compounds (Pukalski and Latowski, 2022).

Carotenoids are organic compounds with a tetraterpene structure and have 40 carbon atoms; their antioxidant activity consists mainly of the formation of adducts with free radicals derived from linolenic acid and the reduction of $Fe^{3+}$ to $Fe^{2+}$ (Im et al., 2014). The conjugated double bonds are responsible for the colors of these molecules (yellow, orange, red, and purple); they absorb between 440 and 490 nm (Rao and Rao, 2007). The high number of conjugated double bonds makes these molecules more reactive and able to function as antioxidants that contribute to protection against cardiovascular diseases, as they inhibit low-density lipoprotein (LDL)-cholesterol oxidation, prevent atherosclerosis development, reduce photosensitivity, stimulate the immune system response, and scavenge peroxyl radicals, avoiding lipid oxidation and subsequent degradation (Ribeiro et al., 2011). The main compound responsible for the biological activity of carotenoids is β-carotene, which is a precursor of vitamin A when consumed by humans, but there are also lycopene, lutein, zeaxanthin, cryptoxanthin, and phytoene, which are not degraded during cooking or baking (Thane and Reddy, 1997). Sharma and Gautam (2015) reported the content of carotenoids, including β-carotene and lycopene, in 20 species of wild edible mushrooms, indicating that *P. cystidiosus* presented 0.79 μg/100 g. It has been reported that there are different amounts of carotenoids in the fruiting body of *P. ostreatus*, fresh mushrooms (100 g) presented total polyphenols and total flavonoids with values of 151–300 and 35–98 mg of catechin equivalents, respectively, as well as 2.1–5.5 mg of vitamin C, 0.548–1.004 mg of β-carotene, 0.336–0.450 mg of lycopene, and 40.46–55.16 mg of tocopherols, with high antioxidant activity by ABTS, DPPH, and ferric reducing/antioxidant power (FRAP) methods. In blanched mushrooms, the content of all the components mentioned above was reduced, including antioxidant activity (Jaworska et al., 2015). Mishra et al. (2014) reported that methanolic extracts of *P. sajor-caju* pileus have higher antioxidant activity than the stipe and present higher lycopene content than β-carotene.

Another molecule of great importance in mushrooms is glucans, which have been attributed a role in antioxidant activity. However, it has been reported that polysaccharides of the glucan type have a low capacity to eliminate free radicals *in vitro* but can stimulate free radical activity in murine macrophages, increasing the level of ROS associated with immunomodulation (Tsiapali et al., 2001). Bobek et al. (2001) showed that *P. ostreatus* glucan (pleuran) had an effect on antioxidant activity and reduced precancerous lesions in the colon using rats as a model. Recently, the antioxidant activity of β-glucan from *P. eryngii* was analyzed, where the antioxidant effect was concentration-dependent; to obtain the $IC_{50}$, 0.52 mg/mL for the inhibition of the DPPH radical, 1.15 mg/mL for the superoxide radical, 0.89 mg/mL for the hydroxyl radical, and 2.83 mg/mL for nitric oxide were required (Petraglia et al., 2023). The free radical scavenging capacity may be related to hydrogen and hydroxyl groups in the polysaccharide chain, acting with a mechanism similar to that of phenolic compounds. In this sense, a polysaccharide was extracted from *P. ostreatus* that contained mainly β-glucan and α-glucan, presenting antioxidant activity (DPPH) and, in a cellular model, its protective function against oxidative damage induced by $H_2O_2$ was corroborated (Barbosa et al., 2020). Liu et al. (2019) obtained a polysaccharide from *P. citrinopileatus*; it was a typical pyranose with a β-type glycosidic bond and showed antioxidant, anti-inflammatory, and antifibrotic effects against $CCl_4$-induced acute liver damage, which could be a potential source of natural hepatic antioxidants. Another essential molecule in fungi is trehalose, a non-reducing disaccharide of glucose, which plays an essential physiological role not only as a reserve carbohydrate but also as a thermoprotector, protecting proteins

**TABLE 6.4**
**Antioxidant Activity in Some *Pleurotus* spp.**

| Species | Mushroom Tissue | Extraction Solvent | Antioxidant Activity and Method Used | References |
|---|---|---|---|---|
| *P. citrinopileatus* | Fruiting body | Methanol | 58.65~90.46%, DPPH | Alam et al., 2011b |
| | Fruiting body | Acetone | 52.56~92.65%, DPPH | |
| | Fruiting body | Hot water | 43.47~95.34%, DPPH | |
| | Fruiting body | Ethanol 70% | 78.14%, chelating activity | Khatun et al., 2015 |
| | Fruiting body | Methanol | 10.13 μmol/g, FRAP; 50%, DPPH | Yıldız et al., 2017 |
| | Fruiting body | 80% acetone | 39.84%. DPPH; 1.27 mg/mL, reducing power | Yin et al., 2019 |
| | Fruiting body | Aqueous | 85.95%. DPPH; 0.6 mg/mL, reducing power | |
| | Fruiting body | Aqueous | 78.14%, ABTS; 81.07%, DPPH | Acosta-Urdapilleta et al., 2020 |
| | Fruiting body | Hot water | 96.3%, ABTS; 61.24%, DPPH | |
| | Fruiting body | Methanol/acetone | 74.93% ABTS; 98.36% DPPH | |
| | Fruiting body | Methanol | 91.55%, DPPH | Krakowska et al., 2020 |
| | Fruiting body | Methanol | 2.42 mg trolox/g, DPPH; 5 mg quercetin/g, ABTS | Diamantis et al., 2024 |
| | Fruiting body | Polysaccharides | 97.22%, ABTS; 89.88%, DPPH; 4.24 mmol/L, FRAP | Xiao et al., 2024 |
| | Mycelium | Methanol | 72.44%, DPPH | Krakowska et al., 2020 |
| | Mycelium | Hot water | 64.12%, DPPH | Özdal et al., 2019 |
| *P. cystidiosus* | Fruiting body | Ethanol 95% | 50.00%, DPPH and FRAP | Babu et al., 2014 |
| | Mycelium | Ethanol 80% | 73.77%, DPPH | Garcia et al., 2020 |
| | Mycelium | Ethanol 80% | 79.01%, DPPH | |
| | Mycelium | Ethanol 80% | 58.95%, DPPH | |
| | Mycelium | Ethanol 80% | 89.63%, DPPH | |
| *P. eryngii* | Fruiting body | Methanol | 50.00%, DPPH | Reis et al., 2012 |
| | Fruiting body | Methanol | 50.00%, DPPH | |
| | Fruiting body | Ethanol | 50.00%, DPPH and chelating activity | Lin et al., 2014 |
| | Fruiting body | Ethanol | 50.00%, DPPH and chelating activity | |
| | Fruiting body | Ethanol | 50.00%, DPPH and chelating activity | |
| | Fruiting body | Methanol | 91.84%, DPPH | Krakowska et al., 2020 |
| | Fruiting body with added Zn | Not indicated | 50.00%, DPPH | Gąsecka et al., 2016 |
| | Fruiting body | Not indicated | 50.00%, DPPH | |
| | Fruiting body | Hot water | 48.85%, DPPH | Özdal et al., 2019 |
| | Fruiting body | Methanol | 69.01%, DPPH | Krakowska et al., 2020 |
| | Fruiting body | Aqueous | 77.95%, ABTS; 49.4%, DPPH | Acosta-Urdapilleta et al., 2020 |
| | Fruiting body | Hot water | 60.34%, ABTS; 24.59%, DPPH | |
| | Fruiting body | Methanol/acetone | 68.57% ABTS; 87.91% DPPH | |
| | Mycelium | Ethanol | 68.01%, DPPH | Dundar et al., 2013 |
| *P. djamor* | Fruiting body | Ethanol 95% | 50.00%, DPPH and FRAP | Babu et al., 2014 |
| | Fruiting body | Methanol | 87.15%, DPPH | Krakowska et al., 2020 |
| | Fruiting body | Methanol | 40%, DPPH; 65%, ABTS | Cruz-Moreno et al., 2023 |
| | Fruiting body | Methanol | 65%, DPPH; 75%, ABTS | |
| | Mycelium | Methanol | 85.95%, DPPH | Krakowska et al., 2020 |
| | Mycelium | Not indicated | 20.80%, DPPH; 28.5% ABTS | Morales-Flores et al., 2022 |
| *P. djamor* var. *roseus* | Fruiting body | Petroleum ether | 72.21%, DPPH | Jegadeesh et al., 2014 |
| | Fruiting body | Hexane | 83.83%, DPPH | |
| | Fruiting body | Chloroform | 81.89%, DPPH | |
| | Fruiting body | Ethyl acetate | 78.28%, DPPH | |
| | Fruiting body | Methanol | 86.89%, DPPH | |
| | Fruiting body | Aqueous | 74.52%, ABTS; 57.14%, DPPH | Acosta-Urdapilleta et al., 2020 |
| | Fruiting body | Hot water | 53.50%, ABTS; 33.17%, DPPH | |
| | Fruiting body | Methanol/acetone | 66.75% ABTS; 79.73% DPPH | |
| *Pleurotus ferulaginis* | Mycelium | Not indicated | 62.4%, DPPH; 48.3% ABTS | Morales-Flores et al., 2022 |
| *P. flabellatus* | Fruiting body | Ethanol 65% | 50.00%, DPPH and ABTS | Pumtes et al., 2016 |

**TABLE 6.4** ***(Continued)***
**Antioxidant Activity in Some *Pleurotus* spp.**

| Species | Mushroom Tissue | Extraction Solvent | Antioxidant Activity and Method Used | References |
|---|---|---|---|---|
| | Fruiting body | Ethanol 80% | 50.00%, DPPH and ABTS | |
| | Fruiting body | Ethanol 95% | 50.00%, DPPH and ABTS | |
| | Fruiting body | Methanol/ chloroform | 24.9 mg of TE/g extract | Stastny et al., 2022 |
| | Mycelium | Not indicated | 78.60%, DPPH and 83.70%, ABTS | Morales-Flores et al., 2022 |
| *P. levis* | Fruiting body | Ethanol 32% | 50.00%, DPPH and ABTS | Adebayo et al., 2018 |
| *P. ostreatus* | Fruiting body | Not indicated | 50.00%, DPPH | Gąsecka et al., 2016 |
| | Fruiting body with added Zn | | 50.00%, DPPH | |
| | Fruiting body | Ethanol 32% | 50.00%, DPPH and ABTS | Adebayo et al., 2018 |
| | Fruiting body | Methanol | 61.98%, DPPH | Krakowska et al., 2020 |
| | Fruiting body | Water | 50.00%, DPPH and Chelating activity | Effiong et al., 2024 |
| | Fruiting body | Methanol | 50.00%, DPPH and Chelating activity | |
| | Fruiting body | Etthanol | 50.00%, DPPH and Chelating activity | |
| | Fruiting body | Chloroform | 50.00%, DPPH and Chelating activity | |
| | Fruiting body | n-Hexane | 50.00%, DPPH and Chelating activity | |
| | Fruiting body | Aqueous | 72.70%, ABTS; 54.0%, DPPH | Acosta-Urdapilleta et al., 2020 |
| | Fruiting body | Hot water | 82.42%, ABTS; 38.66%, DPPH | |
| | Fruiting body | Methanol/acetone | 74.48% ABTS; 94.77% DPPH | |
| | Fruiting body | Methanol/ chloroform | 9.0 mg of TE/g extract | Stastny et al., 2022 |
| | Fruiting body | Methanol/ chloroform | 4.3 mg of TE/g extract | |
| | Mycelium | Methanol | 50.00%, DPPH | Reis et al., 2012 |
| | Mycelium | Methanol | 50.00%, DPPH | |
| | Mycelium | Ethanol | 71.29%, DPPH | Dundar et al., 2013 |
| | Mycelium | Methanol | 60.23%, DPPH | Krakowska et al., 2020 |
| | Mycelium | Not indicated | 43.20%, DPPH; 43.80 ABTS | Morales-Flores et al., 2022 |
| | Not indicated | Not indicated | 59.50%, DPPH; 16.00%, ABTS | Vargas-Sánchez et al., 2018 |
| *P. pulmonarius* | Not indicated | Not indicated | 25.67%, DPPH | Nguyen et al., 2016 |
| | Fruiting body | Ethanol 32% | 50.00%, DPPH and ABTS | Adebayo et al., 2018 |
| | Fruiting body | Ethanol 32% | 50.00%, DPPH and ABTS | |
| | Fruiting body | Water | 50.00%, DPPH, ABTS and FRAP | Contato et al., 2020 |
| | Fruiting body | Methanol | 78.79%, DPPH | Krakowska et al., 2020 |
| | Fruiting body | Aqueous | 79.35%, ABTS; 48.17%, DPPH | Acosta-Urdapilleta et al., 2020 |
| | Fruiting body | Hot water | 70.53%, ABTS; 35.88%, DPPH | |
| | Fruiting body | Methanol/acetone | 63.76% ABTS; 96.99% DPPH | |
| | Fruiting body | Methanol/ chloroform | 8.5 mg of TE/g extract | Stastny et al., 2022 |
| | Mycelium | Methanol/ chloroform | 50.00%, DPPH, ABTS and FRAP | Contato et al., 2020 |
| | Mycelium | Not indicated | 49.68%, DPPH | Krakowska et al., 2020 |
| *P. sajor-caju* | Not indicated | Ethanol 95% | 50.00%, DPPH, ABTS and FRAP | Babu et al., 2014 |
| | Mycelium | Ethanol | 62.82%, DPPH | Dundar et al., 2013 |
| | Mycelium | Hot water | 69.67%, DPPH | Özdal et al., 2019 |
| *P. tuber-regium* | Fruiting body | Etthanol 32% | 50.00%, DPPH and ABTS | Adebayo et al., 2018 |

*ABTS:* 2,2-azino-bis (ácido 3-etilbenzotia zolina-6-sulfónico); DPPH: 2,2-difenil-1-picrilhidrazilo; FRAP: the ferric reducing *antioxidant* power.

and cell membranes from inactivation or denaturation under different stress conditions, and can alleviate high-temperature-induced damage in *P. ostreatus* (Yan et al., 2021). Table 6.4 lists studies where different methods have determined the antioxidant activity of some fungi of the *Pleurotus* genus.

The antioxidant effect of mushrooms of the genus *Pleurotus* has been analyzed in aquatic vertebrate models since, currently, thermal stress and the disturbance of the pH of the marine environment can cause an overproduction of free radicals, oxidation of lipids, proteins, and DNA, as well as the alteration of antioxidant enzymes and the depletion of endogenous antioxidants in fish, which causes oxidative liver damage, decreasing their growth and performance (Qiang et al., 2017). The state of defense has been improved by supplementing with a natural source of antioxidants, an alternative convenient to protect fish from stress (Ahmed et al., 2017a). Ahmed et al. (2017b) suggested that severe deterioration of the aquatic habitat may compromise the homeostasis of antioxidant metabolism in fish, so they used *P. pulmonarius* culture waste as a source of antioxidants (glucans) to supplement feed for *Oreochromis niloticus* (Nile tilapia). The inclusion of 5 g/kg of the extract in the diet was shown to be an antioxidant agent; it presented total phenols (77.84 mg GAE/g), and the $IC_{50}$ values for the DPPH radical, lipid peroxidation, and metal chelation were 0.890, 0.814, and 4.14 mg/mL, respectively. Therefore, using fungal antioxidants is a beneficial alternative for tilapia cultivation, in addition to mitigating pH-induced stress, which improved the growth rate under stress conditions (Ahmed et al., 2017a). Uluköy et al. (2016) showed that supplementation of fish diets with *P. ostreatus* extract at concentrations of 1% and 2% significantly improved hematological parameters and modulated the immune response against *Lactococcus garvieae* in *Oncorhynchus mykiss*. On the other hand, Sattanathan et al. (2018) indicated that the survival rate of *Gibelion catla* was higher in the group fed with *P. sajor-caju* extract (0.5% and 0.75%), growth and immune response were also improved, and the mortality rate against *Aeromonas hydrophila* was decreased. Mycelium, fruiting bodies, polysaccharides, and fungal extracts have been used in aquaculture as growth promoters and immunostimulants, improving the activity of digestive enzymes and antimicrobials, and generally improving health status, making them an alternative in aquaculture, which is one of the leading food production sectors in the world (Mohan et al., 2022).

Consuming foods rich in antioxidants plays a protective role in human health due to the reduction of oxidative damage that results in increased generation of free radicals (Gąsecka et al., 2016). Typically, mushroom biomass and bioactive molecules are obtained through solid culture; however, the production of fruiting bodies requires a long time, even months. Therefore, research is being carried out in liquid culture to obtain biomolecules from mycelium in less time (a few weeks). In addition, the extraction and purification of these compounds would be facilitated, and depending on the cultivation conditions, their yield could be increased. Since species of the *Pleurotus* genus are artificially cultivated quite easily (Mukhopadhyay and Guha, 2015), mycelium is an alternative for obtaining mushroom-derived products (mushroom extracts and tablets) (Özdal et al., 2019). Mushroom mycelium is known to be an attractive alternative source of protein due to its rapid growth in sustainable substrates and promising nutritional and organoleptic properties (van Dam et al., 2024). Dundar et al. (2013) indicated that mycelium extracts present antioxidant activity through the capacity of hydrogen donation; they are efficient superoxide scavengers and exhibit chelating activity since the compounds present interfered in the formation of the ferrous and ferrozine complex, capturing the ferrous ion before the ferrozine.

## 6.6 CONCLUSION AND FUTURE PROSPECTS

Mushrooms of the *Pleurotus* genus are among the most produced and consumed in the world due to their appreciated flavor and excellent nutritional quality. Since they possess biological activities, these mushrooms could be considered nutraceutical foods. It has been observed that the antioxidant activity of these mushrooms is due to various molecules, the concentration of which varies depending on the extraction method, the species or strain of mushrooms, the growth conditions (culture), the cellular age of the sporocarp or mycelium, conservation, and so on. In addition, most of the studies have been carried out *in vitro*, so there is still much research to be done to determine the conditions under which antioxidant activity increases and to validate its effect in *in vivo* systems. However, with the evidence currently available, the consumption of these mushrooms should be encouraged, either as food or as a dietary supplement.

## REFERENCES

Acosta-Urdapilleta, M. L., E. Villegas, A. Estrada-Torres, M. Téllez-Téllez, and G. Díaz-Godínez. 2020. Antioxidant activity and proximal chemical composition of fruiting bodies of mushroom, *Pleurotus* spp. produced on wheat straw. *Journal of Environmental Biology* 41(5): 1075–1081. http://doi.org/10.22438/jeb/41/5/MRN-1307

Adebayo, E. A., D. Martínez-Carrera, P. Morales, et al. 2018. Comparative study of antioxidant and antibacterial properties of the edible mushrooms *Pleurotus levis*, *P. ostreatus*, *P. pulmonarius* and *P. tuber-regium*. *International Journal of Food Science and Technology* 53(5): 1316–1330. https://doi.org/10.1111/ijfs.13712

Adin, C. A. 2021. Bilirubin as a therapeutic molecule: challenges and opportunities. *Antioxidants* 10(10): 1536. https://doi.org/10.3390/antiox10101536

Adwas, A. A., A. Elsayed, A. E. Azab, and F. A. Quwaydir. 2019. Oxidative stress and antioxidant mechanisms in human body. *Journal of Applied Biotechnology and Bioengineering* 6(1): 43–47. https://doi.org/10.15406/jabb.2019.06.00173

Aguilar, T. A. F., B. C. H. Navarro, and J. A. M. Pérez. 2016. Endogenous antioxidants: a review of their role in oxidative stress. In: *A Master Regulator of Oxidative Stress, the Transcription Factor Nrf2*, ed. J. A. Morales-Gonzalez, A. Morales-Gonzalez, and E. O. Madrigal-Santillan, 3–20. IntechOpen Press. https://doi.org/10.57772/657151rd

Ahmed, M., N. Abdullah, H. M. Yusof, A. S. Shuib, and S. A. Razak. 2017a. Influence of raw polysaccharide extract from mushroom stalk waste on growth and pH perturbation induced-stress in Nile tilapia, *Oreochromis Niloticus*. *Aquaculture* 468: 60–70. https://doi.org/10.1016/j.aquaculture.2016.09.043

Ahmed, M., N. Abdullah, H. M. Yusof, A. S. Shuib, and S. A. Razak. 2017b. Improvement of growth and antioxidant status in Nile tilapia, *Oreochromis niloticus*, fed diets supplemented with mushroom stalk waste hot water extract. *Aquaculture Research* 48(3): 1146–1157. https://doi.org/10.1111/are.12956

Alam, N., K. N. Yoon, and T. S. Lee. 2011a. Evaluation of the antioxidant and antityrosinase activities of three extracts from *Pleurotus nebrodensis* fruiting bodies. *African Journal of Biotechnology* 10: 2978–2986. https://doi.org/10.5897/AJB10.2660

Alam, N., K. N. Yoon, K. R. Lee, et al. 2011b. Assessment of antioxidant and phenolic compound concentrations as well as xanthine oxidase and tyrosinase inhibitory properties of different extracts of *Pleurotus citrinopileatus* fruiting bodies. *Mycobiology* 39(1): 12–19. https://doi.org/10.4489/MYCO.2011.39.1.012

Andreadi, A., A. Bellia, N. Di Daniele, et al. 2022. The molecular link between oxidative stress, insulin resistance, and type 2 diabetes: a target for new therapies against cardiovascular diseases. *Current Opinion in Pharmacology* 62: 85–96. https://doi.org/10.1016/j.coph.2021.11.010

Arbaayah, H. H., and K. Y. Umi. 2013. Antioxidant properties in the oyster mushrooms (*Pleurotus* spp.) and split gill mushroom (*Schizophyllum commune*) ethanolic extracts. *Mycosphere* 4(4): 661–673. https://doi.org/10.5943/mycosphere/4/4/2

Babu, D. R., M. Pandey, and G. N. Rao. 2014. Antioxidant and electrochemical properties of cultivated *Pleurotus* spp. and their sporeless/low sporing mutants. *Journal of Food Science and Technology* 51(11): 3317–3324. https://doi.org/10.1007/s13197-012-0822-9

Barbosa, J. R., M. M. S. Freitas, L. C. Oliveira, et al. 2020. Obtaining extracts rich in antioxidant polysaccharides from the edible mushroom *Pleurotus ostreatus* using binary system with hot water and supercritical CO2. *Food Chemistry* 330: 127173. https://doi.org/10.1016/j.foodchem.2020.127173

Bazinet, L., and A. Doyen. 2017. Antioxidants, mechanisms, and recovery by membrane processes. *Critical Reviews in Food Science and Nutrition* 57(4): 677–700. https://doi.org/10.1080/10408398.2014.912609

Bell, V., C. R. P. G. Silva, J. Guina, and T. H. Fernandes. 2022. Mushrooms as future generation healthy foods. *Frontiers in Nutrition* 9: 1050099. https://doi.org/10.3389/fnut.2022.1050099

Bengu, A. S., C. H. Yilmaz, I. Turkekul, and H. Işik. 2019. The levels of minerals and vitamins in some wild and cultured mushroom species from Turkey. *Journal of Pharmacy and Pharmacology* 7: 207–211. https://doi.org/10.17265/2328-2150/2019.04.007

Birben, E., U. M. Sahiner, C. Sackesen, S. Erzurum, and O. Kalayci. 2012. Oxidative stress and antioxidant defense. *World Allergy Organization Journal* 5(1): 9–19. https://doi.org/10.1097/WOX.0b013e3182439613

Bobek, P., V. Nosalova, and S. Cerna. 2001. Effect of pleuran (beta-glucan *Pleurotus ostreatus*) in diet or drinking fluid on colitis rats. *Food Nahrung* 45(5): 360. https://doi.org/10.3390/nu13041333

Bokoch, G. M., and U. G. Knaus. 2003. NADPH oxidases: not just for leukocytes anymore. *Trends in Biochemical Sciences* 28(9): 502–508. https://doi.org/10.1016/S0968-0004(03)00194-4

Borodina, I., L. C. Kenny, C. M. McCarthy, et al. 2020. The biology of ergothioneine, an antioxidant nutraceutical. *Nutrition Research Reviews* 33(2): 190–217. https://doi.org/10.1017/S0954422419000301

Chang, S. T., and S. P. Wasser. 2012. The role of culinary-medicinal mushrooms on human welfare with a pyramid model for human health. *International Journal of Medicinal Mushrooms* 14(2): 95–134. https://doi.org/10.1615/IntJMedMushr.v14.i2.10

Chaudhary, P., P. Janmeda, A. O. Docea, et al. 2023. Oxidative stress, free radicals and antioxidants: potential crosstalk in the pathophysiology of human diseases. *Frontiers in Chemistry* 11: 1158198. https://doi.org/10.3389/fchem.2023.1158198

Cohen, R., L. Persky, and Y. Hadar. 2002. Biotechnological applications and potential of wood-degrading mushrooms of the genus *Pleurotus*. *Applied Microbiology and Biotechnology* 58: 582–594. http://dx.doi.org/10.1007/s00253-002-0930-y

Contato, A. G., F. D. Inácio, C. A. V. de Araújo, et al. 2020. Comparison between the aqueous extracts of mycelium and basidioma of the edible mushroom *Pleurotus pulmonarius*: chemical composition and antioxidant analysis. *Journal of Food Measurement and Characterization* 14: 830–837. https://doi.org/10.1007/s11694-019-00331-0

Cruz-Moreno, B. A., A. A. F. Pérez, J. F. García-Trejo, S. A. Pérez-García, and C. Gutiérrez-Antonio. 2023. Identification of secondary metabolites of interest in *Pleurotus djamor* using *Agave tequilana* bagasse. *Molecules* 28(2): 557. https://doi.org/10.3390/molecules28020557

Diamantis, I., M. Dedousi, E. M. Melanouri, et al. 2024. Impact of spent mushroom substrate combined with hydroponic leafy vegetable roots on *Pleurotus citrinopileatus* productivity and fruit bodies biological properties. *Microorganisms* 12(9): 1807. https://doi.org/10.3390/microorganisms12091807

Dissanayake, D. P., D. T. U. Abeytunga, N. S. Vasudewa, and W. D. Ratnasooriya. 2009. Inhibition of lipid peroxidation by extracts of *Pleurotus ostreatus*. *Pharmacognosy Magazine* 5(19): 266–271. https://phcog.com/article/sites/default/files/PhcogMag-5-19-266_0.pdf

Dong, J., M. Zhang, L. Lu, et al. 2012. Nitric oxide fumigation stimulates flavonoid and phenolic accumulation and enhances antioxidant activity of mushroom. *Food Chemistry* 135(3):1220–1225. https://doi.org/10.1016/j.foodchem.2012.05.055

Droge, W. 2002. Free radicals in the physiological control of cell function. *Physiological Reviews* 82(1): 47–95. https://doi.org/10.1152/physrev.00018.2001

Dubost, N. J., R. B. Beelman, D. Peterson, and D. J. Royse. 2006. Identification and quantification of ergothioneine in cultivated mushrooms by liquid chromatography-mass spectroscopy. *International Journal of Medicinal Mushrooms* 8: 215–222. https://doi.org/10.1615/IntJMedMushr.v8.i3.30

Dundar, A., V. Okumus, S. Ozdemir, and A. Yildiz. 2013. Antioxidant properties of cultured mycelia from four *Pleurotus* species produced in submerged medium. *International Journal of Food Properties* 16(5): 1105–1116. https://doi.org/10.1080/10942912.2011.576793

Effiong, M. E., C. P. Umeokwochi, I. S. Afolabi, and S. N. Chinedu. 2024. Comparative antioxidant activity and phytochemical content of five extracts of *Pleurotus ostreatus* (oyster mushroom). *Scientific Reports* 14(1): 3794. https://doi.org/10.1038/s41598-024-54201-x

El-Enshasy, H. A., S. Selvamani, D. J. Dailin, et al. 2018. Antioxidant compounds of the edible mushroom *Pleurotus ostreatus*. *International Journal of Biotechnology for Wellness Industries* 7: 1–14. https://doi.org/10.6000/1927-3037.2018.07.01

Erjavec, J., J. Kos, M. Ravnikar, T. Dreo, and J. Sabotič. 2012. Proteins of higher fungi-from forest to application. *Trends in Biotechnology* 30(5): 259–273. https://doi.org/10.1016/j.tibtech.2012.01.004

Estévez, M., P. Kylli, E. Puolanne, R. Kivikari, and M. Heinonen. 2008. Oxidation of skeletal muscle myofibrillar proteins in oil-in-water emulsions: interaction with lipids and effect of selected phenolic compounds. *Journal of Agricultural and Food Chemistry* 56(22): 10933–10940. https://doi.org/10.1021/jf801784h

Feeney, M. J., J. Dwyer, C. M. Hasler-Lewis, et al. 2014. Mushrooms and health summit proceedings. *Journal of Nutrition* 144: 1128S–1136S. https://doi.org/10.3945/jn.114.19072

Ferreira, I. C., L. Barros, and R. Abreu. 2009. Antioxidants in wild mushrooms. *Current Medicinal Chemistry* 16(12): 1543–1560. https://doi.org/10.2174/092986709787909587

Fulgoni, V. L., and S. Agarwal. 2021. Nutritional impact of adding a serving of mushrooms on usual intakes and nutrient adequacy using National Health and Nutrition Examination Survey 2011–2016 data. *Food Science and Nutrition* 9: 1504–1511. https://doi.org/10.1002/fsn3.2120

Garcia, K., C. J. Garcia, R. Bustillos, and R. M. Dulay. 2020. Mycelial biomass, antioxidant, and mycoactives of mycelia of abalone mushroom *Pleurotus cystidiosus* in liquid culture. *Journal of Applied Biology and Biotechnology* 8(2): 94–97. https://doi.org/0.7324/JABB.2020.80215

Gąsecka, M., M. Mleczek, M. Siwulski, and P. Niedzielski. 2016. Phenolic composition and antioxidant properties of *Pleurotus ostreatus* and *Pleurotus eryngii* enriched with selenium and zinc. *European Food Research and Technology* 242: 723–732. https://doi.org/10.1007/s00217-015-2580-1

Gil-Ramírez, A., C. Pavo-Caballero, E. Baeza, et al. 2016. Mushrooms do not contain flavonoids. *Journal of Functional Foods* 25: 1–13. https://doi.org/10.1016/j.jff.2016.05.005

González, A., C. Nobre, L. S. Simões, et al. 2021. Evaluation of functional and nutritional potential of a protein concentrate from *Pleurotus ostreatus* mushroom. *Food Chemistry* 346(1): 128884. https://doi.org/10.1016/j.foodchem.2020.128884

González-Palma, I., H. B. Escalona-Buendía, E. Ponce-Alquicira, et al. 2016. Evaluation of the antioxidant activity of aqueous and methanol extracts of *Pleurotus ostreatus* in different growth stages. *Frontiers in Microbiology* 7: 1099. https://doi.org/10.3389/fmicb.2016.01099

Halliwell, B. 1994. Free radicals, antioxidants, and human disease: curiosity, cause, or consequence? *Lancet* 344(8924): 721–724. https://doi.org/10.1016/s0140-6736(94)92211-x

Halliwell, B. 2006. Reactive species and antioxidants redox biology is a fundamental theme of aerobic life. *Plant Physiology* 141: 312–322. https://doi.org/10.1104/pp.106.077073

Halliwell, B., I. K. Cheah, and R. M. Y. Tang. 2018. Ergothioneine a diet-derived antioxidant with therapeutic potential. *FEBS Letters* 592: 3357–3366. https://doi.org/10.1002/1873-3468.13123

Halliwell, B., J. M. C. Gutteridge, and C. E. Cross. 1992. Free radicals, antioxidants, and human disease: where are we now? *Journal of Laboratory and Clinical Medicine* 119(6): 598–620.

Hamza, A., A. Mylarapu, K. V. Krishna, and D. S. Kumar. 2024. An insight into the nutritional and medicinal value of edible mushrooms: a natural treasury for human health. *Journal of Biotechnology* 381(10): 86–99. https://doi.org/10.1016/j.jbiotec.2023.12.014

Hawksworth, D. L. 1995. *Biodiversity: Measurement and Estimation* (Vol. 345). Springer Science and Business Media. 140 pp.

Heleno, S. A., A. Martins, M. J. Queiroz, C. F. R. I. Ferreira. 2015. Bioactivity of phenolic acids: metabolites versus parent compounds: a review. *Food Chemistry* 73: 501–513. https://doi.org/10.1016/j.foodchem.2014.10.057

Herforth, A., M. Arimond, C. Alvarez-Sanchez, J. Coates, K. Christianson, and E. Muehlhoff. 2019. A global review of food-based dietary guidelines. *Advances in Nutrition* 10: 590–605. https://doi.org/10.1093/advances/nmy130

Hilber, O. 1982. Die Gattung *Pleurotus* (Fr.) Kummer unter besonderer Beruecksichtigung des Pleurotus eryngii-Formen komplexes. In: *Bibliotheca Mycologica*, ed. J. Cramer Vaduz, 87. Cornell University.

Ighodaro, O. M., and O. A. Akinloye. 2018. First line defense antioxidants-superoxide dismutase (SOD), catalase (CAT) and glutathione peroxidase (GPX): their fundamental role in the entire antioxidant defense grid. *Alexandria Journal of Medicine* 54(4): 287–293. https://doi.org/10.1016/j.ajme.2017.09.001

Im, K. H., T. K. Nguyen, D. B. Shin, K. R. Lee, and T. S. Lee. 2014. Appraisal of antioxidant and anti-inflammatory activities of various extracts from the fruiting bodies of *Pleurotus florida*. *Molecules* 19: 3310–3326. https://doi.org/10.3390/molecules19033310.

Islam, M. N., A. Rauf, F. I. Fahad, et al. 2021. Superoxide dismutase: an updated review on its health benefits and industrial applications. *Critical Reviews in Food Science and Nutrition* 62(26): 7282–7300. https://doi.org/10.1080/10408398.2021.1913400

Jakubczyk, K., K. Dec, J. Kałduńska, D. Kawczuga, J. Kochman, and K. Janda. 2020. Reactive oxygen species-sources, functions, oxidative damage. *Polski Merkuriusz Lekarski* 48(284): 124–127.

Jaworska, G., K. Pogoń, E. Bernaś, and A. Duda-Chodak. 2015. Nutraceuticals and antioxidant activity of prepared for consumption commercial mushrooms *Agaricus bisporus* and *Pleurotus ostreatus*. *Journal of Food Quality* 38(2): 111–122. https://doi.org/10.1111/jfq.12132

Jayakumar, T., P. A. Thomas, and P. Geraldine. 2009. *In-vitro* antioxidant activities of an ethanolic extract of the oyster mushroom, *Pleurotus ostreatus*. *Innovative Food Science & Emerging Technologies* 10(2): 228–234. https://doi.org/10.1016/j.ifset.2008.07.002

Jayakumar, T., P. A. Thomas, J. R. Sheu, and P. Geraldine. 2011. *In-vitro* and *in-vivo* antioxidant effects of the oyster mushroom *Pleurotus ostreatus*. *Food Research International* 44(4): 851–861. https://doi.org/10.1016/j.foodres.2011.03.015

Jeena, G. S., H. Punatha, O. Prakash, M. Chandra, and K. P. S. Kushwaha. 2016. Study on in vitro antioxidant potential of some cultivated *Pleurotus* species (Oyster mushroom). *Indian Journal of Natural Products and Resources* 5(1): 56–61. https://doi.org/10.56042/ijnpr.v5i1.12457

Jegadeesh, R., L. Hariprasath, K. Kumaresan, and N. Raaman. 2014. *In vitro* antioxidant and antibacterial activities of fractionized extracts of edible mushroom *Pleurotus djamor* var. *roseus*. *Journal of Academia and Industrial Research* 3(5): 202–208.

Jegadeesh, R., H. Lakshmanan, J. Kab-Yeul, V. Sabaratnam, and N. Raaman. 2018. Cultivation of pink oyster mushroom *Pleurotus djamor* var. *roseus* on various agro-residues by low cost technique. *Journal of Mycopathological Research* 56(3): 213–220.

Jomova, K., R. Raptova, S. Y. Alomar, et al. 2023. Reactive oxygen species, toxicity, oxidative stress, and antioxidants: chronic diseases and aging. *Archives in Toxicology* 97: 2499–2574. https://doi.org/10.1007/s00204-023-03562-9

Kabel, A. M. 2014. Free radicals and antioxidants: role of enzymes and nutrition. *World Journal of Nutrition and Health* 2(3): 35–38. https://doi.org/10.12691/jnh-2-3-2

Kakon, A. J., M. B. K. Choudhury, and S. Saha. 2012. Mushroom is an ideal food supplement. *Journal of Dhaka National Medical College and Hospital* 18(1): 58–62.

Kalaras, M. D., J. P. Richie, A. Calcagnotto, and R. B. Beelman. 2017. Mushrooms: a rich source of the antioxidants ergothioneine and glutathione. *Food Chemistry* 233: 429–433. https://doi.org/10.1016/j. foodchem.2017.04.109

Kaur, M., S. Sharma, R. Kaur, J. Kaur, and A. Singh. 2023. Effect of UV-B irradiation on bioconversion of ergosterol to vitamin D2 and its impact on nutritional properties of oyster mushroom (*Pleurotus florida*). *International Journal of Food Science and Technology* 58(10): 5114–5120. https://doi.org/10.1111/ijfs.16610

Khatun, S., A. Islam, U. Cakilcioglu, P. Guler, and N. C. Chatterjee. 2015. Nutritional qualities and antioxidant activity of three edible oyster mushrooms (*Pleurotus* spp.) *NJAS-Wageningen Journal of Life Sciences* 72–73: 1–5. https://doi.org/10.1016/j.njas.2012.03.003

Kim, M. Y., P. Seguin, J. K. Ahn, et al. 2008. Phenolic compound concentration and antioxidant activities of edible and medicinal mushrooms from Korea. *Journal of Agricultural and Food Chemistry* 56(16): 7265–7270. https://doi.org/10.1021/jf8008553

Kirk, P. M., P. F. Cannon, J. C. David, and J. A. Stalpers. 2008. *Ainsworth & Brisby's Dictionary of the Fungi*. CAB International, Wallingford. 759 pp.

Krakowska, A., P. Zięba, A. Włodarczyk, et al. 2020. Selected edible medicinal mushrooms from *Pleurotus* genus as an answer for human civilization diseases. *Food Chemistry* 327(15): 127084. https://doi.org/10.1016/j.foodchem.2020.127084

Krishnamurthy, H. K., I. Rajavelu, M. Pereira, et al. 2024. Inside the genome: understanding genetic influences on oxidative stress. *Frontiers in Genetics* 15: 1397352. https://doi.org/10.3389/fgene.2024.1397352

Krzyczkowski, W., E. Malinowska, P. Suchocki, J. Kleps, M. Olejnik, and F. Herold. 2009. Isolation and quantitative determination of ergosterol peroxide in various edible mushroom species. *Food Chemistry* 113(1): 351–355. https://doi.org/10.1016/j.foodchem.2008.06.075

Kumar, S., and A. K. Pandey. 2013. Chemistry and biological activities of flavonoids: an overview. *The Scientific World Journal* 2013: 162750. https://doi.org/10.1155/2013/16275

Lapornik, B., M. Prosek, and A. G. Wondra. 2005. Comparison of extracts prepared from plant by-products using different solvents and extraction time. *Journal of Food Engineering* 71: 214–222. https://doi.org/10.1016/j.jfoodeng.2004.10.03

Lavelli, V., C. Proserpio, F. Gallotti, M. Laureati, and E. Pagliarini. 2018. Circular reuse of bio-resources: the role of *Pleurotus* spp. in the development of functional foods. *Food and Function* 9(3): 1353–1372. https://doi.org/10.1039/C7FO01747B

Liang, C. H., K. J. Ho, L. Y. Huang, C. H. Tsai, S. Y. Lin, and J. L. Mau. 2013. Antioxidant properties of fruiting bodies, mycelia, and fermented products of the culinary-medicinal king oyster mushroom, *Pleurotus eryngii* (higher Basidiomycetes), with high ergothioneine content. *International Journal of Medicinal Mushrooms* 15(3): 267–275. https://doi.org/10.1615/IntJMedMushr.v15.i3.40

Liao, X., Z. Zhu, S. Wu, et al. 2020. Preparation of antioxidant protein hydrolysates from *Pleurotus geesteranus* and their protective effects on H2O2 oxidative damaged PC12 cells. *Molecules* 25(22): 5408. https://doi.org/10.3390/molecules25225408

Lin, J. T., C. W. Liu, Y. C. Chen, et al. 2014. Chemical composition, antioxidant and anti-inflammatory properties for ethanolic extracts from *Pleurotus eryngii* fruiting bodies harvested at different time. *LWT-Food Science and Technology* 55(1): 374–382. https://doi.org/10.1016/j.lwt.2013.08.023

Liu, X., H. Pang, Z. Gao, H. Zhao, J. Zhang, and L. Jia. 2019. Antioxidant and hepatoprotective activities of residue polysaccharides by *Pleurotus citrinipileatus*. *International Journal of Biological Macromolecules* 131: 315–322. https://doi.org/10.1016/j.ijbiomac.2019.03.074

Maiorino, M., V. Bosello, G. Cozza, A. Roveri, S. Toppo, and F. Ursini. 2011. Glutathione peroxidase–4. In: *Selenium: Its Molecular Biology and Role in Human Health*, ed. L. H. Dolph, M. J. Berry, and V. N. Gladyshev, 181–195. Springer Press.

Mapari, S. A., U. Thrane, and A. S. Meyer. 2010. Fungal polyketide azaphilone pigments as future natural food colorants? *Trends in Biotechnology* 28(6): 300–307. https://doi.org/10.1016/j.tibtech.2010.03.004

Matsugo, S., T. Sakamoto, K. Wakame, Y. Nakamura, K. Watanabe, and T. Konishi. 2022. Mushrooms as a resource for Mibyou-Care functional food; the role of basidiomycetes-x (shirayukidake) and its major components. *Nutraceuticals* 2(3): 132–149. https://doi.org/10.3390/nutraceuticals2030010

Medihi, N. I., Z. A. Haiyee, R. Sukor, and S. Raseetha. 2024. Exploring the functional properties and nutritional values of colored oyster mushrooms species (*Pleurotus*, Agaricomycetes): a review. *International Journal of Medicinal Mushrooms* 26(6): 25–38. https://doi.org/10.1615/IntJMedMushrooms.2024053563

Milovanović, I., M. Stajic, T. P. Stanojkovic, A. Knežević, and J. Vukojević. 2015. Effects of selenium presence in mycelia of *Ganoderma* species (higher basidiomycetes) on their medicinal properties. *International Journal of Medicinal Mushrooms* 17(1): 11–20. https://doi.org/10.1615/IntJMedMushrooms.v17.i1.20

Mishra, K. K., R. S. Pal, and R. Arunkumar. 2014. Antioxidant activities and bioactive compound determination from caps and stipes of specialty medicinal mushrooms *Calocybe indica* and *Pleurotus sajor-caju* (higher basidiomycetes) from India. *International Journal of Medicinal Mushrooms* 16(6): 555–567. http://dx.doi.org/10.1615/intjmedmushrooms.v16.i6.50.

Mohan, K., K. D. Rajan, T. Muralisankar, R. A. Ganesan, K. Marimuthu, and P. Sathishkumar. 2022. The potential role of medicinal mushrooms as prebiotics in aquaculture: a review. *Reviews in Aquaculture* 14(3): 1300–1332. https://doi.org/10.1111/raq.12651

Morales-Flores, S., J. Cepeda-Negrete, G. M. M. de Oca, A. Ángel-Hernández, J. Hernández-Ruiz, and J. E. Ruiz-Nieto. 2022. *In vitro* molecular identification and characterization of *Pleurotus* spp. strains in Guanajuato, Mexico. *Agrociencia* 56(2): 1–11. https://doi.org/10.47163/agrociencia.v56i2.2780

Mukhopadhyay, R., and A. K. Guha. 2015. A comprehensive analysis of the nutritional quality of edible mushroom *Pleurotus sajor-caju* grown in deproteinized whey medium. *LWT-Food Science and Technology* 61: 339–345. https://doi.org/10.1016/j.lwt.2014.12.055

Nandi, A., L. J. Yan, C. K. Jana, and N. Das. 2019. Role of catalase in oxidative stress-and age-associated degenerative diseases. *Oxidative Medicine and Cellular Longevity* 2019: 1–19. https://doi.org/10.1155/2019/9613090

Nguyen, T. K., K. H. Im, J. Choi, P. G. Shin, and T. S. Lee. 2016. Evaluation of antioxidant, anti-cholinesterase, and anti-inflammatory effects of culinary mushroom *Pleurotus pulmonarius*. *Mycobiology* 44(4): 291–301. https://doi.org/10.5941/MYCO.2016.44.4.291

Niego, A. G., S. Rapior, N. Thongklang, et al. 2021. Macrofungi as a nutraceutical source: promising bioactive compounds and market value. *Journal of Fungi* 7(5): 397. https://doi.org/10.3390/jof7050397

Nowak, R., N. Nowacka-Jechalke, W. Pietrzak, and U. Gawlik-Dziki. 2022. A new look at edible and medicinal mushrooms as a source of ergosterol and ergosterol peroxide-UHPLC-MS/MS analysis. *Food Chemistry* 369: 130927. https://doi.org/10.1016/j.foodchem.2021.130927

Özdal, M., Ö. Gülmez, Ö. G. Özdal, and Ö. F. Algur. 2019. Antibacterial and antioxidant activity of mycelial extracts of different *Pleurotus* species. *Food and Health* 5(1): 12–18. https://doi.org/10.3153/FH19002

Palacios, I., M. Lozano, C. Moro, et al. 2011. Antioxidant properties of phenolic compounds occurring in edible mushrooms. *Food Chemistry* 128(3): 674–678. https://doi.org/10.1016/j.foodchem.2011.03.085.

Petraglia, T., T. Latronico, A. Fanigliulo, A. Crescenzi, G. M. Liuzzi, and R. Rossano. 2023. Antioxidant activity of polysaccharides from the edible mushroom *Pleurotus eryngii*. *Molecules* 28(5): 2176. https://doi.org/10.3390/molecules28052176

Philippoussis, A. 2009. Production of mushrooms using agro-industrial residues as substrates. In: *Biotechnology for Agro-industrial Residues Processing*, ed. P. S. Nigam, and A. Pandey, 163–196. Dordrecht Press.

Phillips, J. M., S. L. Ooi, and S. C. Pak. 2022. Health-promoting properties of medicinal mushrooms and their bioactive compounds for the COVID-19 era-an appraisal: do the pro-health claims measure up? *Molecules* 27(7): 72302. https://doi.org/10.3390/molecules27072302

Plays, M., S. Müller, and R. Rodriguez. 2021. Chemistry and biology of ferritin. *Metallomics* 13(5): mfab021. https://doi.org/10.1093/mtomcs/mfab021

Pukalski, J., and D. Latowski. 2022. Secrets of flavonoid synthesis in mushroom cells. *Cells* 11(19): 3052. https://doi.org/10.3390/cells11193052

Pumtes, P., K. Rojsuntornkitti, T. Kongbangkerd, and N. Jittrepotch. 2016. Effects of different extracting conditions on antioxidant activities of *Pleurotus flabellatus*. *International Food Research Journal* 23(1): 173.

Puttaraju, N. G., S. U. Venkateshaiah, S. M. Dharmesh, S. M. N. Urs, and R. Somasundaram. 2006. Antioxidant activity of indigenous edible mushrooms. *Journal of Agricultural and Food Chemistry* 54(26): 9764–9772. https://doi.org/10.1021/jf0615707

Qiang, J., Y. F. Tao, J. He, P. Xu, J. W. Bao, and Y. L. Sun. 2017. miR-122 promotes hepatic antioxidant defense of genetically improved farmed tilapia (GIFT, *Oreochromis niloticus*) exposed to cadmium by directly targeting a metallothionein gene. *Aquatic Toxicology* 182: 39–48. https://doi.org/10.1016/j.aquatox.2016.11.009

Rao, A. V., and L. G. Rao. 2007. Carotenoids and human health. *Pharmacological Research* 55(3): 207–216. https://doi.org/10.1016/j.phrs.2007.01.012

Rašeta, M. J., M. S. Rakić, E. V. Čapelja, and M. V. Karaman. 2022. Update on research data on the nutrient composition of mushrooms and their potentials in future human diets. In: *Edible Fungi: Chemical Composition, Nutrition and Health Effect*, ed. D. Stojković and L. Barros, 27–67. Royal Society of Chemistry Press. https://doi.org/10.1039/9781839167522-00027

Reis, F. S., A. Martins, L. Barros, and I. C. Ferreira. 2012. Antioxidant properties and phenolic profile of the most widely appreciated cultivated mushrooms: a comparative study between in vivo and in vitro samples. *Food and Chemical Toxicology* 50(5): 1201–1207. https://doi.org/10.1016/j.aquatox.2016.11.009

Reverberi, M., S. Zjalic, A. Ricelli, et al. 2008. Modulation of antioxidant defense in *Aspergillus parasiticus* is involved in aflatoxin biosynthesis: a role for the ApyapA gene. *Eukaryotic Cell* 7(6): 988–1000. https://doi.org/10.1128/ec.00228-07

Ribeiro, B., P. G. D. Pinho, P. B. Andrade, et al. 2011. Do bioactive carotenoids contribute to the color of edible mushrooms? *Open Chemical and Biomedical Methods Journal* 4: 14–18. https://doi.org/10.2174/1875038901104010014

Roupas, P., J. Keogh, M. Noakes, C. Margetts, and P. Taylor. 2012. The role of edible mushrooms in health: evaluation of the evidence. *Journal of Functional Foods* 4(4): 687–709. https://doi.org/10.1016/j.jff.2012.05.003

Royse, D. J., J. Baars, and Q. Tan. 2017. Current overview of mushroom production in the world. In: *Edible and Medicinal Mushrooms: Technology and Applications,* ed C. D. Cunha, A. Pardo-Gimenez, 5–13. Wiley Press. https://doi.org/10.1002/9781119149446.ch2

Sastre, J., F. V. Pallardo, and J. Viña. 2005. Glutathione. In: *Reactions, Processes: Oxidants and Antioxidant Defense Systems*, ed. T. Grune, 240–229. Springer Press.

Sattanathan, G., V. Shyamala, B. Deepa, and R. Keerthiga. 2018. Dietary administration of *Pleurotus sajor-caju* mushroom extract on growth, immune response and disease resistance in *Gibelion catla* against *Aeromonas hydrophila*. *International Journal of Life Sciences* 7(3): 142–150. https://doi.org/10.5958/2319-1198.2018.00020.9

Schwab, L., L. Goroncy, S. Palaniyandi, et al. 2014. Neutrophil granulocytes recruited upon translocation of intestinal bacteria enhance graft-versus-host disease via tissue damage. *Nature Medicine* 20: 648–654. https://doi.org/10.1038/nm.3517

Sharma, E., R. Bairwa, P. Lal, et al. 2024. Edible mushrooms trending in food: nutrigenomics, bibliometric, from bench to valuable applications. *Heliyon* 10(17): e36963. https://doi.org/10.1016/j.heliyon.2024.e36963

Sharma, S. K., and N. Gautam. 2015. Chemical, bioactive, and antioxidant potential of twenty wild culinary mushroom species. *BioMed Research International* 2015: 346508. https://doi.org/10.1155/2015/346508

Shnyreva, A. A., E. Y. Kozhevnikova, A. V. Barkov, and A. V. Shnyreva. 2017. Solid-state cultivation of edible oyster mushrooms, *Pleurotus* spp. under laboratory conditions. *Advances in Microbiology* 7(2): 125–136. https://doi.org/10.4236/aim.2017.72010

Shukla, S., and A. K. Jaitly. 2011. Morphological and biochemical characterization of different oyster mushroom (*Pleurotus* spp.). *Journal of Phytology* 3(8): 18–20. https://core.ac.uk/download/pdf/236018496.pdf

Sies, H. 1985. Oxidative stress: introductory remarks. In: *Oxidative Stress*, ed. H. Sies, 1–8. Academic Press.

Sies, H. 2018. On the history of oxidative stress: concept and some aspects of current development. *Current Opinion in Toxicology* 7: 122–126. https://doi.org/10.1016/j.cotox.2018.01.002

Sies, H., and D. P. Jones. 2007. Oxidative stress. In: *Encyclopedia of Stress*, ed. G. Fink, 45–48. Elsevier Press.

Singer, R. 1986. *The Agaricales in Modern Taxonomy*. Koeltz Scientific Books, Koenigstein, Universidad de California. 981 pp.

Smith, D. F., and A. Casadevall. 2019. The role of melanin in fungal pathogenesis for animal hosts. In: *Fungal Physiology and Immunopathogenesis*, ed. M. L. Rodrigues, 1–30. Springer Press. https://doi.org/10.1007/82_2019_173

Stamets, P. 1993. *Growing Gourmet and Medicinal Mushrooms*. Ten Speed Press. Crown Publishing Group, New York. 544 pp.

Stastny, J., P. Marsik, J. Tauchen, et al. 2022. Antioxidant and anti-inflammatory activity of five medicinal mushrooms of the genus *Pleurotus*. *Antioxidants* 11(8): 1569. https://doi.org/10.3390/antiox11081569

Suwannarach, N., J. Kumla, K. Sujarit, T. Pattananandecha, C. Saenjum, and S. Lumyong. 2020. Natural bioactive compounds from fungi as potential candidates for protease inhibitors and immunomodulators to apply for coronaviruses. *Molecules* 25(8): 1800. https://doi.org/10.3390/molecules25081800

Tan, R. Y., Z. Ilham, W. A. A Wan, et al. 2024. Mushroom oils: a review of their production, composition, and potential applications. *Heliyon* 10(11): e31594. https://doi.org/10.1016/j.heliyon.2024.e31594

Thane, C., and S. Reddy. 1997. Processing of fruit and vegetables: effect on carotenoids. *Nutririon and Food Science* 97: 58–65. https://doi.org/10.1108/00346659710161858

Toledo, A. V., M. E. E. Franco, S. M. Y. Lopez, M. I. Troncozo, M. C. N. Saparrat, and P. A. Balatti. 2017. Melanins in fungi: types, localization and putative biological roles. *Physiological and Molecular Plant Pathology* 99: 2–6. https://doi.org/10.1016/j.pmpp.2017.04.004

Toyokuni, S. 2016. The origin and future of oxidative stress pathology: from the recognition of carcinogenesis as an iron addiction with ferroptosis-resistance to non-thermal plasma therapy. *Pathology International* 66(5): 245–259. https://doi.org/10.1111/pin.12396

Tsiapali, E., S. Whaley, J. Kalbfleisch, H. E. Ensley, I. W. Browder, and D. L. Williams. 2001. Glucans exhibit weak antioxidant activity, but stimulate macrophage free radical activity. *Free Radical Biology and Medicine* 30(4): 393–402. https://doi.org/10.1016/S0891-5849(00)00485-8

Turunen, M., J. Olsson, and G. Dallner. 2004. Metabolism and function of coenzyme Q. *Biochimica et Biophysica Acta (BBA)-Biomembranes* 1660(1–2): 171–199. https://doi.org/10.1016/j.bbamem.2003.11.012

Uluköy, G., E. Baba, and C. Öntaş. 2016. Effect of oyster mushroom, *Pleurotus ostreatus*, extract on hemato-immunological parameters of Rainbow trout, *Oncorhynchus mykiss*. *Journal of the World Aquaculture Society* 47(5): 676–684. https://doi.org/10.1111/jwas.12318

Valko, M., D. Leibfritz, J. Moncol, M. Cronin, M. Mazur, and J. Telser. 2007. Free radicals and antioxidants in normal physiological functions and human disease. *The International Journal of Biochemistry and Cell Biology* 39(1): 44–84. https://doi.org/10.1016/j.biocel.2006.07.001

Valko, M., C. J. B. Rhodes, J. Moncol, M. M. Izakovic, and M. Mazur. 2006. Free radicals, metals and antioxidants in oxidative stress-induced cancer. *Chemico-Biological Interactions* 160(1): 1–40. https://doi.org/10.1016/j.cbi.2005.12.009

van Dam, L., P. Cruz-Morales, N. R. Valerón, et al. 2024. GastronOmics: edibility and safety of mycelium of the oyster mushroom *Pleurotus ostreatus*. *Current Research in Food Science* 9: 100866. https://doi.org/10.1016/j.crfs.2024.100866

Vargas-Sánchez, R. D., G. R. Torrescano-Urrutia, F. J. Ibarra-Arias, J. J. Portillo-Loera, F. G. Ríos-Rincón, and A. Sánchez-Escalante. 2018. Effect of dietary supplementation with *Pleurotus ostreatus* on growth performance and meat quality of Japanese quail. *Livestock Science* 207: 117–125. https://doi.org/10.1016/j.livsci.2017.11.015

Venturella, G., M. L. Gargano, and R. Compagno. 2015. The genus *Pleurotus* in Italy. *Flora Mediterranea* 25 (Special Issue): 143–156. https://doi.org/10.7320/FlMedit25SI.143

Walter, W. M., and A. E. Purcell. 1979. Evaluation of several method for analysis of sweet potato phenolic. *Journal of Agricultural and Food Chemistry* 27: 942–946. https://doi.org/10.1021/jf60225a030

Wang, Y., S. M. Gapstur, B. D. Carter, et al. 2018. Untargeted metabolomics identifies novel potential biomarkers of habitual food intake in a cross-sectional study of postmenopausal women. *Journal of Nutrition* 148: 932–943. https://doi.org/10.1093/jn/nxy027

Weidinger, A., and A. V. Kozlov. 2015. Biological activities of reactive oxygen and nitrogen species: oxidative stress versus signal transduction. *Biomolecules* 5(2): 472–484. https://doi.org/10.3390/biom5020472

Willett, W., J. Rockstrom, B. Loken, et al. 2019. Food in the anthropocene: the EAT-Lancet commission on healthy diets from sustainable food systems. *Lancet* 393: 447–492. https://doi.org/10.1016/S0140-6736(18)31788-4

Woldegiorgis, A. Z., D. Abate, G. D. Haki, and G. R. Ziegler. 2014. Antioxidant property of edible mushrooms collected from Ethiopia. *Food Chemistry* 157: 30–36. https://doi.org/10.1016/j.foodchem.2014.02.014

Xiao, X., Y. Li, X. Li, et al. 2024. Identification, nutrient composition, and evaluation of a wild *Pleurotus citrinopileatus* strain (X21156) from Tibet for antioxidant and cytotoxic activities. *Horticulturae* 10(4): 377. https://doi.org/10.3390/horticulturae10040377

Yadav, D., and P. S. Negi. 2021. Bioactive components of mushrooms: processing effects and health benefits. *Food Research International* 148: 110599. https://doi.org/10.1016/j.foodres.2021.110599

Yan, Z. Y., M. R. Zhao, C. Y. Huang, L. J. Zhang, and J. X. Zhang. 2021. Trehalose alleviates high-temperature stress in *Pleurotus ostreatus* by affecting central carbon metabolism. *Microbial Cell Factories* 20: 1–11. https://doi.org/10.1186/s12934-021-01572-9

Yin, C. M., X. Fan, C. Liu, et al. 2019. The antioxidant properties, tyrosinase and α-glucosidase inhibitory activities of phenolic compounds in different extracts from the golden oyster mushroom, *Pleurotus citrinopileatus* (Agaricomycetes). *International Journal of Medicinal Mushrooms* 21(9): 865–874. https://doi.org/10.1615/IntJMedMushrooms.2019031857

Yıldız, S., A. Yılmaz, Z. Can, C. Kılıç, and Ü. C. Yıldız, 2017. Total phenolic, flavonoid, tannin contents and antioxidant properties of *Pleurotus ostreatus* and *Pleurotus citrinopileatus* cultivated on various sawdust. *Gıda: The Journal of Food* 42(3): 315–323. https://doi.org/10.15237/gida.GD16099

You, S., Y. Ma, B. Yan, et al. 2022. The promotion mechanism of prebiotics for probiotics: a review. *Frontiers in Nutrition* 9: 1000517. https://doi.org/10.3389/fnut.2022.1000517

# 7 Antimicrobial Activity of *Pleurotus* spp.

*D. Matijašević[1] and A. Sknepnek[2]

[1] Institute of Molecular Genetics and Genetic Engineering, University of Belgrade, Belgrade, Serbia

[2] Faculty of Agriculture, Institute of Food Technology and Biochemistry, University of Belgrade, Belgrade, Serbia

## 7.1 INTRODUCTION

One of the most important discoveries of the last century that changed medicine was certainly the discovery of the antibiotic penicillin in 1928 by the Scottish bacteriologist Alexander Fleming. However, the first penicillin was not used clinically on a large scale until the end of the Second World War and thereafter saved the lives of millions of people (Gaynes, 2017). The discovery of penicillin was groundbreaking as other researchers, inspired by this discovery, started to develop a number of new antibiotics (such as streptomycin, chloramphenicol, erythromycin, vancomycin, and many others) between the 1950s and 1970s. During this golden era, more than 20 classes of antibiotics were discovered from environmental bacteria and fungi. Since then, no new antibiotic classes have been discovered (Aminov, 2010; Durand et al., 2019).

In recent years, the problem of increasing resistance of microorganisms to existing antibiotics has emerged. The term antimicrobial resistance (AMR) refers to the resistance of microorganisms (including bacteria, viruses, fungi, and parasites) to an antimicrobial drug that was originally successful in treating infections caused by these microorganisms but became ineffective over time (Sknepnek et al., 2018; Dadgostar, 2019). Several international organizations (the US Centers for Disease Control and Prevention (CDC), the World Health Organization (WHO), the United Nations, and many others) consider AMR to be a major threat to public health systems (Durand et al., 2019). In 2019, bacterial infections caused an estimated 4.95 million deaths worldwide, of which 1.27 million were directly attributable to these infections. In addition to the high mortality and disability rates, these infections also have a serious financial impact. According to World Bank estimates, bacterial infections could lead to annual gross domestic product losses of 1 to 3.4 trillion dollars by 2030. Without the development of new therapies, the death rate from untreatable infections is expected to increase more than tenfold by 2050 (Dash et al., 2024). The 2024 WHO Bacterial Priority Pathogens List covers 24 pathogens from 15 families of antibiotic-resistant bacterial pathogens that are prioritized due to their impact on public health. In the list, these pathogens are divided into groups with critical (carbapenem-resistant *Acinetobacter baumannii*, various pathogens of the order Enterobacterales and rifampicin-resistant *Mycobacterium tuberculosis*), high (fluoroquinolone-resistant *Salmonella* ser. *typhi*, non-typhoidal *Salmonella* and *Shigella* spp., *Pseudomonas aeruginosa* and *Staphylococcus aureus*, resistant to last-line antibiotics, vancomycin-resistant *Enterococcus faecium* and third-generation cephalosporin and/or fluoroquinolone-resistant *Neisseria gonorrhoeae*), and medium priority (Group A and B *Streptococci*, *Streptococcus pneumoniae*, and *Haemophilus influenzae*) (WHO, 2024).

For all these reasons, new antimicrobial substances that can be isolated from various natural sources such as medicinal plants, marine organisms, prokaryotes, fungi, and algae are constantly being tested. To date, the number of therapeutically active compounds from microbial sources exceeds the number discovered from all other sources. Due to the renewed interest in traditional medicine, more and more research is being conducted to find antimicrobial agents from mushrooms (Moussa et al., 2022).

It is estimated that there are over 140,000 species of mushrooms on earth, of which only about 15,000 are known to science, and only a small percentage have been studied (Miletić et al., 2020). However, only about 200 to 300 species are commonly consumed as food worldwide. Mushrooms of the genus *Pleurotus*, known as "oyster mushrooms," are among the most commonly consumed mushrooms due to their flavor, aroma, and nutritional value. Today, at least 40 species are known, some of which (i.e., *Pleurotus florida*, *Pleurotus tuber-regium*, *Pleurotus pulmonarius*, *Pleurotus ostreatus*, *Pleurotus sajor-caju*, *Pleurotus citrinopileatus* and *Pleurotus eryngii*) are of great medicinal importance (Golak-Siwulska et al., 2018). In particular, a wide range of mycochemicals have been found in the fruiting bodies, mycelium, or culture broth of *Pleurotus* sp. The most appreciated compounds that have been shown to have a beneficial effect on human health and high nutritional value are high-molecular-weight (HMW) compounds, including polysaccharides, proteins, peptides, polysaccharopeptides, and lipids, and low molecular weight (LMW) compounds, i.e., lectins, phenolic compounds, terpenoids, and alkaloids. Their content, composition, and bioactivity depend on various internal and external conditions, i.e., species, substrate composition, cultivation and postharvest conditions, and final processing and storage (Carrasco-González et al., 2017; Sknepnek et al., 2018; Matijašević et al., 2022; Sknepnek and Miletić, 2022). This review presents a comprehensive overview of the antimicrobial properties of oyster mushrooms (*Pleurotus* spp.), with a particular focus on the wide range of

* Corresponding author: danka.matijasevic@imgge.bg.ac.rs



DOI: 10.1201/9781003527572-7

bioactive compounds contributing to these effects. By examining the current body of scientific literature, the review highlights the mechanisms through which these compounds exert antimicrobial action. Ultimately, the review aims to underscore the therapeutic potential of *Pleurotus* species as natural antimicrobial agents and to explore their promising applications in food preservation, pharmaceutical formulations, and the development of functional foods.

## 7.2 BIOACTIVE CONSTITUENTS OF *PLEUROTUS* SPP.

### 7.2.1 Proteins

Mushrooms are known as a valuable source of non-meat proteins. Some species of the genus *Pleurotus* contain all essential and non-essential amino acids and may also have unique bioactive properties. The protein content in the genus *Pleurotus* varies between 11% and 42%, depending on the species, with the highest protein content found in *P. ostreatus*, *P. sajor-caju*, *Pleurotus sapidius*, *Pleurotus geesteranus*, and so on (Khan and Tania, 2012; Carrasco-González et al., 2017). In addition to these differences, significant variations have also been observed due to the physical and chemical conditions of the culture medium, its composition, the time of harvest, and the size of the pileus. Although many of the proteins of *Pleurotus* spp. have not yet been identified or characterized, it is known that there are proteins that exhibit various antimicrobial activities, including bioactive peptides, lectins, ribonucleases, laccases, ribosome inactivating proteins (RIP), and many others (Xu et al., 2011). Lectins belong to the most extensively studied group of bioactive proteins from mushrooms. They are dimeric proteins or glycoproteins that bind to carbohydrates on the cell surface and cause agglutination of cells, reducing their ability to cause infection, growth, and proliferation (Deepalakshmi and Mirunalini, 2014). Another group of proteins, laccases, have enzymatic activity and belong to a subclass of multicopper oxidases. They are a widely distributed group of enzymes with an important role in physiological and ecological processes such as pathogenesis, in which they contribute to the virulence of microorganisms; morphogenesis, in which they influence structural changes in organisms; and immunogenesis, in which they modulate host immune responses (Xu et al., 2011).

### 7.2.2 Carbohydrates

Carbohydrates in mushrooms are the major fractions of dry matter, which include monosaccharides, disaccharides, sugar alcohols, oligosaccharides, polysaccharides, glycoproteins, and hemicelluloses. Mushroom polysaccharides are natural macromolecular polymers that can be obtained from fruiting bodies, mycelium, or as extracellular polysaccharides in the culture broth. Of the bioactive components of mushrooms, polysaccharides have been studied most intensively. The cell wall of mushrooms consists of a rigid fibril structure of cellulose and chitin and another layer in the form of a matrix composed of α-glucans, β-glucans, and glycoproteins (Leong et al., 2021). The glucans take up the largest part of the cell wall (50%–60%), glycoproteins are present at 20% to 30% and chitin at 10% to 20% (Rodrigues Barbosa et al., 2020). Glucans are divided into (1→3)-α-glucans and (1→3)-, (1→6)-β-glucans based on the glycosidic bonds. The most appreciated β-glucans of *Pleurotus* spp. are essentially composed of glucose monomers, but they can have different anomeric configurations, degrees of branching (several side chains, linkages of one or more monosaccharides at different positions), molecular weights, and chain conformations, which have an overall impact on their bioactivity (Rodrigues Barbosa et al., 2020). The highest total carbohydrate content in *Pleurotus* spp. usually varies between 40% and 60%, depending on the species (Khan and Tania, 2012; Deepalakshmi and Mirunalini, 2014); but in some cases (in *P. eryngii*), even 75% carbohydrates have been found (Carrasco-González et al., 2017).

To reach these metabolites of interest bound in the cell wall, extraction techniques have been developed and are constantly being improved to break the outer layer of the cell wall and facilitate the penetration of the solvent into the inner layer without damaging or altering the polysaccharides, while improving yield and purity and maintaining or even improving bioactivity (Sharma et al., 2021). For the extraction of polysaccharides, an aqueous extraction technique based on a compatible polarity of water and polysaccharide is usually used. Pretreatment with ethanol or acetone facilitates the separation of polysaccharides from non-polar compounds (lipids, terpenes, phenols) but is not mandatory. Increasing the temperature or time during hot water extraction can improve extraction efficiency, but special care must be taken to preserve sensitive compounds that contribute to bioactivity (Rodrigues Barbosa et al., 2020). A further increase in temperature can lead to degradation of the β-glucans and/or disruption of the triple helix structure, which subsequently reduces bioactivity (Leong et al., 2021). New techniques have also been developed, such as microwave, ultrasound-assisted, subcritical water extraction, pressurized hot water extraction, and deep eutectic solvent extraction, and so on. These techniques mostly use environmentally friendly "green" solvents, but some of them have shown that a long extraction time can lead to degradation of the polysaccharides and consequently to reduced bioactivity. Therefore, careful optimization of the process is required to avoid or minimize the presence of unwanted and interfering substances (Rodrigues Barbosa et al., 2020).

### 7.2.3 Lipids

Mushrooms in general, including *Pleurotus* species, contain low concentrations of lipids, depending on the species, ranging from 0.2% to 8%. Some essential fatty acids are present in *Pleurotus* mushrooms; among the monounsaturated fatty acids, oleic acid dominates, while the major polyunsaturated fatty acid is linoleic acid. Lipid composition and content vary depending on the type of substrate and growth conditions such as nutrients, temperature, and oxygen (Khan and Tania, 2012). Some researchers have shown that lipids play a crucial role in antimicrobial activity.

### 7.2.4 Terpenes

Terpenes are secondary metabolites that represent a diverse and large group with mostly lipophilic nature and are predominantly derived from higher plants but can also be synthesized by microorganisms and animals. These compounds consist of isoprene units (five carbon atoms each) linked together according to the "head-tail" arrangement and are divided into several classes depending on the number of isoprene units. The terpenes with antimicrobial activity detected in mushrooms are predominantly sesquiterpene or diterpene derivatives (Masyita et al., 2022; Matijašević and Sknepnek, 2024).

### 7.2.5 Phenolic Compounds

For a long time, plants were appreciated as the main producers of phenolic compounds, but mushrooms have gained attention for this unique ingredient due to their role as functional foods and natural preservatives (Moussa et al., 2022). Phenolic compounds can range from simple molecules to complex polymers but are all characterized by having one or more aromatic rings (C6) and one or more hydroxyl groups. Mushrooms are rich in various phenolic compounds such as phenolic acids, flavonoids, tocopherols, tannins, and others. The phenolic acids found in *Pleurotus* species are caffeic acid, gallic acid, ferulic acid, *p*-coumaric acid, cinnamic acid, and homogentisic acid, while the most frequently detected flavonoids are myricetin, rutin, and quercetin. The concentration of phenolic compounds has been found to be closely related to antimicrobial activity against pathogenic microorganisms. In addition, the phenolic content of mushrooms is influenced by the extraction solvent. In the literature, high levels are reported in aqueous, ethanolic (EtOH), methanolic (MeOH), and acetonitrile extracts, while moderate concentrations are found in ethyl acetate extracts and low in hexane extracts (Dulay et al., 2017). In general, Gram-negative bacteria are more resistant to phenolic compounds than Gram-positive species, possibly due to the complex outer lipopolysaccharide barrier in Gram-negative bacteria, which limits permeability for many compounds into the cell. Phenolic compounds act as antimicrobial agents by inhibiting protein, nucleic acid, and cell wall synthesis (Abdelshafy et al., 2022).

## 7.3 HISTORY OF BASIDIOMYCETES ANTIMICROBIAL SUBSTANCES DISCOVERY, WITH SPECIAL EMPHASIS ON *PLEUROTUS* SP.

Since the discovery of penicillin, much of the attention has focused on the identification and isolation of antimicrobial substances from the division of Ascomycota, while other divisions of the Fungi kingdom have been somewhat neglected. In the 1940s, extensive work was carried out on a wide range of antibiotics from Basidiomycetes. Wilkins and Haris investigated the bacteriostatic properties of extracts from the sporophores of 722 species spread over 96 genera of the Basidiomycetes and found that about 20% of the species produced such substances. The work also included five species of the genus *Pleurotus*. Under the conditions tested, only *Pleurotus palmatus* (now classified as *Rhodotus palmatus*) and *Pleurotus sapidus* were active against *S. aureus*, while no species was active against *Escherichia coli*. In summary, they considered the possibility of producing antibacterial substances from Basidiomycetes to be extremely promising (Wilkins and Harris, 1944).

In a similar study, Robbins et al. (1945) tested the antibacterial activity of more than 300 wood-decaying fungi and found that 35% of the fungi produced inhibition zones for *S. aureus*. According to their results, none of the *Pleurotus* species tested (*Pleurotus corticatus*, *P. ostreatus*, *P. sapidus*, and *Pleurotus ulmarius* [now classified as *Hypsizygus ulmarius*]) were active against *S. aureus* or *E. coli*. The same authors found that *Pleurotus griseus* (now classified as *Hohenbuehelia grisea*) showed moderate to strong activity against *S. aureus*. Furthermore, the culture liquid obtained from *P. griseus* was active at a dilution of 1:1000, while *Penicillium notatum*, with which Fleming began his study, was active against *S. aureus* at a dilution of 1:800 under the same conditions. In addition, the authors drew the following conclusions: (a) the inhibition of *S. aureus* exceeded that of *E. coli*, (b) the activity was not limited to certain genera, (c) the medium on which the fungus grew had a strong influence on the formation of antibiotic material, and (d) the test method influenced the antibiotic activity.

The search for antibiotic substances from Basidiomycetes was also the focus of Robbins et al. (1946), who tested around 300 species from 42 genera and found that two-thirds of the species inhibited the growth of *S. aureus*. They found that, in most cases, the activity was inversely proportional to the growth rate and hypothesized that secondary metabolites have antibacterial effects.

The following year, the same group of authors (Robbins et al., 1947) published their work with *P. griseus*. The authors extracted the active substance with chloroform ether and obtained amber-colored crystals. The crystalline antibiotic substance was named pleurotin and was active against *S. aureus* at a concentration of 1 μg/mL. The chemical and physical properties demonstrated led to the conclusion that it was a new substance, different from all other antibiotics. Pleurotin was only active against Gram-positive bacteria such as *S. aureus*, *Bacillus mycoides*, *Bacillus subtilis*, and *Aliivibrio fischeri* (formerly *Photobacterium fischeri*). The toxicity test showed that pleurotin is not toxic when administered intravenously to white mice. This work is considered one of the earliest documented studies on the antimicrobial activity of *Pleurotus* species. In later decades, renewed interest in pleurotin led to the realization that it is actually a naphthoquinone antibiotic (Shipley et al., 2006).

The interest of Kavanagh et al. (1951) in species of the genus *Pleurotus* led to the isolation of a white, crystalline, antibacterial substance called pleuromutilin from *Pleurotus mutilus* (now known as *Clitopilus scyphoides*) and *Pleurotus passeckerianus* (synonym *Clitopilus passeckerianus*). Further analyses revealed that pleuromutilin is easily soluble in water and thermostable. Antibacterial tests showed its activity against all strains of *Staphylococcus*, *Streptococcus*, *Enterococcus*, *B. subtilis*, 80% of *E. coli* strains, and 60% of strains of *Vibrio cholerae*, and had no effect against one strain of *Bacillus cereus*. Pleuromutilin showed no antifungal activity at a concentration of 0.5 mg/mL. Since pleuromutilin was not toxic to mice, it was administered immediately after infection of the mice with *Streptococcus haemolyticus*. After 7 days, 63% of mice treated with pleuromutilin, all penicillin-treated mice, and 20% of untreated control mice survived. Today, pleuromutilin is extracted from certain species of mushrooms, including *P. mutilus* and *Clitopilus*. The elucidation of the structure in 1962 and further attempts to improve the antimicrobial activity led to the development of the first veterinary pleuromutilin, tiamulin, which was approved in 1979. In 2007, the first pleuromutilin antibiotic, retapamulin, was approved for short-term topical treatment in humans (Novak, 2011). Another pleuromutilin, lefamulin (BC-3781), is approved for intravenous (IV) and oral use and is the first systemic antibiotic in a new class of antibiotics with favorable clinical data in the treatment of adults with community-acquired bacterial pneumonia (CABP) in over 15 years. Lefamulin offers a promising intravenous and oral monotherapy alternative to fluoroquinolones or macrolides for the treatment of patients with CABP, including those at higher risk of poor outcomes due to advanced age or comorbidities (File et al., 2021, 2019). After almost a decade of large-scale screening projects, the focus shifted from Basidiomycetes to the easier-to-cultivate *Streptomyces* bacteria. This trend continued for the next nearly 50 years until the early 2000s, when new studies were conducted to identify antimicrobial compounds from Basidiomycetes (Suay et al., 2000; Tambekar et al., 2006).

The need to find compounds with a different structure and new mechanisms of action has meant that difficulties such as the slow growth rate of Basidiomycetes and the low yield of products obtained have been ignored, once putting them on the back burner (Suay et al., 2000).

In a comprehensive screening program conducted in Spain in the 2000s, the antimicrobial potential of 317 isolates of Basidiomycetes, representing 204 species and divided into 44 families, was investigated against a range of multidrug-resistant (MDR) clinical pathogens. Under the experimental conditions, 40% of the isolates showed antibacterial activity, while 20% exhibited antifungal properties. Of the *Pleurotus* species tested, two *P. eryngii* strains were active only against *B. subtilis*, while the third *P. eryngii* strain inhibited the growth of *B. subtilis*, β-lactam- and vancomycin-resistant *Enterococcus faecium*, methicillin-resistant *S. aureus* (MRSA), and *Aspergillus fumigatus* (Suay et al., 2000).

## 7.4 *PLEUROTUS* SPP. WITH ANTIMICROBIAL ACTIVITY

### 7.4.1 *Pleurotus* spp. Extracts with Antimicrobial Activity against Gram-positive Bacteria

Numerous extracts of different *Pleurotus* strains have been described as having antibacterial activity against Gram-positive bacteria (Table 7.1). The most studied mushroom of the genus *Pleurotus* is *P. ostreatus*, which has a broad spectrum of antimicrobial activity. The MeOH extract of the fruiting body of *P. ostreatus* showed high antibacterial activity against *B. subtilis*, *S. aureus* (Chowdhury et al., 2015), and *Bacillus thuringiensis* (Younis et al., 2015) and moderate activity against *B. cereus* and *Micrococcus luteus* (Chaiharn et al., 2018). Of the tested extracts (EtOH, MeOH, and aqueous), the aqueous extract showed the highest and broadest activity against all tested Gram-positive bacteria (*B. subtilis*, *B. megaterium*, *B. thuringiensis*, *S. aureus*, and *Streptococcus pyogenes*) (Younis et al., 2015). Due to its high polarity, water extracts easily dissolve water-soluble small (phenolic compounds) and high-molecular-weight compounds (saccharides, polysaccharides, and polysaccharide–protein complexes) (Shen et al., 2017). When EtOH and ethyl acetate were used as extraction solvents, *B. cereus* was found to be the most sensitive bacterium, with a minimal inhibitory concentration (MIC) value of 1.5 mg/mL, and was comparable to amoxicillin (MIC = 1.25 mg/mL) (Chaiharn et al., 2018). *Bacillus cereus*, a spore-forming bacterium widely distributed in nature, is mainly associated with food poisoning but also with severe and potentially fatal infections of the non-gastrointestinal tract (Bottone, 2010). As most isolates produce a potent β-lactamase, the choice of treatment is quite complicated, and any agent that helps reduce bacterial survival may be of great importance. The ability of *P. ostreatus* acetone and ethyl acetate extracts to improve oral health was investigated by Han et al. (2015), as they showed antibacterial activity on *Streptococcus sobrinus*, *Streptococcus mutans*, *Streptococcus ratti*, *Streptococcus sanguinis*, *Streptococcus anginosus*, and *Actinomyces naeslundii*, which are implicated in gingivitis and dental caries and contribute significantly to tooth decay. In line with these findings, Gangwar et al. (2024) found that bioactive proteins extracted from *P. ostreatus* inhibit dental pathogens, so their inclusion in oral care products (e.g. gargle, toothpaste) holds great potential. According to Adebayo et al. (2018), the hydroalcoholic extract of *P. ostreatus* inhibited the growth of *Listeria monocytogenes* at a concentration of 73 µg/mL, while concentrations between 8 and 73 µg/mL exerted inhibition on *Streptococcus agalactiae*.

In contrast to the fruiting body, the aqueous extract of the mycelium of *P. ostreatus* showed low-to-moderate inhibition of *S. aureus* and *S. pyogenes* (Younis et al., 2015). However, the EtOH extract showed an inhibitory effect on all tested Gram-positive bacteria, with the highest capacity against *Sarcina lutea* (Kalyoncu et al., 2010). The

**TABLE 7.1**
***Pleurotus* spp. Extracts with Antimicrobial Activity Against Gram-positive Bacteria**

| Mushroom | Microorganism | Extract | Results | References |
|---|---|---|---|---|
| *Pleurotus ostreatus* (Jacq. Ex Fr.) P. Kumm. (FB) | *B. cereus* | Aqueous, Hexane, Ethyl acetate, Methanol, Ethanol | IZD = 5–14 mm<br>MIC = 1.5 mg/mL | Hearst et al., 2009; Okafor et al., 2017; Chaiharn et al., 2018; Sutthisa and Anujakkawan, 2023) |
| | *B. subtilis* | Aqueous, Methanol, Ethyl acetate, Ethanol, Hydroalcoholic, Acetone | IZD = 5–16 mm<br>MIC = 0.073–7 mg/mL<br>MBC = 146 µg/mL | Hearst et al., 2009; Chowdhury et al., 2015; Adebayo et al., 2018; Han et al., 2015; Younis et al., 2015 |
| | *B. megaterium* | Water, Methanol | IZD = 7.5–8.5 mm | Akyuz et al., 2010; Younis et al., 2015 |
| | *B. thuringiensis* | Water, Ethanol, Methanol | IZD = 7–16 mm | Younis et al., 2015 |
| | *Bacillus* sp. (an) | Methanol | IZD = 7 mm | Karaman et al., 2010 |
| | *S. aureus* | Ethyl acetate, Methanol, Water, Chloroform, Hydroalcoholic, Ethanol,<br>Methanol:chloroform:water = 2:2:1 | IZD = 5–25 mm<br>MIC = 0.073–6 mg/mL | Akyuz et al., 2010; Karaman et al., 2010; Han et al., 2015; Chowdhury et al., 2015; Adebayo et al., 2018; Chaiharn et al., 2018; Hamad et al., 2022; Younis et al., 2015 |
| | *S. pyogenes* | Water | IZD = 13 mm | Younis et al., 2015 |
| | *S. agalactiae* | Hexane, Hydroalcoholic | IZD = 12 mm<br>MIC = 8–73 µg/mL<br>MBC = 146 µg/mL | Okafor et al., 2017; Adebayo et al., 2018 |
| | *S. sobrinus* (oral) | Acetone, Ethyl acetate | IZD = 9–10 mm | Han et al., 2015 |
| | *S. mutans* (oral) | | IZD = 9–10 mm | |
| | *S. ratti* (oral) | | IZD = 9–10 mm | |
| | *S. sanguinis* (oral) | | IZD = 9–10 mm | |
| | *S. anginosus* (oral) | | IZD = 6–10 mm | |
| | *A. naeslundii* (oral) | | IZD = 6–8 mm | |
| | *L. monocytogenes* | Hydroalcoholic | MIC = 73 µg/mL | Adebayo et al., 2018 |
| | *M. luteus* | Ethyl acetate, Methanol, Chloroform.<br>Methanol:chloroform:water = 2:2:1 | IZD = 6.3–12 mm<br>MIC = 3 mg/mL | Karaman et al., 2010; Chaiharn et al., 2018; Hamad et al., 2022 |
| *Pleurotus ostreatus* (M) | *B. cereus* | Ethanol, Hot water | IZD = 7–16 mm | Kalyoncu et al., 2010; Özdal et al., 2019 |
| | *B. subtilis* | Ethanol | IZD = 12 mm | Kalyoncu et al., 2010 |
| | *S. aureus* | Ethanol, Water | IZD = 5–24 mm | Kalyoncu et al., 2010; Younis et al., 2015 |
| | *S. pyogenes* | Water | IZD = 5 mm | Younis et al., 2015 |
| | *S. lutea* | Ethanol | IZD = 30 mm | Kalyoncu et al., 2010 |
| *Pleurotus ostreatus* (CB) | *S. aureus* | Water | IZD = 12 mm | Younis et al., 2015 |
| | *S. pyogenes* | | IZD = 17 mm | |
| | *B. subtilis* | | IZD = 12 mm | |
| | *B. thuringiensis* | | IZD = 17 mm | |
| *Pleurotus sajor-caju* (FB) | *S. aureus* | Aqueous, Ethanol, Methanol, Ether, Benzene, Xylene, Acetone | IZD = 7.5–25 mm | Tambekar et al., 2006; Akyuz et al., 2010; Gogavekar et al., 2014 |
| | *S. epidermidis* | Aqueous, Ethanol, Methanol, Ether, Benzene, Xylene, Acetone | IZD = 14–17 mm | Tambekar et al., 2006 |
| | *S. agalactiae* | Hexane | IZD = 12 mm | Okafor et al., 2017 |
| | *B. cereus* | Hexane, Methanol | IZD = 13–18 mm | Okafor et al., 2017; Gogavekar et al., 2014 |
| | *M. luteus* | Methanol | IZD = 23 mm | Gogavekar et al., 2014 |
| *Pleurotus sajor-caju* (M) | *B. cereus* | Hot water extract | IZD = 10.2 mm | Özdal et al., 2019 |
| *Pleurotus eryngii* (FB) | *S. aureus* | Methanol | IZD = 7.5–12 mm | Akyuz et al., 2010 |
| var. *eryngii* | *B. megaterium* | | IZD = 8.5–9 mm | |
| var. *ferulae* | | | | |
| *Pleurotus eryngii* (M) | *S. aureus* | Ethanol, Methanol | IZD = 8–15 mm | Suay et al., 2000; Kalyoncu et al., 2010 |

**TABLE 7.1** ***(Continued)***
***Pleurotus* spp. Extracts with Antimicrobial Activity Against Gram-positive Bacteria**

| Mushroom | Microorganism | Extract | Results | References |
|---|---|---|---|---|
| | *B. subtilis* | Methanol | IZD < 15 mm | Suay et al., 2000 |
| | *B. cereus* | Hot water | IZD = 9.8 mm | Özdal et al., 2019 |
| *Pleurotus pulmonarius* (FB) | *B. cereus* | Ethyl acetate, Aqueous, Hexane | IZD = 7.1–15 mm<br>MIC = 1.25 mg/mL | Okafor et al., 2017; Chaiharn et al., 2018 |
| | *B. subtilis* | Methanol, Hydroalcoholic | IZD = 23 mm<br>MIC = 0.068–3 mg/mL | Ramesh and Pattar, 2010; Adebayo et al., 2018 |
| | *M. luteus* | Ethyl acetate, Aqueous | IZD = 6.3–8 mm<br>MIC = 1.75 mg/mL | Chaiharn et al., 2018 |
| | *L. monocytogenes* | Hydroalcoholic | MIC = 68–126 μg/mL | Adebayo et al., 2018 |
| | *S. aureus* | Ethyl acetate, Aqueous, Methanol, Hydroalcoholic | IZD = 6.1–18 mm<br>MIC = 0.063–2.5 mg/mL | Ramesh and Pattar, 2010; Chaiharn et al., 2018; Adebayo et al., 2018 |
| | *S. agalactiae* | Hexane, Hydroalcoholic | IZD = 11 mm<br>MIC = 63–137 μg/mL | Okafor et al., 2017; Adebayo et al., 2018 |
| *Pleurotus squarrosolus* (FB) | *B. cereus* | Ethanol, Hot water, Cold water | IZD = 4.44–9.76 mm | Nwachukwu and Uzoeto, 2010 |
| | *S. aureus* | | IZD = 0.86–11.71 mm<br>MIC = 25–50 mg/mL | |
| | *S. pneumonia* | | IZD = 5.12–6.75 mm | |
| *Pleurotus populinus* (FB) | *B. cereus* | Hexane | IZD = 11 mm | Okafor et al., 2017 |
| | *S. agalactiae* | | IZD = 10 mm | |
| *Pleurotus flabelatus* (FB) | *S. aureus* | Methanol | IZD = 17.6 mm<br>MIC = 62.5–83.3 mg/mL | Prastiyanto et al., 2020; Pandey et al., 2021 |
| *Pleurotus flabelatus* (M) | *S. aureus* | Hot water, Hot alkali | MIC = 2.5–5 mg/mL | Vunduk et al., 2019 |
| *Pleurotus flabelatus* (CB) | *S. aureus* | Exopolysaccharide extract | MIC = 2.5 mg/mL | |
| *Pleurotus citrinopileatus* (M) | *B. cereus* | Hot water | IZD = 9.6 mm | Özdal et al., 2019 |
| *Pleurotus florida* (M) | *B. cereus* | Hot water, Aqueous | IZD = 8.8 mm | Özdal et al., 2019; Illuri et al., 2022 |
| | *B. subtilis* | | IZD = 30 mm | |
| | *S. aureus* | | IZD = 37 mm | |
| | *S. mutans* | | IZD = 32 mm | |
| *Pleurotus djamor* (FB) | *S. aureus* | Hot water, Acetonitrile | IZD = 8–9.4 mm | Dulay et al., 2017; Otali et al., 2024 |
| *Pleurotus djamor* (M) | *S. aureus* | Aqueous | IZD = 28 mm | Illuri et al., 2022 |
| | *S. mutans* | | IZD = 30 mm | |
| | *B. subtilis* | | IZD = 28 mm | |
| *Pleurotus tuber-regium* (FB) | *B. subtilis* | Hydroalcoholic | MIC = 48 μg/mL<br>MBC = 48 μg/mL | Adebayo et al., 2018 |
| | *L. monocytogenes* | | MIC = 24 μg/mL<br>MBC = 48 μg/mL | |
| | *S. aureus* | | MIC = 12 μg/mL<br>MBC = 48 μg/mL | |
| | *S. agalactiae* | | MIC = 6 μg/mL<br>MBC = 48 μg/mL | |
| *Pleurotus levis* (FB) | *B. subtilis* | Hydroalcoholic | MIC = 3.3 μg/mL | |
| | *L. monocytogenes* | | MIC = 106.6 μg/mL | |
| | *S. aureus* | | MIC = 26.6 μg/mL | |
| | *S. agalactiae* | | MIC = 13.3 μg/mL | |

*The antibacterial activity is expressed in MIC/MBC (minimal inhibitory/bactericidal concentrations), IZD (internal zone diameter); FB–fruiting body, M–mycelium, CB–culture broth; AN-animal origin.*

antibacterial potential of the culture broth from the submerged fermentation of *P. ostreatus* was also the focus of the study. The water extract of the culture broth produced a 17 mm zone against *S. pyogenes* and *B. thuringiensis*, followed by 12 mm for *S. aureus* and *B. subtilis*, while no inhibition was observed only against *B. megaterium* (Younis et al., 2015).

In addition to *P. ostreatus*, other species of the genus *Pleurotus* have also been tested for the production of antibacterial agents. In the study by Tambekar et al. (2006), the effect of different solvents used for the extraction of fruiting bodies of *P. sajor-caju* on antibacterial activity was investigated. The extracts of ether, MeOH, and xylene showed the highest antibacterial activity against *S. aureus*, while the acetone extract showed the lowest activity. The high activity of the MeOH extract against *S. aureus* has also been demonstrated elsewhere (Akyuz et al., 2010; Gogavekar et al., 2014). *Staphylococcus epidermidis* was most sensitive to EtOH extract, while the smallest zone of inhibition was achieved with aqueous and xylene extracts. This illustrates how solvent extraction influences both the type and concentration of compounds in the final product, which ultimately affects the antimicrobial spectrum of activity. The MeOH extract of the same mushroom was also active against *M. luteus* and *B. cereus* (Gogavekar et al., 2014), while the latter reacted only moderately to the hexane extract (Okafor et al., 2017). The moderate activity was also observed when *B. cereus* was treated with a hot water extract (HWE) from the mycelium of *P. sajor-caju* (Özdal et al., 2019).

Similar to *P. ostreatus*, the growth of *B. subtilis* and *S. aureus* was also strongly inhibited by the MeOH extract of the fruiting body of *P. pulmonarius* (Ramesh and Pattar, 2010). The bacterium that was most sensitive to the aqueous extract was *B. cereus* (MIC = 1.25 mg/mL), followed by *M. luteus* (MIC = 1.75 mg/mL), while *S. aureus* was the most resistant, with the highest MIC of 2.5 mg/mL (MIC values for amoxicillin were in the range of 1.25–2.75 mg/mL) (Chaiharn et al., 2018). On the other hand, the growth of *S. aureus* was strongly inhibited by the hydroalcoholic extract of *P. pulmonarius* (MIC = 63–68 µg/mL), followed by *L. monocytogenes* and *B. subtilis* (MIC = 68–126 µg/mL), while slightly higher concentrations were required to inhibit the growth of *S. agalactiae* (MIC = 63–137 µg/mL) (Adebayo et al., 2018). In addition, the ethyl acetate extract showed the lowest antibacterial activity against all tested Gram-positive bacteria (Chaiharn et al., 2018).

The fruiting body of *P. eryngii* (MeOH extract) showed antimicrobial properties against *S. aureus* and was active against *B. subtilis* and *E. faecium* (Suay et al., 2000). Additionally, moderate activity against *B. cereus* was observed with HWE of the same mushroom (Özdal et al., 2019).

When comparing the antibacterial growth activities in EtOH, HWE, and cold-water extract from fruiting bodies of *Pleurotus squarrosolus*, the EtOH extract showed the highest inhibitory activity against *B. cereus*, cold-water extract against *S. aureus*, and HWE against *S. pneumoniae* (Nwachukwu and Uzoeto, 2010). The efficiency of HWE depends on the temperature and time applied and can be a double-edged sword—higher temperature can increase extraction efficiency but can also degrade sensitive compounds. *Pleurotus squarrosolus* extracted at 100 °C for 4 h showed better antibacterial activity compared to extraction at 28 °C for 36 h (Shen et al., 2017).

The hexane extract of *Pleurotus populinus* showed moderate activity against *B. cereus* and *S. agalactiae* (Okafor et al., 2017). A high concentration of *Pleurotus flabelatus* MeOH extract was required to inhibit the growth of *S. aureus* (MIC = 62.5 mg/mL) (Prastiyanto et al., 2020), which is consistent with the results of Pandey et al. (2021) who reported MIC = 83.3 mg/mL. However, HWE and hot alkali extract of its mycelium were active against *S. aureus* (MIC = 2.5 and 5 mg/mL, respectively) and *E. faecalis* (MIC = 10 and 20 mg/mL, respectively) (Vunduk et al., 2019). Similar results were demonstrated for *Pleurotus djamor*, whose HWE of the fruiting body showed low to moderate activity against *S. aureus* (internal zone diameter or IZD = 8 mm) (Otali et al., 2024), while the aqueous extract of the mycelium showed significant antimicrobial activity (IZD = 28–30 mm) against *S. aureus*, *B. subtilis*, and *S. mutans* (Illuri et al., 2022). The broad-spectrum antimicrobial activity of the aqueous extract was also demonstrated for the mycelium of *P. florida*, which proved to be very effective against *B. subtilis*, *S. aureus*, and *S. mutans*. However, HWE from the mycelium of *P. citrinopileatus* showed moderate activity against *B. cereus* (Özdal et al., 2019).

When comparing hydroalcoholic extracts of *P. tuber-regium* and *Pleurotus levis*, Adebayo et al. (2018) found that *B. subtilis* was very sensitive to *P. levis*, while lower MIC values were obtained for *P. tuber-regium* for all other strains tested (*L. monocytogenes*, *S. aureus*, and *S. agalactiae*).

### 7.4.2 *Pleurotus* spp. Extracts with Antimicrobial Activity against Gram-Negative Bacteria

Extensive studies were conducted to test the antibacterial activity of the fruiting body, mycelium, and culture broth of *Pleurotus* spp. against Gram-negative bacteria (Table 7.2). Among the solvents tested for the extraction of bioactive compounds with antibacterial properties from the fruiting body of *P. ostreatus*, the aqueous extract showed the strongest activity against *P. aeruginosa* (Hearst et al., 2009; Younis et al., 2015), followed by MeOH (Chowdhury et al., 2015; Younis et al., 2015), while EtOH and hexane fractions showed the weakest activity (Okafor et al., 2017; Sutthisa and Anujakkawan, 2023). Adebayo et al. (2018) investigated the activity of hydroalcoholic extracts of two strains of *P. ostreatus* (CP-50, country of origin: USA and Ukraine; and CP-800, country of origin: Nigeria) and found that no inhibitory activity was observed for strain CP-50 even at the highest concentration tested, while strain CP-800 was bacteriostatic at 73 µg/mL and bactericidal at 146 µg/mL against the clinical isolate *P. aeruginosa*. *Enterobacter aerogenes* and *Enterobacter cloacae*, which are considered

**TABLE 7.2**
***Pleurotus* spp. Extracts with Antimicrobial Activity Against Gram-negative Bacteria**

| Mushroom | Microorganism | Extract | Results | References |
|---|---|---|---|---|
| *Pleurotus ostreatus* (Jacq. Ex Fr.) P. Kumm. (FB) | *P. aeruginosa* | Aqueous, Hexane, Ethanol, Methanol, Hydroalcoholic | IZD = 7–20 mm<br>MIC = 0.073–150 mg/mL<br>MBC = 146 μg/mL | Hearst et al., 2009; Okafor et al., 2017; Younis et al., 2015; Chowdhury et al., 2015; Adebayo et al., 2018; Sutthisa and Anujakkawan, 2023 |
| | *E. aerogenes* | Ethyl acetate, Ethanol | IZD = 6.5–8 mm<br>MIC = 12.5 mg/mL | Chaiharn et al., 2018 |
| | *E. cloacae* | Ethanol | IZD = 9.6 mm | Sutthisa and Anujakkawan, 2023 |
| | *E. coli* | Methanol, Chloroform, Water, Hydroalcoholic, Ethyl acetate, Acetone, Methanol:chloroform:water = 2:2:1 | IZD = 6–23 mm<br>MIC = 0.073–6 mg/mL<br>MBC = 146 μg/mL | Akyuz et al., 2010; Karaman et al., 2010; Han et al., 2015; Chowdhury et al., 2015; Younis et al., 2015; Adebayo et al., 2018; Hamad et al., 2022; Sutthisa and Anujakkawan, 2023 |
| | *E. coli* O157:H7 | Ethyl acetate, Methanol | IZD = 6.3–9 mm<br>MIC = 5.5 mg/mL | Chaiharn et al., 2018 |
| | *P. vulgaris* | Ethyl acetate, Ethanol | IZD = 6.1–8.5 mm<br>MIC = 5.25 mg/mL | |
| | *P. mirabilis* | Methanol | IZD = 9.2 mm | Karaman et al., 2010 |
| | *K. pneumonia* | Methanol, Water | IZD = 8–19 mm<br>MIC = 7 mg/mL | Akyuz et al., 2010; Younis et al., 2015; Chowdhury et al., 2015 |
| | *S. flexneri* | Chloroform | IZD = 9.5 mm | Karaman et al., 2010 |
| | *S. dysenteriae* | Water, Methanol | IZD = 6–13 mm | Younis et al., 2015 |
| | *S. enterica* | | IZD = 6–8 mm | |
| | *S.* Typhimurium | Ethyl acetate, Methanol, Ethanol | IZD = 6.5–7.1 mm<br>MIC = 17.5 mg/mL | Chaiharn et al., 2018 |
| | *S.* Typhi | Methanol, Ethanol, Hydroalcoholic | IZD = 9.2–11.5 mm<br>MIC = 0.125–5 mg/mL | Chowdhury et al., 2015; Adebayo et al., 2018; Sutthisa and Anujakkawan, 2023 |
| | *S. marcescens* | Ethanol | IZD = 10.11 mm<br>MIC = 150 mg/mL | Sutthisa and Anujakkawan, 2023 |
| | *Stenotrophomonas sp.* | Hydroalcoholic | MIC = 36–63 μg/mL<br>MBC = 36 μg/mL | Adebayo et al., 2018 |
| | *A. actinomycetemcomitans* | Acetone | IZD = 9–10 mm | Han et al., 2015 |
| | *A. vitis* | Hexane | IZD = 11 mm | Okafor et al., 2017 |
| *Pleurotus ostreatus* (M) | *P. aeruginosa* | Hot water | IZD = 9.4 mm | Özdal et al., 2019 |
| | *K. oxytoca* | | IZD = 10.8 mm | |
| | *H. pylori* | | IZD = 12.4 mm | |
| | *X. campestris* | | IZD = 10.7 mm | |
| | *K. pneumonia* | Water | IZD = 5 mm | Younis et al., 2015 |
| | *S. dysenteriae* | | IZD = 5 mm | |
| | *E. coli* | Ethanol | IZD = 18 mm | Kalyoncu et al., 2010 |
| | *S.* Typhimurium | | IZD = 12 mm | |
| | *P. vulgaris* | | IZD = 8 mm | |
| | *E. cloacae* | | IZD = 20 mm | |
| | *E. aerogenes* | | IZD = 8 mm | |
| *Pleurotus ostreatus* (CB) | *E. coli* | Water | IZD = 12 mm | Younis et al., 2015 |
| | *K. pneumoniae* | | IZD = 12 mm | |
| | *P. aeruginosa* | | IZD = 13 mm | |
| | *S. dysenteriae* | | IZD = 13 mm | |
| | *S. enterica* | | IZD = 13 mm | |
| *Pleurotus sajor-caju* (FB) | *E. aerogenes* | Aqueous, Ethanol, Methanol, Ether, Benzene, Xylene, Acetone, Ethyl acetate | IZD = 6.1–19 mm | Tambekar et al., 2006; Chaiharn et al., 2018 |

*(Continued)*

**TABLE 7.2 *(Continued)***
***Pleurotus* spp. Extracts with Antimicrobial Activity Against Gram-negative Bacteria**

| Mushroom | Microorganism | Extract | Results | References |
|---|---|---|---|---|
| | *E. coli* | Aqueous, Ethanol, Methanol, Ether, Benzene, Xylene, Acetone | IZD = 12–21 mm | Tambekar et al., 2006; Gogavekar et al., 2014 |
| | *K. pneumonia* | Aqueous, Ethanol, Methanol, Ether, Benzene, Xylene, Acetone | IZD = 12–19 mm | Tambekar et al., 2006 |
| | *P. vulgaris* | | IZD = 13–19 mm | |
| | *P. aeruginosa* | Aqueous, Ethanol, Methanol, Ether, Benzene, Xylene, Acetone, Hexane | IZD = 11–19 mm | Tambekar et al., 2006; Okafor et al., 2017 |
| | *S.* Typhi | Aqueous, Ethanol, Methanol, Ether, Benzene, Xylene, Acetone | IZD = 14–17 mm<br>MIC = 100 μg/mL<br>$IC_{50}$ = 25 μg/mL | Tambekar et al., 2006; Kandasamy et al., 2020 |
| | *S.* Typhimurium | | IZD = 12–29 mm | Tambekar et al., 2006; Gogavekar et al., 2014 |
| | *A. vitis* | Hexane | IZD = 12 mm | Okafor et al., 2017 |
| *Pleurotus sajor-caju* (M) | *P. aeruginosa* | Hot water | IZD = 9 mm | Özdal et al., 2019 |
| | *K. oxytoca* | | IZD = 10.3 mm | |
| | *H. pylori* | | IZD = 7.3 mm | |
| | *X. campestris* | | IZD = 9.7 mm | |
| *Pleurotus eringii* (FB) var. eryngii var. ferulae | *E. coli* | Methanol | IZD = 8–8.5 mm | Akyuz et al., 2010 |
| *Pleurotus eringii* (M) | *E. coli* | Ethanol | IZD = 8 mm | Kalyoncu et al., 2010 |
| | *P. aeruginosa* | Hot water | IZD = 8.8 mm | Özdal et al., 2019 |
| | *K. oxytoca* | | IZD = 15.2 mm | |
| | *H. pylori* | | IZD = 8.4 mm | |
| | *X. campestris* | | IZD = 12.1 mm | |
| *Pleurotus pulmonarius* (FB) | *E. aerogenes* | Ethyl acetate, Aqueous | IZD = 6.2–8.5 mm<br>MIC = 10 mg/mL | Chaiharn et al., 2018 |
| | *E. coli* O157:H7 | | IZD = 6.3–8 mm<br>MIC = 5.25 mg/mL | |
| | *P. vulgaris* | | IZD = 6.3–8 mm<br>MIC = 5.25 mg/mL | |
| | *S.* Typhimurium | | IZD = 6.3–9 mm<br>MIC = 15.5 mg/mL | |
| | *E. coli* | Methanol, Hexane, Hydroalcoholic | IZD = 3–12 mm<br>MIC = 126 μg/mL | Ramesh and Pattar, 2010; Okafor et al., 2017; Adebayo et al., 2018 |
| | *P. aeruginosa* | | IZD = 10–19 mm<br>MIC = 0.126–2 mg/mL | |
| | *S.* Typhi | Hydroalcoholic | MIC = 126–137 μg/mL | Adebayo et al., 2018 |
| | *Stenotrophomonas spp.* | | MIC = 63–137 μg/mL | |
| | *S. dysenteriae* | Hexane | IZD = 2 mm | Okafor et al., 2017 |
| | *A. vitis* | | IZD = 11 mm | |
| *Pleurotus squarrosolus* (FB) | *P. aeruginosa* | Hot water | IZD = 3.41 mm<br>MIC = 50 mg/mL | Nwachukwu and Uzoeto, 2010 |
| | *E. coli* | Ethanol, Hot water | IZD = 1.7–2.1 mm<br>MIC = 50 mg/mL | |
| | *K. pneumonia* | Ethanol, Hot water, Cold water | IZD = 0.5–6.14 mm<br>MIC = 50 mg/mL | |
| | *P. mirabilis* | | IZD = 1.9–3.1 mm<br>MIC = 50 mg/mL | |
| | *S.* Typhi | | IZD = 1.04–6.45 mm<br>MIC = 12.5–50 mg/mL | |
| *Pleurotus populinus* (FB) | *P. aeruginosa* | Hexane | IZD = 10 mm | Okafor et al., 2017 |
| | *E. coli* | | IZD = 3 mm | |
| | *S. dysenteriae* | | IZD = 3 mm | |
| | *A. vitis* | | IZD = 10 mm | |

**TABLE 7.2 *(Continued)***
***Pleurotus* spp. Extracts with Antimicrobial Activity Against Gram-negative Bacteria**

| Mushroom | Microorganism | Extract | Results | References |
|---|---|---|---|---|
| *Pleurotus flabelatus* (FB) | *P. aeruginosa* | Methanol | IZD = 15 mm<br>MIC = 62.5 mg/mL | Pandey et al., 2021 |
| | *P. vulgaris* | | IZD = 16.6 mm<br>MIC = 52.08 mg/mL | |
| | *P. mirabilis* | | IZD = 17.6 mm<br>MIC = 26.03 mg/mL | |
| | *S. marcescens* | | IZD = 11.3 mm<br>MIC = 83.3 mg/mL | |
| | *M. catarrhalis* | | IZD = 13.3 mm<br>MIC = 52.08 mg/mL | |
| | *S. flexneri* | | IZD = 14.3 mm<br>MIC = 52.08 mg/mL | |
| *Pleurotus flabelatus* (M) | *E. coli* | Hot water, Hot alkali | MIC = 5–20 mg/mL<br>MBC = 20 mg/mL | Vunduk et al., 2019 |
| | *P. aeruginosa* | | MIC = 5–20 mg/mL<br>MBC = 20 mg/mL | |
| *Pleurotus flabelatus* (CB) | *E. coli* | Exopolysaccharide extract | MIC = 10 mg/mL<br>MBC = 20 mg/mL | |
| | *P. mirabilis* | | MIC = 10 mg/mL | |
| | *P. aeruginosa* | | MIC = 10 mg/mL | |
| *Pleurotus citrinopileatus* (M) | *P. aeruginosa* | Hot water | IZD = 8.1 mm | Özdal et al., 2019 |
| | *K. oxytoca* | | IZD = 9.8 mm | |
| | *H. pylori* | | IZD = 7.3 mm | |
| | *X. campestris* | | IZD = 11.8 mm | |
| *Pleurotus florida* (M) | *P. aeruginosa* | Hot water, Aqueous | IZD = 9.7–38 mm | Özdal et al., 2019; Illuri et al., 2022 |
| | *E. coli* MDR | Aqueous | IZD = 36 mm | Illuri et al., 2022 |
| | *K. pneumonia* | | IZD = 32 mm | |
| | *V. cholera* | | IZD = 45 mm | |
| | *S.* Typhi | | IZD = 32 mm | |
| | *K. oxytoca* | Hot water | IZD = 11.6 mm | Özdal et al., 2019 |
| | *H. pylori* | | IZD = 13 mm | |
| | *X. campestris* | | IZD = 9.8 mm | |
| *Pleurotus djamor* (FB) | *E. coli* | Hot water | IZD = 19.3 mm | Otali et al., 2024 |
| | *P. aeruginosa* | | IZD = 23.3 mm | |
| *Pleurotus djamor* (M) | *E. coli* MDR | Aqueous | IZD = 30 mm | Illuri et al., 2022 |
| | *K. pneumonia* | | IZD = 30 mm | |
| | *P. aeruginosa* | | IZD = 36 mm | |
| | *V. cholera* | | IZD = 39 mm | |
| | *S.* Typhi | | IZD = 32 mm | |
| *Pleurotus tuber-regium* (FB) | *E. coli* | Hydroalcoholic | MIC = 48 μg/mL<br>MBC = 96 μg/mL | Adebayo et al., 2018 |
| | *Stenotrophomonas spp.* | | MIC = 24 μg/mL<br>MBC = 48 μg/mL | |
| | *S.* Typhi | | MIC = 24 μg/mL<br>MBC = 24 μg/mL | |
| | *P. aeruginosa* | | MIC = 12 μg/mL<br>MBC = 24 μg/mL | |
| *Pleurotus levis* (FB) | *Stenotrophomonas spp.* | | MIC = 106.6 μg/mL | |

*The antibacterial activity is expressed in MIC/MBC (minimal inhibitory/bactericidal concentrations), IZD (internal zone diameter); FB–fruiting body, M–mycelium, CB–culture broth; MDR- multidrug-resistant.*

important opportunistic and MDR nosocomial pathogens responsible for various infections in humans, were sensitive to the EtOH extract of *P. ostreatus*, while the ethyl acetate extract showed low activity (Chaiharn et al., 2018; Sutthisa and Anujakkawan, 2023). Furthermore, no inhibitory activity against *E. aerogenes* was found for the MeOH extract (Chaiharn et al., 2018). Most studies found no activity of EtOH and MeOH extracts of *P. ostreatus* against *E. coli* (Karaman et al., 2010; Han et al., 2015; Younis et al., 2015; Sutthisa and Anujakkawan, 2023), while Chowdhury et al. (2015) and Akyuz et al. (2010) reported positive activity. However, these discrepancies can be attributed to the different origin of the mushrooms, the bacterial strains (American Type Culture Collection—ATCC or clinical isolate), and the concentrations used. In addition, water extract showed the highest activity (Younis et al., 2015), followed by hydroalcoholic extract (Adebayo et al., 2018), while ethyl acetate, chloroform, and acetone showed moderate-to-low activity (Karaman et al., 2010; Han et al., 2015; Sutthisa and Anujakkawan, 2023). The growth of *E. coli* O157:H7 was inhibited by ethyl acetate and MeOH extracts (Chaiharn et al., 2018). Other genera within the *Enterobacteriaceae* family showed variable sensitivity to *P. ostreatus* extracts. The ethyl acetate extract was effective in inhibiting the growth of *Proteus vulgaris* and *Salmonella* ser. *typhimurium* (Chaiharn et al., 2018), while no activity was observed in clinical isolates of *S. typhi* and *Serratia marcescens* (Sutthisa and Anujakkawan, 2023), whereas the EtOH extract was active against all four bacteria, with *S. marcescens* being the most sensitive (Chaiharn et al., 2018; Sutthisa and Anujakkawan, 2023). In addition, the EtOH extract showed no antibacterial activity against *Klebsiella pneumoniae*, *Shigella dysenteriae*, and *Salmonella enterica* (Younis et al., 2015). Of the extracts tested, MeOH showed the broadest spectrum of activity, as it inhibited the growth of *Proteus mirabilis* (Karaman et al., 2010), *K. pneumoniae* (Akyuz et al., 2010; Chowdhury et al., 2015), *S. typhi* (Chowdhury et al., 2015), *S. dysenteriae*, *S. enterica* (Younis et al., 2015) and *S. typhimurium* (Chaiharn et al., 2018), while *Shigella flexneri* grew uninhibited (Karaman et al., 2010). The latter were sensitive to chloroform extract (Karaman et al., 2010). According to Younis et al. (Younis et al., 2015), the water extract was effective against *K. pneumoniae*, *S. dysenteriae* and *S. enterica*. The acetone extract of the fruiting body of *P. ostreatus* was not only effective against Gram-positive bacteria associated with dental caries, but also inhibited the growth of *Aggregatibacter actinomycetemcomitans*, the most common pathogen causing periodontitis (Han et al., 2015). When the mycelium of *P. ostreatus* was tested, the EtOH extract proved to be a good antibacterial agent, as it had an inhibitory effect on five (*E. coli*, *S. typhimurium*, *P. vulgaris*, *E. cloacae*, and *E. aerogenes*) out of six Gram-negative bacteria tested (Kalyoncu et al., 2010). Similar activity was obtained when water was used for extraction. *Pseudomonas aeruginosa*, *K. pneumoniae*, *Klebsiella oxytoca*, *S. dysenteriae*, *Helicobacter pylori*, and *Xanthomonas campestris* were sensitive to the extract (Younis et al., 2015; Özdal et al., 2019). The culture broth of *P. ostreatus* inhibited all five tested pathogenic bacteria belonging to the *Enterobacteriaceae* family, with an inhibition zone of 12–13 mm (Younis et al., 2015).

Tambekar et al. (2006) found that in most cases better antibacterial activity was obtained when aqueous, MeOH, xylene, and EtOH extracts were used, while ether, benzene, and especially acetone extracts of the fruiting body of *P. sajor-caju* exhibited the weakest zones of inhibition against the tested Gram-negative bacteria (*E. coli*, *P. aeruginosa*, *K. pneumoniae*, *S. typhi*, *P. vulgaris*, *E. aerogenes*, and *S. typhimurium*). This research was also supported by the study of Gogavekar et al. (2014), who found high activity of the MeOH extract against *E. coli* and especially *S. typhimurium*.

The pathogen *H. pylori*, which is recognized as a major cause of chronic gastritis and a risk factor for gastric malignancies, was inhibited by mycelial HWE of *Pleurotus* spp. in decreasing order: *P. florida* > *P. ostreatus* > *P. eryngii* > *P. sajor-caju* > *P. citrinopileatus*. The clinical isolate *K. oxytoca*, which has been reported to be resistant to a wide range of antibiotics, was the most sensitive of the bacteria tested, and the strongest inhibition was observed with the application of *P. eryngii* extract, followed by *P. florida*, *P. ostreatus*, *P. sajor-caju*, and *P. citrinopileatus*, while the weakest zone of inhibition was found with *P. aeruginosa* and ranged between 8.1 and 9.7 mm. The chemical composition revealed the highest phenolic content in *P. ostreatus*, followed by *P. sajor-caju*, while the highest flavonoid content was found in *P. sajor-caju* and *P. florida* (Özdal et al., 2019).

Chaiharn et al. (2018) compared the antibacterial properties of the ethyl acetate and aqueous extracts of the fruiting body of *P. pulmonarius* and found that all tested Gram-negative bacteria were more sensitive to the aqueous extract, with the lowest MIC of 5.25 mg/mL being achieved against *E. coli* O157:H7 and *P. vulgaris*. The MeOH, hexane, and hydroalcoholic extracts of the same mushroom inhibited the growth of *E. coli* and *P. aeruginosa* (Ramesh and Pattar, 2010; Okafor et al., 2017; Adebayo et al., 2018). Additionally, the hexane extract was effective against *S. dysenteriae* (Okafor et al., 2017); while the hydroalcoholic extract showed activity against *S. typhi* and *Stenotrophomonas* spp. (Adebayo et al., 2018).

Nwachukwu and Uzoeto (2010) tested the activity of EtOH, HWE, and cold-water extract of the fruiting body of *P. squarrosolus* and found that *K. pneumoniae* was most sensitive to the EtOH extract, while HWE and cold-water extract were most active against *S. typhi*.

The hexane extract of *P. populinus* was effective against *P. aeruginosa*, *E. coli* and *S. dysenteriae* (Okafor et al., 2017). The MeOH extract of *P. flabelatus* was able to inhibit the growth of *P. aeruginosa*, *P. vulgaris*, *P. mirabilis*, *S. marcescens*, *S. flexneri*, and *Moraxella catarrhalis*, with *P. mirabilis* being the most sensitive (MIC = 26.03 mg/mL) and *S. marcescens* being the most resistant (MIC = 83.3 mg/mL) (Pandey et al., 2021). The water and alkali mycelium extracts of the same mushroom showed microbicidal activity against clinical isolates of *E. coli* and *P. aeruginosa*, while the ATCC strains were found to be more resistant. When the culture broth was tested, it was found to have an even broader spectrum of activity, as it was also active against clinical isolates of *P. mirabilis* (Vunduk et al., 2019).

Illuri et al. (2022) tested the antibacterial potential of the aqueous mycelial extract of *P. florida* and *P. djamor* and found that *P. florida* was active against most bacteria and had a broader zone of inhibition. In addition, *V. cholerae*, the causative agent of the human intestinal disease cholera, which leads to severe diarrhea, was most sensitive to *P. florida*, followed by *P. aeruginosa* and MDR *E. coli*.

Among the *Pleurotus* spp. tested, the hydroalcoholic extract of the fruiting body of *P. tuber-regium* (from Nigeria) was the one with the strongest antimicrobial activity. The MIC values showed that *P. aeruginosa* was the most sensitive, followed by *S. typhi* and *Stenotrophomonas*, while the highest concentration was required to inhibit the growth of *E. coli*. Furthermore, microbicidal activity against all tested bacteria was only obtained for this mushroom (Adebayo et al., 2018).

## 7.5 *PLEUROTUS* SPP. EXTRACTS WITH ANTIFUNGAL ACTIVITY

Different extracts obtained from *Pleurotus* spp. were described in literature as possessing antifungal activity (Table 7.3). The literature review shows that extracts of *Pleurotus* spp. have mainly been tested against various *Candida* species, in particular *Candida albicans*. *Candida albicans*, a polymorphic fungus, is a member of the normal human microbiome. In most people, it exists as a harmless, lifelong commensal. However, under certain conditions, *C. albicans* can become pathogenic and cause infections ranging from mild skin infections to severe, life-threatening systemic infections (Mayer et al., 2013). Akyuz et al. (2010) found no activity of MeOH extracts from wild (collected from different parts of Turkey) and cultured *P. ostreatus* against *C. albicans*, while Chowdhury et al. (2015) and Younis et al. (2015) reported positive activity. The bioactive compounds extracted from *P. ostreatus* using the solvent mixture methanol, chloroform, and water (in a ratio of 2:2:1) were found to be effective in suppressing the growth of *C. albicans* (Hamad et al., 2022). Of the extracts tested, the highest sensitivity was observed when *C. albicans* was treated with the water extract, while EtOH was the least effective (Younis et al., 2015). Somewhat opposite results were observed for the mycelial extracts, as EtOH appears to be promising against *C. albicans*, while moderate activity was observed for water (Kalyoncu et al., 2010; Younis et al., 2015). The anticandidal activity of the culture broth was even higher than that of the mycelium (Younis et al., 2015).

MeOH and EtOH extracts of the mycelium of *P. eryngii* were effective in inhibiting the growth of *C. albicans* (Kalyoncu et al., 2010; Akyuz et al., 2010), as were the EtOH and water extracts of the fruiting body of *P. squarrosolus* (Nwachukwu and Uzoeto, 2010).

The MeOH extract of the fruiting body of *P. pulmonarius* showed higher activity against *C. albicans* (IZD = 24 mm, MIC = 1 mg/mL) than that of the standard nystatin tested (IZD = 19 mm, MIC = 2 mg/mL) (Ramesh and Pattar, 2010). The aqueous extracts from the mycelium of *P. florida* and *P. djamor* also showed antifungal activity against *C. albicans* (Illuri et al., 2022).

The highest activity against *Candida dubliniensis* was demonstrated for the water extract of the fruiting body of *P. ostreatus* (IZD = 17 mm), while EtOH and water extracts of the mycelium and culture broth were less effective (IZD = 4–7 mm) (Younis et al., 2015). The growth of *Candida glabrata* was inhibited by MeOH extracts of the fruiting body of *P. sajor-caju*, *P. eryngii*, and *P. ostreatus* (Akyuz et al., 2010).

Of the other yeasts tested, the aqueous extract from the mycelium of *P. florida* was very effective in inhibiting the growth of *Scheffersomyces stipites* (formerly *Pichia stipites*), while the extract from the mycelium of *P. djamor* showed less activity (Illuri et al., 2022). The MeOH extract of *P. ostreatus* showed activity against *Saccharomyces cerevisiae* (Chowdhury et al., 2015), while *Geotrichum candidum*, a yeast-like organism that is pathogenic to certain plant tissues, and *Cryptococcus humicola* were sensitive to water, MeOH, and EtOH extracts of the same mushroom. *Trichosporon cutaneum*, which is responsible for superficial hair infections and allergic pneumonitis, was strongly inhibited by the water extract of the fruiting body of *P. ostreatus* and moderately inhibited by its mycelium, while MeOH and EtOH extracts were ineffective (Younis et al., 2015).

Two dermatophytes (pathogenic fungi), *Epidermophyton* spp. (which can cause skin diseases such as athlete's foot, itch, or ringworm) and *Trichophyton* spp., were found to be sensitive to the MeOH extracts of *P. eryngii* and *P. ostreatus*, while *P. sajor-caju* had no effect on their growth (Akyuz et al., 2010). A comprehensive screening of the antifungal activity of the fruiting body, broth, and mycelia of *P. ostreatus* was performed (Younis et al., 2015). According to the results, the water extract of the fruiting body showed the broadest activity with the highest inhibition zone values, as it acted on 16 out of 20 fungal species, while MeOH and EtOH extracts as well as the culture broth and mycelia acted on 7–8 out of 20 species. Specifically, the water extract strongly inhibited the growth of *Fusarium moniliforme* (plant pathogen), *A. fumigatus* (one of the main causative agents of invasive aspergillosis), *Aspergillus flavus* (human pathogen, allergen, and mycotoxin producer—aflatoxin), *Drechslera rostrata*, and *Curvularia clavata* (human pathogen for skin infections), had a moderate effect on *Alternaria chlamydospore*, while lower activity was observed against *Penicillium expansum* (global postharvest pathogen), *Trichoderma viride*, *Aspergillus niger*, *Aspergillus terreus* (pathogen of aspergillosis with poor response to recommended therapies), and *Eupenicillium ludwigii*.

**TABLE 7.3**
***Pleurotus* spp. Extracts with Antifungal Activity**

| Mushroom | Microorganism | Extract | Results | References |
|---|---|---|---|---|
| *Pleurotus ostreatus* (Jacq. Ex Fr.) P. Kumm. (FB) | *C. albicans* | Methanol, Ethanol, Water, Methanol:chloroform:water = 2:2:1 | IZD = 1–30 mm<br>MIC = 4 mg/mL | Chowdhury et al., 2015; Younis et al., 2015; Hamad et al., 2022 |
| | *C. dubliniensis* | Water, Ethanol | IZD = 4–17 mm | Younis et al., 2015 |
| | *C. glabrata* | Methanol | IZD = 8–15.5 mm | Akyuz et al., 2010 |
| | *C. humicola* | Water, Ethanol, Methanol | IZD = 6–30 mm | Younis et al., 2015 |
| | *G. candidum* | | IZD = 6–25 mm | |
| | *T. viride* | | IZD = 3–12 mm | |
| | *T. cutaneum* | Water | IZD = 30 mm | |
| | *A. flavus* | | IZD = 23 mm | |
| | *A. niger* | | IZD = 12 mm | |
| | *A. terreus* | | IZD = 14 mm | |
| | *P. expansum* | | IZD = 8 mm | |
| | *E. ludwigii* | Water, Methanol | IZD = 7–15 mm | |
| | *D. rostrata* | | IZD = 6–23 mm | |
| | *A. chlamydospora* | | IZD = 3–19 mm | |
| | *F. moniliforme* | | IZD = 3–25 mm | |
| | *C. clavate* | Water, Ethanol | IZD = 3–23 mm | |
| | *A. fumigatus* | | IZD = 3–25 mm | |
| | *F. oxysporum* | Methanol:chloroform: water = 2:2:1 | GI = 47% | Hamad et al., 2022 |
| | *F. solani* | | GI = 28% | |
| | *R. solani* | | GI =21% | |
| | *S. cerevisiae* | Methanol | IZD = 15 mm<br>MIC = 4 mg/mL | Chowdhury et al., 2015 |
| | *Trichophyton* sp. (dermatophytes) | | IZD = 8–11.5 mm | Akyuz et al., 2010 |
| | *Epidermophyton* sp. (dermatophytes) | | IZD = 8–8.5 mm | |
| *Pleurotus ostreatus* (M) | *C. albicans* | Ethanol, Water | IZD = 3–12 mm | Kalyoncu et al., 2010; Younis et al., 2015 |
| | *C. dubliniensis* | Water | IZD = 6 mm | Younis et al., 2015 |
| | *C. humicola* | | IZD = 10 mm | |
| | *T. cutaneum* | | IZD = 15 mm | |
| | *G. candidum* | | IZD = 3 mm | |
| | *P. expansum* | | IZD = 3 mm | |
| | *C. clavate* | | IZD = 6 mm | |
| | *F. moniliforme* | | IZD = 3 mm | |
| *Pleurotus ostreatus* (CB) | *C. albicans* | | IZD = 12 mm | |
| | *C. dubliniensis* | | IZD = 7 mm | |
| | *A. flavus* | | IZD = 11 mm | |
| | *A. fumigatus* | | IZD = 7 mm | |
| | *A. niger* | | IZD = 5 mm | |
| | *A. terreus* | | IZD = 11 mm | |
| | *T. viride* | | IZD = 12 mm | |
| | *C. clavate* | | IZD = 5 mm | |
| | *F. moniliforme* | | IZD = 5 mm | |
| | *T. harzianum* | Culture filtrate | GI = 6.4–11.6% | Owaid et al., 2017 |
| | *Verticillium* sp. | | GI = 0.9–6.3% | |
| | *Pythium* sp. | | GI = 1.8–5.7% | |
| *Pleurotus sajor-caju* (FB) | *C. glabrata* | Methanol | IZD = 8.5 mm | Akyuz et al., 2010 |
| *Pleurotus eryngii* (FB) var. eryngii var. ferulae | *C. albicans* | | IZD = 7.5–8.5 mm | |

**TABLE 7.3 *(Continued)***
***Pleurotus* spp. Extracts with Antifungal Activity**

| Mushroom | Microorganism | Extract | Results | References |
|---|---|---|---|---|
| | *C. glabrata* | | IZD = 8.5–11.5 mm | |
| | *Epidermophyton* sp. (dermatophytes) | | IZD = 8–10 mm | |
| | *Trichophyton* sp. (dermatophytes) | | IZD = 9.5 mm | |
| *Pleurotus eryngii* (M) | *C. albicans* | Ethanol | IZD = 10 mm | Kalyoncu et al., 2010 |
| | *A. fumigatus* | Methanol | hazy/very hazy inhibition zone | Suay et al., 2000 |
| *Pleurotus pulmonarius* (FB) | *C. albicans* | Methanol | IZD = 24 mm<br>MIC = 1 mg/mL | Ramesh and Pattar, 2010 |
| *Pleurotus squarrosolus* (FB) | *C. albicans* | Ethanol, Cold water | IZD = 6.9–7.1 mm<br>MIC = 50 mg/mL | Nwachukwu and Uzoeto, 2010 |
| *Pleurotus florida* (M) | *C. albicans* | Aqueous | IZD = 30 mm | Illuri et al., 2022 |
| | *P. stipitis* | | IZD = 34 mm | |
| | *A. niger* | | GI = 79.2% | |
| | *A. terreus* | | GI = 82% | |
| *Pleurotus djamor* (M) | *C. albicans* | | IZD = 29 mm | |
| | *P. stipitis* | | IZD = 27 mm | |
| | *A. niger* | | GI = 86% | |
| | *A. terreus* | | GI = 82% | |
| *Pleurotus salmoneostramineus* | *T. harzianum* | Culture filtrate | GI = 4.4% | Owaid et al., 2017 |
| | *Verticillium* sp. | | GI = 12.3% | |
| | *Pythium* sp. | | GI = 6.8% | |
| *Pleurotus cornucopiae* | *T. harzianum* | | GI = 6.0% | |
| | *Verticillium* sp. | | GI = 1.3% | |
| | *Pythium* sp. | | GI = 4.6% | |

*The antifungal activity is expressed in MIC (minimal inhibitory concentrations), IZD (internal zone diameter), GI (growth inhibition) value. FB, fruiting body; M, mycelium, CB, culture broth.*

The extract showed no activity against *Penicillium roqueforti, Botrytis cinerea, Mucor rouxii*, and *Syncephalastrum racemosus*.

Illuri et al. (2022) tested the antifungal activity of the aqueous mycelial extract of *P. djamor* and *P. florida* using the biomass inhibition method and found that the highest inhibition (86%) was elicited by *P. djamor* on *A. niger*. Both mushrooms were significantly inhibitory to *A. terreus* (82%) at the highest concentration tested. Using the same assay, the polar extract of *P. ostreatus* was tested for its antifungal activity against three plant pathogenic fungi and showed the highest growth inhibition value against *Fusarium oxysporum* (47%), followed by *Fusarium solani* (28%) and *Rhizoctonia solani* (21%) (Hamad et al., 2022).

The percentage inhibition of different *Pleurotus* spp. (*P. ostreatus, Pleurotus cornucopiae* var. *citrinopileatus*, and *Pleurotus salmoneostramineus*) culture filtrates was tested against pathogenic fungi: *Verticillium* sp., *Trichoderma harzianum* (main cause of large losses in the cultivation of oyster mushrooms and white button mushrooms), and *Pythium* sp. The highest inhibition rate of 12.3%, followed by 11.6%, was observed for *P. salmoneostramineus* and *P. ostreatus* against *Verticillium* sp. and *T. harzianum*, respectively, while the highest growth inhibition (GI) value of 6.8% against *Pythium* sp. was observed for *P. salmoneostramineus* (Owaid et al., 2017).

## 7.6 ISOLATED COMPOUNDS FROM *PLEUROTUS* SPP. WITH REPORTED ANTIBACTERIAL/ANTIFUNGAL ACTIVITY AND THEIR MODE OF ACTION

Mushroom extracts have been studied mainly for their antimicrobial activity, while the compounds responsible for the activity remain largely unknown, although some of them have been described in the literature. Glucosylceramide, a sphingolipid molecule isolated from the fruiting body of the golden oyster mushroom *P. citrinopileatus* (synonym: *P. cornucopiae, P. cornucopiae* var. *citrinopileatus*), showed activity against *E. coli* and *S. aureus* with an $IC_{50}$ value of 275.1 μM (200 μg/mL) and 323.2 μM (235 μg/mL), respectively (Meng et al., 2012).

Of the 11 bacterial species tested, only *P. aeruginosa* ($IC_{50}$ = 51 μM), *Pseudomonas fluorescens* ($IC_{50}$ = 186 μM), and *S. aureus* ($IC_{50}$ = 34 μM) were sensitive to the ribonuclease (antibacterial protein) isolated from *P. sajor-caju* (Ngai and Ng, 2004).

A monomeric 17-kDa hemolysin, termed eryngeolysin, isolated from fresh fruiting bodies of the mushroom *P. eryngii*, showed antibacterial activity against *Bacillus* sp. ($IC_{50}$ values were in the range of 25 to 175 μM) but not against other Gram-positive and Gram-negative bacterial species. In addition, pore formation has been described as a mechanism of action of eryngeolysin on bacteria (Ngai and Ng, 2006)

A hexane–dichloromethane extract of *P. ostreatus* containing *p*-anisaldehyde showed antimicrobial activity against *B. subtilis, P. aeruginosa, F. oxysporum*, and *A. niger* (Okamoto et al., 2002).

As mentioned earlier, the diterpene antibiotic pleuromutilin was the first antibacterial drug discovered in the research of Kavanagh et al. (1951). Subsequent improvements led to the development of a number of new derivatives (more than 1200) that inhibit prokaryotic protein synthesis through a unique binding to the peptidyl transferase center of the 50S ribosomal subunit, preventing the binding of transfer RNA (Novak, 2011; File et al., 2019). While pleuromutilin itself exhibits Gram-positive activity (*S. aureus*, MRSA, MSSA, β-hemolytic *Streptococcus* spp., *S. pyogenes, S. agalactiae*, MDR *S. pneumoniae*), further modifications contribute significantly to lower MICs, although they have only a limited impact on the overall antimicrobial spectrum. In addition, pleuromutilins cover some fastidious Gram-negative pathogens, including MDR *Haemophilus influenzae* and *Moraxella catarrhalis*, and are also highly effective against atypical pathogens such as *Chlamydia* spp., *Mycoplasma* spp., and *Legionella* spp. The lack of activity against *E. faecalis* and Enterobacteriaceae is probably related to the presence of efflux pumps, which are common in these organisms (Novak, 2011). Pleuromutilins are the only class of antibiotics derived from macromycetes on the market (Hamers et al., 2021).

Detailed phytochemical analyses of the MeOH extract of *Pleurotus eous* by gas chromatography–mass spectrometry (GC–MS) confirmed the presence of 26 compounds with antibacterial and antifungal activity (such as various metabolites from the class of esters, alkanes, and organometallic compounds). The scanning electron micrographs showed that the cell membrane of the *S. aureus* cells treated with the extract was completely ruptured due to the adhesion of the extract molecules to the negatively charged surface of the cell wall and membrane. In addition, the results showed a significant decrease in membrane integrity of the treated *S. aureus* (42.67%) and a significant increase in membrane permeability (60.32%) (Dash et al., 2024). These results are consistent with previous studies indicating that treatment with mushroom extracts can damage bacterial cytoplasmic membranes, suggesting their potential targets (Matijaševic et al., 2016; Chan and Chong, 2022). In the MeOH extract of *P. sajor-caju*, four different compounds were identified by GC–MS that could be responsible for possible antimicrobial activities: β-sitosterol, cholestanol, 1,5-dibenzoylnaphthalene, and 1,2-benzenedicarboxylic acid (Gogavekar et al., 2014). In a similar study, the chemical profile of the polar extract of *P. ostreatus* using the GC–MS technique revealed the presence of antifungal and antibacterial compounds such as ethyl iso-allocholate (steroid derivative), hexadecane (CAS, alkanes), and pentacosane (aliphatic hydrocarbon) (Hamad et al., 2022).

Sub-fraction A of the ethyl acetate extract of *P. giganteus* showed antifungal activity on all tested *Candida* spp. with MIC values ranging from 2.0 to 10.3 μg/mL. GC–MS characterization showed the presence of large amounts of fatty acid methyl esters, that is., methyl and ethyl palmitate, methyl and ethyl oleate, methyl stearate, and methyl linoleate (Phan et al., 2013).

Pleurostrin, a 7 kDa peptide, an antifungal peptide isolated from the fruiting bodies of *P. ostreatus* by aqueous extraction and further purified by ion-exchange chromatography followed by affinity chromatography and gel filtration, showed activity at a concentration of 15.6 μM against phytopathogenic fungi such as *F. oxysporum, Mycosphaerella arachidicola*, and *Physalospora piricola* with GI of 20%, 45%, and 63%, respectively (Chu et al., 2005). These pathogens are difficult to control and can infect a range of crops (e.g., apples, pears, tomatoes, cucumbers, peanuts). Therefore, the compounds that act on these toxin producers could have important applications in the food industry. Another antifungal peptide with a molecular mass of 10 kDa, isolated from the fruiting bodies of *P. eryngii* and named eryngin, inhibited mycelial growth of *F. oxysporum* ($IC_{50}$ = 1.35 μM) and *M. arachidicola* ($IC_{50}$ = 3.5 μM). The N-terminal amino acid sequence of eryngin showed only slight similarity to the antifungal protein of the mushroom *Lyophyllum shimeji* (Wang and Ng, 2004).

A 12 kDa ribonuclease (an enzyme family) isolated from the fruiting bodies of *P. sajor-caju* inhibited mycelial growth of the fungi *F. oxysporum* and *M. arachidicola* with $IC_{50}$ values of 95 and 7 μM, respectively (Ngai and Ng, 2004)

The secondary metabolites of *Pleurotus* spp., 10-oxo-trans-8-decenoic acid (ODA) and 1-octen-3-ol, inhibit the growth of *P. expansum*, a common food spoilage organism, at concentrations >1.25 mM (Assaf et al., 1995; Okull et al., 2003; Gogavekar et al., 2014). The activity was more pronounced at lower pH values, suggesting that the undissociated carboxyl of ODA is important for inhibition (Okull et al., 2003).

## 7.7 *PLEUROTUS* SPP. EXTRACTS AND BIOACTIVE COMPOUNDS WITH ANTIVIRAL ACTIVITY

Viruses are the causative agents of around 60% of epidemic infectious diseases worldwide, leading to high mortality rates and economic problems. The effectiveness of crude extracts or isolated and purified compounds from edible and medicinal mushrooms in suppressing viruses has attracted the attention of many researchers. Various molecules such as polysaccharides, sterols, phenolic compounds, terpenoids, laccase, and lectins have been investigated for their antiviral activity, and various mechanisms of action have been proposed. Mushroom extracts and their compounds could slow down viral infection directly by inhibiting virus replication, targeting various viral proteins, and neutralizing host receptors to inhibit virus uptake and prevent viral entry, or indirectly by improving immunity, eliminating inflammation, and regulating intestinal function (Guo et al., 2021; Rangsinth et al., 2021; Arunachalam et al., 2022). In most studies, the selectivity index (SI) is calculated to assess the safety of compounds to be used as antiviral agents. Higher SI values indicate higher efficacy and safety, and preferably the antiviral agent would target the virus at a low concentration and the host cell at a high concentration. According to Doğan et al. (2018), an SI value above 4 is considered a reliable indicator of the potential antiviral activity of an extract.

The antiviral effect of extracts and bioactive compounds of *Pleurotus* spp. against socially important viral infections was investigated in detail (Table 7.4). Seven strains of *P. pulmonarius* (collected in the Novosibirsk region, Russia) were tested for their antiviral activity against the pathogenic strain of influenza virus A (H1N1 pdm09). The authors found that EtOH extracts showed more pronounced activity compared to aqueous extracts (Vlasenko et al., 2020). The study by Krupodorova et al. (2014) also found that mycelia of *P. ostreatus* and *P. eryngii* exhibited anti-influenza activity.

The results on antiviral activity against the herpes simplex virus (HSV) are somewhat contradictory. The study, which investigated the antiviral activity of five inedible and five edible mushroom species from Turkey against HSV-1, concluded that the crude MeOH and aqueous *P. ostreatus* extracts showed no effect (Doğan et al., 2018). In contrast, Santoyo et al. (2012) found that the aqueous and MeOH extracts of the fruiting body of *P. ostreatus* showed strong antiviral properties against HSV-1 based on the SI (15.15 and 10.65, respectively) and $IC_{50}$ values (26.7 and 94.2 µg/mL). In addition, the HMW fraction of the aqueous extract showed even more pronounced activity with an SI value of 163.4 and an $IC_{50}$ value of 4.8 µg/mL, and the activity correlated well with the presence of β-glucans. In addition, the polysaccharide fraction was able to prevent HSV-1 infection, impair virus adsorption, and inhibit intracellular replication, so it can be used to prevent and also to treat viral infections. According to Elhusseiny et al. (2021b), a slight activity of the aqueous extract of the fruiting body of *P. ostreatus* against HSV-2 was observed (SI = 2.0). A similar study found that other species of the genus *Pleurotus* also showed low antiviral activity against HSV-2 (*Pleurotus columbinus*, SI = 0.87, and *P. sajor-caju*, SI = 0.74). In contrast, the mycelia of *P. ostreatus* showed high anti-HSV-2 activity with an SI of 80.6, while no antiherpetic activity was found for *P. eryngii* (Krupodorova et al., 2014). Zhang et al. (2004) showed that sulfated β-glucans from the sclerotium of *P. tuber-regium* exhibited potent antiviral activity against enveloped HSV that was comparable to the control, acyclovir. The authors speculated that negative charges on the polymer chain might interact with the positively charged glycoproteins on the surface of HSV, preventing the latter from infecting the host cells, or that an extended chain conformation of a glucan molecule would promote the formation of additional interactions (hydrogen bonds and van der Waals forces) between the glucans and virus particles. The hypothesis was supported by the results, which showed that the simultaneous addition of sulfated β-glucans with HSV-1 and HSV-2 was more effective than their addition 1 h after infection.

The laccase purified from the fruiting body of *P. ostreatus* was tested for its activity against the hepatitis C virus (anti-HCV). According to the results, incubation of laccase with human peripheral blood mononuclear cells and hepatoma cells (HepG2) and subsequent addition of HCV did not protect the cells from virus entry and infection, while direct interaction between the laccase and the virus molecules (at concentrations of 2.0 and 2.5 mg/mL) completely blocked virus entry. In addition, laccase was able to inhibit HCV replication in infected HepG2 cells in a concentration-dependent manner (EL-Fakharany et al., 2010).

The aqueous extracts of *P. columbinus* and *P. ostreatus* showed strong activity against human adenovirus type 7, Ad7 (SI = 4.6 and SI = 4.5, respectively), while the activity of *P. sajor-caju* was about three times lower (SI = 1.4) (Elhusseiny et al., 2021a, 2021b).

Considerable antiviral activity with an SI of 4.34 against SARS-CoV-2 was observed for the aqueous extract of *P. ostreatus*. Further docking analysis of certain metabolites against the SARS-CoV-2 main protease (protease Mpro—the main enzyme involved in viral replication) revealed that small biomolecules such as kaempferol, ascorbic acid, and catechin, and macromolecules such as superoxide dismutase and catalase, were able to effectively bind to the Mpro (Elhusseiny et al., 2022). A previous study also confirmed the anti-SARS-CoV-2 activity of kaempferol through multiple mechanisms, such as binding to angiotensin-converting enzyme 2 (ACE2) and 3C-like protease (3CL pro), also known as major protease (Mpro), thereby inhibiting viral adsorption, invasion, and replication. In addition, kaempferol could suppress the release of cytokines, attenuate excessive immune responses, and eliminate inflammation (Huang et al., 2020; Rangsinth et al., 2021).

The antiviral activity of various extracts (HWE, EtOH, hexane-soluble extract (HSE), ethyl acetate-soluble extract (ESE), and aqueous-soluble extract (ASE)) of *P. giganteus* against dengue virus serotype 2 (DENV-2) was investigated by Ellan et al. (2019) using three different assays (simultaneous (s), attachment (a), and penetration (p) assay). The ASE extract showed a very pronounced anti-DENV2 activity with high SI values (SIs = 15.7; SIa = 11.5; SIp = 32.3), which were higher than HWE (SIs = 12.4; SIp = 2.5) and the positive control ribavirin (SIs = 10.9; SIp = 4.3). EtOH,

**TABLE 7.4**
***Pleurotus* spp. with Antiviral Agents Against Various Viruses and Mode of Action**

| Virus | Mushroom | Extract/Compound | Results (Target/Activity) | Reference |
|---|---|---|---|---|
| Influenza A (H1N1 pdm09) | *P. pulmonarius*, *P. ostreatus* (M), *P. eryngii* (M) | Aqueous, Ethanol | SI = 0.3–7.5 | Krupodorova et al., 2014; Vlasenko et al., 2020 |
| Human immunodeficiency virus (HIV) | *P. citrinopileatus*, *P. eryngii*, *P. nebrodensis*, *P. abalonus*, *P. ostreatus* | Lectin, Laccase, Nebrodeolysin, Polysaccharide–peptide complex LB-1b, Polysaccharide LA, Ubiquitin-like protein | Reverse transcriptase (RT) inhibitor ($IC_{50}$ = 2.4 pM–12.5 μM) | Wang and Ng, 2000, 2006; Li et al., 2008; Lv et al., 2009; Wang et al., 2011; Li et al., 2012 |
| SARS-CoV- 2 (Severe Acute Respiratory Syndrome – Coronavirus-2) | *P. ostreatus* | Aqueous | Blocking viral replication by binding to the Mpro (SI = 4.34) | Elhusseiny et al., 2022 |
| Respiratory syncytial virus (RSV) | *P. tuber-regium* | Sulfated β-glucans | Binding to the viral particles ($IC_{50}$ = 10–20 μg/mL) | Zhang et al., 2004 |
| Hepatitis C virus (HCV) | *P. ostreatus* | Laccase | Direct interaction between HCV and the laccase led to a complete inhibition of virus entry; inhibition of HCV replication in infected cells | EL-Fakharany et al., 2010 |
| Human adenovirus type 7 (Ad7) | *P. columbinus*, *P. sajor-caju* *P. ostreatus* | Aqueous | SI = 1.4–4.6 | Elhusseiny et al., 2021a, 2021b |
| Herpes simplex virus type 1 and/or type 2 (HSV-1, HSV-2) | *P. columbinus*, *P. sajor-caju*, *P. ostreatus*, *P. ostreatus* (M), *P. ostreatus*, *P. tuber-regium* | Aqueous, Methanol, Polysaccharide fraction, Sulfated β-glucans | SI = 0.74–163.38 Binding to the viral particles (HSV-1, $IC_{50}$ = 1.9–2.5 μg/mL; HSV-2, $IC_{50}$ = 0.38–1.25 μg/mL) | Zhang et al., 2004; Santoyo et al., 2012; Krupodorova et al., 2014; Elhusseiny et al., 2021a, 2021b |
| Dengue virus serotype 2 (DENV-2) | *P. giganteus* | Hot aqueous extract Aqueous-soluble extract | SI = 2.5–32.3 | Ellan et al., 2019 |
| Infectious bursal disease virus (IBDV) | *P. djamor*, *P. sajor-caju*, *P. citrinopileatus* | Methanol | Moderate to strong antiviral activities | Kidukuli et al., 2010 |
| Pox virus | *P. djamor*, *P. sajor-caju*, *P. citrinopileatus* | | | |
| Human Cytomegalovirus (HCMV) | *P. ostreatus* | Methanol | SI = 6.24 | Roy et al., 2020 |

*SI, selectivity index = $CC_{50}/IC_{50}$, where $CC_{50}$ is concentration of the drug which causes a cytotoxic effect in 50% of the cells, and $IC_{50}$ is the minimum concentration of the drug that inhibits virus-induced cytopathic changes by 50%.*

HSE, and ESE showed no anti-dengue activity, indicating a higher polarity of bioactive compounds active against DENV-2. In addition, ASE significantly reduced the expression of dengue envelope proteins (ENV) and non-structural protein 5 (NS5) in the simultaneous and penetration experiments. The authors also found that pretreatment with the mushroom extract did not protect the cells from DENV-2 infection, so they concluded that the addition of the extract prior to infection would be of no benefit.

Of the three *Pleurotus* spp. tested, the MeOH extracts of *P. djamor* showed strong antiviral activity against infectious bursal disease virus (IBDV) and pox virus, while *P. citrinopileatus* and *P. sajor-caju* showed moderate antiviral activities. In addition, all species showed protease inhibitory properties, promoting their potential use as therapeutics, especially in individuals infected with human immunodeficiency virus (HIV, the causative agent of AIDS) (Kidukuli et al., 2010).

A lectin isolated and purified from the fruiting body of *P. citrinopileatus* showed excellent HIV-1 reverse transcriptase (RT) activity ($IC_{50}$ = 0.93 μM) compared to many other fungal lectins (Li et al., 2008). The activity was concentration-dependent, as at 0.12 μM, 0.6 μM, and 3 μM the lectin inhibited HIV-1 reverse transcriptase by 10%, 43.7%, and 90.3%, respectively. In a similar study, a novel polysaccharide–peptide complex (LB-1b) from the fruiting bodies of the edible *Pleurotus abalonus* inhibited HIV-1 RT with an $IC_{50}$ of 12.50 μM, and the inhibition was 40% at 0.63 μM (Li et al., 2012). Laccase purified from the fruiting body of *P. eryngii* showed an inhibitory activity against HIV-1 reverse transcriptase with $IC_{50}$ of 2.2 μM (Wang and Ng, 2006) and the polysaccharide LA from *P. abalonus* inhibited HIV-1 reverse transcriptase with $IC_{50}$ of 10.4 μg/mL (i.e., 87 nM) (Wang et al., 2011).

A hemolysin (nebrodeolysin, a monomeric protein) isolated from the edible mushroom *Pleurotus nebrodensis* showed anti-HIV-1 activity in human T lymphocytes (CEM). However, none of the other DNA and RNA viruses were inhibited at sub-toxic concentrations (Lv et al., 2009).

A novel ubiquitin-like protein, ULP (a glycoprotein), from the mushroom *P. ostreatus* induced a dose-dependent inhibition of HIV-1 reverse transcriptase, which was further enhanced after succinylation and sulfation. Inhibition of more than 90% of HIV-1 RT was observed for ULP, succinylated ULP, and sulfated ULP at 5 mg/mL, while at 0.5 mg/mL ULP, succinylated ULP, and sulfated ULP showed inhibition of 3.5%, 81.8%, and 30%, respectively. Even a weak inhibitory effect of ULP on the glycohydrolases involved in HIV infection was observed (Wang and Ng, 2000). The MeOH extract of the fruiting body of *P. ostreatus* showed strong activity against human cytomegalovirus (HCMV) with SI value of 6.24. In addition, the crude extract showed 100% inhibition of HCMV replication at 180 µg/mL. The time-dependent antiviral response of the mushroom extract showed a percentage inhibition of 14 and 70 in 24 and 36 hpi (hour post-infection), respectively, while in 48 to 72 hpi a strong reduction in viral load (80%–90%) was achieved compared to ganciclovir (50%–55%). The authors concluded that the extract is able to effectively inhibit the propagation of HCMV at an early stage of viral replication (Roy et al., 2020).

## 7.8 MYCOSYNTHESIS OF NANOPARTICLES WITH ANTIMICROBIAL ACTIVITY

Fungi contain various extracellular enzymes that make them suitable as bio-reducing and stabilizing agents for the synthesis of nanoparticles (NP). In contrast to chemically synthesized NPs, biological synthesis meets the requirements for a green synthesis approach, as it is non-toxic, environmentally friendly, simple, and inexpensive, and could be suitable for application in biomedicine and the development of active food packaging. With this approach, it is possible to produce NPs with the desired properties, i.e., shape, size, and surface, which are crucial for achieving certain functional properties. In fungal cells or on the cell membrane, enzymatic reduction occurs through the activity of the enzyme reductase (Khandel and Shahi, 2018; Bhardwaj et al., 2020). There are two possible routes for this process: intracellular and extracellular. In the intracellular approach, NP synthesis takes place in the fungal cell, i.e., fungal mycelium is incubated in the presence of the metal precursor when the reduction takes place. In the extracellular approach, an aqueous filtrate containing fungal molecules is used to treat a metal precursor, and the metal ions are reduced and adsorbed on the cell surface. The extracellular approach is predominantly used, as it is more effective and less expensive—no downstream process is required. However, the NPs formed in the intracellular technique are smaller and suitable for the production of composite films (Bhardwaj et al., 2020).

Intracellular and extracellular synthesis of different metal ion NPs with antimicrobial properties (Ag, Au, $TiO_2$, ZnO, ZnS, Se, Fe) was performed by *Pleurotus* spp. They have been shown to be suitable for the reduction of metal NPs as a rich source of chemical components (alkaloids, saponins, steroids, tannins, flavonoids, and anthraquinones) that play a crucial role in the reduction process. The most commonly used species for the mycosynthesis of NPs are *P. ostreatus* (28%), *P. sajor-caju* (21%), *P. florida* (17%), but other species have also been successfully used for the synthesis of NPs (Owaid et al., 2017). When positively charged metal NPs come into direct contact with negatively charged components of the microbial cell membrane (peptidoglycans, proteins, porins), they bind to each other and cause membrane damage, leakage of intracellular contents, and eventually cell death (Wang et al., 2017). Metallic NPs can also act on microbial cells by allowing reactive oxygen species, i.e., •OH and $•O_2$, to act on proteins, enzymes, DNA, or ribosomes, causing their deactivation or damage and ultimately cell death (Bhardwaj et al., 2020).

In the research of Bhat et al. (2011), AgNPs were synthesized by extracellular approach using *P. florida* as a bioreductant. Significant antimicrobial properties against *S. aureus*, *S. typhi*, *Providencia alcalifaciens*, and *P. mirabilis* were observed (MIC values ranging from 20 µg/mL to 2.5 µg/mL). The AgNPs produced by *P. ostreatus* showed anti-*Candida* activity against *C. albicans*, *C. tropicalis*, *C. glabrata*, *C. parapsilosis*, and *C. krusei* with MIC values between 5 µg/mL and 28 µg/mL (Yehia and Al-Sheikh, 2014). As part of a strategy to combat MDR bacteria, the antimicrobial potential of AgNPs is enhanced by their conjugation with glucans from *Pleurotus* fungi, i.e., glucan from *P. florida* was produced with silver nanoparticles and thus showed improved antimicrobial activity (MIC = 40 µg/mL) against *K. pneumoniae* and proved to be safer for medical use due to the attenuated oxidizing properties of silver (Irshad et al., 2020; Hamers et al., 2021).

## 7.9 APPLICATION OF *PLEUROTUS* SPP. WITH ANTIMICROBIAL PROPERTIES

### 7.9.1 Application of *Pleurotus* spp. for Food Preservation and Safety

Technologies and methods for food preservation have been developed with regard to the microbiological safety of food. Preservatives are food additives that can be ingredients or mixtures of different ingredients added to food in small amounts during production or processing to ensure food safety and extend the shelf life of a product (Torres-Martínez et al., 2022; Matijašević and Sknepnek, 2024). In some cases, the use of preservatives can have undesirable side effects. Synthetic compounds (benzoates, sorbates, sulfites, nitrates), which are widely used in food preservation, can cause various allergic reactions, such as abdominal pain, urticaria, asthma, and so on., and can even be life-threatening. Concerns about the health risks associated with the use of synthetic preservatives, as well as the need to ensure food safety, have increased interest in the use of natural food preservatives. This has led to the definition of bio-preservation as a novel preservation method. The aim is to ensure food safety by using preservatives derived from natural sources. Shelf-life extension can be achieved by bio-preserving various foods (meat, fruits, and vegetables) and by applying bioactive packaging (Pisoschi et al., 2018; Rakasivi and Chin, 2022).

As mentioned above, mushrooms are composed of a variety of bioactive compounds, such as HMW and LMW components, which have antimicrobial properties against various microorganisms and can therefore be used as natural preservatives to ensure food safety. Furthermore, the antimicrobial components of mushrooms are expected to have additional preventive and/or therapeutic effects, such as immune-boosting and antioxidant effects, which are in line with consumers' desire to consume foods with health-promoting effects (Shen et al., 2017; Pisoschi et al., 2018; Sknepnek et al., 2021). In addition, other food properties, such as color stability, taste, aroma, texture, and so on, can also be influenced (Pisoschi et al., 2018). The phenolic compound *p*-anisaldehyde from *P. ostreatus* has been shown to act as a preservative in food and also imparts a sweet flavor (Shen et al., 2017). The application of *P. sajour-caju* and cinnamon powder (*Cinnamomum cassia*) at concentrations of 0.1% to 0.2% to chicken sausages affected product quality and shelf life during storage. Bacterial growth in the sausages was significantly inhibited during the 35-day storage period compared to the control. In addition, the powders of natural origin synergistically reduced lipid oxidation in the product and, together with the delayed microbial spoilage, led to an extended shelf life of the sausages. The recommended concentration of *P. sajor-caju* was up to 0.1%, as higher concentrations had a negative effect on the textural properties of the meat product (Rakasivi and Chin, 2022). In beef patties, *P. ostreatus* flour was used at concentrations of 2.5% or 5% to reduce fat and salt. In addition to affecting the sensory properties of the product, its nutritional profile was also improved, while the cell number of *P. aeruginosa* in the product was reduced after 13 days of storage. It was recommended to use 2.5% mushroom flour in beef patties to achieve the desired functional properties and consumer acceptance (Cerón-Guevara et al., 2020).

Ready-to-eat (RTE) foods are a recognized food safety risk, as they are often contaminated with foodborne pathogens, usually due to poor hygiene, improper preparation techniques, or inadequate storage. Antibiotic resistance in microorganisms from RTE foods is a common situation that poses a significant risk to public health (Abalkhail, 2023). The secondary metabolites of *P. ostreatus* in monoculture and co-cultures with *Lactobacillus fermentum* and *S. cerevisiae* were investigated against pathogenic microorganisms isolated from 100 RTE foods. It was found that 80%–90% of *S. aureus* from RTE foods showed resistance to erythromycin and nitrofurantoin. More than 46% of *S. enterica* isolates showed resistance to one or more antibiotics. In addition, 28.6% to 100% of *P. fluorescens* isolated from RTE foods were resistant to more than one antibiotic tested. The metabolites of *P. ostreatus* showed antibiotic activity against several antibiotic-resistant microorganisms, i.e., *S. enterica*, *S. dysenteriae*, *E. coli*, *K. pneumoniae*, *P. fluorescens*, *B. cereus*, *L. monocytogenes*, *S. aureus*, *Penicillium chrysogenum*, *A. flavus*, and *A. niger*, with zones of inhibition ranging from 9.10 mm to 14.30 mm. However, the antimicrobial activity was enhanced when a co-culture of *P. ostreatus* and *S. cerevisiae* was used, and in some cases was even stronger when co-cultures of *L. fermentum*, *P. ostreatus*, and *S. cerevisiae* were used. The inhibition zones reached 23.7 mm, which was significantly higher than the activity of ciprofloxacin (21.4 mm) (Bamisi et al., 2024). Co-cultured microorganisms are able to increase their activity due to several secondary metabolites produced by their synergistic metabolic pathways (Kapoore et al., 2022). In monocultured *P. ostreatus*, 15 metabolites were detected by GC–MS, of which linolelaedic acid methyl ester dominated. It is known as a regulator of macrophage differentiation and exhibits anti-cancer properties. In co-culture with *S. cerevisiae*, 23 metabolites were detected, while the addition of *L. fermentum* increased the number of metabolites to 27. These microbial metabolites have great potential to be used as natural preservatives to minimize spoilage of various foods, including RTE foods (Bamisi et al., 2024).

### 7.9.2 Application of *Pleurotus* spp. in Active Food Packaging Materials

Food packaging materials are showing a growing trend toward the development of sustainable, biodegradable materials that are predominantly made from natural raw materials, such as polysaccharides and proteins, often in combination with other natural components (Liu et al., 2024). Nowadays, packaging materials are being developed that have additional capabilities besides the tendency to be biodegradable. They are being developed as (i) active packaging with the ability to enhance food protection and extend shelf life, (ii) intelligent packaging that monitors and gives real-time information to consumers about product safety and quality, and (iii) smart packaging that uses advanced digital technologies to transmit product information throughout the supply chain (Sani et al., 2024).

In the development of active packaging materials, antioxidant or antimicrobial additives are added to inhibit undesirable chemical reactions in foods and thus improve their quality or to protect foods from microbial growth and extend their shelf life. Due to their natural antimicrobial activity, chitin and chitosan can be used to produce biodegradable films with antimicrobial properties. The activity can be further enhanced when combined with bioactive compounds such as polysaccharides, proteins, or other additives with functional properties (Sani et al., 2024). Chitin is a widely used biopolymer that can be extracted from arthropods, fungi, and some algae. After its deacetylation, chitin is converted into chitosan, and both have great potential for use in various industries, including agriculture, food industry, wastewater treatment, biomedicine, and so on. Polysaccharides and glycoproteins are components that predominate in the cell walls of fungi, but chitin is most responsible for the rigidity of the cell wall, even though it is only present in small amounts. Chitin is traditionally extracted from crustacean shells, which requires the use of high temperatures and strong alkaline solutions. In addition, production is limited by geographical location and seasonal fluctuations. These shortcomings are overcome by switching to chitosan production by fungi, which is more environmentally friendly and can be organized by continuous biofermentation (Camele et al., 2024). A polysaccharide extract from the stems of *P. ostreatus* (POP) incorporated into chitosan at a range of 0.25% to 0.75% improved the activity of the films by increasing the antibacterial activity against *E. coli*, with the best effect occurring at a POP concentration of 0.5%. In addition, a very important property for food packaging materials, tensile strength, was improved after the incorporation of POP into chitosan (Liu et al., 2024). One way to improve the functional properties of biodegradable packaging films is to achieve intramolecular cross-linking. This can be achieved by chemical, physical, or enzymatic methods. A common enzyme for the cross-linking of protein and polysaccharide matrices is laccase. Fungal laccases can be obtained from microorganisms, plants, or insects and play an important role in nature as a lignin-degrading enzyme, which is important for carbon recycling. Laccase from *P. ostreatus* has been extensively studied due to its high activity and stability. It has been used to improve various material properties, such as mechanical strength and barrier properties. Laccase coatings have also shown great potential to improve antimicrobial properties (Sani et al., 2024). Enzymatic methods for modifying chitosan are proving to be more sustainable than chemical methods and are also more environmentally friendly and effective. Chitosan polymers modified by the enzymatic activity of laccase from *P. ostreatus* with cinnamic acids showed improved polymer thermostability and improved antibacterial activity of chitosan against the bacterial plant wilt pathogen *Ralstonia solanacearum* (Yang et al., 2024).

### 7.9.3 Application of *Pleurotus* spp. as Biopesticides

Plant diseases are a major cause of global crop losses, accounting for 40% to 45% of total food production losses and damage during postharvest storage. These losses have a significant economic impact, amounting to 550 billion US dollars annually. To prevent the occurrence of diseases and the resulting losses, 3.5 million tons of pesticides are used worldwide every year. The extensive use of synthetic chemical fungicides is of great concern due to their negative impact on human health, the environment, and ecosystems, as well as the development of resistance in plant pathogens (Acay et al., 2024). As an alternative, the primary and secondary metabolites of mushrooms and their derivatives have been recognized as a promising replacement for commercial phytochemicals as an environmentally friendly and sustainable solution to control plant diseases (Faugeron-Girard et al., 2020).

The water-soluble polysaccharide PFP1–1 from *Pleurotus ferulae* has the potential to be used as a biofungicide for plant protection and, at the same time, as a fertilizer to promote plant growth. It showed significant antifungal activity against the fungal pathogen *Rhizoctonia solani*. The mycelial growth of *R. solani* was inhibited in a dose-dependent manner and reached 70% at a concentration of 1000 mg/L, while cucumber seeds infected with *R. solani* resisted the infection by up to 22.92%. Of importance for the expression of the antifungal activity of PFP-1 are the high content of mannose and the presence of fucose, which are known to express antimicrobial activity (Yang et al., 2024).

An environmentally friendly nanomaterial containing *P. eryngii* extract (PE), lecithin, and chitosan, i.e., *P. eryngii*-lecithin-chitosan (PEELCN), was developed. PEELCN and PE alone showed antifungal activity against *Verticillium dahliae*, *Bipolaris sorokiniana*, and *Neoscytalidium dimidiatum*. *Verticillium dahliae* is the causative agent of Verticillium wilt mainly on olive and cotton plants, but also on other species. PEELCN achieved an inhibition of up to 75% after 48 hours, while PE also achieved 70% inhibition. *B. sorokiniana* causes diseases in wheat and affects its production, yield, and grain quality. The inhibition of *B. sorokiniana* by PE was 44% after 2 days. *Neoscytalidium dimidiatum* is a fungus that poses a threat to both humans and over 126 plant species. PEELCN inhibited its growth by 50% after 7 days, while PE achieved a higher inhibition rate (63.3%) (Acay et al., 2024). In the study by Gangwar et al. (2024), chitin from *P. eryngii* was used to produce chitosan by deacetylation. It showed significant antibacterial activity against the pathogenic bacteria *E. coli*, *X. campestris*, *Clavibacter michigenesis*, and *P. fluorescens*, with the lowest MIC and minimal bactericidal concentration (MBC) values for *C. michigenesis* (0.375 mg/mL and 0.75 mg/mL, respectively). In addition, antifungal activity was also observed against *P. expansum*, *B. cinerea*, and *Monilia laxa*, with the lowest MIC value for *B. cinerea* (0.375 mg/mL) and the lowest minimal fungicidal concentration value for *M. laxa* (0.75 mg/mL). *In vivo*, *Prunus domestica* (plum fruit) previously infected with *M. laxa*, a brown rot pathogen, was sprayed with chitosan, and a coating was formed. The chitosan coating on the fruit reduced disease symptoms by 46.4% in the preventive treatment and by 52.6% in the curative treatment. In addition to protection against fungal diseases, the coating also improved the weight retention and firmness of the fruit. Therefore, the chitosan coating of *P. eryngii* can serve as an environmentally friendly biopesticide and extend the shelf life of the fruit. The amino groups in the chitosan polymer are responsible for its cationic properties and increased reactivity toward microbial cells, causing their disruption through increased permeability of the lipid cell membrane and, thus, final cell death (Gangwar et al., 2024). The antimicrobial activity of chitosan also depends on its molecular weight. When LMW chitosan is used against fungal cells, it can penetrate the cells and bind DNA and RNA, affecting fungal growth and viability. HMW chitosan forms a layer on the cell wall of microorganisms, preventing nutrients and oxygen from entering the cell, which can lead to lethal consequences (Yang et al., 2024).

One of the very important secondary metabolites of *Pleurotus* spp. is pleurotin, an antibiotic metabolite of the naphthoquinone class. Pleurotin has been shown to be effective against various bacteria and fungi. In the control of bean leaf spot disease caused by *Alternaria alternata*, pleurotin acted as a generator of peroxidase and chitinase activity, which in turn led to increased plant resistance to the pathogenic fungi (Awad and Hassan, 2023). Peroxidase is an enzyme whose activity is usually increased when a plant is exposed to biotic stress factors. Its increased activity allows the plant being able to degrade pathogenic enzymes, preventing their ability to invade the plant tissue (Faugeron-Girard et al., 2020; Awad and Hassan, 2023). In addition, pleurotin increased chitinase activity in plants, which improved their ability to degrade the hyphal wall of fungi and their subsequent death (Awad and Hassan, 2023). Similar biological activity was observed in pleuran complex cell wall extract (CCWE) obtained by alkaline extraction of *P. ostreatus* mushroom. The pleuran CCWE consisted predominantly of β-glucans and proteins (40%), while a lower proportion of α-glucans, peptides, free amino acids, monosaccharides, polysaccharides, and minerals was found. Peroxidase activity was increased when 350 mg/L pleuran CCWE was sprayed on tomato leaves to control *B. cinerea* infection. In addition, pretreatment of *Brachypodium distachyon* with pleuran CCWE significantly reduced the development of the pathogenic fungus *Fusarium graminearum*. Although the previous treatments were validated under controlled conditions, the effect of pleuran CCWE was also demonstrated in the field on wheat (*Triticum aestivum*) and vine (*Vitis vinifera*). It showed the ability to limit the symptoms of plant diseases caused by pathogens (Faugeron-Girard et al., 2020).

## 7.10 CONCLUSION AND FUTURE PERSPECTIVE

The current and comprehensive information provided in this review shows that oyster mushrooms, not only the basidiomata but also the mycelia and the fermentation broth, exhibit strong antimicrobial activities due to the presence of various myco-constituents. Furthermore, the activity against Gram-negative organisms was found to be comparable to that of Gram-positive organisms, emphasizing the uniqueness of *Pleurotus* extracts and metabolites and their ability to overcome the challenging membranous nature of Gram-negative microbes. In addition, the studies demonstrated the antimicrobial activity of *Pleurotus* spp. against pathogenic microorganisms, including MDR and contaminating microorganisms. However, most studies focused on screening the activity of mushroom extracts rather than isolated compounds. In addition, much of the research has traditionally relied on conventional extraction methods. Furthermore, it is of great importance to investigate antimicrobial activities after processing operations when mushrooms are utilized as foods or medicines. Treatments such as freezing, drying, thermal treatment, or alkali or acid processing can influence the stability and activity of bioactive components. To fully exploit the enormous potential of mushrooms, several requirements need to be met in the near future:

1. Advanced extraction techniques should be intensified to obtain more potent *Pleurotus* extracts rich in bioactive compounds.
2. Although a number of compounds have already been defined, further identification of specific molecules, together with structure determination, should be a priority.
3. Further research to elucidate the mode of action of the various metabolites isolated from *Pleurotus* spp., as well as preclinical and clinical trials, is required to elucidate these effects and ensure their safety.

In the age of "omics," all requirements would be much easier to investigate, so that *Pleurotus* spp. could be upgraded to an innovative antimicrobial food bio-ingredient, an adequate alternative to conventional drugs, or, more generally, to increase its therapeutic importance.

## REFERENCES

Abalkhail, A. 2023. Frequency and antimicrobial resistance patterns of foodborne pathogens in ready-to-eat foods: an evolving public health challenge. *Applied Sciences* 13(23): 12846. https://doi.org/10.3390/app132312846

Abdelshafy, A. M., T. Belwal, Z. Liang, et al. 2022. A comprehensive review on phenolic compounds from edible mushrooms: occurrence, biological activity, application and future prospective. *Critical Reviews in Food Science and Nutrition* 62(22): 6204–6224. https://doi.org/10.1080/10408398.2021.1898335

Acay, H., I. G. Güney, A. Yildirim, S. Derviş, and E. Dereli. 2024. Green synthesis of *Pleurotus eryngii*-derived nanomaterials for phytopathogen control. *Chemistry and Biodiversity* 21(12): e202401972. https://doi.org/10.1002/cbdv.202401972

Adebayo, E. A., D. Martínez-Carrera, P. Morales, et al. 2018. Comparative study of antioxidant and antibacterial properties of the edible mushrooms *Pleurotus Levis, P. Ostreatus, P. Pulmonarius* and *P. Tuber-Regium*. *International Journal of Food Science and Technology* 53(5): 1316–1330. https://doi.org/10.1111/ijfs.13712

Akyuz, M., A. N. Onganer, P. Erecevit, and S. Kirbag. 2010. Antimicrobial activity of some edible mushrooms in the Eastern and Southeast Anatolia Region of Turkey. *Gazi University Journal of Science* 23(2): 125–130.

Aminov, R. I. 2010. A brief history of the antibiotic era: lessons learned and challenges for the future. *Frontiers in Microbiology* 1: 134. https://doi.org/10.3389/fmicb.2010.00134

Arunachalam, K., S. P. Sasidharan, and X. Yang. 2022. A concise review of mushrooms antiviral and immunomodulatory properties that may combat against COVID–19. *Food*

*Chemistry Advances* 1: 100023. https://doi.org/10.1016/j.focha.2022.100023

Assaf, S., Y. Hadar, and C. G. Dosoretz. 1995. Biosynthesis of 13-hydroperoxylinoleate, 10-0x0-8-decenoic acid, and l-octen-3-01 from linoleic acid by a mycelial-pellet homogenate of *Pleurotus pulmonarius*. *Journal of Agricultural and Food Chemistry* 43.

Awad, M. A., and A. A. K. Hassan. 2023. Evaluation of the Pleurotin efficacy extracted from *Pleurotus* Spp. in controlling leaf spot disease caused by *Alternaria alternata* on broad bean. *Euphrates Journal of Agricultural Science* 15(2): 120–130.

Bamisi, O. E., C. O. Ogidi, and B. J. Akinyele. 2024. Antimicrobial metabolites from probiotics, *Pleurotus ostreatus* and their co-cultures against foodborne pathogens isolated from ready-to-eat foods. *Annals of Microbiology* 74(1): 34. https://doi.org/10.1186/s13213-024-01776-5

Bhardwaj, K., A. Sharma, N. Tejwan, et al. 2020. *Pleurotus* macrofungi-assisted nanoparticle synthesis and its potential applications: a review. *Journal of Fungi* 6(4): 351. https://doi.org/10.3390/jof6040351

Bhat, R., R. Deshpande, S. V. Ganachari, D. S. Huh, and A. Venkataraman. 2011. Photo-irradiated biosynthesis of silver nanoparticles using edible mushroom *Pleurotus florida* and their antibacterial activity studies. *Bioinorganic Chemistry and Applications* 2011: 650979. https://doi.org/10.1155/2011/650979

Bottone, E. J. 2010. *Bacillus cereus*, a volatile human pathogen. *Clinical Microbiology Reviews* 23(2): 382–398. https://doi.org/10.1128/CMR.00073-09

Camele, I., A. A. Mohamed, A. A. Ibrahim, and H. S. Elshafie. 2024. Biochemical characterization and disease control efficacy of *Pleurotus eryngii*-derived chitosan—an *in vivo* study against *Monilinia laxa*, the causal agent of plum brown rot. *Plants* 13(18): 2598. https://doi.org/10.3390/plants13182598

Carrasco-González, J. A., S. O. Serna-Saldívar, and J. A. Gutiérrez-Uribe. 2017. Nutritional composition and nutraceutical properties of the *Pleurotus* fruiting bodies: potencial use as food ingredient. *Journal of Food Composition and Analysis* 58: 69–81. https://doi.org/10.1016/j.jfca.2017.01.016

Cerón-Guevara, M. I., E. Rangel-Vargas, J. M. Lorenzo, et al. 2020. Effect of the addition of edible mushroom flours (*Agaricus bisporus* and *Pleurotus ostreatus*) on physicochemical and sensory properties of cold-stored beef patties. *Journal of Food Processing and Preservation* 44(3): e14351. https://doi.org/10.1111/jfpp.14351

Chaiharn, M., W. S Phutdhawong, D. Amornlerdpison, and W. Phutdhawong. 2018. Antibacterial, antioxidant properties and bioactive compounds of Thai cultivated mushroom extracts against food-borne bacterial strains. *Chiang Mai Journal of Science* 45(4): 1713–1727. http://epg.science.cmu.ac.th/ejournal/

Chan, Y. S., and K. P. Chong. 2022. Bioactive compounds of *Ganoderma boninense* inhibited methicillin-resistant *Staphylococcus aureus* growth by affecting their cell membrane permeability and integrity. *Molecules* 27(3): 838. https://doi.org/10.3390/molecules27030838

Chowdhury, H. M. H., K. Kubra, and R. R. Ahmed. 2015. Screening of antimicrobial, antioxidant properties and bioactive compounds of some edible mushrooms cultivated in Bangladesh. *Annals of Clinical Microbiology and Antimicrobials* 14(1): 8. https://doi.org/10.1186/s12941-015-0067-3

Chu, K. T., L. Xia, and T. B. Ng. 2005. Pleurostrin, an antifungal peptide from the oyster mushroom. *Peptides* 26(11): 2098–2103. https://doi.org/10.1016/j.peptides.2005.04.010

Dadgostar, P. 2019. Antimicrobial resistance: implications and costs. *Infection and Drug Resistance* 12: 3903–3910. https://doi.org/10.2147/IDR.S234610

Dash, P., B. Kar, M. Gochhi, et al. 2024. Antimicrobial properties of the edible pink oyster mushroom, *Pleurotus eous*: *in-vivo* and *in-vitro* studies. *Microbial Pathogenesis* 196: 106915. https://doi.org/10.1016/j.micpath.2024.106915

Deepalakshmi, K., and S. Mirunalini. 2014. Propolis: an enrichment material for mycelium development of oyster mushroom (*Pleurotus ostreatus*). *Journal of Biochemical Technology* 2(5): 718–726.

Doğan, H. H., S. Karagöz, and R. Duman. 2018. *In vitro* evaluation of the antiviral activity of some mushrooms from Turkey. *International Journal of Medicinal Mushrooms* 20(3): 201–212. https://doi.org/10.1615/IntJMedMushrooms.2018025468

Dulay, R. M. R., L. A. Miranda, J. S. Malasaga, et al. 2017. Antioxidant and antibacterial activities of acetonitrile and hexane extracts of *Lentinus tigrinus* and *Pleurotus djamour*. *Biocatalysis and Agricultural Biotechnology* 9: 141–144. https://doi.org/10.1016/j.bcab.2016.12.003

Durand, G. A., D. Raoult, and G. Dubourg. 2019. Antibiotic discovery: history, methods and perspectives. *International Journal of Antimicrobial Agents*: 53(4): 371–382. https://doi.org/10.1016/j.ijantimicag.2018.11.010

EL-Fakharany, E. M., B. M. Haroun, T. B. Ng, and E-R. M. Redwan. 2010. Oyster mushroom laccase inhibits Hepatitis C virus entry into peripheral blood cells and hepatoma cells. *Protein & Peptide Letters* 17(8): 1031–1039. https://doi.org/10.2174/092986610791498948

Elhusseiny, S. M., T. S. El-Mahdy, M. F. Awad, et al. 2021a. Antiviral, cytotoxic, and antioxidant activities of three edible Agaricomycetes mushrooms: *Pleurotus columbinus*, *Pleurotus sajor-caju*, and *Agaricus bisporus*. *Journal of Fungi* 7(8): 645. https://doi.org/10.3390/jof7080645

Elhusseiny, S. M., T. S. El-Mahdy, M. F. Awad, et al. 2021b. Proteome analysis and *in vitro* antiviral, anticancer and antioxidant capacities of the aqueous extracts of *Lentinula edodes* and *Pleurotus ostreatus* edible mushrooms. *Molecules* 26(15): 4623. https://doi.org/10.3390/molecules26154623

Elhusseiny, S. M., T. S. El-Mahdy, N. S. Elleboudy, et al. 2022. *In vitro* anti SARS-CoV-2 activity and docking analysis of *Pleurotus ostreatus*, *Lentinula edodes* and *Agaricus bisporus* edible mushrooms. *Infection and Drug Resistance* 15: 3459–3475. https://doi.org/10.2147/IDR.S362823

Ellan, K., R. Thayan, J. Raman, et al. 2019. Anti-viral activity of culinary and medicinal mushroom extracts against dengue virus serotype 2: an *in-vitro* study. *BMC Complementary and Alternative Medicine* 19(1): 260. https://doi.org/10.1186/s12906-019-2629-y

Faugeron-Girard, C., V. Gloaguen, R. Koçi, et al. 2020. Use of a *Pleurotus ostreatus* complex cell wall extract as elicitor of plant defenses: from greenhouse to field trial. *Molecules* 25(5): 1094. https://doi.org/10.3390/molecules25051094

File, T. M. Jr., E. Alexander, L. Goldberg, et al. 2021. Lefamulin efficacy and safety in a pooled phase 3 clinical trial population with community-acquired bacterial pneumonia and common clinical comorbidities. *BMC Pulmonary Medicine* 21(1): 154. https://doi.org/10.1186/s12890-021-01472-z

File, T. M., L. Goldberg, A. Das, et al. 2019. Efficacy and Safety of intravenous-to-oral lefamulin, a Pleuromutilin antibiotic, for the treatment of community-acquired bacterial Pneumonia: the phase III lefamulin evaluation against Pneumonia (LEAP 1) trial. *Clinical Infectious Diseases* 69(11): 1856–1867. https://doi.org/10.1093/cid/ciz090

Gangwar, R., M. M. Salem, V. K. Maurya, et al. 2024. Exploring time-killing and biofilm inhibition potential of bioactive proteins extracted from two varieties of *Pleurotus ostreatus*. *Frontiers in Microbiology* 15: 1456358. https://doi.org/10.3389/fmicb.2024.1456358

Gaynes, R. 2017. The discovery of Penicillin—new insights after more than 75 years of clinical use. *Emerging Infectious Diseases* 23(5): 849–853. https://doi.org/10.3201/eid2305.161556

Gogavekar, S. S., S. A. Rokade, R. C. Ranveer, et al. 2014. Important nutritional constituents, flavour components, antioxidant and antibacterial properties of *Pleurotus sajor-caju*. *Journal of Food Science and Technology* 51(8): 1483–1491. https://doi.org/10.1007/s13197-012-0656-5

Golak-Siwulska, I., A. Kałuzewicz, T. Spizewski, M. Siwulski, and K. Sobieralski. 2018. Bioactive compounds and medicinal properties of oyster mushrooms (*Pleurotus* sp.). *Folia Horticulturae* 30(2): 11–21. https://doi.org/10.2478/fhort-2018-0012

Guo, Y., X. Chen, and P. Gong. 2021. Classification, structure and mechanism of antiviral polysaccharides derived from edible and medicinal fungus. *International Journal of Biological Macromolecules* 183: 1753–1773. https://doi.org/10.1016/j.ijbiomac.2021.05.139

Hamad, D., H. El-Sayed, W. Ahmed, H. Sonbol, and M. A. H. Ramadan. 2022. GC-MS analysis of potentially volatile compounds of *Pleurotus ostreatus* polar extract: *in vitro* antimicrobial, cytotoxic, immunomodulatory, and antioxidant activities. *Frontiers in Microbiology* 13: 834525. https://doi.org/10.3389/fmicb.2022.834525

Hamers, V., C. Huguet, M. Bourjot, and A. Urbain. 2021. Antibacterial compounds from mushrooms: a lead to fight ESKAPEE pathogenic bacteria? *Planta Medica* 87(5): 351–367. https://doi.org/10.1055/a-1266-6980

Han, S. R., K. H. Kim, K. O. Lim, and T. J. Oh. 2015. Biological activity analysis of different solvent extracts from *Pleurotus ostreatus*. *Indian Journal of Science and Technology* 8(26): 1–8. https://doi.org/10.17485/ijst/2015/v8i26/80610

Hearst, R., D. Nelson, G. McCollum, et al. 2009. An examination of antibacterial and antifungal properties of constituents of Shiitake (*Lentinula edodes*) and oyster (*Pleurotus ostreatus*) mushrooms. *Complementary Therapies in Clinical Practice* 15(1): 5–7. https://doi.org/10.1016/j.ctcp.2008.10.002

Huang, Y. F., C. Bai, F. He, Y. Xie, and H. Zhou. 2020. Review on the potential action mechanisms of chinese medicines in treating Coronavirus disease 2019 (COVID-19). *Pharmacological Research* 158: 104939. https://doi.org/10.1016/j.phrs.2020.104939

Illuri, R., M. Eyini, M. Kumar, et al. 2022. Bio-prospective potential of *Pleurotus djamor* and *Pleurotus florida* mycelial extracts towards Gram positive and Gram negative microbial pathogens causing infectious disease. *Journal of Infection and Public Health* 15(2): 297–306. https://doi.org/10.1016/j.jiph.2021.10.012

Irshad, A., N. Sarwar, H. Sadia, et al. 2020. Silver nanoparticles: synthesis and characterization by using glucans extracted from *Pleurotus ostreatus*. *Applied Nanoscience (Switzerland)* 10(8): 3205–3214. https://doi.org/10.1007/s13204-019-01103-4

Kalyoncu, F., M. Oskay, H. Sag, F. Erdog, and A. U. Tamer. 2010. Antimicrobial and antioxidant activities of mycelia of 10 wild mushroom species. *Journal of Medicinal Food* 13(2): 415–419. https://doi.org/10.1089/jmf.2009.0090

Kandasamy, S., S. Chinnappan, S. Thangaswamy, S. Balakrishnan, and A. Y. Z. Khalifa. 2020. Assessment of antioxidant, antibacterial activities and bioactive compounds of the wild edible mushroom *Pleurotus sajor-caju*. *International Journal of Peptide Research and Therapeutics* 26(3): 1575–1581. https://doi.org/10.1007/s10989-019-09969-2

Kapoore, R. V., G. Padmaperuma, S. Maneein, and S. Vaidyanathan. 2022. Co-culturing microbial consortia: approaches for applications in biomanufacturing and bioprocessing. *Critical Reviews in Biotechnology* 42(1): 46–72. https://doi.org/10.1080/07388551.2021.1921691

Karaman, M., E. Jovin, R. Malbaša, M. Matavuly, and M. Popović. 2010. Medicinal and edible lignicolous fungi as natural sources of antioxidative and antibacterial agents. *Phytotherapy Research* 24(10): 1473–1481. https://doi.org/10.1002/ptr.2969

Kavanagh, F., A. Hervey, and W. J. Robbins. 1951. Antibiotic substances from basidiomycetes. *Proceedings of the National Academy of Sciences* 37(9): 570–574. https://doi.org/10.1073/pnas.37.9.570

Khan, M. A., and M. Tania. 2012. Nutritional and medicinal importance of *Pleurotus* mushrooms: an overview. *Food Reviews International* 28(3): 313–329. https://doi.org/10.1080/87559129.2011.637267

Khandel, P., and S. K. Shahi. 2018. Mycogenic nanoparticles and their bio-prospective applications: current status and future challenges. *Journal of Nanostructure in Chemistry* 8: 369–391. https://doi.org/10.1007/s40097-018-0285-2

Kidukuli, A. W., Z. H. Mbwambo, H. M. Malebo, C. A. Mgina, and M. J. Mihale. 2010. *In vivo* antiviral activity, protease inhibition and brine shrimp lethality of selected Tanzanian wild edible mushrooms. *Journal of Applied Sciences* 31: 1887–1894.

Krupodorova, T., S. Rybalko, and V. Barshteyn. 2014. Antiviral activity of basidiomycete mycelia against influenza type A (serotype H1N1) and herpes simplex virus type 2 in cell culture. *Virologica Sinica* 29(5): 284–290. https://doi.org/10.1007/s12250-014-3486-y

Leong, Y. K., F. C. Yang, and J. S. Chang. 2021. Extraction of polysaccharides from edible mushrooms: emerging technologies and recent advances. *Carbohydrate Polymers* 251: 117006. https://doi.org/10.1016/j.carbpol.2020.117006

Li, N., L. Li, J. C. Fang, et al. 2012. Isolation and identification of a novel polysaccharide-peptide complex with antioxidant, anti-proliferative and hypoglycaemic activities from the Abalone mushroom. *Bioscience Reports* 32(3): 221–228. https://doi.org/10.1042/BSR20110012

Li, Q. H. Liu, H. X. Wang, and T. B. Ng. 2008. A novel lectin with potent antitumor, mitogenic and HIV-1 reverse transcriptase inhibitory activities from the edible mushroom *Pleurotus citrinopileatus*. *Biochimica et Biophysica Acta—General Subjects* 1780(1): 51–57. https://doi.org/10.1016/j.bbagen.2007.09.004

Liu, T., Q. Tang, H. Lei, et al. 2024. Preparation, physicochemical and biological evaluation of chitosan *Pleurotus ostreatus* polysaccharides active films for food packaging. *International Journal of Biological Macromolecules* 254: 127470. https://doi.org/10.1016/j.ijbiomac.2023.127470

Lv, H., Y. Kong, Q. Yao, et al. 2009. Nebrodeolysin, a novel hemolytic protein from mushroom *Pleurotus nebrodensis* with apoptosis-inducing and anti-HIV-1 effects. *Phytomedicine* 16(2–3): 198–205. https://doi.org/10.1016/j.phymed.2008.07.004

Masyita, A., R. Mustika Sari, W. Astuti, et al. 2022. Terpenes and terpenoids as main bioactive compounds of essential oils, their roles in human health and potential application as natural food preservatives. *Food Chemistry: X* 13: 100217. https://doi.org/10.1016/j.fochx.2022.100217

Matijaševic, D., M. Pantic, B. Rašković, et al. 2016. The antibacterial activity of coriolus versicolor methanol extract and its effect on ultrastructural changes of *Staphylococcus aureus* and *Salmonella enteritidis*. *Frontiers in Microbiology* 7: 1226. https://doi.org/10.3389/fmicb.2016.01226

Matijašević, D., M. Pantić, N. Stanisavljević, et al. 2022. *Coriolus versicolor* mushroom grown on selenium-rich zeolitic tuff as a potential novel food supplement. *Food Technology and Biotechnology* 60(1): 67–79. https://doi.org/10.17113/ftb.60.01.22.7172

Matijašević, D., and A. Sknepnek. 2024. Development and emergence of *Ganoderma*-based industry: a global perspective. In: *Ganoderma: Cultivation, Chemistry, and Medicinal Applications*, ed. K. Acharya, and S. Khatua, 2: 176–202. CRC Press.

Mayer, F. L., D. Wilson, and B. Hube. 2013. *Candida albicans* pathogenicity mechanisms. *Virulence* 4(2): 119–128. https://doi.org/10.4161/viru.22913

Meng, T. X., H. Ishikawa, K. Shimizu, S. Ohga, and R. Kondo. 2012. A glucosylceramide with antimicrobial activity from the edible mushroom *Pleurotus citrinopileatus*. *Journal of Wood Science* 58(1): 81–86. https://doi.org/10.1007/s10086-011-1213-y

Miletić, D., M. Pantić, A. Sknepnek, I. Vasiljević, M. Lazović, and M. Nikšić. 2020. Influence of selenium yeast on the growth, selenium uptake and mineral composition of *Coriolus versicolor* mushroom. *Journal of Basic Microbiology* 60(4): 331–340. https://doi.org/10.1002/jobm.201900520

Moussa, A. Y., S. Fayez, H. Xiao, and B. Xu. 2022. New insights into antimicrobial and antibiofilm effects of edible mushrooms. *Food Research International* 162(Pt A): 111982. https://doi.org/10.1016/j.foodres.2022.111982

Ngai, P. H. K., and T. B. Ng. 2004. A ribonuclease with Antimicrobial, antimitogenic and antiproliferative activities from the edible mushroom *Pleurotus sajor-caju*. *Peptides* 25(1): 11–17. https://doi.org/10.1016/j.peptides.2003.11.012

Ngai, P. H. K., and T. B. Ng. 2006. A hemolysin from the mushroom *Pleurotus eryngii*. *Applied Microbiology and Biotechnology* 72(6): 1185–1191. https://doi.org/10.1007/s00253-006-0406-6

Novak, R. 2011. Are Pleuromutilin antibiotics finally fit for human use? *Annals of the New York Academy of Sciences* 1241(1): 71–81. https://doi.org/10.1111/j.1749-6632.2011.06219.x

Nwachukwu, E., and H. O. Uzoeto. 2010. Antimicrobial activity of some local mushrooms on pathogenic isolates. *Journal of Medicinal Plants Research* 4(23): 2460–2465. https://doi.org/10.5897/jmpr10.154

Okafor, D. C., N. C. Onuegbu, N. E. Odimegwu, et al. 2017. Antioxidant and antimicrobial activities of oyster mushroom. *American Journal of Food Science and Technology* 5(2): 64–69. https://doi.org/10.12691/ajfst-5-2-6

Okamoto, K., S. Narayama, A. Katsuo, I. Shigematsu, and H. Yanase. 2002. Biosynthesis of p-Anisaldehyde by the white-rot basidiomycete *Pleurotus ostreatus*. *Journal of Bioscience and Bioengineering* 93(2): 207–210. https://doi.org/10.1263/jbb.93.207

Okull, D. O, R. B. Beelman, and H. Gourama. 2003. Antifungal activity of 10-oxo-trans-8-decenoic acid and 1-octen-3-ol against *Penicillium expansum* in potato dextrose agar medium. *Journal of Food Protection* 66: 1503–1505.

Otali, C. C., C. S. O. Otoikhian, T. Onuoha, C. S. Akpeji, and B., O. Bosah. 2024. Antibacterial activities of *Pleurotus ostreatus* and *Pleurotus djamor* against selected bacterial pathogens. *Bima Journal of Science and Technology* 8(2B): 397–402. https://doi.org/10.56892/bima.v8i2B.746

Owaid, N. M., S. S. S. Al Saeedi, I. A. A. Abed, P. Shahbazi, and V. Sabaratnam. 2017. Antifungal activities of some *Pleurotus* species (higher Basidiomycetes). *Agricultural Technology and Biological Sciences* 14(3): 215–224. http://wjst.wu.ac.th

Özdal, M., Ö. Gülmez, Ö. G. Özdal, and Ö. F. Algur. 2019. Antibacterial and antioxidant activity of mycelial extracts of different *Pleurotus* species. *Food and Health* 5(1): 12–18. https://doi.org/10.3153/fh19002

Pandey, A. T., I. Pandey, P. Kerkar, and M. P. Singh. 2021. Antimicrobial activity and mycochemical profile of methanol extract from *Pleurotus flabellatus*. *Vegetos* 34(3): 619–629. https://doi.org/10.1007/s42535-021-00242-w

Phan, C. W., G. S. Lee, I. G. Macreadie, et al. 2013. Lipid constituents of the edible mushroom, *Pleurotus giganteus* demonstrate anti-candida activity. *Natural Product Communications* 8(12): 1763–1765. https://doi.org/10.1177/1934578x1300801228

Pisoschi, A. M., A. Pop, C. Georgescu, V. Turcuş, N. K. Olah, and E. Mathe. 2018. An overview of natural antimicrobials role in food. *European Journal of Medicinal Chemistry* 143: 922–935. https://doi.org/10.1016/j.ejmech.2017.11.095

Prastiyanto, M. E., R. M. Rukmana, D. K. Saraswati, S. Darmawati, E. T. W. Maharani, and Y. Tursinawati. 2020. Anticancer potential of methanolic extracts from *Pleurotus* species on Raji cells and antibacterial activity against methicillin-resistant *Staphylococcus aureus*. *Biodiversitas* 21(12): 5644–5649. https://doi.org/10.13057/biodiv/d211221

Rakasivi, K. G. J., and K. B. Chin. 2022. Evaluation of product quality of chicken sausages with added cinnamon (*Cinnamomum cassia*) and mushroom (*Pleurotus sajur-caju*) powders at various concentrations. *Journal of Food Processing and Preservation* 46(11): e16958. https://doi.org/10.1111/jfpp.16958

Ramesh, C., and M. G. Pattar. 2010. Antimicrobial properties, antioxidant activity and bioactive compounds from six wild edible mushrooms of Western Ghats of Karnataka, India. *Pharmacognosy Research* 2(2): 107–112. https://doi.org/10.4103/0974-8490.62953

Rangsinth, P., C. Sillapachaiyaporn, S. Nilkhet, T. Tencomnao, A. T. Ung, and S. Chuchawankul. 2021. Mushroom-derived bioactive compounds potentially serve as the inhibitors of SARS-CoV-2 main protease: an *in silico* approach. *Journal of Traditional and Complementary Medicine* 11(2): 158–172. https://doi.org/10.1016/j.jtcme.2020.12.002

Robbins, F. K., and A. Hervey. 1947. Antibiotic substances from Basidiomycetes: I. *Pleurotus griseus*. *Proceedings of the National Academy of Sciences of the United States of America* 33(6): 171–176. https://doi.org/10.1073/pnas.33.6.171

Robbins, W. J., A. Hervey, R. W. Davidson, R. Ma, and W. C. Robbins. 1945. Torrey Botanical Society a survey of some wood-destroying and other fungi for antibacterial activity. *Source: Bulletin of the Torrey Botanical Club* 72(2): 165–190

Robbins, W. J., F. Kavanagh, and A. Hervey. 1946. Production of antibiotic substances by Basidiomycetes. *Annals of the New York Academy of Sciences* 48(2): 67–72.

Rodrigues Barbosa, J., M. M. dos Santos Freitas, L. H. da Silva Martins, and R. N. de Carvalho. 2020. Polysaccharides of mushroom *Pleurotus* spp.: new extraction techniques, biological activities and development of new technologies. *Carbohydrate Polymers* 229: 115550. https://doi.org/10.1016/j.carbpol.2019.115550

Roy, D., S. Ansari, A. Chatterjee, A. Luganini, S. K. Ghosh, and N. Chakraborty. 2020. *In vitro* search for antiviral activity against human Cytomegalovirus from medicinal mushrooms *Pleurotus* sp. and *Lentinus* sp. *Journal of Antivirals & Antiretrovirals* 12(3): 201. https://doi.org/10.35248/1948-5964.20.12.201

Sani, M. A., R. Priyadarshi, W. Zhang, A. Khezerlou, and J. W. Rhim. 2024. Innovative application of laccase enzyme in food packaging. *Trends in Food Science and Technology*. 151: 104623. https://doi.org/10.1016/j.tifs.2024.104623

Santoyo, S., A. C. Ramirez-Anguiano, L. Aldars-Garcia, G. Reglero, and C. Soler-Rivas. 2012. Antiviral activities of *Pleurotus ostreatus* extracts and polysaccharide fractions against herpes simplex virus type 1. *Journal of Food and Nutrition Research* 51(4): 225–235.

Sharma, A., A. Sharma, and A. Tripathi. 2021. Biological activities of *Pleurotus* spp. polysaccharides: a review. *Journal of Food Biochemistry* 45(6): e13748. https://doi.org/10.1111/jfbc.13748

Shen, H. S., S. Shao, J. C. Chen, and T. Zhou. 2017. Antimicrobials from mushrooms for assuring food safety. *Comprehensive Reviews in Food Science and Food Safety* 16(2):316–329. https://doi.org/10.1111/1541-4337.12255

Shipley, S. M., A. L. Barr, S. J. Graf, R. P. Collins, T. G. McCloud, and D. J. Newman. 2006. Development of a process for the production of the anticancer lead compound Pleurotin by fermentation of *Hohenbuehelia atrocaerulea*. *Journal of Industrial Microbiology and Biotechnology* 33(6): 463–468. https://doi.org/10.1007/s10295-006-0089-0

Sknepnek, A., and D. Miletić. 2022. Application of mushrooms in beverages. In: *Fungal Biotechnology Prospects and Avenues*, ed. S. K. Deshmukh, K. R. Sridhar, and S. M. Badalyan, 1st ed., 280–309. CRC Press.

Sknepnek, A., M. Pantić, D. Matijašević, D. Miletić, S. Lević, and M. Nikšić. 2018. Novel Kombucha beverage from Lingzhi or Reishi medicinal mushroom, *Ganoderma lucidum*, with antibacterial and antioxidant effects. *International Journal of Medicinal Mushrooms* 20(3): 243–258. https://doi.org/10.1615/IntJMedMushrooms.2018025833

Sknepnek, A., S. Tomić, D. Miletić, et al. 2021. Fermentation characteristics of novel *Coriolus versicolor* and *Lentinus edodes* Kombucha beverages and immunomodulatory potential of their polysaccharide extracts. *Food Chemistry* 342: 128344. https://doi.org/10.1016/j.foodchem.2020.128344

Suay, I., F. Arenal, F. J. Asensio, A. Basilio, et al. 2000. Screening of Basidiomycetes for antimicrobial activities. *Antonie van Leeuwenhoek* 78(2): 129–139. https://doi.org/10.1023/a:1026552024021

Sutthisa, W., and S. Anujakkawan. 2023. Antibacterial potential of oyster mushroom (*Pleurotus ostreatus* (Jacq. Ex Fr.) P. Kumm.) extract against pathogenic bacteria. *Journal of Pure and Applied Microbiology* 17(3): 1907–1915. https://doi.org/10.22207/JPAM.17.3.56

Tambekar, D. H., T. P. Sonar, M. V. Khodke, and B. S. Khante. 2006. The novel antibacterials from two edible mushrooms: *Agaricus bisporus* and *Pleurotus sajor caju*. *International Journal of Pharmacology* 2(5): 584–587.

Torres-Martínez, B. D. M., R. D. Vargas-Sánchez, G. R. Torrescano-Urrutia, et al. 2022. *Pleurotus* genus as a potential ingredient for meat products. *Foods* 11(6): 779. https://doi.org/10.3390/foods11060779

Vlasenko, V. A., T. N. Ilyicheva, T. V. Teplyakova, et al. 2020. Antiviral activity of total polysaccharide fraction of water and ethanol extracts of *Pleurotus pulmonarius* against the nfluenza A virus. *Current Research in Environmental and Applied Mycology* 10(1): 224–235. https://doi.org/10.5943/CREAM/10/1/22

Vunduk, J., W. A. A. Q. I. Wan-Mohtar, S. A. Mohamad, et al. 2019. Polysaccharides of *Pleurotus flabellatus* strain mynuk produced by submerged fermentation as a promising novel tool against adhesion and biofilm formation of foodborne pathogens. *LWT* 112: 108221. https://doi.org/10.1016/j.lwt.2019.05.119

Wang, C. R., T. B. Ng, L. Li, et al. 2011. Isolation of a polysaccharide with antiproliferative, hypoglycemic, antioxidant and HIV-1 reverse transcriptase inhibitory activities from the fruiting bodies of the Abalone mushroom *Pleurotus abalonus*. *Journal of Pharmacy and Pharmacology* 63(6): 825–832. https://doi.org/10.1111/j.2042-7158.2011.01274.x

Wang, H., and T. B. Ng. 2004. Eryngin, a novel antifungal peptide from fruiting bodies of the edible mushroom *Pleurotus eryngii*. *Peptides* 25(1): 1–5. https://doi.org/10.1016/j.peptides.2003.11.014

Wang, H. X., and T. B. Ng. 2000. Isolation of a novel ubiquitin-like protein from *Pleurotus ostreatus* mushroom with anti-human immunodeficiency virus, translation-inhibitory, and ribonuclease activities. *Biochemical and Biophysical Research Communications* 276(2): 587–593. https://doi.org/10.1006/bbrc.2000.3540

Wang, H. X., and T. B. Ng. 2006. Purification of a laccase from fruiting bodies of the mushroom *Pleurotus eryngii*. *Applied Microbiology and Biotechnology* 69(5): 521–525. https://doi.org/10.1007/s00253-005-0086-7

Wang, L., C. Hu, and L. Shao. 2017. The antimicrobial activity of nanoparticles: present situation and prospects for the future. *International Journal of Nanomedicine* 12: 1227–1249. https://doi.org/10.2147/IJN.S121956

WHO. 2024. WHO Bacterial Priority Pathogens List, 2024. Bacterial Pathogens of Public Health Importance to Guide Research, Development and Strategies to Prevent and Control Antimicrobial Resistance.

Wilkins, W. H., and G. C. M. Harris. 1944. Investigation into the production of bacteriostatic substances by fungi VI. Examination of the larger Basidiomycetes. *Annals of Applied Biology* 31(4): 261–270. https://doi.org/10.1111/j.1744-7348.1944.tb06737.x

Xu, X., H. Yan, J. Chen, and X. Zhang. 2011. Bioactive proteins from mushrooms. *Biotechnology Advances* 29(6): 667–674. https://doi.org/10.1016/j.biotechadv.2011.05.003

Yang, X., Y. Gao, M. Reyimu, et al. 2024. Structural analysis of *Pleurotus ferulae* polysaccharide and its effects on plant fungal disease and plant growth. *International Journal of Biological Macromolecules* 282 (Pt 5): 137396. https://doi.org/10.1016/j.ijbiomac.2024.137396

Yehia, R. S., and H. Al-Sheikh. 2014. Biosynthesis and characterization of silver nanoparticles produced by *Pleurotus ostreatus* and their anticandidal and anticancer activities. *World Journal of Microbiology and Biotechnology* 30(11): 2797–2803. https://doi.org/10.1007/s11274-014-1703-3

Younis, A. M., F. S. Wu, and H. H. El Shikh. 2015. Antimicrobial activity of extracts of the oyster culinary medicinal mushroom *Pleurotus ostreatus* (higher Basidiomycetes) and identification of a new antimicrobial compound. *International Journal of Medicinal Mushrooms* 17(6): 579–590. https://doi.org/10.1615/IntJMedMushrooms.v17.i6.80

Zhang, M., P. C. K. Cheung, V. E. C. Ooi, and L. Zhang. 2004. Evaluation of sulfated fungal β-glucans from the sclerotium of *Pleurotus tuber-regium* as a potential water-soluble anti-viral agent. *Carbohydrate Research* 339(13): 2297–2301.

# 8 Hepatoprotective, Antidiabetic, and Prebiotic Potentials of Different Oyster Mushrooms

*Monika Choudhary and Sandipta Ghosh**
School of Biological and Environmental Sciences, Shoolini University of Biotechnology and Management Sciences, Solan, India

## 8.1 INTRODUCTION

The worldwide increase in liver disease, diabetes, and intestinal disorders (diseases primarily caused by contemporary lifestyle factors, like diet, stress, and environmental exposure) poses an increasing threat to human health (Flandroy et al., 2018; Asrani et al., 2019). A vital component of the immune system, the liver is positioned to identify and eradicate infections from the digestive system. It is the primary barrier against infections and contains the greatest number of phagocytic cells. The liver normally reacts to harmless foreign substances, such as food antigens, by maintaining an anti-inflammatory or tolerant immune state. Since insufficient immunity can result in cancer or chronic infections, while excessive inflammation can damage tissue, it is crucial to strike a balance between immunity and tolerance. Through interactions among different immune cells in the liver, this balance is preserved (Kubes and Jenne, 2018). The liver's role becomes even more important when considering diabetes, a condition estimated to impact hundreds of millions of people globally. Diabetes mellitus is a non-communicable disease that affects both developed and developing nations. The liver is one of the many bodily systems impacted by this metabolic illness. Studies have shown that storage of fat can cause damage in the liver, which can further increase blood sugar levels. Insulin resistance is the primary cause of hyperglycemia, which impacts the metabolism of proteins, fats, and carbohydrates and can result in non-alcoholic fatty liver disease, which can then lead to cirrhosis, non-alcoholic steatohepatitis, and ultimately hepatocellular carcinomas (Mohamed et al., 2016). By promoting digestion, nutrient absorption, and immunological function, the gut microbiota contributes to human health in ways that go beyond liver and metabolic health (Zhang, 2022). An imbalance in gut bacteria is caused by poor eating habits, stress, and a lack of fiber. This imbalance is frequently connected to diseases like inflammatory bowel disease and irritable bowel syndrome (Yan et al., 2022; Aziz et al., 2023). Prebiotics are indigestible fibers that help the growth of health-promoting microorganisms, which, in turn, helps restore the balance of the microbiota. Common prebiotics that withstand breakdown in the upper gastrointestinal tract and reach the colon, where they promote the growth of good bacteria, include inulin, fructooligosaccharides, and mannanoligosaccharides (Adhikari and Kim, 2017; Mercola et al., 2017; Zoumpopoulou et al., 2018). By supporting liver function, regulating blood sugar levels, and promoting a balanced gut microbiome, hepatoprotective, antidiabetic, and prebiotic approaches collectively offer a synergistic strategy to enhance resilience against common health conditions and improve overall quality of life (Khatua et al., 2023; Saha et al., 2024).

Mushrooms are regarded as a rich source of essential amino acids, which are the building blocks of vital proteins, as well as phytonutrients like polysaccharides, dietary fibers, and other micronutrients. Generally speaking, mushrooms have a broad range of pharmacological qualities, such as antidiabetic, antioxidant, antiviral, antibacterial, osteoprotective, nephroprotective, and hepatoprotective effects. Their *in vitro* and *in vivo* tests demonstrate the potent biological and therapeutic properties of both wild edible and medicinal mushrooms. The multifunctional properties of mushroom extracts, along with the distinct bioactivities of their constituent compounds, present a wide range of applications, particularly in the restoration and healing of human organs and cells. Because of their rich composition and previously mentioned qualities, mushrooms are used to make pharmaceuticals and nutraceuticals (Anusiya et al., 2021). *Pleurotus*, a genus of edible mushrooms known for their medicinal properties, has shown promise in helping to mitigate the effects of liver diseases, diabetes, and gut issues (Singh et al., 2023; Shamim et al., 2023). Studies suggest that bioactive compounds in *Pleurotus*, such as polysaccharides and phenolic acids, can reduce oxidative stress and inflammation, improving liver health and potentially offering a natural, affordable supplement (Riaz et al., 2022). In diabetes, *Pleurotus* mushrooms have been shown to improve insulin sensitivity and lower blood glucose levels (BGL), offering a natural supplement to conventional diabetes management (De et al., 2012; Balaji et al., 2020). Additionally, *Pleurotus* mushrooms are rich in dietary fiber

* Corresponding author: sandiptaghosh@shooliniuniversity.com

DOI: 10.1201/9781003527572-8

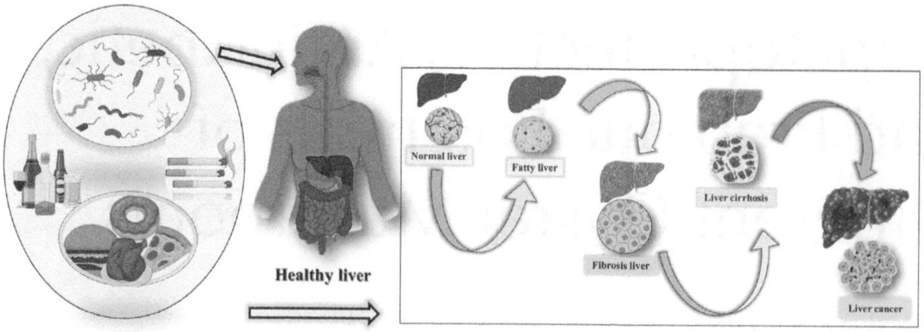


**FIGURE 8.1** Unhealthy lifestyle choices lead to liver damage, starting with fatty liver and culminating in liver cancer.

and prebiotics, which support gut health by promoting a balanced microbiome and improving digestion (Yin et al., 2020). In this chapter, *Pleurotus* mushrooms are presented as a natural, affordable, and sustainable option with significant potential to complement existing treatments for liver diseases, diabetes, and gut disorders, offering a holistic approach to managing these prevalent health conditions.

## 8.2 HEPATOPROTECTIVE PROPERTIES OF *PLEUROTUS* SP.

Clinical disorders or conditions resulting in liver dysfunction are referred to as liver diseases. Both acute and chronic diseases can be broadly classified into two categories. Acute liver disease manifests rapidly and is frequently transient, whereas chronic liver disease is a chronic illness that usually lasts longer than six months. Fibrosis and cirrhosis may eventually develop in chronic cases as a result of the liver going through cycles of parenchymal damage and regeneration (Crawford, 2007).

Roughly 2 million deaths globally are attributed to liver disease every year, with cirrhosis and liver cancer accounting for 3.5% of these deaths. Liver cancer is the 16th most common cause of death, whereas cirrhosis is ranked eleventh. With approximately 2 billion drinkers worldwide and 75 million individuals with alcohol-use disorders, alcohol consumption is a main risk factor that increases the risk of liver disease. Moreover, 400 million adults have diabetes, and 2 billion adults are overweight or obese; these conditions are major causes of non-alcoholic fatty liver disease (Figure 8.1). There is still a high prevalence of viral hepatitis, and drug-induced liver damage is increasing. Despite being a common treatment, liver transplants only cover less than 10% of the world's need. The majority of liver diseases are avoidable, which presents a significant opportunity to improve the state of public health (Asrani et al., 2019).

### 8.2.1 Bioactive Compounds Responsible for Hepatoprotective Effects

In order to address liver diseases, both dietary management and medicinal plant-based treatments are essential. This highlights the significance of bioactive compounds in prevention and treatment strategies. The goal of dietary management is to control the consumption of bioactive substances found in both plant and animal products, such as fatty acids, phytosterols, and alkaloids. These substances have the ability to affect several metabolic pathways, improving liver function and possibly delaying the onset of liver disease (Kumar et al., 2018) (Figure 8.2).

Rich in polysaccharides, phenols, tocopherols, flavonoids, carotenoids, glycosides, ergothioneine, and ascorbic acid, mushrooms serve as an abundant natural source of antioxidants. Specifically, polysaccharides decrease tissue lipid peroxidation and strengthen the body's antioxidant enzymes. With less chance of promoting oxidation, they are safer and more potent as macromolecules than small molecular antioxidants. Numerous species of *Pleurotus* are abundant in bioactive substances like β-glucans, phenolic compounds, and vitamin C, which can boost the action of antioxidant enzymes, such as superoxide dismutase and catalase. By lowering oxidative stress and averting hepatic cell necrosis, these enzymes are essential for liver protection (Fu et al., 2009) (Table 8.1).

Zhang et al. (2016) investigated the hepatoprotective properties of *Pleurotus eryngii* mycelia-isolated intracellular polysaccharides (IPS) and its fractions, IPS-1 and IPS–2. By lowering serum albumin and triglyceride levels and raising bilirubin and serum enzyme activities, IPS-2 demonstrated better hepatic protection. It reduced liver structural damage and enhanced antioxidant status as well. Analysis of the monosaccharide composition and bond types indicated that IPS and its fractions may serve as good candidates for use in natural remedies and functional foods to prevent acute liver damage.

In the study by Xu et al. (2017), hyperlipidemic mice fed a high-fat, high-cholesterol diet were used to evaluate the hepatoprotective and antioxidant properties of enzymatically extracted mycelial zinc polysaccharides (En-MZPS) from *P. eryngii* var. *tuoliensis*. En-MZPS reduced hepatic lipid levels and improved serum enzyme and lipid levels. It also enhanced the activity of antioxidant enzymes, including superoxide dismutase (SOD), glutathione peroxidase (GSH-Px), catalase (CAT), and total antioxidant capacity (T-AOC), while decreasing lipid peroxidation. *In vitro*, En-MZPS inhibited hydroxyl radicals by 59.98% and 2,2-diphenyl-1-picrylhydrazyl (DPPH) radicals by 37.01%.

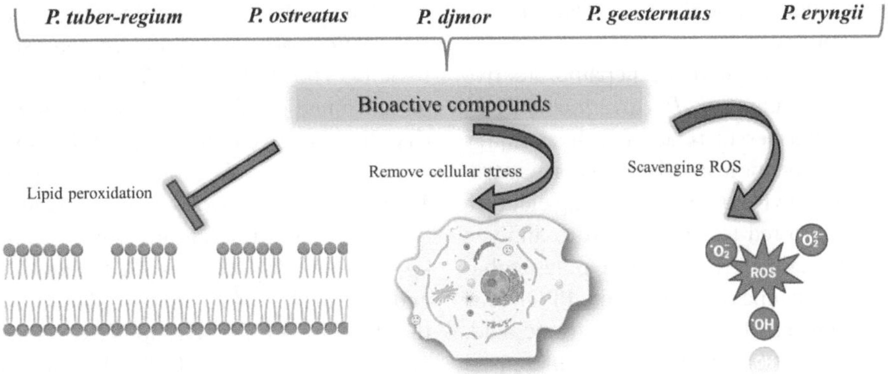


**FIGURE 8.2** Different species of *Pleurotus* mushrooms can support liver health by providing bioactive compounds that protect against liver damage.

**TABLE 8.1**
**Bioactive Compounds of *Pleurotus* spp. Responsible for Hepatoprotection**

| Species | Bioactive Compound | Role in Hepatoprotection | Reference |
|---|---|---|---|
| *Pleurotus citrinopileatus* | Polysaccharide-peptides | Protective effects of on hepatoprotective and gut microbiota | Huang et al., 2022 |
| *Pleurotus djamor* | Polysaccharides, Phenolic compounds | Preventing the $CCl_4$-induced acute liver damage | Zhang et al., 2016 |
| *Pleurotus eryngii* | Phenolic compounds, Ergothioneine | Protects liver cells from oxidative stress | Borodina et al., 2020 |
| *Pleurotus ostreatus* | Polysaccharides (β-glucans), Lovastatin | Reduces oxidative stress, lowers cholesterol, immune modulation | Golak-Siwulska et al., 2018 |
| | Ergothioneine, Lovastatin | Reduces hepatic inflammation, lowers cholesterol levels, preventing fatty liver and hepatic damage | Saad et al., 2022; Parnian-Khajehdizaj et al., 2024 |
| *Pleurotus sajor-caju* | Polysaccharides (β-glucans) | Reduces oxidative stress, supports liver detoxification | Glamočlija et al., 2018 |

The findings suggest that En-MZPS could be used as a functional food or conventional drug for preventing hyperlipidemia and non-alcoholic fatty liver disease.

Song et al. (2018) examined the effects of acidic- and enzymatic-hydrolyzed mycelium polysaccharides (AcMPS and EnMPS) from *Pleurotus geesteranus* on mice with alcoholic liver disease (ALD) in terms of their antioxidant and hepatoprotective properties. Histopathological analysis supported the results, which indicated that both polysaccharides improved inflammatory symptoms, decreased lipid accumulation, mitigated oxidative stress, and alleviated alcoholic hepatitis. L-Rhamnose, D-Ribose, L-Arabinose, D-Glucose, D-Mannose, and D-Galactose were all present in AcMPS (84% yield), whereas L-Rhamnose, L-Arabinose, D-Galactose, and D-Glucose were all present in EnMPS (90% yield). The polysaccharides with (1→3)- and (1→6)-linkages were recognized as β-pyranoside polysaccharides. According to the study, AcMPS and EnMPS may be used as complementary and alternative medicines to prevent ALD.

Osman et al. (2019) evaluated the hepatoprotective benefits of an oyster mushroom crude phenolic-rich extract (CPRE) on albino rats' liver damage caused by carbon tetrachloride ($CCl_4$). Reduced activity of antioxidant enzymes and increased markers of liver impairment were observed following exposure to $CCl_4$. Activities of SOD and GSH-Px were elevated, and liver damage markers were significantly reduced upon CPRE administration. Results of the study indicate that CPRE has potent antioxidant properties and can defend the liver from $CCl_4$-induced damage. This suggests that CPRE may find application in food products as a natural antioxidant and antibacterial agent.

Zhu et al. (2019) investigated hepatoprotective effects of polysaccharides from *Pleurotus ostreatus* (POP) on $CCl_4$-induced acute liver injury (ALI) in rats. POP treatment led to reduced levels of alanine transaminase and aspartate transaminase, increased antioxidant enzyme levels (SOD, catalase, and GSH-Px), and decreased malondialdehyde (MDA) levels in both blood and liver. TUNEL and immunohistochemistry analyses indicated a significant reduction in apoptotic cell rates and expression of pro-apoptotic markers (Bax, cytochrome C, caspase-3, caspase-9) and an increase in the anti-apoptotic marker Bcl–2. Metabolite analysis identified 11 key metabolites related to energy, fatty acid, amino acid, and gut bacteria metabolism as biomarkers. The study concluded that POP effectively protects against $CCl_4$-induced ALI, primarily through antioxidant mechanisms that regulate metabolic disorders and mitigate liver mitochondrial apoptosis.

In 2020, researchers investigated the hepatoprotective properties and structural features of *P. ostreatus* (PPOP) phosphorylated polysaccharides in relation to liver injury induced by $CCl_4$ in Kunming mice. Fourier transform infrared spectroscopy (FT-IR), environmental scanning electron microscopy, and the Congo red test were used to analyze PPOP, which was synthesized from natural polysaccharides (POP). Based on its chemical composition, PPOP was found to be mainly composed of xylose, galacturonic acid, and rhamnose. *In vivo* tests showed that PPOP increased antioxidant enzyme activity and decreased cholesterol and liver enzyme levels. According to the study, PPOP may be a viable treatment for liver damage since it has greater hepatoprotective effects than POP (Duan et al., 2020).

Abdel-Monem et al. (2020) discussed the need for substitute chemopreventive agents in light of Egypt's rising liver disease incidence. In rats exposed to diethylnitrosamine (DEN) and $CCl_4$, the study assessed the hepatoprotective effects of an exopolysaccharide-peptide (PSP) complex from *P. ostreatus*. Liver injury markers such as alanine aminotransferase (ALT), aspartate aminotransferase (AST), and alkaline phosphatase were markedly elevated following DEN/$CCl_4$ treatment, also indicating significant oxidative stress. PSP, on the other hand, counteracted these effects and restored liver integrity and function. The findings imply that PSP derived from *P. ostreatus* may provide noteworthy antioxidant and hepatoprotective advantages, thereby warranting additional clinical testing in patients with liver disease.

Zhang et al. (2020) used alkali precipitation to prepare a protein extract from *P. ostreatus* (POPE), and then used pepsin enzymatic hydrolysis to isolate a particular fraction (POPEP-III) with a molecular weight of 3000–5000 Da. POPEP-III demonstrated noteworthy antioxidant characteristics, such as the ability to scavenge hydrogen peroxide and DPPH, chelate iron, and inhibit lipid peroxidation. After $CCl_4$ treatment, it dramatically decreased levels of ALT and AST in the serum and MDA in the liver in mice. In addition, POPEP-III prevented $CCl_4$-induced liver histological damage and enhanced activities of SOD and GSH-Px. These findings suggest that POPEP-III is an effective hepatoprotectant against oxidative damage.

Dkhil et al. (2020) investigated the antioxidant and hepatoprotective effects of *P. ostreatus* extract (PE) in cadmium-intoxicated rats. Cadmium exposure caused significant liver damage, oxidative stress, and increased cadmium accumulation. PE treatment improved histological damage, reduced cadmium levels, and mitigated oxidative stress by normalizing GSH, CAT, MDA, and ROS levels. Additionally, PE downregulated the expression of Nrf2-mRNA, which was upregulated due to cadmium toxicity. The study concluded that PE possesses strong antioxidant and anti-inflammatory properties, effectively protecting the liver against cadmium-induced hepatotoxicity.

Okolo et al. (2021) examined the hepatoprotective potential of *Pleurotus tuber-regium* against $CCl_4$-induced liver toxicity. The study utilized *in vitro* antioxidant assays and gas chromatography–mass spectrometry (GC–MS) to identify the bioactive compounds in the mushroom. Sixty rats were divided into six groups (10 per group) and treated over 13 weeks. The control group received olive oil (3 mL/kg) twice weekly, while Group II was exposed to $CCl_4$ (3 mL/kg) twice weekly. Groups III, IV, and V received 100, 200, and 500 mg/kg of *P. tuber-regium* mixed with feed daily, in addition to $CCl_4$. Group VI received 500 mg/kg of *P. tuber-regium* without $CCl_4$. Various liver function tests and histopathological evaluations were conducted. The analysis identified 10 antioxidant compounds in *P. tuber-regium*. CCl4 significantly increased liver enzyme levels, bilirubin, and MDA, indicating liver damage. However, *P. tuber-regium* treatment reduced these markers and protected against necrosis, edema, and vein congestion caused by $CCl_4$. Overall, *P. tuber-regium* showed significant hepatoprotective effects, suggesting its potential as a natural remedy for liver protection.

Chandana et al. (2023) used both *in vitro* and *in vivo* techniques to evaluate *P. ostreatus* antioxidant effects. Rats were placed into six groups and given a methanolic extract (MEPO), with one group receiving CCl4 to induce liver damage. For 21 days, other groups received treatments consisting of different dosages of silymarin (100 mg/kg) or MEPO (200, 400, and 800 mg/kg). With $IC_{50}$ values of 72.1 versus 66.3 for silymarin, MEPO demonstrated noteworthy antioxidant activity. Treatment lowered MDA levels while raising catalase and superoxide dismutase levels. According to the results, MEPO may offer protection against hepatic injury caused by $CCl_4$.

## 8.3 ANTIDIABETIC POTENTIAL OF *PLEUROTUS* SP.

Diabetes mellitus is a group of chronic metabolic disorders characterized by elevated BGL. These conditions result from impaired glucose metabolism, which may be caused by inadequate insulin secretion, insufficient insulin action, or insulin resistance (Nnemolisa et al., 2024) (Figure 8.3). In many nations, the prevalence of diabetes mellitus in children is rising. Although there are obvious signs of regional variations in trends, the annual increase is thought to be about 3% overall. An estimated 79,000 children worldwide are diagnosed with type 1 diabetes each year. Evidence suggests that the incidence is rising more rapidly in some low-prevalence European countries, with the most significant increases observed among young children compared to other regions. Additionally, data from high-incidence European nations suggest that the trend of rising incidence appears to be leveling off, which could be encouraging for the future (Patterson et al., 2014). Historically, the natural world has provided a wealth of medicinal compounds. Numerous contemporary medicines have been

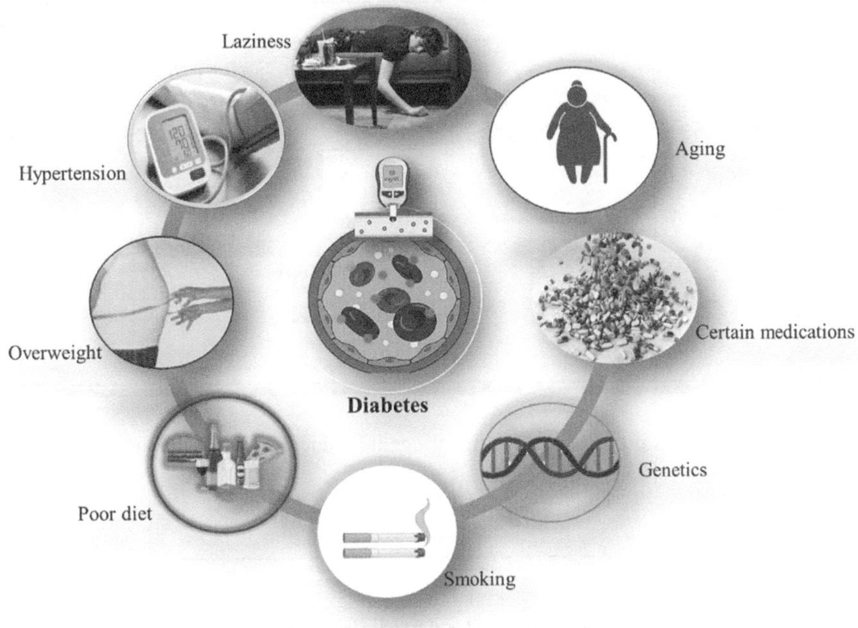


**FIGURE 8.3** Key risk factors and complications of diabetes.

derived from natural sources, frequently drawing inspiration from their application in traditional medicine (Bhushan and Kulshreshtha, 2018).

Interest in natural ingredients has increased dramatically in the last few centuries, as they are being explored more and more as potential nutraceuticals to replace synthetic compounds, whose application is frequently limited because of their negative side effects (El Omari et al., 2019). Mushrooms are widely recognized macro-fungi featuring distinct fruiting bodies that can be either hypogeous (underground) or epigeous (above ground) (Yadav and Negi, 2021). The *Pleurotus* genus ranks among the most widely produced mushroom species globally (Torres-Martínez et al., 2022). Studies have shown that extracts from *P. ostreatus* (Ravi et al., 2013), *Pleurotus giganteus* (Paravamsivam et al., 2016), and other species can lower BGL in diabetic models, making them promising candidates for dietary interventions and functional foods aimed at managing diabetes.

### 8.3.1 Bioactive Compounds Responsible for Antidiabetic Effects

The wide variety of bioactive metabolites originating in plants and fungi, especially mushrooms, makes them important sources of natural compounds used in drugs (Cragg et al., 1997). Bioactive compounds found in mushrooms include polysaccharides, particularly β-glucans and glycoproteins, as well as secondary metabolites such as terpenoids, acids, alkaloids, polyphenolic compounds, lactones, sesquiterpenes, sterols, and nucleotide analogues, vitamins, and metal-chelating agents. Because of these biologically active ingredients, mushrooms are advantageous as hepatoprotective, immune-boosting, antiviral, anticancer, and cholesterol-lowering agents (Kumar et al., 2021) (Figure 8.4). The abundance of bioactive compounds found in *Pleurotus* species is drawing attention, as they may have the potential to regulate blood sugar levels. These mushrooms contain polysaccharides, like β-glucans and ergosterol, that increase glucose metabolism and insulin sensitivity (Chowdhury et al., 2020; Arunachalam et al., 2022) (Table 8.2).

*Pleurotus florida* is an excellent source of extractable phytochemicals that exhibit inhibitory effects on key enzymes, specifically α-amylase and α-glucosidase, which are associated with Type 2 diabetes mellitus. *In vitro* assays play a critical role in evaluating the antidiabetic potential of drugs during the initial screening phase, especially when there is a need to assess a large number of prospective therapeutic candidates. Additionally, these tests can provide valuable insights into the mechanisms of action of the therapeutic agents (LoPiparo et al., 2008).

Ravi et al. (2013) evaluated *P. ostreatus* antidiabetic properties in mice with normal and alloxan-induced diabetes. Four groups of mice were used in the study to test the ethanol extracts from *P. ostreatus* fruiting bodies: normal control, diabetic control, diabetic mice treated with glibenclamide, and diabetic mice treated with *P. ostreatus* extract. After 15 days, the extract significantly improved lipid profiles by lowering serum cholesterol, triglycerides, and low-density lipoprotein (LDL) while increasing high-density lipoprotein (HDL), and significantly lowered blood glucose. Additionally, it prevented diabetic mice from losing weight

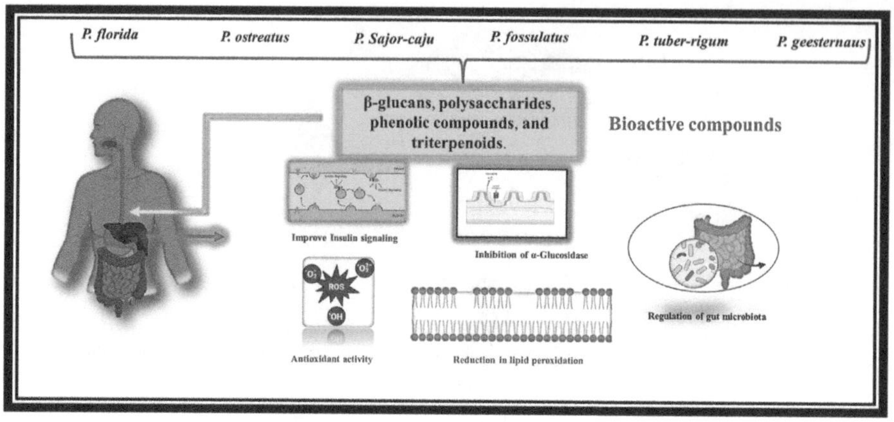


**FIGURE 8.4** Different species of *Pleurotus* promote insulin signaling and reduce oxidative damage for improved health.

**TABLE 8.2**
**Bioactive Compounds of *Pleurotus* spp. Contributing to Antidiabetic Effects**

| Species | Bioactive Compound | Antidiabetic Effects | Reference |
|---|---|---|---|
| *Pleurotus citrinopileatus* | Water soluble polysaccharide | Reduced fasting blood glucose level in diabetic mice | Hu et al., 2006 |
| *Pleurotus djamor* | Zinc polysaccharide | Reduced blood glucose in mice | Zhang et al., 2015 |
| *Pleurotus eryngii* | Polysaccharides, Ergosterol | Improved glucose tolerance, enhanced insulin secretion | Chowdhury et al., 2020 |
| *Pleurotus florida* | Phenols, flavonoids, saponins, tannins and terpenoid | Controlled diabetics | Prabu et al., 2017 |
| *Pleurotus ostreatus* | Ergothioneine | Reduced blood sugar | Aslam et al., 2021 |
| | Polysaccharides (β-glucans) | Improved insulin sensitivity, reduced blood glucose level, and enhanced glucose metabolism | Arunachalam et al., 2022 |
| *Pleurotus sajor-caju* | β-glucan rich polysaccharide | Activated AMPK pathway, which regulates lipid and glucose metabolism in 3T3-L1 cells. | Kanagasabapathy et al., 2014 |
| *Pleurotus tuber-regium* | Extracellular polysaccharide | Controlled progression of diabetes | Huang et al., 2014 |

and enhanced kidney function, indicating the potential of *P. ostreatus* as a therapeutic agent for managing diabetes.

Ganeshpurkar et al. (2014) investigated the antihyperglycemic effects of the polysaccharide fraction of *Pleurotus florida* (PFP) in streptozotocin-induced diabetic models. Acute toxicity studies, following standard guideline 423, showed that PFP was nontoxic up to 4000 mg/kg. Doses of 200 and 400 mg/kg were tested, resulting in a significant reduction in blood glucose, serum cholesterol, urine glucose, triglycerides, and ketones compared to controls. PFP also decreased MDA and nitric oxide (NO) levels while restoring catalase, superoxide dismutase, and reduced glutathione concentrations. The findings suggest that PFP may help manage hyperglycemia and hypercholesterolemia in type 2 diabetes and could be considered an adjunct to first-line therapies.

Paravamsivam et al. (2016) evaluated the effects of *Pleurotus giganteus* on adipocyte differentiation and glucose uptake in 3T3-L1 cells. Methanol extracts of the mushroom were fractionated to obtain an ethyl acetate extract, which was tested using 3–(4,5-dimethylthiazol-2-yl)-2,5-diphenyltetrazolium bromide (MTT), Oil Red O, and fluorescence glucose uptake assays. The ethyl acetate extract significantly enhanced adipogenic differentiation and glucose uptake in 3T3-L1 adipocytes by upregulating sterol regulatory element-binding protein, peroxisome proliferator-activated receptor γ (PPARγ), and the PI3K/Akt pathway. It also increased the expression of glucose transporters Glut1 and Glut4. The findings suggest that *P. giganteus* ethyl acetate extract has insulin-sensitizing effects and could be a potential adjuvant for managing type 2 diabetes.

Johnny and Okon (2013) investigated the antidiabetic effects of ethanolic extract of *P. ostreatus* on alloxan-induced diabetic rats. The median lethal dose ($LD_{50}$) was found to be 3,472.14 mg/kg. BGL were measured over 7 and 15 hours after rats received single intraperitoneal doses of 380.0, 760.0, and 1140.0 mg/kg body weight. With outcomes similar to the prescribed medication metformin, the extract dramatically decreased BGL, reaching its maximum reduction after five hours. This study supports the traditional use of *P. ostreatus* in the management of diabetes.

Onuekwuzu et al. (2019) studied the effects of an aqueous extract of *Pleurotus tuber-regium* sclerotia ("usu" milk) on body weight, fasting blood glucose, organ weights, lipid profiles, and atherogenic indices in normal and alloxan-induced diabetic rabbits. Alloxan was used to induce diabetes, and the extract was administered at 100, 200, and 300 mg/kg doses, with metformin serving as the standard medication. Gas chromatography analysis identified 31 flavonoids, the main constituents of which were luteolin (24.3%), isoquercetin (28.5%), quercetin (18.8%), and kaempferol (11.3%). Blood glucose, triglycerides, very low-density lipoprotein (VLDL), LDL, and non-HDL cholesterol were all significantly reduced by the extract, while HDL cholesterol and several atherogenic indices improved. The results suggest that the extract may possess hypoglycemic, lipid-lowering, and cardioprotective properties, which could explain its traditional use in the treatment of diabetes mellitus.

Dubey et al. (2020) evaluated the antidiabetic effects of *Pleurotus fossulatus* aqueous extract (PFA-extract) on liver and kidney function in diabetic rats. Five groups of albino Wistar rats were used in the study: two test groups received PFA-extract at 250 mg/kg (D1) and 500 mg/kg (D2), and the other groups were diabetic and normal. The drug control group used metformin. PFA-extract improved body weight recovery (97.9%) and significantly lowered blood glucose levels (53%) in the D2 group. In the D2 group, there was a noticeable improvement in lipid profiles, liver function, and kidney function (creatinine, blood urea nitrogen, and uric acid), with values returning to nearly normal ranges. According to these results, PFA-extract shows promise as a natural diabetes treatment and as a means of preserving kidney and liver health.

Krishna et al. (2023) evaluated antidiabetic and antioxidant properties of extracts from eight *Pleurotus* mushroom species. The methanolic extract of *P. florida* was found to have the highest concentrations of total phenolics, tannins, and flavonoids. It also demonstrated the strongest antioxidant activity in multiple tests, such as ferric reducing antioxidant power (FRAP), 2,2′-azino-bis(3-ethylbenzothiazoline-6-sulfonic acid) (ABTS), and DPPH. Significant β-carotene protection and hydroxyl radical scavenging were also demonstrated by *P. florida*. Methanolic extract at 1000 μg/mL outperformed the benchmark medication acarbose (96.54%) in antidiabetic assays, achieving 99.02% α-amylase inhibition. The results imply that *P. florida* bioactive compounds have strong *in vitro* antidiabetic and antioxidant properties.

Nnemolisa et al. (2024) explored the chemical composition and therapeutic potential of *P. ostreatus* methanol extract (MEPO) using *in vitro* and *in silico* methods. Gas chromatography-flame ionization detector (GC-FID) analysis identified 13 compounds, including ephedrine, oxalate, rutin, naringin, and kaempferol. DPPH and FRAP assays were used to assess antioxidant activity, which revealed moderate activity ($IC_{50}$ = 732.41 mg/mL for DPPH). Compared to α-amylase ($IC_{50}$ = 1074.05 mg/mL), MEPO showed stronger inhibition of α-glucosidase ($IC_{50}$ = 246.58 mg/mL). Using molecular docking, it was shown that rutin and naringin could form strong hydrogen bonds with important residues related to the activities of α-glucosidase, α-amylase, and phosphoenolpyruvate (PEP) carboxykinase. According to the study, MEPO contains a variety of bioactive substances that may have antidiabetic effects.

Das et al. (2024) evaluated six hybrid mushrooms, developed by protoplast fusion of *Calocybe indica* and *Pleurotus sajor-caju*, for antioxidant, antimicrobial, and antidiabetic activities. The hybrids exhibited robust antioxidant activity along with a high content of phenolics and flavonoids. Six major compounds were identified by UHPLC, and the extracts showed antibacterial activity against *Pseudomonas aeruginosa*, *Bacillus subtilis*, *Klebsiella pneumoniae*, and *Staphylococcus aureus*. Furthermore, APS-4 demonstrated significant α-galactosidase and α-glucosidase inhibition properties, indicating potential antidiabetic qualities.

## 8.4 PREBIOTIC POTENTIAL OF *PLEUROTUS* SP.

In recent decades, there has been growing awareness regarding the impact of food on health. Numerous epidemiological studies have indicated that functional foods can lower the risk of cancer, promote cardiovascular health, enhance immune function, and improve gastrointestinal health (Al-Sheraji et al., 2013). Mushrooms boost immune function and gut health by inhibiting harmful bacteria and promoting beneficial ones, especially in the colon. This enhances immunity and supports a balanced gut microbiome for overall health (Yin et al., 2020) (Figure 8.5). Mushroom-derived D-glucans are resistant to digestion by human enzymes, making them part of the dietary fiber category. These compounds pass through the small intestine without being digested or absorbed, reaching the large intestine intact. Once there, they serve as a source of energy and carbon for colonic bacteria, contributing to gut microbiota metabolism (Holscher, 2017). Colonic fermentation of mushroom polysaccharides can stimulate the growth of beneficial microorganisms while reducing the presence of opportunistic pathogens. This process indirectly or directly enhances immune responses by modulating the gut microbiota. The resulting shift in microbial populations promotes the proliferation of organisms that produce immune-regulating compounds, such as short-chain fatty acids (SCFAs), which play a key role in supporting immune function (Ma et al., 2017; Tang and Lu, 2019).

### 8.4.1 POLYSACCHARIDES AND FIBERS AS PREBIOTIC AGENTS IN *PLEUROTUS* SP.

Mushrooms are promising sources of prebiotics due to their diverse polysaccharide content, including hemicellulose, xylans, α- and β-glucans, chitin, mannans, and galactans. Notably, β-(1→3)-D-glucans, along with their peptide/protein derivatives (polysaccharide-peptide/protein complexes)

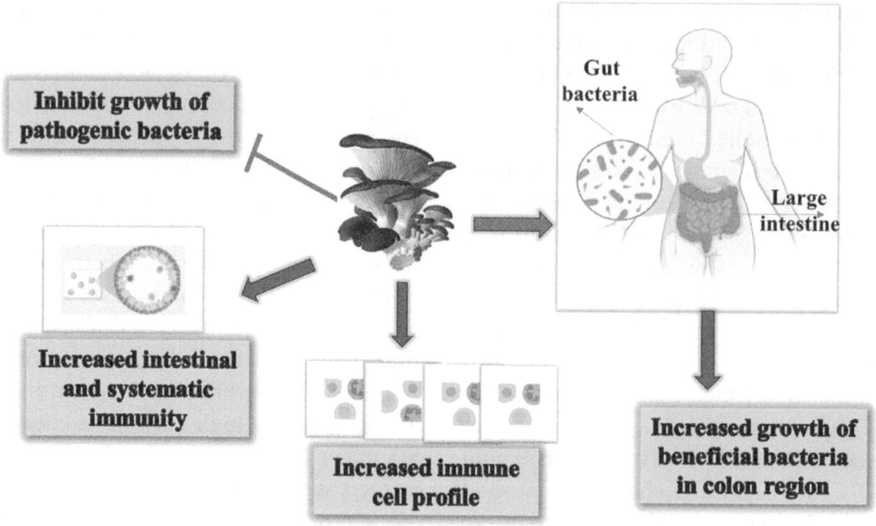


**FIGURE 8.5** Roles of mushrooms in supporting gut and immune health.

and proteoglycans, serve as key prebiotics that play crucial roles in immunomodulation or exhibit antitumor activities (Singdevsachan et al., 2016) (Table 8.3).

*In vitro* studies have demonstrated that polysaccharides from *P. ostreatus* and *P. eryngii* promote the growth of probiotic bacteria, particularly *Lactobacillus*. Notably, the growth rate of *Enterococcus* was observed to be lower compared to that of *Lactobacillus*. This indicates that these mushroom-derived polysaccharides preferentially support the proliferation of beneficial *Lactobacillus* species, which may enhance gut health (Synytsya et al., 2009). The primary bioactive compound in *P. ostreatus* is β-glucan [(1,3)/(1,6)-glucans], which is key to its prebiotic activity (Hu et al., 2022). This mushroom is particularly rich in β-glucan, which may be bound to proteins, enhancing its functionality. Another important component is chitin and its derivatives, which modulate gut microflora by providing energy and a carbon source for probiotics, facilitating their digestion and metabolism. These compounds contribute to the overall prebiotic effects of *P. ostreatus*, supporting gut health and beneficial microbial growth (Guan et al., 2022).

Moumita and Das (2022) investigated five edible mushrooms *P. sajor-caju, P. ostreatus, P. florida, Agaricus bisporus*, and *Lentinus edodes*—for their prebiotic content and bioactive compounds. They measured the amounts of β-glucan, inulin, eritadenine, flavonoids, and polyphenols using biochemical techniques and HPLC. *Pleurotus florida* was found to have the highest levels of insulin and β-glucan, indicating that it is a rich source of prebiotics. Its extract also yielded a higher prebiotic activity score than commercial inulin due to its high eritadenine content and strong compatibility with *Enterococcus faecium*. Probiotics and these mushrooms were used to formulate synbiotic microcapsules, which had antihypertensive and cholesterol-lowering properties.

Freitas et al. (2018) investigated the use of spent coffee grounds (SCGs), a by-product of coffee brewing, as a growth substrate for beneficial edible mushrooms. This approach holds promise for mitigating environmental impacts, reducing production costs, and enhancing consumer health. The study focused on the cultivation of *Pleurotus citrinopileatus* and *Pleurotus salmoneo-stramineus*, revealing that SCGs facilitated a maximum production yield of 25.1% (w/w) for *P. citrinopileatus*. Both ethanolic and aqueous extracts of the cultivated mushrooms, with or without SCGs, demonstrated significant prebiotic potential, effectively promoting the growth of *Bifidobacterium animalis* compared to fructooligosaccharides (FOS).

Kerezoudi et al. (2021) explored the potential of *Ganoderma lucidum* and *P. ostreatus* mushrooms as prebiotics in healthy and osteopenic women. The study involved fermenting mushroom powder and extracts with fecal inocula from both groups and analyzing gut microbiota composition and SCFA production. The outcomes demonstrated that the mushroom treatments increased osteoblastic activity, decreased RANKL levels, and improved gut microbiota. According to the research, mushrooms high in β-glucan may improve bone health by altering gut microbiota and increasing the synthesis of SCFAs.

Soodpakdee et al. (2022) investigated the nutritional and prebiotic enhancement of Riceberry rice through germination and fermentation with *P. ostreatus* mycelium. Over four days, β-aminobutyric acid (GABA) levels peaked on day three. Germinated Riceberry (GR) showed moderate prebiotic activity against probiotics compared to commercial agents, while ungerminated Riceberry (R) had no prebiotic effects. Mycelium fermentation further boosted GR's prebiotic potential. The study suggests that GRM could be used in rice-based products, especially for the elderly, to improve nutritional value and digestive health.

**TABLE 8.3**
**Polysaccharides and Fibers Found in *Pleurotus* Species, Highlighting their Prebiotic Effects on Gut Microbiota and Immune Modulation**

| Source (*Pleurotus* Species) | Polysaccharide/Fiber | Type | Prebiotic Function | Reference |
|---|---|---|---|---|
| *Pleurotus ostreatus* | Xylan | Hemicellulose | Encourages beneficial bacteria of the colon to grow | Synytsya et al., 2009 |
| | Mannan | Polysaccharide | Enhances beneficial microbial activity in the colon | Törős et al., 2010 |
| | Glucuronoxylomannan | | Immunomodulatory, supports beneficial gut flora | Giavasis, 2014 |
| | Beta-glucans | | Stimulates growth of beneficial bacteria, enhances immune response | Van et al., 2016; Vetvicka et al., 2019 |
| *Pleurotus tuber- regium* | β-glucans | | Promote the growth of beneficial probiotic bacteria, such as lactic acid bacteria and bifidobacteria, to improve human gut health. | Singdevsachan et al., 2016 |
| *Pleurotus ostreatus, Pleurotus eryngii* | Beta-glucans | | Stimulates growth of beneficial bacteria, enhances immune response | Van et al., 2016; Vetvicka et al., 2019 |
| | Chitin | | Promotes gut microbiota balance and supports colonic health | Liang et al., 2021 |
| | Galactomannan | | Enhances gut microbiota activity, supports immune function | Törős et al., 2023 |
| | Hemicellulose | Dietary Fiber | Promotes intestinal health by supporting microbiota diversity | Törős et al., 2010; Ma et al., 2017 |
| | Pectin | | Fermented by gut bacteria, promotes short-chain fatty acid production | Nakahara et al., 2020 |
| | Cellulose | | Aids in digestion and supports growth of gut microbes | Dong et al., 2024 |
| | Fructooligosaccharids (FOS) | Oligosaccharide | Increases beneficial bacteria such as *Bifidobacteria* and *Lactobacillus* | Saxami et al., 2021 |

Setyawan et al. (2023) evaluated the prebiotic potential of white oyster mushroom extract. In this study, alkaline and hot water extraction techniques were used to isolate β-glucan from the mushroom. Subsequently, the β-glucan content was quantified using an assay kit and examined in glucose-free media to evaluate its uptake by the pathogen *Escherichia coli* and the probiotic *Lactobacillus plantarum*. The β-glucan content from the *P. ostreatus* water extract was found to be 37.15 g/100 g, with a prebiotic activity score of 0.91, a prebiotic index of 1.42, and other favorable results. The extract may be a novel prebiotic because it promoted the synthesis of organic acids like lactic acid.

Törős et al. (2023) reported that *P. ostreatus* contains bioactive compounds, including polysaccharides like glucan and chitin, which act as prebiotics by promoting beneficial gut bacteria and potentially reducing antibiotic resistance. These compounds, along with secondary metabolites, exhibited antimicrobial properties that may help prevent harmful bacterial growth.

Yadav et al. (2024) investigated the prebiotic potential of chitosan extracted from *P. ostreatus* mushroom waste (MCH). When supplemented with probiotic strains (*Lactobacillus casei*, *L. plantarum*, *L. helveticus*, and *L. rhamnosus*) at 1%, MCH exhibited prebiotic activity scores of 0.73 to 0.82. It promoted the growth of *Lactobacillus* strains, increased SCFA production, particularly in *L. rhamnosus*, and improved the adhesion efficiency of probiotics. These results suggest that MCH can serve as an effective prebiotic, supporting gut health and microbiota balance.

Thikham et al. (2024) investigated the extraction of β-glucan from *P. pulmonarius* by-products (caps and stems) using pulsed electric field technology and evaluated their prebiotic properties. The greatest β-glucan content was produced by PEF treatment at 7.5 kV/cm, yielding 4.18% from the cap and 14.10% from the stem; treatment at 10 kV/cm produced 21.39% from the stem and 2.34% from the cap. PEF performed better than conventional hot water extraction, and the samples had substantial structural damage, as shown by scanning electron microscopy. The extracts demonstrated their prebiotic potential by encouraging the growth of lactic acid bacteria and synthesis of SCFAs. According to this study, PEF improves the β-glucan extraction efficiency from *P. pulmonarius* by-products, potentially providing prebiotic benefits.

## 8.5 CONCLUSION

In conclusion, the increasing amount of data shows that *Pleurotus* species have substantial therapeutic potential for treating a variety of illnesses, such as diabetes, gut health, and liver protection. Polysaccharides, phenolic compounds, flavonoids, and β-glucans are among the bioactive substances found in these mushrooms that have significant hypoglycemic, anti-inflammatory, and antioxidant properties. However, there are some shortcomings that need to be addressed. First, numerous studies have demonstrated that the bioactive compounds in *Pleurotus* species can reduce liver damage caused by oxidative stress, inflammation, and toxic agents. These mushrooms are promising natural options for the prevention and treatment of liver disease because of their capacity to reduce pro-apoptotic markers, alter antioxidant pathways, and maintain liver histology. Nevertheless, more clinical trials and pharmacokinetic research are necessary to fully confirm their therapeutic efficacy in humans. Second, *Pleurotus* species have demonstrated promising antidiabetic properties, driven by their bioactive compounds that contribute to hypoglycemic, antioxidant, and lipid-lowering effects. They are promising candidates for development as functional foods or nutraceuticals for diabetes management because of their cardiovascular benefits and general safety profile. However, standardizing extract formulations and carrying out thorough clinical trials to validate their therapeutic viability should be the main goals of future research. Third, *Pleurotus* species have significant prebiotic potential, improving gut health by influencing immune function and encouraging the growth of beneficial bacteria. Their abundance of chitin, β-glucans, and other prebiotic fibers promotes a healthy gut microbiota, aids in the synthesis of SCFAs, and lowers inflammation. Their antimicrobial qualities also support digestive health, highlighting their potential as nutraceuticals or functional foods. To fully explore the range of their health benefits, future research should focus on more detailed studies of their prebiotic properties, as very few articles related to prebiotic activity have been reported. Furthermore, while *P. ostreatus* has been thoroughly investigated for all the activities reported here, other species have been poorly studied. Thus, ongoing research could lead to the development of novel natural remedies and functional foods that offer long-term alternatives to traditional pharmaceutical methods, ultimately improving public health outcomes.

## REFERENCES


Abdel-Monem, N. M., M. A. El-Saadani, A. S. Daba, S. R. Saleh, and E. Aleem. 2020. Exopolysaccharide-peptide complex from oyster mushroom (*Pleurotus ostreatus*) protects against hepatotoxicity in rats. *Biochemistry and Biophysics Reports* 24: 100852. https://doi.org/10.1016/j.bbrep.2020.100852

Adhikari, P. A., and W. K. Kim. 2017. Overview of prebiotics and probiotics: focus on performance, gut health and immunity–a review. *Annals of Animal Science* 17(4): 949–966. https://doi.org/10.1515/aoas-2016-0092

Al-Sheraji, S. H., A. Ismail, M. Y. Manap, S. Mustafa, R. M. Yusof, and F. A. Hassan. 2013. Prebiotics as functional foods: a review. *Journal of Functional Foods* 5: 1542–1553. https://doi.org/10.1016/j.jff.2013.08.009

Anusiya, G., U. Gowthama Prabu, N. V. Yamini, et al. 2021. A review of the therapeutic and biological effects of edible and wild mushrooms. *Bioengineered* 12(2): 11239–11268. https://doi.org/10.1080/21655979.2021.2001183

Arunachalam, K., P. S. Sreeja, and X. Yang. 2022. The antioxidant properties of mushroom polysaccharides can potentially mitigate oxidative stress, beta-cell dysfunction and insulin resistance. *Frontiers in Pharmacology* 13: 874474. https://doi.org/10.3389/fphar.2022.874474

Aslam, T., U. Khalil, N. Mehmood, S. Sohail, A. Shahid, M. Nadeem, and H. Rana. 2021. Bioactive compounds derived from *Agaricus* species and bioactivity for type 2 diabetes mellitus. *Global Scientific Journal* 9(9): 1996–2016.

Asrani, S. K., H. Devarbhavi, J. Eaton, and P. S. Kamath. 2019. Burden of liver diseases in the world. *Journal of Hepatology* 70(1): 151–171. https://doi.org/10.1016/j.jhep.2018.09.014

Aziz, T., A. A. Khan, A. Tzora, C. Voidarou, and I. Skoufos. 2023. Dietary implications of the bidirectional relationship between the gut microflora and inflammatory diseases with special emphasis on irritable bowel disease: current and future perspective. *Nutrients* 15(13): 2956. https://doi.org/10.3390/nu15132956

Balaji, P., R. Madhanraj, K. Rameshkumar, et al. 2020. Evaluation of antidiabetic activity of *Pleurotus pulmonarius* against streptozotocin-nicotinamide induced diabetic Wistar albino rats. *Saudi Journal of Biological Sciences* 27(3): 913–924. https://doi.org/10.3390/nu15132956

Bhushan, A., and M. Kulshreshtha. 2018. The medicinal mushroom *Agaricus bisporus*: review of phytopharmacology and potential role in the treatment of various diseases. *Journal of Nature and Science of Medicine* 1(1): 4–9. https://doi.org/10.4103/JNSM.JNSM_1_18

Borodina, I., L. C. Kenny, C. M. McCarthy, et al. 2020. The biology of ergothioneine, an antioxidant nutraceutical. *Nutrition Research Reviews* 33(2): 190–217. https://doi.org/10.4103/JNSM.JNSM_1_18

Chandana, L., and D. A. Bhikshapathi. 2023. Free radical scavenging activity of *Pleurotus ostreatus* against $CCL_4$-induced hepatic damage in Wister rats. *International Journal of Pharmacy and Pharmaceutical Sciences* 15(12): 17–22. https://dx.doi.org/10.22159/ijpps.2023v15i12.49478

Chowdhury, P., and S. Paul. 2020. The potential role of mushrooms in the prevention and treatment of diabetes: a review. *Journal of Biologically Active Products from Nature* 10(5): 429–454. https://doi.org/10.1080/22311866.2020.1831958

Cragg, G. M., D. J. Newmann, and K. M. Snader. 1997. Natural products in drug discovery and development. *Journal of Natural Products* 60: 52–60. https://doi.org/10.1021/np9604893

Crawford, J. M. 2007. Basic mechanism in hepatopathology. In: *Pathology of the Liver*, 5th ed., ed. N. M. Roderick, D. A. Burt, B. Portmann, and L. D. Ferrell, 119–146. Elsevier, Churchill Livingstone.

Das, P., A. De, A. Das, and A. Samanta. 2024. Assessment of antioxidant, antidiabetic, and antimicrobial properties of hybrid mushroom strains derived through intergeneric hybridization between *Pleurotus sajor-caju* and *Calocybe indica*.

*Bioactive Carbohydrates and Dietary Fibre* 32: 100450. https://doi.org/10.1016/j.bcdf.2024.100450

De Silva, D. D., S. Rapior, K. D. Hyde, and A. H. Bahkali. 2012. Medicinal mushrooms in prevention and control of diabetes mellitus. *Fungal Diversity* 56: 1–29.

Dkhil, M. A., M. S. Diab, M. S. Lokman, H. El-Sayed, H. M. Aljawdah, S. Al-Quraishy, and A. A. Bauomy. 2020. Hepato-protective effect of *Pleurotus ostreatus* extracts in cadmium-intoxicated rats. *Journal of King Saud University–Science* 32(8): 3432–3436. https://doi.org/10.1016/j.jksus.2020.09.029

Dong, H., H. Zhuang, C. Yu, X. Zhang, and T. Feng. 2024. Interactions between soluble dietary fibers from three edible fungi and gut microbiota. *International Journal of Biological Macromolecules* 278: 134685. https://doi.org/10.1016/j.ijbiomac.2024.134685

Duan, Z., Y. Zhang, C. Zhu, Y. Wu, B. Du, and H. Ji. 2020. Structural characterization of phosphorylated *Pleurotus ostreatus* polysaccharide and its hepatoprotective effect on carbon tetrachloride-induced liver injury in mice. *International Journal of Biological Macromolecules* 162: 533–547. https://doi.org/10.1016/j.ijbiomac.2020.06.107

Dubey, S., C. Yadav, A. Bajpeyee, and M. P. Singh. 2020. Effect of *Pleurotus fossulatus* aqueous extract on biochemical properties of liver and kidney in streptozotocin-induced diabetic rat. *Diabetes, Metabolic Syndrome and Obesity* 13: 3035–3046.

El Omari, N., K. Sayah, S. Fettach, et al. 2019. Evaluation of in vitro antioxidant and antidiabetic activities of *Aristolochia longa* extracts. *Evidence-Based Complementary and Alternative Medicine* 2019(1): 7384735. https://doi.org/10.1155/2019/7384735

Flandroy, L., T. Poutahidis, G. Berg, et al. 2018. The impact of human activities and lifestyles on the interlinked microbiota and health of humans and of ecosystems. *Science of the Total Environment* 627: 1018–1038. https://doi.org/10.1016/j.scitotenv.2018.01.288

Freitas, A. C., M. B. Antunes, D. Rodrigues, et al. 2018. Use of coffee by-products for the cultivation of *Pleurotus citrinopileatus* and *Pleurotus salmoneo-stramineus* and its impact on biological properties of extracts thereof. *International Journal of Food Science and Technology* 53(8): 1914–1924. https://doi.org/10.1111/ijfs.13778

Fu, H. Y., D. E. Shieh, and C. T. Ho. 2009. Antioxidant and free radical scavenging activities of edible mushrooms. *Journal of Food Lipids* 9: 35–43. https://doi.org/10.1111/j.1745-4522.2002.tb00206.x

Ganeshpurkar, A., S. Kohli, and G. Rai. 2014. Antidiabetic potential of polysaccharides from the white oyster culinary-medicinal mushroom *Pleurotus florida* (higher Basidiomycetes). *International Journal of Medicinal Mushrooms* 16(3): 207–217. https://doi.org/10.1615/IntJMedMushr.v16.i3.10

Giavasis, I. 2014. Bioactive fungal polysaccharides as potential functional ingredients in food and nutraceuticals. *Current Opinion in Biotechnology* 26: 162–173. https://doi.org/10.1016/j.copbio.2014.01.010

Glamočlija, J., M. Kostić, and M. Soković. 2018. Antimicrobial and hepatoprotective activities of edible mushrooms. In: *Biology of Macrofungi*, ed. B. Singh, A. Lallawmsanga, and Passari, 81–113. Springer, Cham.

Golak-Siwulska, I., A. Kałużewicz, T. Spiżewski, M. Siwulski, and K. Sobieralski. 2018. Bioactive compounds and medicinal properties of oyster mushrooms (sp.). *Folia Horticulturae* 30(2): 191–201. https://doi.org/10.2478/fhort-2018-0012

Guan, Z., and Q. Feng. 2022. Chitosan and chitooligosaccharide: the promising non-plant-derived prebiotics with multiple biological activities. *International Journal of Molecular Sciences* 23: 6761. https://doi.org/10.3390/ijms23126761

Holscher, H. 2017. Dietary fiber and prebiotics and the gastrointestinal microbiota. *Gut Microbes* 8(2): 172–184.

Hu, S. H., J. C. Wang, J. L. Lien, E. T. Liaw, and M. Y. Lee. 2006. Antihyperglycemic effect of polysaccharide from the fermented broth of *Pleurotus citrinopileatus*. *Applied Microbiology and Biotechnology* 70(1): 107–113.

Hu, Y., J. Xu, Y. Sheng, J. Liu, H. Li, M. Guo, W. Xu, Y. Luo, K. Huang, and X. He. 2022. *Pleurotus ostreatus* ameliorates obesity by modulating the gut microbiota in obese mice induced by high-fat diet. *Nutrients* 14: 1868. https://doi.org/10.3390/nu14091868

Huang, H. Y., M. Korivi, H. T. Yang, C. C. Huang, Y. Y. Chaing, and Y. C. Tsai. 2014. Effect of *Pleurotus tuber-regium* polysaccharides supplementation on the progression of diabetes complications in obese-diabetic rats. *Chinese Journal of Physiology* 57(4): 198–208. https://doi.org/10.4077/CJP.2014.BAC245

Huang, Y., Y. Gao, X. Pi, S. Zhao, and W. Liu. 2022. In vitro hepatoprotective and human gut microbiota modulation of polysaccharide-peptides in *Pleurotus citrinopileatus*. *Frontiers in Cellular and Infection Microbiology* 12: 892049. https://doi.org/10.3389/fcimb.2022.892049

Johnny, I., and J. Okon. 2013. Antidiabetic effect of *Pleurotus ostreatus* (Jacq. ex Fr) Kumm. mushroom on alloxan-induced diabetic rat. *Indian Journal of Pharmaceutical and Biological Research* 1(2): 31.

Kanagasabapathy, G., K. H. Chua, S. N. A. Malek, S. Vikineswary, and U. R. Kuppusamy. 2014. AMP-activated protein kinase mediates insulin-like and lipo-mobilizing effects of β-glucan-rich polysaccharides isolated from *Pleurotus sajor-caju* (Fr.), Singer mushroom, in 3T3-L1 cells. *Food Chemistry* 145: 198–204. https://doi.org/10.1016/j.foodchem.2013.08.051

Kerezoudi, E. N., E. K. Mitsou, K. Gioti, et al. 2021. Fermentation of *Pleurotus ostreatus* and *Ganoderma lucidum* mushrooms and their extracts by the gut microbiota of healthy and osteopenic women: potential prebiotic effect and impact of mushroom fermentation products on human osteoblasts. *Food and Function* 12(4): 1529–1546. https://doi.org/10.1039/D0FO02581J

Khatua, S., S. Bose, L. M. C. Cordeiro, and K. Acharya. 2023. The emergence of mushrooms as novel resources of potential prebiotics: an updated view. In: *Applied Mycology for Agriculture and Food Industrial Applications*, ed. S. K. Singh, D. Kumar, Md. Shamim, and R. Sharma, 271–314. Apple Academic Press and CRC Press. https://doi.org/10.1201/9781003369868

Krishna, K. V., J. M. Murugan, H. Khan, et al. 2023. Exploring the therapeutic potential of edible *Pleurotus* mushroom species for oxidative stress and diabetes management. *Journal of King Saud University–Science* 35(9): 102926. https://doi.org/10.1016/j.jksus.2023.102926

Kubes, P., and C. Jenne. 2018. Immune responses in the liver. *Annual Review of Immunology* 36(1): 247–277. https://doi.org/10.1016/j.jksus.2023.102926

Kumar, K., R. Mehra, R. P. Guiné, M. J. Lima, N. Kumar, R. Kaushik, N. Ahmed, A. N. Yadav, and H. Kumar. 2021. Edible mushrooms: a comprehensive review on bioactive compounds with health benefits and processing aspects. *Foods* 10(12): 2996. https://doi.org/10.3390/foods10122996

Kumar, R., S. Gautam, K. D. Singh, P. Kumar, N. Haque, V. Yadav, R. Kumar, R. P. Diwakar, Rishikant, and S. B. Kumar. 2018. Pharmacological properties of *Boerhavia diffusa*: a review. *International Journal of Chemical Studies* 6(4): 72–80.

Liang, J., M. Zhang, X. Wang, Y. Ren, T. Yue, Z. Wang, and Z. Gao. 2021. Edible fungal polysaccharides, the gut microbiota, and host health. *Carbohydrate Polymers* 273: 118558. https://doi.org/10.1016/j.carbpol.2021.118558

LoPiparo, E., H. Scheib, N. Frei, G. Williamson, M. Grigorov, and C. J. Chou. 2008. Flavonoids for controlling starch digestion: structural requirements for inhibiting human α-amylase. *Journal of Medicinal Chemistry* 51: 3555–3561. https://doi.org/10.1021/jm800115x

Ma, G., B. M. Kimatu, L. Zhao, W. Yang, F. Pei, and Q. Hu. 2017. *In vivo* fermentation of a *Pleurotus eryngii* polysaccharide and its effects on fecal microbiota composition and immune response. *Food & Function* 8(5): 1810–1821. https://doi.org/10.1039/C7FO00341B

Mercola, J. 2017. *Fat for Fuel Ketogenic Cookbook: Recipes and Ketogenic Keys to Health from a World-Class Doctor and an Internationally Renowned Chef.* Hay House, Inc. (November 14).

Mohamed, J., A. N. Nafizah, A. H. Zariyantey, and S. Budin. 2016. Mechanisms of diabetes-induced liver damage: the role of oxidative stress and inflammation. *Sultan Qaboos University Medical Journal* 16(2): e132. 10.18295/squmj.2016.16.02.002

Moumita, S., and B. Das. 2022. Assessment of the prebiotic potential and bioactive components of common edible mushrooms in India and formulation of synbiotic microcapsules. *LWT* 156: 113050. https://doi.org/10.1016/j.lwt.2021.113050

Nakahara, D., C. Nan, K. Mori, M. Hanayama, H. Kikuchi, S. Hirai, and Y. Egashira. 2020. Effect of mushroom polysaccharides from *Pleurotus eryngii* on obesity and gut microbiota in mice fed a high-fat diet. *European Journal of Nutrition* 59: 3231–3244.

Nnemolisa, S. C., C. C. Chukwurah, S. C. Edeh, et al. 2024. Antidiabetic and antioxidant potentials of *Pleurotus ostreatus*-derived compounds: an in vitro and in silico approach. *Food Chemistry Advances* 4: 100639. https://doi.org/10.1016/j.focha.2024.100639

Okolo, K. O., and O. E. Orisakwe. 2021. *In vitro* antioxidants and hepatoprotective effects of *Pleurotus tuber-regium* on carbon tetrachloride–treated rats. *Journal of Basic and Clinical Physiology and Pharmacology* 32(2): 67–78.

Onuekwuzu, I. M., I. C. Chidinma, and I. J. Chigozie. 2019. Anti-diabetic effect of a flavonoid and sitosterol-rich aqueous extract of *Pleurotus tuberregium* sclerotia in alloxan-induced diabetic rabbits. *Endocrine, Metabolic and Immune Disorders—Drug Targets* 19(8): 1148–1156. https://doi.org/10.2174/1871530319666190206213843

Osman, A., and A. O. Toliba. 2019. Hepatoprotective effects of crude phenolic-rich extract from oyster mushroom (*Pleurotus ostreatus*). *Egyptian Journal of Food Science* 47(2): 157–164. https://doi.org/10.21608/ejfs.2019.16213.1017

Paravamsivam, P., C. K. Heng, S. N. Malek, V. Sabaratnam, and U. R. Kuppusamy. 2016. Giant oyster mushroom *Pleurotus giganteus* (Agaricomycetes) enhances adipocyte differentiation and glucose uptake via activation of PPARγ and glucose transporters 1 and 4 in 3T3-L1 cells. *International Journal of Medicinal Mushrooms* 18(9): 821–831. https://doi.org/10.1615/IntJMedMushrooms.v18.i9.60

Parnian-Khajehdizaj, N., E. Mainer Pardos, S. B. Machek, and H. Nobari. 2024. Association between *Pleurotus ostreatus* consumption and more optimal sports performance: a narrative review. *Journal of Food Biochemistry* 2024(1): 2610415. https://doi.org/10.1155/2024/2610415

Patterson, C., L. Guariguata, G. Dahlquist, G. Soltész, G. Ogle, and M. Silink. 2014. Diabetes in the young—a global view and worldwide estimates of numbers of children with type 1 diabetes. *Diabetes Research and Clinical Practice* 103(2): 161–175. https://doi.org/10.1016/j.diabres.2013.11.005

Prabu, M., and R. Kumuthakalavalli. 2017. Antidiabetic potential of the oyster mushroom *Pleurotus florida* (Mont.) Singer. *International Journal of Current Pharmaceutical Research* 9(4): 51–54. https://doi.org/10.22159/ijcpr.2017v9i4.20765

Ravi, B., R. E. Renitta, M. L. Prabha, R. Issac, and S. Naidu. 2013. Evaluation of antidiabetic potential of oyster mushroom (*Pleurotus ostreatus*) in alloxan-induced diabetic mice. *Immunopharmacology and Immunotoxicology* 35(1): 101–109.

Riaz, S., A. Ahmad, R. Farooq, M. Ahmed, M. Shaheryar, and M. Hussain. 2022. Edible mushrooms, a sustainable source of nutrition, biochemically active compounds and its effect on human health. In: *Current Topics in Functional Food*, ed. N. Shiomi and A. Savitskaya. IntechOpen, Rijeka, Croatia. https://doi.org/10.5772/intechopen.102694

Saad, A. M., M. Z. Sitohy, B. A. Omar, M. T. El-Saadony, and B. Sitohy. 2022. Mushroom; chemistry, bioactive components, and application. In: *Current Topics in Functional Food*, ed. N. Shiomi, and A. Savitskaya. IntechOpen, Rijeka, Croatia. https://doi.org/10.5772/intechopen.104182

Saha, A., P. Jana, and S. Khatua. 2024. A protective role of *Ganoderma* sp. on cardiovascular disease, atherosclerosis and hypertension. In: *Ganoderma: Cultivation, Chemistry and Medicinal Applications, Volume 2*, ed. K. Acharya, and S. Khatua, 102–126. CRC Press. https://doi.org/10.1201/9781003490258-6

Saxami, G., E. N. Kerezoudi, E. K. Mitsou, et al. 2021. Fermentation supernatants of *Pleurotus eryngii* mushroom ameliorate intestinal epithelial barrier dysfunction in lipopolysaccharide-induced Caco-2 cells via upregulation of tight junctions. *Microorganisms* 9(10): 2071. https://doi.org/10.3390/microorganisms9102071

Setyawan, R. H., I. Saskiawan, N. Widhyastuti, K. Kasirah, and M. Mulyadi. 2023. Potensi prebiotic dari ekstrak jamur tiram putih (*Pleurotus ostreatus*). *Berita Biologi* 22(1): 51–59. https://doi.org/10.55981/beritabiologi.2023.806

Shamim, M. Z., A. K. Mishra, T. Kausar, et al. 2023. Exploring edible mushrooms for diabetes: unveiling their role in prevention and treatment. *Molecules* 28(6): 2837. https://doi.org/10.3390/molecules28062837

Singdevsachan, S. K., P. Auroshree, J. Mishra, B. Baliyarsingh, K. Tayung, and H. Thatoi. 2016. Mushroom polysaccharides as potential prebiotics with their antitumor and immunomodulating properties: a review. *Bioactive Carbohydrates and Dietary Fibre* 7(1): 1–4. https://doi.org/10.1016/j.bcdf.2015.11.001

Singh, V., S. Bhatia, and R. Shri. 2023. Bioactive constituents from mushrooms for management of diabetes: a scientific corroboration. In: *Therapeutic Mushrooms for Diabetes Mellitus*, 67–108. Apple Academic Press.

Song, X., Z. Liu, J. Zhang, et al. 2018. Antioxidative and hepatoprotective effects of enzymatic and acidic-hydrolysis of *Pleurotus geesteranus* mycelium polysaccharides on alcoholic liver diseases. *Carbohydrate Polymers* 201: 75–86. https://doi.org/10.1016/j.carbpol.2018.08.058

Soodpakdee, K., J. Nacha, N. Rattanachart, A. Owatworakit, and S. Chamyuang. 2022. Fermentation with *Pleurotus ostreatus* enhances the prebiotic properties of germinated riceberry rice. *Frontiers in Nutrition* 9: 839145. https://doi.org/10.1016/j.carbpol.2018.08.058

Synytsya, A., K. Míčková, A. Synytsya, I. Jablonský, J. Spěváček, V. Erban, et al. 2009. Glucans from fruit bodies of cultivated mushrooms *Pleurotus ostreatus* and *Pleurotus eryngii*: structure and potential prebiotic activity. *Carbohydrate Polymers* 76: 548–556. https://doi.org/10.1016/j.carbpol.2008.11.021

Tang, C., and Z. Lu. 2019. Health promoting activities of probiotics. *Journal of Food Biochemistry* 43(8): e12944.

Thikham, S., T. Jeenpitak, K. Shoji, S. Phongthai, P. Therdtatha, A. Yawootti, and W. Klangpetch. 2024. Pulsed electric field-assisted extraction of mushroom β-glucan from *Pleurotus pulmonarius* by-product and study of prebiotic properties. *International Journal of Food Science and Technology* 59(6): 3939–3949. https://doi.org/10.1111/ijfs.17144

Törős, G., H. El-Ramady, J. Prokisch, F. Velasco, X. Llanaj, D. H. Nguyen, and F. Peles. 2023. Modulation of the gut microbiota with prebiotics and antimicrobial agents from *Pleurotus ostreatus* mushroom. *Foods* 12(10): 2010. https://doi.org/10.3390/foods12102010

Torres-Martínez, B. D., R. D. Vargas-Sánchez, G. R. Torrescano-Urrutia, M. Esqueda, J. G. Rodríguez-Carpena, J. Fernández-López, J. A. Perez-Alvarez, and A. Sánchez-Escalante. 2022. *Pleurotus* genus as a potential ingredient for meat products. *Foods* 11(6): 779. https://doi.org/10.3390/foods11060779

Van Doan, H., S. Doolgindachbaporn, and A. Suksri. 2016. Effects of *Eryngii* mushroom (*Pleurotus eryngii*) and *Lactobacillus plantarum* on growth performance, immunity and disease resistance of *Pangasius* catfish (*Pangasius bocourti*, Sauvage 1880). *Fish Physiology and Biochemistry* 42: 1427–1440.

Vetvicka, V., O. Gover, M. Karpovsky, H. Hayby, O. Danay, N. Ezov, Y. Hadar, and B. Schwartz. 2019. Immune-modulating activities of glucans extracted from *Pleurotus ostreatus* and *Pleurotus eryngii*. *Journal of Functional Foods* 54: 81–91. https://doi.org/10.1016/j.jff.2018.12.034

Xu, N., Z. Gao, J. Zhang, H. Jing, S. Li, Z. Ren, S. Wang, and L. Jia. 2017. Hepatoprotection of enzymatic-extractable mycelia zinc polysaccharides by *Pleurotus eryngii* var. *tuoliensis*. *Carbohydrate Polymers* 157: 196–206. https://doi.org/10.1016/j.carbpol.2016.09.082

Yadav, D., and P. S. Negi. 2021. Bioactive components of mushrooms: processing effects and health benefits. *Food Research International* 148: 110599. https://doi.org/10.1016/j.carbpol.2016.09.082

Yadav, D., K. H. Prashanth, and P. S. Negi. 2024. Low molecular weight chitosan from *Pleurotus ostreatus* waste and its prebiotic potential. *International Journal of Biological Macromolecules* 267: 131419. https://doi.org/10.1016/j.ijbiomac.2024.131419

Yan, J., L. Wang, Y. Gu, H. Hou, T. Liu, Y. Ding, and H. Cao. 2022. Dietary patterns and gut microbiota changes in inflammatory bowel disease: current insights and future challenges. *Nutrients* 14(19): 4003. https://doi.org/10.3390/nu14194003

Yin, C., G. D. Noratto, X. Fan, Z. Chen, F. Yao, D. Shi, and H. Gao. 2020. The impact of mushroom polysaccharides on gut microbiota and its beneficial effects to host: a review. *Carbohydrate Polymers* 250: 116942. https://doi.org/10.1016/j.carbpol.2020.116942

Zhang, C., S. Li, J. Zhang, C. Hu, G. Che, M. Zhou, and L. Jia. 2016. Antioxidant and hepatoprotective activities of intracellular polysaccharide from *Pleurotus eryngii* SI–04. *International Journal of Biological Macromolecules* 91: 568–577. https://doi.org/10.1016/j.ijbiomac.2016.05.104

Zhang, J., G. Meng, C. Zhang, et al. 2015. The antioxidative effects of acidic-, alkalic- and enzymatic-extractable mycelium zinc polysaccharides by *Pleurotus djamor* on liver and kidney of streptozocin-induced diabetic mice. *BMC Complementary and Alternative Medicine* 15: 440.

Zhang, L., Y. Lu, X. Feng, Q. Liu, Y. Li, J. Hao, Y. Wang, Y. Dong, and H. D. Wang. 2020. Hepatoprotective effects of *Pleurotus ostreatus* protein hydrolysates yielded by pepsin hydrolysis. *Catalysts* 10(6): 595. https://doi.org/10.3390/catal10060595

Zhang, P. 2022. Influence of foods and nutrition on the gut microbiome and implications for intestinal health. *International Journal of Molecular Sciences* 23(17): 9588. https://doi.org/10.3390/ijms23179588

Zhu, B., Y. Li, T. Hu, and Y. Zhang. 2019. The hepatoprotective effect of polysaccharides from *Pleurotus ostreatus* on carbon tetrachloride-induced acute liver injury in rats. *International Journal of Biological Macromolecules* 131: 1–9. https://doi.org/10.1016/j.ijbiomac.2019.03.043

Zoumpopoulou, G., M. Kazou, V. Alexandraki, A. Angelopoulou, K. Papadimitriou, B. Pot, and E. Tsakalidou. 2018. Probiotics and prebiotics: an overview on recent trends. In: *Probiotics and Prebiotics in Animal Health and Food Safety*, ed. D. Di Gioia, and B. Biavati. Springer, Cham. https://doi.org/10.1007/978-3-319-71950-4_1

# 9 Anti-Inflammatory and Anticancer Activities of *Pleurotus* spp.

*Jasmina Ćilerdžić and Milica Galić

Faculty of Biology, University of Belgrade, Belgrade, Serbia

## 9.1 INTRODUCTION

### 9.1.1 Inflammation: A Basis of Numerous Diseases

Inflammation is one of the most important evolutionarily acquired responses of an organism to injury, toxins, or infection, which eliminate stimuli, repair damaged tissues, and promote recovery through controlled cell proliferation (Kotas and Medzhitov, 2015). However, there are two types of inflammation—acute and chronic—which differ in several important characteristics: triggering patterns, duration, magnitude, outcomes, age dependence, and biomarkers (Furman et al., 2019). Contrary to acute inflammation, which is a short-term response (lasting from several hours to a few days) to trauma or infections (bacterial, viral, etc.) and is typically resolved once the threat is eliminated, chronic inflammation is a prolonged (months to years), low-grade inflammatory response that has significant hazardous effects on human health. While acute inflammation induces healing, trigger removal, and tissue repair, systemic chronic inflammation (SCI) leads to continuous tissue injury and repair cycles, causing many collateral damages in the organism. SCI is characterized by the activation of immune components that can be distinct from those activated during an acute immune response and also have no canonical standard biomarkers as is the case in acute inflammation (IL-6, TNF-α, IL-1β, CRP). Finally, acute inflammation is not age-related, contrary to SCI. This condition can be triggered by a plethora of endogenous and exogenous stimuli continuously activating the response of the immune system (Figure 9.1).

The low-grade inflammation during the aging process without active infection has been defined as "inflammaging" which is a cause of increased mortality in the aging population. It appears that various processes associated with aging disrupt the balance between pro-inflammatory and anti-inflammatory signaling, resulting in a persistent state of inflammation (Baechle et al., 2023). There is numerous evidence for the cyclic relation between chronic inflammation and other hallmarks of aging, leading to the further progression of aging and age-dependent diseases. In this state, immune cells like macrophages, T-cells, and neutrophils continuously release pro-inflammatory cytokines, reactive oxygen species (ROS), and other molecules that cause tissue damage over time. On the other hand, one of the most important discoveries of the 21st century is that inflammatory processes are involved in the initiation and genesis of diseases that dominate as disabling and morbidity factors (cardiovascular diseases, cancer, metabolic, neurodegenerative, and autoimmune disorders) and account for more than 50% of all deaths worldwide (Furman et al., 2019). The first indication of a linkage between inflammation and cancer was observed even in the 19th century, and to date it has been shown that an inflammatory microenvironment is the essential basis of all tumors. Considering that chronic inflammation is the main trigger of the leading causes of mortality today, the importance of prevention and treatment of this condition is absolutely justified. Numerous studies have confirmed the great potential of fungal extracts or purified compounds to prevent or reduce inflammation, but additional studies are needed for the development of novel fungi-based nutraceuticals and pharmaceuticals (Kushairi et al., 2020; Bassi et al., 2024).

### 9.1.2 Cancer: A Growing Cause of Mortality

After cardiovascular diseases, cancer remains the leading cause of death globally, with an increasing incidence (Nagai and Kim, 2017). According to the World Health Organization, approximately 10 million people worldwide died from cancer in 2020, accounting for about 17% of all deaths annually. Therefore, understanding and preventing the triggers of the disease, early recognition of symptoms, and exploring future treatment options are key aspects in the fight against this global issue. The etiology of cancer is complex and encompasses a range of factors contributing to cancer's prevalence (genetic predisposition, exposure to carcinogens, unhealthy lifestyles, certain infections, stress, etc.), but metabolic alterations related to the aging process are one of the main risk factors. Therefore, if humans lived long enough, they would all eventually develop cancer (Figure 9.2).

However, genetic factors play a crucial role, so mutations in certain genes, such as BRCA1 and BRCA2, significantly increase the risk of developing breast and ovarian cancer. In addition, exposure to carcinogenic substances, such as tobacco smoke, environmental pollutants, and certain chemicals, induces DNA changes that result in tumor formation. Lifestyle factors, such as poor nutrition, physical inactivity, and obesity, are also associated with an increased risk of cancer. The role of bacterial infections (e.g., *Helicobacter pylori*) and viruses, such as the human papillomavirus (HPV), hepatitis B, and C, has been proven to trigger

* Corresponding author: simonicj@bio.bg.ac.rs



DOI: 10.1201/9781003527572-9

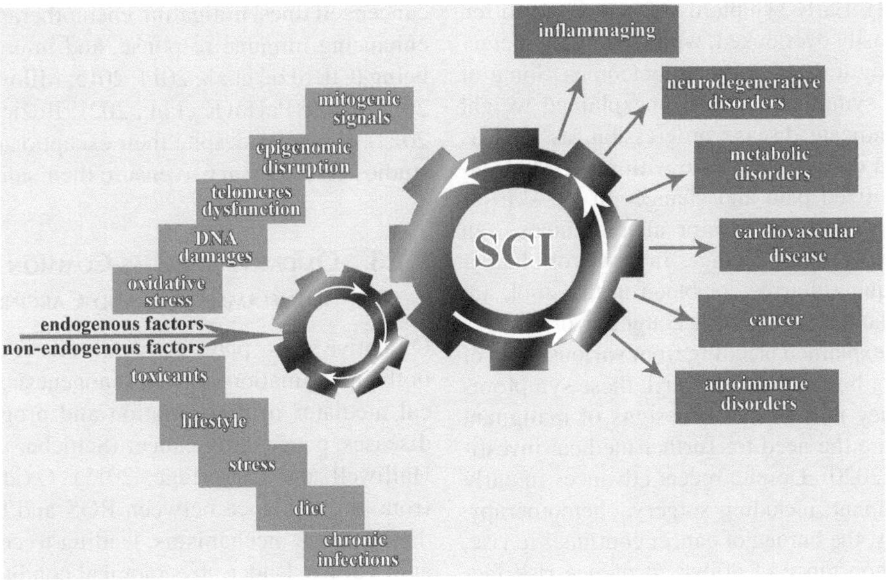


**FIGURE 9.1** The triggers and outcomes of systemic chronic inflammation.

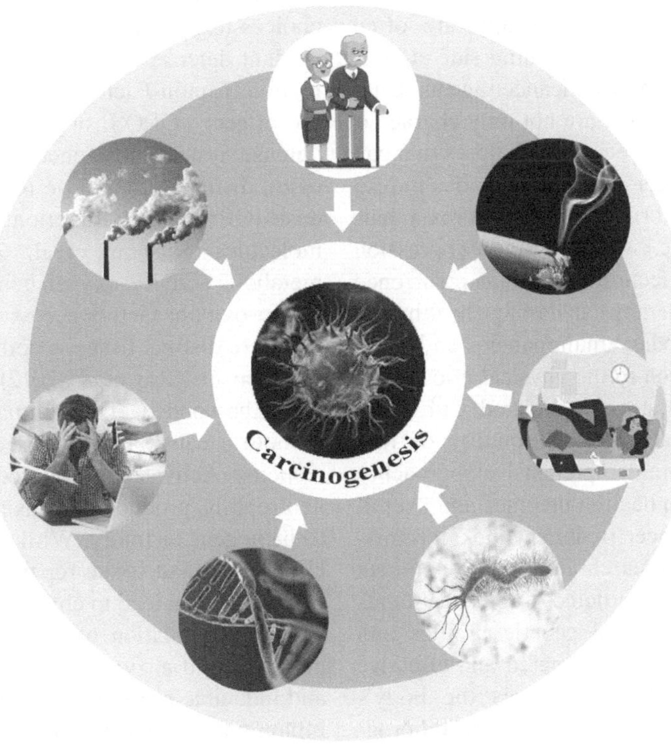


**FIGURE 9.2** The main triggers of cancer.

certain types of cancer. The elderly population is particularly susceptible to developing cancer, highlighting cell senescence as a tumor promoter acting via inflammatory mechanisms (Grivennikov et al., 2010). Moreover, there are numerous common triggers for cancer and inflammation, which is itself often a key precursor to cancer development. Namely, SCI promotes cellular damage, genetic mutations, and immune evasion, which facilitate the development of cancer through the secretion of pro-inflammatory cytokines (Coussens and Werb, 2002). The most common cancer types include lung cancer (the most common and deadliest cancer globally), breast cancer (one of the most common cancers in post-reproductive women), colorectal cancer (the third most common and fourth deadliest cancer), cervical cancer (the second most common cancer in women), and prostate cancer (common in older men population) (Bray et al., 2018;

Mishra et al., 2021). Early symptoms of cancer are often subtle and can be easily overlooked, while their early detection can significantly impact treatment outcomes. Some of the most common symptoms include unexplained weight loss, which may indicate disease progression, as well as chronic fatigue that does not improve with rest. Additionally, persistent localized pain and changes in the skin or mucous membranes, such as new or altered moles, can serve as early warning signs. Changes in bowel or bladder habits, such as frequent diarrhea or blood in the stool, are also important indicators. Persistent coughing or hoarseness, along with unexplained bleeding from various parts of the body, should not be ignored. Although these symptoms are not specific, they may be the first signs of malignant diseases and indicate the need for further medical investigation (Koo et al., 2020). Despite recent advances in early detection and treatment, including surgery, chemotherapy, and immunotherapy, the burden of cancer continues to rise, highlighting the importance of efforts to reduce risk factors, enhance screening programs, and develop novel therapies for the fight against cancer as a global health challenge in the current era (Bray et al., 2018).

The conventional procedures in cancer treatment, including surgery, chemotherapy, and radiotherapy, are often associated with significant risks and potential side effects, so there is an evident need for new anticancer agents. Chemotherapeutic agents, for instance, are not only characterized by poor bioavailability and require high doses that can induce drug resistance in cancer cells, but they also induce systemic toxic effects (in the form of fatigue, anorexia, hair loss, vomiting, headaches, etc.), while immunosuppression increases susceptibility to infection and tumor recurrence (Mishra et al., 2021). Radiotherapy can damage healthy tissue, which can lead to secondary malignancies and long-term complications. In addition to the physical effects, the emotional and psychological stress associated with these treatments can further impact patients' quality of life, highlighting the need for more comprehensive approaches to oncology that address both health outcomes and overall well-being. The future of cancer treatment looks promising due to advancements in research and the development of new therapies. Personalized medicine, where treatments are developed based on individual genetic profiles and tumor characteristics, enables more targeted and effective therapies. Immunotherapy, which harnesses the body's immune system to fight cancer, has shown significant progress, particularly in the treatment of certain cancer types. Targeted therapies that act directly on the mechanisms that drive tumor growth are also gaining popularity. Combining different therapies, including immunotherapy and chemotherapy, may enhance treatment efficacy and overcome drug resistance. However, an immense number of scientific reports suggest plant or fungal preparations as promising alternative or auxiliary agents to conventional therapies. The extracts and purified compounds (polysaccharides, polyphenols, alkaloids, etc.) of numerous plant and fungal species have been shown to be cytotoxic against various cancer cell lines, mitigating chemotherapy-induced toxicity, enhancing immune response, and improving overall well-being (Ćilerdžić et al., 2014, 2015; Milovanović et al., 2014, 2015; Oalđe Pavlović et al., 2021; Božić Nedeljković et al., 2022). However, despite their exceptional potential, further studies are necessary to ensure their safety and efficacy.

### 9.1.3 Oxidative Stress as Common Trigger of Inflammation and Carcinogenesis

Oxidative stress plays a pivotal role in the pathogenesis of both inflammation and carcinogenesis, acting as a critical mediator in the initiation and progression of various diseases, particularly cancer (Schieber and Chandel, 2014; Halliwell and Gutteridge, 2015). Oxidative stress arises from an imbalance between ROS and the body's antioxidant defense mechanisms, leading to cell damage (Reuter et al., 2010). Under physiological conditions, ROS (including $O^{2-}$, $H_2O_2$, •OH, etc.) are generated as by-products of cellular metabolism, but their accumulation due to environmental factors, such as ultraviolet (UV) and ionizing radiation, xenobiotics, and heavy metals, or lifestyle choices (e.g., smoking and poor diet) can overcome antioxidant defenses and promote tissue damage and inflammation (Limón-Pacheco and Gonsebatt, 2009). To mitigate the effects of ROS, organisms have evolved antioxidant defense systems that encompass enzymes such as superoxide dismutase, catalase (CAT), and glutathione peroxidase (GPx), which function to neutralize these reactive molecules (Kushairi et al., 2020). Due to variations in metabolic activity, oxygen consumption, and susceptibility to pro-oxidant factors across different tissues and organs, there are distinct tissue-specific adaptations of antioxidant mechanisms (Ray et al., 2012).

In the context of inflammation, oxidative stress serves as both a cause and a consequence of the inflammatory response. Activated immune cells, such as macrophages and neutrophils, produce ROS as part of the immune response to pathogens or injury. While ROS play a role in pathogen elimination and tissue repair, their persistent or excessive production can lead to chronic inflammation, characterized by the upregulation of pro-inflammatory cytokines, chemokines, and enzymes such as cyclooxygenase-2 (COX-2) and inducible nitric oxide synthase (iNOS). This persistent inflammatory state is a key factor in the development of various chronic diseases and disorders such as arthritis, diabetes, and cancer (Limón-Pacheco and Gonsebatt, 2009).

The link between oxidative stress and carcinogenesis is particularly well-documented. Namely, ROS-induced damage of macromolecules, including DNA, proteins, and lipids, can lead to mutations, chromosomal instability, and altered signaling pathways that promote tumorigenesis (O'Byrne and Dalgleish, 2001; Ray et al., 2012). Oxidative DNA damage is particularly associated with the activation of oncogenes and the inactivation of tumor suppressor genes, both of which contribute to the initiation and progression of cancer. Furthermore, chronic inflammation, triggered by oxidative stress, creates a tumor-promoting microenvironment by facilitating angiogenesis, suppressing immune surveillance, and enhancing cell proliferation and survival. The sustained oxidative damage thus not only initiates carcinogenesis but also promotes the progression and metastasis of established tumors (Sosa et al., 2013). Since oxidative stress plays a central role in the pathophysiology of inflammation and carcinogenesis (through a complex interplay of cellular damage, immune dysregulation, and genetic mutations) understanding the mechanisms by which oxidative stress contributes to these processes provides important insights into potential therapeutic targets as an indirect strategy for prevention and treatment of chronic inflammation and various cancers (Reuter et al., 2010; Ćilerdžić et al., 2013). When the body's internal antioxidant system is insufficient to combat oxidative stress, supplementary antioxidants may be required (Sarmadi and Ismail, 2010). Natural antioxidants, derived from plant, fungal, or microbial sources, offer several advantages over conventional synthetic antioxidants in terms of efficacy, safety, and broader health benefits. Unlike synthetic antioxidants, which may cause toxicity or adverse effects with prolonged use, natural antioxidants are generally considered safer and less prone to harmful side effects because they are often bioavailable and metabolized by the body in ways that are more aligned with natural physiological processes. Furthermore, natural antioxidants, such as polyphenols, flavonoids, carotenoids, and vitamins, often exhibit multiple bioactive properties beyond antioxidant activity, including anti-inflammatory, anticancer, and immunomodulatory effects, thus offering a more holistic approach to health promotion. Strategies to address oxidative stress therapeutically include (i) limiting ROS production, (ii) neutralizing existing ROS, and (iii) targeting metal-protein interactions that exacerbate oxidative stress—while plant/fungal extracts or their purified compounds have shown considerable capacity in all of these patterns (Barnham et al., 2004).

## 9.2 NATURAL MYCOCHEMICALS AS PROMISING ALTERNATIVE TO CONVENTIONAL MEDICINE

Mushrooms have been part of the human nutrition and traditional medicine since ancient times, while their diverse use is gaining more and more importance in modern science (Nowakowski et al., 2021; Stajić et al., 2023; Ray et al., 2024). Modern studies have just confirmed the justification of using mushrooms in the traditional medicine of ancient, primarily eastern civilizations. Therefore, it is well known that certain macromycetes, such as reishi (*Ganoderma lucidum*), shiitake (*Lentinula edodes*), and maitake (*Grifola frondosa*), possess significant medicinal potential thanks to the presence of polysaccharides, triterpenes, proteins, ergosterol, β-glucan, lovastatin, triterpenes, and other bioactive components (Patel and Goyal, 2012). Owing to the abundance of bioactive compounds, numerous other fungal species have been shown to possess antitumor, anti-inflammatory, antioxidative, antimicrobial, immunomodulatory, hypoglycemic, hypolipidemic, neuroprotective, and antiviral effects, making them a wide area of research for enhancing health and the development of new therapeutic agents. In addition to their medicinal properties, mushrooms have no toxicity or side effects, which makes them promising alternatives or auxiliaries to conventional therapeutics (Mishra et al., 2021). Since mycochemicals not only strengthen the immune system but also directly affect tumor growth and reduce inflammatory processes in the body, they are novel promising agents in the fight against the most important diseases today (Jedinak et al., 2011; Božić Nedeljković et al., 2022).

### 9.2.1 Medicinal Potential of *Pleurotus* spp.

Species of the genus *Pleurotus* (family Pleurotaceae), commonly known as oyster mushrooms, represent the second most commercially important group of edible mushrooms worldwide. The genus *Pleurotus* includes more than 700 species and varieties distributed worldwide, among which are some of the most appreciated mushrooms (Yin et al., 2021; Galić et al., 2022a). Thanks to the unique taste and nutritional composition of *Pleurotus* basidiocarps, they are a well-known gastronomic specialty and functional food. Besides the exceptional content of raw fibers and polysaccharides, *Pleurotus* spp. basidiocarps, due to their amino acid composition, fully satisfy the needs of essential amino acids for adults, placing them at the top in nutritional ratings. The bioactive agents responsible for antioxidant, anticancer, anti-inflammatory, immunomodulatory, antihypertensive, antibacterial, antiviral, anti-nematode, hypolipidemic, anticholesterolemic, hypoglycemic, anti-neurodegenerative, anti-obesity, anti-aging, and many other *Pleurotus* spp. properties (Figure 9.3) include a variety of mycochemicals. These compounds encompass polysaccharides, polysaccharide-peptide complexes, proteins, peptides, triterpenoids, phenols, flavonoids, polyketides, nucleotides, and so on (Deepalakshmi and Mirunalini, 2014). Fruiting bodies of these species have the ability to accumulate various minerals, the most abundant being potassium, phosphorus, magnesium, sodium, calcium, iron, zinc, manganese, and copper. Thus, it has been shown that *P. ostreatus* basidiocarps can provide more than 70% of the daily recommended value of iron and one third of the calcium needs (Carrasco-González et al., 2017). However, the majority of studies have demonstrated the bioactivity of crude extracts or fruiting body powder, as opposed to scarce investigation of the capacity of mycochemicals as isolated compounds. The incorporation of *P. ostreatus* basidiocarp powder into animal food formulations has been shown to potentially reduce serum glucose, total cholesterol, and triglyceride levels in a leptin receptor-deficient mouse model, while simultaneously increasing high-density lipoprotein (HDL)-cholesterol (Kim et al., 2010). A similar reduction

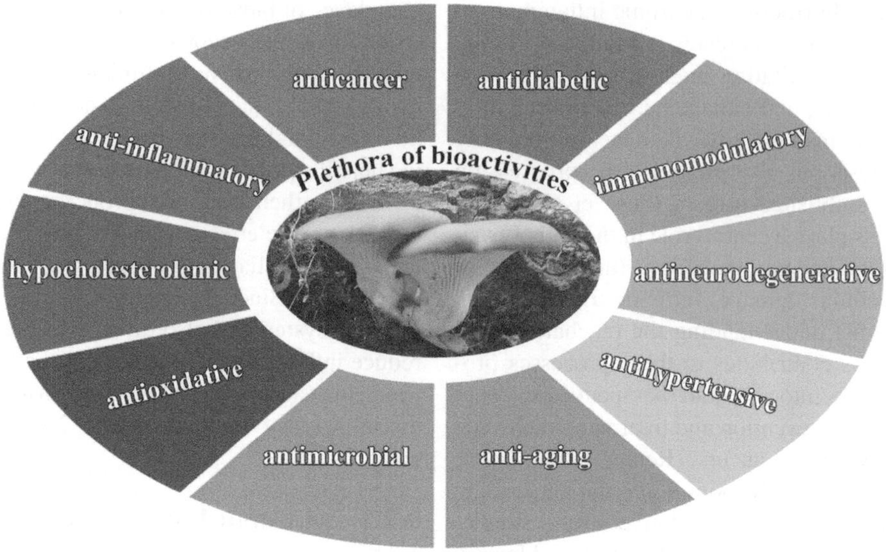


**FIGURE 9.3** Medicinal potential of *Pleurotus* spp.

in the lipid profile was observed in rats on high-cholesterol diets, which is attributed to enhanced excretion of cholesterol and lipids (Deepalakshmi and Mirunalini, 2014). Furthermore, a human trial demonstrated that consuming tomato soup supplemented with dried *P. ostreatus* for three weeks significantly decreased serum triglyceride and oxidized low-density lipoprotein levels compared to a control group consuming only tomato soup. Since numerous studies have shown their bioactivities, the fruiting bodies of these species are also promising therapeutic agents. On the other hand, given their well-developed lignocellulolytic enzyme system and their easy and high-yield cultivation on available lignocellulosic substrates, species of the genus *Pleurotus* are among the most commonly cultivated mushrooms globally (the third place following the genera *Agaricus* and *Lentinula*), with annual production reaching more than 900 000 t (Silveira et al., 2014; Fernandes et al., 2015; Galić et al., 2018; Galić et al., 2021a, 2021b, 2022b; Ćilerdžić et al., 2022). However, only a few *Pleurotus* spp. are commercially cultivated and intensively studied in recent decades. The most widely cultivated species within the genus is *Pleurotus ostreatus*, followed by *Pleurotus pulmonarius* (dos Santos Bazanella et al., 2013). As can be seen from Figure 9.4, the largest number of studies in the last two decades have concerned the medicinal potential of *P. ostreatus*, followed by *P. eryngii*, with approximately the same number of publications on *P. sajor-caju*, *P. florida*, *P. pulmonarius*, and *P. citrinopileatus*, while a significantly smaller number dealt with *P. djamor*.

#### 9.2.1.1 Anti-inflamatory Potential of *Pleurotus* spp. Crude Extracts

Previous studies have reported that polysaccharides, terpenes, phenolic compounds, sterols, fatty acids, and other bioactive metabolites isolated from numerous fungal species are promising anti-inflammatory substances able to reduce the production of inflammatory mediators (Taofiq et al., 2016; Bassi et al., 2024). Owing to the plethora of pharmacologically active molecules, *Pleurotus* spp. is also one of the most promising anti-inflammatory agents. Ethanolic extracts from the basidiocarps of *P. eryngii* showed significant anti-inflammatory activity through nitric oxide (NO) production in murine tumor cells, RAW264.7. Similarly, extracts of *P. ostreatus* basidiocarps showed a significant capacity to reduce intracellular oxidative stress in mammalian cells, as they completely inhibited ROS production in keratinocytes treated with hydrogen peroxide, partially in keratinocytes treated with UV radiation, and also inhibited the production of proinflammatory cytokines such as tumor necrosis factor (TNF)-α, monocyte chemoattractant protein (MCP)-1, interleukin (IL)-1β, IL-6, and vascular endothelial growth factor (VEGF) (Bassi et al., 2024). The same authors revealed the stronger anti-inflammatory capacity of extracts from *P. ostreatus* basidiocarps obtained on a substrate enriched with waste from lavender essential oil production due to modified mycochemical content. These basidiocarps were enriched with several compounds (polyphenols, hydroxy-methyl-coumarin, adenylosuccinic acid, lactones, tryptophan, and cholanic and fatty acid epoxides) extracted with methanol and acetonitrile. This opens numerous possibilities for enriching fruiting bodies with bioactive compounds by modifying the composition of the nutrient medium using lignocellulosic waste, especially since the general trend is toward a green or circular economy. The commercially cultivated basidiocarps of *P. ostreatus* have also shown considerable anti-inflammatory potential. Jedinak et al. (2011) tested the concentrate (water extract at the concentration of the commercial preparation), which significantly reduced lipopolysaccharide (LPS)-dependent production of the pro-inflammatory mediators TNF-α, IL-6, and IL−12. The same preparation also inhibited LPS-induced production of prostaglandin E2 ($PGE_2$)

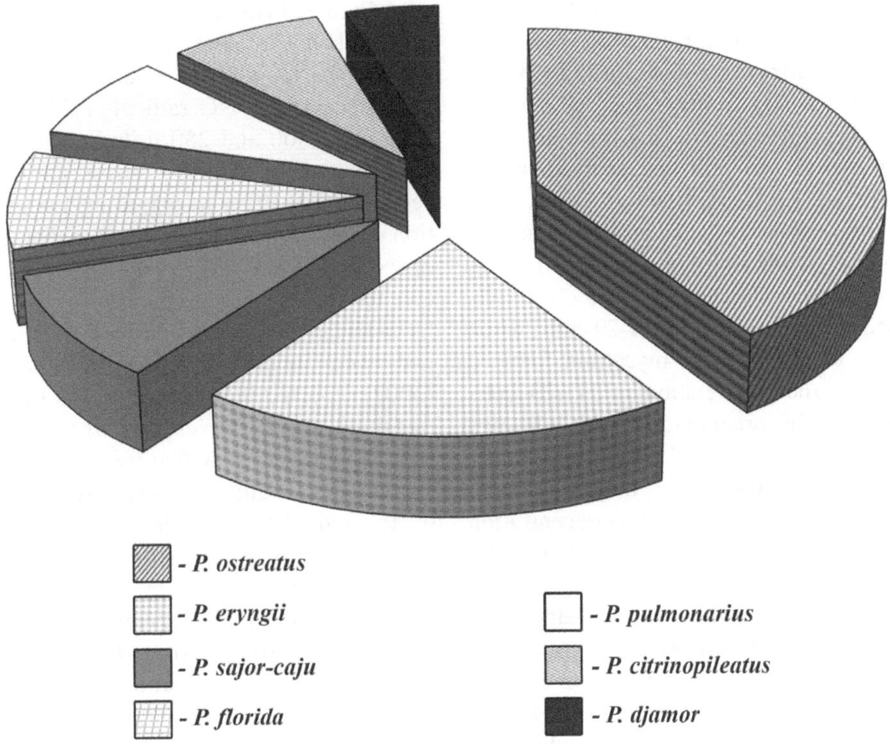


**FIGURE 9.4** The proportion of various *Pleurotus* species in scientific publications on their medicinal properties according to the ScienceDirect database.

and NO by downregulation of COX-2 and iNOS in macrophages and suppressed activation of transcription factors (AP-1 and NF-κB).

### 9.2.1.2. Anticancer Potential of *Pleurotus* spp. Crude Extracts

In several cases, crude extracts of the mushrooms showed higher cytotoxicity toward cancer cell lines compared to the purified compounds, probably due to synergistic effects between them. Numerous studies confirmed the exceptional anticancer potential of various extracts (aqueous, ethanol, methanol, etc.) derived from basidiocarps and mycelia of *Pleurotus* spp. (Jedinak et al., 2008; Milovanović et al., 2014; Mishra et al., 2021). Yang et al. (2018a) reported the significant potential of *P. ferulae* fruiting body ethanol extracts to inhibit proliferation of murine melanoma (B16), human esophageal cancer (Eca-109), human gastric cancer (BGC823), and human cervical cancer (HeLa) cells through induction of apoptosis. Similarly, ethanol extracts of *P. ostreatus* basidiocarps obtained on pure barley and lavender-enriched substrate showed cytotoxicity toward B16F10 human melanoma cell line, decreasing viability by 30%–50% at a concentration of 500 μg/mL (Bassi et al., 2024). The same extracts showed the ability to increase cell survival in HaCaT human keratinocytes threatened by $H_2O_2$ and UV light. The ethanolic extract of *P. ostreatus* basidiocarps also showed antiproliferative and apoptotic effects on breast cancer cells (MCF-7 line) by decreasing mitochondrial membrane potential, inducing DNA fragmentation, and causing oxidative stress, while healthy fibroblastic Vero cells were not affected (Deepalakshmi and Mirunalini, 2016a). Addition of an ethanol extract of *P. ostreatus* to female Sprague-Dawley rats, in which breast cancer was induced by 7,12-dimethylbenz(a)anthracene (DMBA), led to a reduction in tumors and improvement in general condition, resulting in increased body weight, showing that the tested extract did not show toxicity toward healthy cells. Research has also shown that oral administration of this extract in treated animals inhibits estrogen and progesterone receptors as well as COX-2, most likely due to the presence of ergosterol, while lovastatin inhibits angiogenesis and metastasis through inhibition of metalloproteinases (MMP-2 and MMP-9) in metastatic breast cancer cell lines (Deepalakshmi and Mirunalini, 2016b; Tika et al., 2016). The ethanol extract of *P. tuberregium* proved more effective than the water extract in inhibiting the proliferation of human colorectal (HCT-116) and cervical (HeLa) cancer cell lines, even showing significantly higher cytotoxicity compared to Mitomycin C™, a well-known anticancer drug (Maness et al., 2011). Ethanolic extracts of *P. ferulae* induced apoptosis in several cancer cell lines (B16, Eca-109, BGC823, and HeLa) by mitochondrial dysfunction, cleavage of caspase-12, heat shock protein 70, and an increase in phosphorylated c-Jun N-terminal kinase (JNK) levels (Yang et al., 2018b). The ethanolic extract of the basidiocarp of *P. ferulae* tested in the study by Choi et al. (2004) showed significant cytotoxicity toward SiHa and HeLa cervical cancer cell lines (at concentrations as low as 40 μg/mL) and also toward lung cancer cells (at concentrations as low as 10 μg/mL). On the other hand, the methanolic extract of *P. ostreatus* studied by Jedinak and Silva (2008) showed the ability to arrest the cell cycle in several breast cancer cell lines by reducing the expression of key transcription factors and proteins involved in cell cycle regulation (E2F, DP-1, PCNA, CDK4, CDK6), leading to inhibition of tumor cell growth and proliferation. The methanol extract from *P. ostreatus* basidiocarps also inhibited the proliferation of colon cancer cell lines (HCT-116, HT-29) and increased cell cycle arrest in HT-29 cells, while the hot water extract of *P. ostreatus* mycelium induced apoptosis and inhibited the proliferation of HT-29 cells through upregulation of the proapoptotic molecule Bax and cytosolic cytochrome c (Lavi et al., 2006; Deepalakshmi and Mirunalini, 2016a). Additionally, the ethyl acetate extract of *P. ostreatus* mycelial biomass at a concentration of 107.59 μg/mL showed cytotoxicity toward 50% of HeLa cervical cancer cells, which initially exhibited morphological changes, and after 72 h of incubation, 100% apoptosis was detected (Ekowati et al., 2017). However, significantly lower anticancer potential was reported for ethanol extracts of submergedly cultivated mycelia of *P. eryngii*, *P. ostreatus*, and *P. pulmonarius* toward human cervix adenocarcinoma (HeLa) and human colon carcinoma (LS174) cell lines, with half maximal inhibitory concentration ($IC_{50}$) values ranging from 196 μg $mL^{-1}$ to more than 400 μg $mL^{-1}$ (Milovanović et al., 2014). Recent studies have also demonstrated the possible use of *Pleurotus* spp. extracts in the synthesis of CuO nanoparticles with significant anticancer potential, exhibiting an IC50 value of 45.62 μg/mL toward the colon cancer cell line (HT-29) (Manimaran et al., 2023).

#### 9.2.1.3 Antioxidative Activity as the Basis for Anticancer and Anti-inflammatory Ones

The considerable antioxidative potential of basidiocarps, mycelia, and cultivation broth of *Pleurotus* species has been reported by numerous studies, which showed that their bioactive compounds have a great ability to scavenge various free radicals (DPPH•, ABTS•, •OH, •NO, etc.) and mitigate oxidative stress (Milovanović et al., 2014; Ćilerdžić et al., 2015, 2019; Bassi et al., 2024). Those studies have revealed the concentration-dependent antioxidant activity of various extracts (96% ethanol, 70% ethanol, methanol, hot water, and acetonitrile), which significantly differs according to the fungal part used for extraction. It was shown that the extract of submergedly cultivated *P. ostreatus* mycelia reduced the DPPH radical to a greater extent (with a maximum of 89.91%) than the basidiocarp extract, which opens great possibilities for varying the composition of the nutrient medium and cultivation in controlled conditions for a shorter period compared to traditional basidiocarp cultivation (Ćilerdžić et al., 2015). The addition of various plant residues to the cultivation medium significantly affected the antioxidant activity of *P. ostreatus* extracts, as mycelium obtained on wheat bran-enriched medium was a stronger antioxidant agent than mycelium cultivated on non-amended medium. Ethanol and methanol extracts of *P. ostreatus* basidiocarps grown on pure barley and lavender-enriched substrates significantly scavenged NO radicals (70% and 75% at concentrations of 500 and 250 μg/mL), which cause significant damage during oxidative stress and serve as important second messengers in signaling pathways and transduction of inflammatory signals (Bassi et al., 2024). The type of extractant was also revealed as an important parameter for the extracts preparation in several studies. Thus, Ćilerdžić et al. (2019) obtained most active basidiocarps and mycelium extracts using 96% ethanol, while Lee et al. (2007) reported that hot water was the best extractant for metabolites in *P. citrinopileatus* fruiting bodies, yet ethanol was a better extractant for mycelial metabolites. However, an extraordinary antioxidative potential was reported by Tel et al. (2015) for ethyl acetate and methanol extracts of *P. ostreatus*, which, in extremely low concentrations, reduced DPPH radicals by 50% (0.1421 and 0.1921 mg/mL, respectively). Besides phenolics and flavonoids, polysaccharides extracted from *Pleurotus* species, including β-glucans, have been identified as the main contributors to antioxidative activity. These compounds exhibit significant free radical scavenging, metal-chelation, and enzyme modulation effects, which together offer a robust defense against oxidative damage. All of the above-mentioned factors make *Pleurotus* species a promising base for the development of natural antioxidants for health promotion and disease prevention.

## 9.3 BIOLOGICALLY ACTIVE COMPOUNDS OF *PLEUROTUS* SPP.

Extensive studies have shown that *Pleurotus* spp. basidiocarps and mycelium are reservoirs of polysaccharides, phenolic compounds, triterpenes, proteins and proteoglycans, amino acids, sterols, vitamins, and minerals that together contribute to their nutritional and health benefits, such as hypocholesterolemic, antitumor, immunomodulatory, antioxidant, anti-neurodegenerative, and anti-diabetic effects (Milovanović et al., 2014; Sarma et al., 2018; Ćilerdžić et al., 2019; Mishra et al., 2021).

### 9.3.1 Polysaccharides as Anti-inflamatory and Anticancer Agents

Various studies have shown that extracts derived from *Pleurotus* spp. basidiocarps, mycelia, and fermentation broth are rich in polysaccharides, which show a plethora of bioactivities. To date, more than 30 different polysaccharides have been isolated from basidiocarps, mycelia, and fermentation broth of these species, showing antioxidant, immunomodulatory, hypoglycemic, hypolipidemic, anti-aging, hepatoprotective, antitumor, and anti-inflammatory effects (Pandya et al., 2019; Yin et al., 2021; Li et al., 2021). *Pleurotus* spp. polysaccharides are proven to possess great diversity in molecular weight, monosaccharide composition, chain structure, and configuration, all affecting their bioactivity, which can be further enhanced by phosphorylation, sulfonation, and selenization. β-glucan isolated from *P. pulmonarius* was a potent anti-inflammatory agent inhibiting pro-inflammatory cytokines (Smiderle et al., 2008). Similarly, polysaccharides isolated from *P. eryngii* and its sulfonated form significantly reduced cytokine secretion (IL-1β/IL-10, IL-6/IL-10, TNF-α/IL-10) and inhibited ROS production (Li and Shah, 2016). Newly identified polysaccharides (PEP and NPEP) from *P. eryngii* notably suppressed LPS-induced inflammatory responses in RAW264.7 macrophages by modulating the production of NO, PGE2, IL-1β, TNF-α, and IL-6 through inhibition of MAPK and NF-κB signaling pathways. Polysaccharides isolated from *Pleurotus* species also maintain the balance between pro-inflammatory and anti-inflammatory macrophages, which is important for preventing immunodeficiency diseases, chronic inflammation, allergies, and cancer. β-Glucan isolated from *P. citrinopileatus* promoted the differentiation of human CD14+ and THP-1 monocytes into anti-inflammatory M2 macrophages via Dectin-1 and Toll-like receptor (TLR)2. Similarly, polysaccharides isolated from *P. ostreatus* and *P. eryngii* competed for binding to anti-Dectin-1 and anti-CR3, significantly reducing TNF-mediated intestinal inflammation induced by dextran sulfate sodium (Vetvicka et al., 2019). Silveira et al. (2014) found that β-D-glucan extracted from *Pleurotus sajor-caju* exhibited immunomodulatory effects on THP-1 macrophages, attenuated the inflammatory response during the formalin-induced nociceptive phase in mice, and decreased both total leukocyte count and myeloperoxidase activity in response to LPS stimulation.

Glucans isolated from *Pleurotus* spp. have also shown anti-proliferative effect against cancer cell lines without harming the normal cells (Milovanović et al., 2014; Mishra et al., 2021). β-Glucans is isolated from *P. ostreatus* caused a reduction in the volume of existing tumors and the incidence of new tumors in rats with breast cancer induced by 7,12-dimethylbenz[a]anthracene (DMBA), reducing the total number of tumor nodules (Susanti et al., 2018). Polysaccharide isolated from the aqueous extract of *P. ostreatus* fruiting bodies inhibited the proliferation of cervical cancer cell lines (HeLa) at a concentration of 200 μg/mL (Tong et al., 2009). Polysaccharides isolated from *Pleurotus eous* suppressed angiogenesis in MCF-7 human breast cancer cells by inhibiting VEGF expression and increasing caspase 3 and 9, Bax, and phospho-JNK expression, thus decreasing mitochondrial membrane potential and inducing apoptosis (Xu et al., 2015). The acidic fraction of polysaccharides isolated from *P. abalonus* at a concentration of 193 μg/mL caused a loss of mitochondrial membrane potential, an increase in ROS concentration, and a shift in the balance of proapoptotic and antiapoptotic proteins (Bak/Bcl2) toward programmed cell death (Shi et al., 2013). Carboxymethylated β-glucan isolated from the sclerotia of *P. tuberregium* showed similar activity toward MCF-7 cells, inducing apoptosis in 50% of tumor cells at a concentration of 204 μg/mL (Zhang et al., 2006). Due to the induction of humoral and cellular immune responses, β-glucan isolated from *P. sajor-caju*, together with an oncogenic protein produced by human papilloma, could be used as a vaccine against cervical cancer (Roopngam, 2018).

### 9.3.2 Proteins and Peptides as Anti-inflamatory and Anticancer Agents

Numerous studies have demonstrated the significant antitumor potential of proteins derived from mushrooms, which not only have cytotoxic effects against tumor cells but also enhance immune responses (Yang et al., 2018a, 2018b; Jeitler et al., 2020). Additionally, these proteins have been shown to reduce adverse side effects commonly associated with conventional radiotherapy and chemotherapy. Liu et al. (2022) isolated a novel protein—pleuroferin—from *P. ferulae* lanzi basidiocarps and reported its *in vitro* antitumor activity against non-small cell lung cancer with an $IC_{50}$ value of only 190 μg/mL. Additionally, the same protein was reported to induce adenosine monophosphate-activated protein kinase (AMPK)-mediated autophagy and G1-phase cell cycle arrest in A549 lung cancer cells, and the same mechanism was demonstrated in a xenograft mouse model *in vivo* (Liu et al., 2023). The antitumor and immunomodulatory activity of a polypeptide isolated from *P. eryngii* mycelia, which inhibited the proliferation of cervical, breast, and gastric cancer cells at concentrations of 0.05 to 2 mg/mL, was demonstrated by Sun et al. (2017). The same polypeptide promoted macrophage proliferation, TNF-α and IL-6 secretion, and expression of TLR2 and TLR4. The protein extract of *P. ostreatus* basidiocarp studied by Wu et al. (2011) showed an apoptotic effect on the human colorectal adenocarcinoma cell line (SW480) through ROS production, glutathione depletion, and mitochondrial dysfunction, similar to clinical anticancer drugs such as cisplatin. The concanavalin A–like lectin derived from *P. ostreatus* is also a potential anti-inflammatory and antitumor agent, while Cibacron blue affinity protein from the same species showed anticancer capacity in several animal models—Dalton lymphoma (DL)-bearing mice, Sarcoma 180, and B16F0 melanoma tumor-bearing mice (Sarma et al., 2018).

### 9.3.3 Other Bioactive Compounds

Besides polysaccharides, proteins, proteoglycans, and peptides as the main carriers of bioactivity in *Pleurotus* spp. fruiting bodies or mycelium, numerous phenolic compounds, flavonoids, tannins, saponins, alkaloids, mono- and sesquiterpenoids, ergosterols, and fatty acid esters have also been identified as important bioactive constituents (Rahimah et al., 2019). Several phenolic acids (chlorogenic acid, *p*-hydroxybezoic acid, vanillic acid, ferulic acid, sinapic acid, syringic acid, and p-anisic acid) and flavonoids (catechin, epicatechin, rutin, myricetin, hesperidin, and quercetin) have been identified in *Pleurotus* spp. extracts as major agents contributing to its antoxidanttheir antioxidant and anti-inflammatory potential (Lin et al., 2014). Wang et al. (2013) reported the cytotoxicity of terpenoids isolated from *P. cornucopiae* mycelium toward HeLa and HepG2 cancer cells.

## 9.4 CONCLUSIONS AND FUTURE PERSPECTIVES

Given the increasing scientific interest in natural products for inflammation and cancer prevention and treatment, mushrooms have received great scientific attention as a promising alternative to conventional therapeutics. Numerous studies have shown *Pleurotus* spp. to be a promising avenue for the development of novel nutraceutical and therapeutic agents. However, further research, including clinical trials and mechanistic studies, is required to fully elucidate the efficacy, safety, and optimal use of *Pleurotus* spp. in therapeutic applications. Overall, the anti-inflammatory and anticancer potential of *Pleurotus* spp. underscores their value as functional foods and natural products for the prevention and treatment of inflammatory diseases and cancer.

## 9.5 ACKNOWLEDGMENTS

This work was supported by the Ministry of Education and Science of the Republic of Serbia (contract number: 451-03-136/2025-03/200178).

## REFERENCES

Baechle, J. J., N. Chen, P. Makhijani, S. Winer, D. David Furman, and D. A. Winer. 2023. Chronic inflammation and the hallmarks of aging. *Molecular Metabolism* 74: 101755.

Barnham, K. J., C. L. Masters, and A. I. Bush. 2004. Neurodegenerative diseases and oxidative stress. *Nature Reviews Drug Discovery* 3: 205–214.

Bassi, S., M. Benvenuti, S. Mirata, et al. 2024. Enhanced antioxidant and anti-inflammatory activity of the extracts of *Pleurotus ostreatus* edible mushroom grown on *Lavandula angustifolia* residues. *Food Bioscience* 60: 104382.

Božić Nedeljković, B., J. Ćilerdžić, D. Zmijanac, et al. 2022. Immunomodulatory effects of extract of Lingzhi or Reishi medicinal mushroom *Ganoderma lucidum* (Agaricomycetes) basidiocarps cultivated on alternative substrate. *International Journal of Medicinal Mushrooms* 24: 45–59.

Bray, F., J. Ferlay, I. Soerjomataram, R. L. Siegel, L. A. Torre, and A. Jemal. 2018. Global cancer statistics 2018: GLOBOCAN estimates of incidence and mortality worldwide for 36 cancers in 185 countries. *CA: A Cancer Journal for Clinicians* 68: 394–424.

Carrasco-González, J. A., S. O. Serna-Saldívar, and J. A. Gutiérrez-Uribe. 2017. Nutritional composition and nutraceutical properties of the *Pleurotus* fruiting bodies: potential use as food ingredient. *Journal of Food Composition and Analysis* 58: 69–81.

Choi, D., W. S. Cha, S. H. Kang, B. R. Lee. 2004. Effect of *Pleurotus ferulae* extracts on viability of human lung cancer and cervical cancer cell lines. *Biotechnology and Bioprocess Engineering* 9: 356–361.

Ćilerdžić, J., M. Galić, and M. Stajić. 2022. From pomiculture waste to biotechnological raw material: efficient transformation using ligninosomes and cellulosomes from *Pleurotus* spp. *Bioresources and Bioprocessing* 9: 66.

Ćilerdžić J., M. Galić, J. Vukojević, and M. Stajić. 2019. *Pleurotus ostreatus* and *Laetiporus sulphureus* (Agaricomycetes): possible agents against Alzheimer and Parkinson diseases. *International Journal of Medicinal Mushrooms* 21: 275–289.

Ćilerdžić, J., M. Stajić, J. Vukojević, and S. Duletić-Laušević. 2013. Oxidative stress and species of genus *Ganoderma* (Higher Basidiomycetes). *International Journal of Medicinal Mushrooms* 15: 21–28.

Ćilerdžić, J., M. Stajić, J. Vukojević, I. Milovanović, and N. Muzgonja. 2015. Antioxidant and antifungal potential of *Pleurotus ostreatus* and *Agrocybe cylindracea* basidiocarps and mycelia. *Current Pharmaceutical Biotechnology* 16: 179–186.

Ćilerdžić, J., J. Vukojević, M. Stajić, T. Stanojković, and J. Glamočlija. 2014. Biological activity of *Ganoderma lucidum* basidiocarps cultivated on alternative and commercial substrate. *Journal of Ethnopharmacology* 155: 312–319.

Coussens, L. M., and Z. Werb. 2002. Inflammation and cancer. *Nature* 420: 860–867.

Deepalakshmi, K., and S. Mirunalini. 2014. *Pleurotus ostreatus*: an Oyster Mushroom with nutritional and medicinal properties. *Journal of Biochemical Technology* 5: 718–726.

Deepalakshmi, K., and S. Mirunalini. 2016a. Antiproliferative and apoptotic effect of *Pleurotus ostreatus* on human mammary carcinoma cell line (Michigan Cancer Foundation-7). *Cancer Translational Medicine* 2: 95–104.

Deepalakshmi, K., and S. Mirunalini. 2016b. Efficacy of *Pleurotus ostreatus* (Jacq. Ex Fr.) P.kumm. on 7,12-dimethylbenz(a) anthracene induced mammary carcinogenesis infemale Sprague-Dawley rats. *New Horizons in Translational Medicine* 3: 73–82.

dos Santos Bazanella, G. C., D. F. de Souza, R. Castoldi, et al. 2013. Production of laccase and manganese peroxidase by *Pleurotus pulmonarius* in solid-state cultures and application in dye decolorization. *Folia Microbiologica* 58: 641–647.

Ekowati, N., A. Mumpuni, and J. S. Muljowati. 2017. Effectiveness of *Pleurotus ostreatus* extract through cytotoxic test and apoptosis mechanism of cervical cancer cells. *Biosaintifika: Journal of Biology and Biology Education* 9: 148–155.

Fernandes, Â., L. Barros, A. Martins, P. Herbert, and I. C. F. R. Ferreira. 2015. Nutritional characterisation of *Pleurotus ostreatus* (Jacq. ex Fr.) P. Kumm. produced using paper scraps as substrate. *Food Chemistry* 169: 396–400.

Furman, D., J. Campisi, E. Verdin, et al. 2019. Chronic inflammation in the etiology of disease across the life span. *Nature Medicine* 25: 1822–1832.

Galić, M., J. Ćilerdžić, and M. Stajić. 2022a. Anticandidal and anti-Aspergillus effects of *Pleurotus ostreatus* and *Laetiporus sulphureus* extracts. Short Oral Pres 06, lecture 08. 11th International Medicinal Mushroom Conference. Belgrade, Serbia.

Galić, M., J. Ćilerdžić, J. Vukojević, and M. Stajić. 2018. Delignification of the most common agricultural plant residues by *Pleurotus pulmonarius*. 7th Balkan Botanical Congress, Novi Sad, Serbia. Book of abstracts: 154.

Galić, M., M. Stajić, and J. Ćilerdžić. 2021a. Lignocellulose waste valorization by an enzymatic cocktail of *P. eryngii* and *P. pulmonarius*. *Cellulose Chemistry and Technology* 55: 1043–1050.

Galić, M., M. Stajić, and J. Ćilerdžić. 2022b. Dynamics of pretreated wheat straw saccharification by cellulosome of

*Trichoderma viride*. *Cellulose Chemistry and Technology* 56: 1013–1019.

Galić, M., M. Stajić, J. Vukojević, and J. Ćilerdžić. 2021b. Obtaining cellulose-available raw materials by pretreatment of common agro-forestry residues with *Pleurotus* spp. *Frontiers in Bioengineering and Biotechnology* 9: 720473.

Grivennikov, S. I., F. R. Greten, and M. Karin. 2010. Immunity, inflammation, and cancer. *Cell* 140: 883–899.

Halliwell, B., and J. M. C. Gutteridge. 2015. *Free Radicals in Biology and Medicine*, 5th ed. Oxford University Press.

Jedinak, A., S. Dudhgaonkar, Q. L. Wu, J. Simon, and D. Sliva. 2011. Anti-inflammatory activity of edible oyster mushroom is mediated through the inhibition of NF-κB and AP-1 signaling. *Nutrition Journal* 10: 52.

Jedinak, A., and D. Silva. 2008. *Pleurotus ostreatus* inhibits proliferation of human breast and colon cancer cells through p53-dependent as well as p53-independent pathway. *International Journal of Oncology* 33: 1307–1313.

Jeitler, M., A. Michalsen, D. Frings, et al. 2020. Significance of medicinal mushrooms in integrative oncology: A narrative review. *Frontiers in Pharmacology* 11: 580656.

Kim, J. I., M. J. Kang, J. Im, et al. 2010. Effect of king oyster mushroom (*Pleurotus eryngii*) on insulin resistance and dyslipidemia in db/db mice. *Food Science Biotechnology* 19: 239–242.

Koo, M. M., R. Swann, S. McPhail, et al. 2020. Presenting symptoms of cancer and stage at diagnosis: evidence from a cross-sectional, population-based study. *Lancet Oncology* 21: 73–79.

Kotas, M. E., and M. Medzhitov. 2015. Homeostasis, inflammation, and disease susceptibility. *Cell* 160: 816827.

Kushairi, N., N. A. K. A. Tarmizi, C. W. Phan, et al. 2020. Modulation of inflammatory pathways by medicinal mushrooms, with particular relevance to Alzheimer's disease. *Trends in Food Science and Technology* 104: 153–162.

Lavi, I., D. Friesem, S. Geresh, Y. Hadar, and B. Schwartz. 2006. An aqueous polysaccharide extract from the edible mushroom *Pleurotus ostreatus* induces anti-proliferative and proapoptotic effects on HT-29 colon cancer cells. *Cancer Letters* 244: 61–70.

Lee, Y. L., G. W. Huang, Z. C. Liang, and J. L. Mau. 2007. Antioxidant properties of three extracts from *Pleurotus citrinopileatus*. *LWT Food Science and Technology* 40: 823–833.

Li, H., Y. Feng, W. Sun, Y. Kong, and L. Jia. 2021. Antioxidation, anti-inflammation and anti-fibrosis effect of phosphorylated polysaccharides from *Pleurotus djamor* mycelia on adenine-induced chronic renal failure mice. *International Journal of Biological Macromolecules* 170: 652–663.

Li, S. Q., and N. P. Shah. 2016. Anti-inflammatory and anti-proliferative activities of natural and sulphonated polysaccharides from *Pleurotus eryngii*. *Journal of Functional Foods* 23: 80–86.

Limón-Pacheco, J., and M. E. Gonsebatt. 2009. The role of antioxidants and antioxidant-related enzymes in protective responses to environmentally induced oxidative stress. *Mutation Research* 674: 137–147.

Lin, J. T., C. W. Liu, Y. C. Chen, et al. 2014. Chemical composition, antioxidant and anti-inflammatory properties for ethanolic extracts from *Pleurotus eryngii* fruiting bodies harvested at different time. *LWT—Food Science and Technology* 55: 374–382.

Liu, M. H., F. Liu, T. B. Ng, and Z. K. Liu. 2022. Purification and characterization of pleuroferin, a novel protein with in vitro anti-non-small cell lung cancer activity from the mushroom *Pleurotus ferulae lanzi*. *Process Biochemistry* 122: 16–25.

Liu, M. H., F. Liu, T. B. Ng, and Z. K. Liu. 2023. New fungal protein from *Pleurotus ferulae* lanzi induces AMPK-mediated autophagy and G1-phase cell cycle arrest in A549 lung cancer cells. *International Journal of Biological Macromolecules* 244: 125453.

Maness, L., N. Sneed, B. Hardy, J. Yu, M. Ahmedna, and I. Goktepe. 2011. Anti-proliferative effect of *Pleurotus tuberregium* against colon and cervical cancer cells. *Journal of Medicinal Plants Research* 5: 6650–6655.

Manimaran, K., D. H. Y. Yanto, C. Kamaraj, et al. 2023. Eco-friendly approaches of mycosynthesized copper oxide nanoparticles (CuONPs) using *Pleurotus citrinopileatus* mushroom extracts and their biological applications. *Environmental Research* 232: 116319.

Milovanović, I., M. Stajić, J. Ćilerdžić, T. Stanojković, A. Knežević, and J. Vukojević. 2014. Antioxidant, antifungal and anticancer activities of Se-enriched *Pleurotus* spp. mycelium extracts. *Archives of Biological Sciences* 66: 1379–1388.

Milovanović, I., T. Stanojković, M. Stajić, et al. 2015. Effect of selenium enrichment of *Lenzites betulinus* and *Trametes hirsuta* mycelia on antioxidant, antifungal and cytostatics potential. *Current Pharmaceutical Biotechnology* 16: 920–926.

Mishra, V., S. Tomar, P. Yadav, and M. P. Singh. 2021. Promising anticancer activity of polysaccharides and other macromolecules derived from oyster mushroom (*Pleurotus* sp.): an updated review. *International Journal of Biological Macromolecules* 182: 1628–1637.

Nagai, H., and Y. H. Kim. 2017. Cancer prevention from the perspective of global cancer burden patterns. *Journal of Thoracic Disease* 9: 448–451.

Nowakowski, P., R. Markiewicz-Żukowska, J. Bielecka, K. Mielcarek, M. Grabia, and K. Socha. 2021. Treasures from the forest: evaluation of mushroom extracts as anti-cancer agents. *Biomedicine and Pharmacotherapy* 143: 112106.

Oalđe Pavlović, M., S. Kolarević, J. Đorđević, et al. 2021. A Study of phytochemistry, genoprotective activity, and antitumor effects of extracts of the selected Lamiaceae species. *Plants* 10: 2306.

O'Byrne, K. J., and A. G. Dalgleish. 2001. Chronic immune activation and inflammation as the cause of malignancy. *Brazilian Journal of Cancer* 85: 473–483.

Pandya, U., U. Dhuldhaj, and N. S. Sahay. 2019. Bioactive mushroom polysaccharides as antitumor: an overview. *Natural Product Research* 33: 2668–2680.

Patel, S., and A. Goyal. 2012. Recent developments in mushrooms as anti-cancer therapeutics: a review. *3 Biotech* 2: 1–15.

Rahimah, S. B., D. D. Djunaedi, A. Y. Soeroto, and T. Bisri. 2019. The phytochemical screening, total phenolic contents and antioxidant activities in vitro of white oyster mushroom (*Pleurotus ostreatus*) preparations. *Open Access Macedonian Journal of Medical Sciences* 7: 2404–2412.

Ray, P., S. Kundu, and D. Paul. 2024. Exploring the therapeutic properties of Chinese mushrooms with a focus on their anti-cancer effects: a systemic review. *Pharmacological Research—Modern Chinese Medicine* 11: 100433.

Ray, P. D., B. W. Huang, and Y. Tsuji. 2012. Reactive oxygen species (ROS) homeostasis and redox regulation in cellular signaling. *Cell signal* 24: 981–990.

Reuter, S., S. C. Gupta, M. M. Chaturvedi, and B. B. Aggarwal. 2010. Oxidative stress, inflammation, and cancer: how are they linked? *Free Radical Biology and Medicine* 49: 1603–1616.

Roopngam, P. E. 2018. Increased response of human T-lymphocytes by dendritic cells pilsed with HPV16E7 and *Pleurotus sajor-caju*-β-glucan (PBG). *Iranian Journal of Immunology* 15: 246–255.

Sarma, D., A. K. Saha, and B. K. Datta. 2018. Bioactive compounds with special references to anticancer property of oyster mushroom *Pleurotus ostreatus*. *Journal of Pharmacognosy and Phytochemistry* 7: 2694–2698.

Sarmadi, B. H., and A. Ismail. 2010. Antioxidative peptides from food proteins: a review. *Peptides* 31: 1949–1956.

Schieber, M., and N. S. Chandel. 2014. ROS function in redox signaling and oxidative stress. *Current Biology* 24: 453–462.

Shi, X., Y. Zhao, Y. Jiao, T. Shi, and X. Yang. 2013. ROS-dependent mitochondria molecular mechanisms underlying antitumor activity of *Pleurotus abalonus* acidic polysaccharides in human breast cancer MCF-7 cells. *PlosONE* 8: e64266.

Silveira, M. L. L., F. R. Smiderle, C. P. Moraes, et al. 2014. Structural characterization and anti-inflammatory activity of a linear β-d-glucan isolated from *Pleurotus sajor-caju*. *Carbohydrate Polymers* 113: 588–596.

Smiderle, F. R., L. M. Olsen, E. R. Carbonero, et al. 2008. Anti-inflammatory and analgesic properties in a rodent model of a (1→3),(1→6)-linked β-glucan isolated from *Pleurotus pulmonarius*. *Immunopharmacology and Inflammation* 597: 86–91.

Sosa, V., T. Moliné, R. Somoza, R. Paciucci, H. Kondoh, and M. E. Lleonart. 2013. Oxidative stress and cancer: an overview. *Ageing Research Reviews* 12: 376–390.

Stajić, M., M. Galić, and J. Ćilerdžić. 2023. Mushrooms—from traditional remedies to the modern therapeutics; In: *Mushrooms with Therapeutic Potentials (Recent Advances in Research and Development)*, ed. D. C. Agrawal, and M. Dhanasekaran. Springer Nature Singapore.

Sun, Y., X. Hu, and W. Li. 2017. Antioxidant, antitumor and immunostimulatory activities of the polypeptide from *Pleurotus eryngii* mycelium. *International Journal of Biological Macromolecules* 97: 323–330.

Susanti, I., H. Wibowo, N. Siregar, S. Pawiroharsono, and F. D. Sujatna. 2018. Antitumor activity of beta glucan extract from Oyster mushroom (*Pleurotus ostreatus* Jacq. P. Kum) on DMBA-induced breast cancer *in vivo*. *International Journal of PharmTech Research* 11: 190–197.

Taofiq, O., A. Martins, M. F. Barreiro, and I. C. F. R. Ferreira. 2016. Anti-inflammatory potential of mushroom extracts and isolated metabolites. *Trends in Food Science and Technology* 50: 193–210.

Tel, G., M. Ozturk, M. E. Duru, and A. Turkoglu. 2015. Antioxidant and anticholinesterase activities of five wild mushroom species with total bioactive contents. *Pharmaceutical Biology* 53: 824–830.

Tika, L. N., L. S. S. Wikanthi, S. Annur, and R. Murwanti. 2016. Oyster mushroom (*Pleurotus ostreatus*) inhibits migration and metastasis on 4T1 breast cancer cells. *Indonesian Journal of Cancer Chemoprevention* 7: 99–103.

Tong, H., F. Xia, K. Feng, et al. 2009. Structural characterization and *in vitro* antitumor activity of a novel polysaccharide isolated from the fruiting bodies of *Pleurotus ostreatus*. *Bioresource Technology* 100: 1682–1686.

Vetvicka, V., O. Gover, M. Karpovsky, et al. 2019. Immune-modulating activities of glucans extracted from *Pleurotus ostreatus* and *Pleurotus eryngii*. *Journal of Functional Foods* 54: 81–91.

Wang, S., L. Bao, F. Zhao, et al. 2013. Isolation, identification and bioactivity of monoterpenoids and sesquiterpenoids from the mycelia of edible mushroom *Pleurotus cornucopiae*. *Journal of Agricultural and Food Chemistry* 61: 5122–5129.

Wu, J. Y., C. H. Chen, W. H. Chang, et al. 2011. Anti-cancer effects of protein extracts from *Calvatia lilacina*, *Pleurotus ostreatus* and *Volvariella volvacea*. *Evidence-Based Complementary and Alternative Medicine* 2011: 982368.

Xu, J. K., Q. G. Yuan, P. Luo, X. L. Sun, and J. C. Ma. 2015. *Pleurotus eous* polysaccharides suppress angiogenesis and induce apoptosis via ROS-dependent JNK activation and mitochondrial mediated mechanisms in MCF-7 human breast cancer cells. *Bangladesh Journal of Pharmacology* 10: 78–86.

Yang, Y., C. Fu, F. Zhou, et al. 2018a. Chemical composition, antioxidant and antitumor activities of sub-fractions of wild and cultivated *Pleurotus ferulae* ethanol extracts. *PeerJ* 6: e6097.

Yang, Y., P. Yuan, X. Wei, et al. 2018b. Cultivated and wild *Pleurotus ferulae* ethanol extracts inhibit hepatocellular carcinoma cell growth via inducing endoplasmic reticulum stress- and mitochondria-dependent apoptosis. *Scientific Reports* 8: 13984.

Yin, Z., D. Sun-Waterhouse, J. Wang, C. Ma, G. I. N. Waterhouse, and W. Kang. 2021. Polysaccharides from edible fungi *Pleurotus* spp.: advances and perspectives. *Journal of Future Foods* 1–2: 128–140.

Zhang, M., P. C. K. Cheung, L. C. M. Chiu, E. Y. L. Wong, V. E. C. Ooi. 2006. Cell-cycle arrest and apoptosis induction in human breast carcinoma MCF-7 cells by carboxymethylated β-glucan from the mushroom sclerotia of *Pleurotus tuber-regium*. *Carbohydrate Polymers* 66: 455–462.

# 10 Immunostimulatory Potential of *Pleurotus* spp.

## *Mechanisms and Applications*

*Somanjana Khatua**
Molecular and Applied Mycology Laboratory, Department of Botany, Faculty of Science, University of Allahabad, Prayagraj, Uttar Pradesh, India

## 10.1 INTRODUCTION

The immune system functions as the body's frontline defense against a wide array of harmful microorganisms, including bacteria, viruses, fungi, and parasites (Marshall et al., 2018). It achieves this by distinguishing and eliminating foreign pathogens while safeguarding the body's own healthy cells (Khatua et al., 2022). Beyond its role in infection control, the immune system is essential for clearing dead or damaged cells, facilitating tissue repair, and developing immunological memory that enables faster and more robust responses upon repeated exposure to pathogens (Sharma et al., 2024). Additionally, it plays a crucial role in immune surveillance by detecting and eliminating abnormal or potentially malignant cells, thereby maintaining physiological homeostasis (Finn, 2012). Under normal conditions, immune responses are precise and well-regulated (Sollome and Fry, 2015). However, factors such as undernutrition and aging can compromise immune function, resulting in increased vulnerability to diseases (Weyand and Goronzy, 2016; Mitra et al., 2022). Such immune dysregulation may manifest as hypersensitivity and autoimmune diseases (where the body erroneously targets its own tissues) or immunodeficiency (characterized by impaired or absent immune responses). Individuals with immunodeficiency are particularly susceptible to infections and have a heightened risk of cancer due to reduced immune surveillance (Ilham et al., 2023). Thus, enhancing host immunity through safe and nontoxic strategies is increasingly viewed as a valuable approach to disease prevention and management (Kong et al., 2014).

In this context, immunostimulants, natural or synthetic agents that enhance the immune system's defensive capabilities, have garnered significant attention (Khatua and Acharya, 2024). A specialized subcategory, "immunoceuticals," has recently emerged to describe nutraceuticals with immunomodulatory properties. These compounds are particularly promising for managing chronic and complex conditions, such as cancer, autoimmune diseases, and infections, which often necessitate long-term therapeutic interventions. Their relevance has surged, especially in light of the COVID-19 pandemic. As the demand for complementary and integrative medicine grows, immunoceuticals are increasingly recognized for their potential to support conventional therapies and prevent immune-related disorders (Son et al., 2024). In line with this, the global nutraceuticals market, including immunoceuticals, was valued at USD 454.55 billion in 2021 and is projected to reach USD 991.09 billion by 2030 (Tieu et al., 2022). Researchers are continually searching for potent immunostimulants that can effectively enhance immune function and provide therapeutic benefits (Khatua et al., 2022).

Notably, mushrooms, valued for their therapeutic and nutritional properties for thousands of years, have emerged as a key focus in this pursuit of natural immunomodulators (Llauradó et al., 2016). Today, a wide range of bioactive compounds with immunomodulatory properties have been identified from various mushroom species, including β-glucans, lentins, polysaccharides, polysaccharide-peptide complexes, triterpenoids, nucleosides, and other secondary metabolites (Nworu et al., 2015). These compounds, whether in crude extracts or isolated forms, are known to boost immune function by stimulating natural killer (NK) cells, T cells, B cells, and macrophages (Jubori et al., 2018). Both commercially available and non-commercial glucans, including those derived from sources such as zymosan, lentinan, schizophyllan, and maitake D fraction, have demonstrated varying degrees of immunological activity (Devi et al., 2015). Recent clinical studies have revealed that consuming fungi-derived β-glucans can significantly enhance immune responses by increasing the levels of immune cells such as $CD3^+$, $CD4^+$, and $CD8^+$ T-lymphocytes, as well as improving the CD4/CD8 ratio (Chen et al., 2023). Many of these substances, due to their immunostimulatory effects, also show promise as antitumor agents (Nworu et al., 2015).

Numerous studies have highlighted the nutritional and medicinal significance of *Pleurotus* mushrooms, which have been recognized not only as a valuable food source but also for their therapeutic potential (Llauradó et al., 2016). Extracts and bioactive compounds derived from oyster mushrooms have shown effectiveness in managing chronic conditions such as diabetes and cancer, mitigating oxidative stress, and boosting immune function. Like many edible and medicinal fungi, *Pleurotus* species are rich in biological response modifiers (BRMs) that contribute to

* Corresponding author: somanjana@allduniv.ac.in

DOI: 10.1201/9781003527572-10

their immunostimulatory effects. Furthermore, submerged mycelial cultures of these mushrooms offer a promising and sustainable method for developing safe and health-promoting products (Llauradó et al., 2015). One notable compound is pleuran, an insoluble β-(1,3/1,6)-D-glucan isolated from *Pleurotus ostreatus* fruiting bodies. Developed as a dietary supplement, pleuran is used to support immune health and alleviate early symptoms of fatigue in both children and adults. Clinical studies have also indicated that Imunoglukan P4H®, a formulation combining pleuran with vitamin C, may serve as an effective immunostimulant, particularly for elite athletes (Van Steenwijk et al., 2021).

Thus, the present review focuses on the immunostimulatory effects of different oyster mushrooms, emphasizing their potential as natural agents to enhance immune function. By compiling recent findings on their bioactive compounds, mechanisms of action, and therapeutic applications, this review aims to provide valuable insights for researchers, healthcare professionals, and nutraceutical developers seeking effective and safe immunomodulatory strategies.

## 10.2 OVERVIEW OF THE IMMUNE SYSTEM

The immune system is a highly sophisticated and dynamic defense network that protects the body from infections and harmful agents, including bacteria, viruses, fungi, and parasites (Cumming and Peters, 2024). It comprises two primary branches: the innate (non-specific) immune system and the adaptive (specific) immune system, which operate in tandem to maintain host defense (Meng et al., 2023a). The innate immune system serves as the body's first line of defense, providing a rapid but non-specific response to invading pathogens. It includes physical and chemical barriers, such as the skin, mucous membranes, and antimicrobial peptides, as well as cellular components like neutrophils, macrophages, dendritic cells (DCs), and natural killer (NK) cells (Morsink et al., 2020; Michailidou et al., 2022; Pradeu et al., 2024; Wang et al., 2024). These cells detect pathogen-associated molecular patterns (PAMPs) through pattern recognition receptors (PRRs) and act immediately, often without discriminating between pathogen types (Li and Wu, 2021). Subsequently, innate immune cells secrete a variety of cytokines that activate and guide the development of adaptive immune responses (Wang et al., 2024).

The adaptive immune system, while slower to respond, offers highly specific and long-lasting protection. It is mediated by lymphocytes, namely T cells and B cells (Clark and Kupper, 2005; Poon and Farber, 2020). T cells originate in the bone marrow and mature in the thymus, differentiating into cytotoxic T cells, which kill infected or abnormal cells, and helper T cells, which activate other immune cells. B cells, which also originate in the bone marrow, differentiate into plasma cells that produce antibodies targeting specific antigens. The adaptive response also includes memory cells, which provide faster and more robust responses upon re-exposure to the same pathogen (Janeway et al., 2001; Morsink et al., 2020; DeMaio et al., 2022).

The immune system is supported by several lymphoid organs, including the lymph nodes, spleen, and thymus (Gutierrez-Lanz et al., 2022). Lymph nodes act as filtration sites where immune cells interact and initiate responses. The spleen filters the blood, removing damaged cells and pathogens, and acts as a reservoir for immune cells. The thymus is essential for the maturation of functional T cells (Lima-Junior et al., 2022). Together, the innate and adaptive immune systems orchestrate a coordinated defense strategy that provides both immediate and long-term protection (Morsink et al., 2020). The immune system's primary functions include disease management, tissue upkeep, and the elimination of damaged or cancerous cells (Singh et al., 2022).

## 10.3 IMMUNODEFICIENCY AND THE ROLE OF MUSHROOM-DERIVED IMMUNOSTIMULANTS

The immune system can become compromised under a variety of circumstances, leading to impaired defense against infections and disease. Immunodeficiency may result from poor nutrition, therapeutic interventions for cancer, such as chemotherapy or radiation, chronic viral infections like human immunodeficiency virus (HIV), or inherited genetic abnormalities. In these situations, the immune response may be insufficient, overly aggressive, or misdirected, making it necessary to enhance or fine-tune immune activity to restore proper function (Tieu et al., 2022). When immunity is weakened, the coordination between immune components deteriorates, reducing the system's ability to effectively combat harmful agents (Jubori et al., 2018). This dysfunction can be worsened by external and physiological parameters, including aging, environmental toxins, malnutrition, and chronic psychological stress, all of which contribute to reduced immune resilience (Calder et al., 2020).

Immunotherapy, also known as biotherapy, offers a way to support or boost immune activity in these cases. It can involve either stimulating the body's natural immune defenses or administering synthetic immune components, such as engineered proteins, to enhance the body's response to infections or malignancies. This approach is particularly beneficial for patients with weakened immunity due to conditions such as liver disease, advanced HIV/AIDS, cancer, or post-bone marrow transplantation. Such individuals are at greater risk of adverse health outcomes, including acute respiratory failure, a frequent and severe condition that often leads to intensive care unit admission among patients with cancer or immune compromise (Oloke and Adebayo, 2015). Strengthening the immune system in these cases is essential for both preventing new infections and improving responses to ongoing disease. Notably, natural immunostimulants, including bioactive compounds found in medicinal plants and edible mushrooms, have shown potential in enhancing both innate and adaptive immunity, offering supportive therapeutic value (Llauradó et al., 2016).

Mushrooms are widely recognized for their immune-enhancing potential and are considered an excellent dietary option, especially for vegetarians and individuals with weakened immune systems, such as cancer patients and those living with HIV/AIDS (Ching et al., 2021). Unlike animals, mushrooms lack an adaptive immune system, relying entirely on their chemical defenses to guard against the diverse array of pathogens present in their natural habitats (Jędrzejewski et al., 2023). Their immunomodulatory effects influence various immune components, including hematopoietic stem cells, lymphocytes, macrophages, DCs, T cells, and NK cells. A significant portion of these effects is attributed to their rich content of biologically active polysaccharides, such as α- and β-glucans, mannans, galactans, xylans, chitin, and hemicellulose, with β-glucans being the most extensively studied and noted for their strong immunostimulatory activities (Ching et al., 2021). Notable mushroom-derived β-glucans include lentinan from *Lentinula edodes*, grifolan from *Grifola frondosa*, and schizophyllan from *Schizophyllum commune*, all of which have demonstrated the ability to activate macrophages and NK cells, enhancing the body's innate immune defenses (Abdullah et al., 2017). Polysaccharides from *Antrodia camphorata* have shown antitumor effects primarily through immune enhancement, with tumor suppression attributed to the upregulation of immune network activity (Liu et al., 2004). Of the over 270 mushroom species known for their immunotherapeutic potential, approximately 50 have been identified as non-toxic and evaluated in animal studies (Jędrzejewski et al., 2023). The growing volume of scientific literature highlights the increasing interest in the medicinal potential of macrofungi. In the past decade alone, more than 800 articles related to "mushroom immunity" have been indexed in the PubMed database.

Beyond human health, mushroom polysaccharides are increasingly explored in aquaculture as natural immunostimulants. Supplementation in fish diets has been shown to not only activate non-specific immune responses but also improve survival and resistance against bacterial infections (Ching et al., 2021). Among the mushroom genera, *Pleurotus* species have yielded a wide range of immune-enhancing compounds that can stimulate various immune cell populations, contributing to an overall modulation of the immune response (Oloke and Adebayo, 2015).

## 10.4 THE GENUS *PLEUROTUS* AS NATURAL IMMUNE ENHANCERS

Oyster mushrooms (*Pleurotus* spp.) have been an integral part of traditional medicine for centuries, especially in Asian cultures, where they are valued not only as a nutritious food but also for their therapeutic benefits (Adams et al., 2019). Their use in folk medicine, however, extends well beyond Asia, with traditional health practices in Central Europe, South America, and Africa also incorporating these mushrooms for their healing properties (Mariga et al., 2014). Globally, approximately 40 species of *Pleurotus* are known, thriving in both tropical and temperate regions, many of which are now cultivated commercially on a large scale (Tanaka et al., 2015; Raman et al., 2020). Traditionally, these mushrooms have been consumed in various forms (including dried powders, decoctions, infusions, and extracts) often prepared as teas, soups, or herbal remedies to promote vitality and general well-being. For example, *Pleurotus eryngii*, known as king oyster mushrooms, has traditionally been brewed into tonics believed to support immune function and enhance stamina (Mariga et al., 2014).

These mushrooms are a reservoir of biologically active compounds, including β- and α-glucans, lectins, proteoglycans, polysaccharides, and polysaccharide–protein complexes. Notably, β-glucans—the major structural components of the fungal cell wall—are among the most thoroughly studied for their immunomodulatory potential (Mishra et al., 2021). In particular, β-(1,3)-glucans from *Pleurotus* species have demonstrated the ability to modulate both innate and adaptive immune responses, acting as BRMs. These compounds stimulate various immune cells, enhance cytokine production, and support host defense mechanisms, all while maintaining a high safety profile due to their low toxicity (Vetvicka et al., 2019). Such characteristics highlight the promise of *Pleurotus* mushrooms as natural agents for immune support and therapeutic development.

### 10.4.1 Immune-Stimulatory Effect of *Pleurotus* sp. on Macrophage Cells

Polysaccharides derived from fungi, particularly those from *Pleurotus* species, have garnered significant attention for their ability to stimulate immune responses, primarily through macrophage activation. Macrophages, as central components of innate immunity, play a crucial role in processes such as phagocytosis, engulfing pathogens and abnormal cells, and initiating downstream immune signaling. Upon stimulation by *Pleurotus* polysaccharides, macrophages exhibit enhanced production of immune effectors, including nitric oxide (NO), reactive oxygen species (ROS), and key pro-inflammatory cytokines like tumor necrosis factor (TNF)-α, which are vital for host defense and antitumor activity (Devi et al., 2015; Mariga et al., 2014). The increase in TNF-α is often accompanied by elevated NO levels, reflecting a heightened inflammatory response. Activated macrophages also secrete other signaling molecules, such as interleukins (ILs) and interferons (IFN), which contribute to immune regulation and coordination (Kong et al., 2014). Morphological alterations, including membrane ruffling and increased adherence, further signify macrophage activation (Khatua et al., 2021). The immunomodulatory effects of fungal polysaccharides are attributed to their interaction with pattern recognition receptors, particularly Toll-like receptors (TLRs). This interaction activates intracellular signaling cascades involving adaptor proteins like MyD88, leading to NF-κB pathway activation and the expression of inflammatory mediators (Ghosh et al., 2021). Additionally, these polysaccharides engage the MAPK and

PI3K-Akt signaling pathways, which support macrophage activation, cytokine release, and cellular survival (Khatua et al., 2022).

To evaluate the immunostimulatory activity of *P. ostreatus*, researchers have isolated β-glucans from both the mycelium and fruiting bodies and conducted comparative analyses of their structural characteristics and biological functions. Mycelial glucans, composed of glucose, mannose, and fucose in a 3:2:1 molar ratio, contain diverse glycosidic linkages such as (1→2), (1→3), (1→4), and (1→6), while fruiting body glucans primarily consist of glucose linked through (1→3) and (1→6) bonds. *In vitro* assays using RAW264.7 macrophages showed that mycelial glucans slightly outperformed fruiting body glucans in enhancing phagocytosis, NO, and TNF-α production. *In vivo* experiments in lymphoma-bearing mice demonstrated that both glucan types significantly stimulated systemic immunity by increasing lymphocyte and macrophage populations, with tumor inhibition rates of 75% and 71.4%, respectively (Devi et al., 2015). Complementing these findings, Llauradó Maury et al. (2021) reported that a hot-water mycelial extract of *P. ostreatus* triggered M1-like macrophage polarization, characterized by elevated NO, TNF-α, and IL-6 production, along with enhanced phagocytic function mediated through NF-κB activation.

Further evidence of *Pleurotus*-derived polysaccharide immunopotency comes from the characterization of WPOP-N1, an alkali-extracted and chromatographically purified polysaccharide. At concentrations of 200 and 400 μg/mL, WPOP-N1 significantly promoted phagocytic activity in murine peritoneal macrophages. The 400 μg/mL dose led to a threefold increase in NO production and a 3.5-fold increase in TNF-α secretion. This was accompanied by upregulation of TNF-α and iNOS gene expression and activation of NF-κB signaling, as evidenced by p65 phosphorylation and IκB degradation (Kong et al., 2014).

In another comparative study, polysaccharides from *P. ostreatus* grown on a 15% *Astragali radix*–enriched substrate ($P^{15}$OP-I) were contrasted with those from a standard substrate ($P^{0}$OP-I). $P^{15}$OP-I had a slightly lower molecular weight (Mw) (20.2 kDa vs. 21.7 kDa) and exhibited higher concentrations of mannose, glucose, galactose, and arabinose, but less galacturonic acid than $P^{0}$OP-I. Both samples enhanced NO and TNF-α production in RAW264.7 macrophages, with $P^{15}$OP-I displaying a more pronounced immunostimulatory effect. The observed activity was associated with the activation of JNK/MAPK, Erk/MAPK, and NF-κB pathways (Meng et al., 2023b).

Additionally, selenium enrichment of *P. ostreatus* led to the development of Se-POP-21, a novel selenium-bound polysaccharide. Although its exact selenium-binding structure remains unresolved, Fourier transform infrared (FTIR) spectroscopy suggests the presence of Se–O–C linkages. Se-POP-21 significantly enhanced macrophage proliferation, phagocytosis, and the production of NO, ROS, TNF-α, IL-1β, and IL-6, predominantly via the NF-κB signaling axis. These findings suggest that trace element fortification can further augment the immunological properties of fungal polysaccharides (Wang et al., 2022).

Beyond *P. ostreatus*, king oyster mushroom has also been investigated for its immune-enhancing properties. In one study, a water-soluble polysaccharide was isolated from its fruiting bodies using a combination of hot-water extraction, DEAE-cellulose anion-exchange, and gel filtration chromatography. This purified compound, termed WPEP-N-b, had an Mw of approximately 21.4 kDa. Analysis of its sugar composition revealed mannose as the major monosaccharide, followed by galactose and glucose. Functional assays using RAW264.7 macrophages showed that WPEP-N-b significantly enhanced phagocytic activity and NO release at concentrations between 25 and 100 μg/mL after 24 hours. Although it did not increase cell proliferation, treatment with 50 μg/mL led to the degradation of IκB-α and phosphorylation of NF-κB p65, indicating pathway activation. The compound also stimulated the phosphorylation of MAPK pathway proteins (JNK, ERK, and p38) in a time-dependent manner. These immunomodulatory effects were primarily mediated via TLR2, suggesting WPEP-N-b influences immune function by engaging TLR2–NF-κB and MAPK signaling pathways (Yan et al., 2019).

Another study examined the isolation and immunomodulatory effects of a polysaccharide, EPA-1, from *P. eryngii* through a series of extraction and purification steps, including water extraction, alcohol precipitation, DEAE anion-exchange, and Sephadex G-50 chromatography. The resulting polysaccharide, EPA-1, is a neutral compound with an average Mw of approximately 99.7 kDa. It was composed of mannose, glucose, and galactose in a ratio of 2.2:1.0:3.2, with its primary structure consisting of (1→6)-linked galactose residues. In functional assays with RAW264.7 cells, EPA-1 at 100 μg/mL significantly enhanced immune activity. Specifically, it increased NO production and elevated the levels of key pro-inflammatory cytokines compared to untreated controls. Additionally, EPA-1 promoted the phosphorylation of important MAPK pathway proteins, including p38, ERK, and JNK, in a dose-dependent manner. These results suggest that EPA-1 exerts potent immunostimulatory effects, likely through the activation of the MAPK signaling pathway, making it a promising candidate for applications aimed at enhancing immune responses (Xu et al., 2016).

Abreu et al. (2021) extracted polysaccharides from *P. eryngii* using a range of low-cost, water-based methods under different temperature and pressure conditions to assess their immunomodulatory properties. The cold-water extracted fraction (CWEF), predominantly consisting of mannogalactan, exhibited the strongest immune-boosting effects, significantly increasing macrophage secretion of NO, IL-10, and IL-1β. In contrast, the hot-water extracted fraction (HWEF), which contained high levels of linear (1→6)-β-D-glucan, also stimulated cytokine and NO production, albeit to a lesser degree. The autoclave-extracted fraction (AEF), containing a blend of α- and β-glucans, including newly identified linear (1→6)- and (1→3)-β-glucans, was effective only in promoting NO secretion and showed no significant impact on cytokine production. These findings highlight that various extraction methods produce distinct polysaccharide profiles with different immunostimulatory potentials, suggesting a simple and cost-efficient strategy for isolating bioactive components from *P. eryngii.*

A protein identified as PEP, isolated from *P. eryngii*, was found to have an Mw of approximately 63 kDa, which is notably higher than other proteins previously reported from *P. eryngii* or other oyster mushroom species. Structural analysis using FTIR and NMR spectroscopy revealed that PEP primarily adopts an α-helical secondary structure, with arginine, glycine, and serine as its predominant amino acid residues. In terms of functionality, PEP displayed multiple immunostimulatory activities. It promoted macrophage spreading and proliferation while exhibiting low cytotoxicity to macrophage membranes. Additionally, PEP significantly enhanced lysosomal enzyme activity and pinocytosis in macrophages. The protein also stimulated the production of NO and $H_2O_2$ to levels comparable to those induced by lipopolysaccharide (LPS), indicating its potential as a bioactive immunomodulator (Mariga et al., 2014).

In another study, Sun et al. (2017) isolated a bioactive polypeptide from *P. eryngii* mycelium, which was purified through Sephadex-50 chromatography, resulting in two distinct fractions. Both fractions, when tested on macrophages at concentrations ranging from 0.5 to 2 mg/mL, enhanced macrophage proliferation and phagocytosis. A dose-dependent increase in NO and $H_2O_2$ production was also observed. Furthermore, the polypeptides upregulated the expression of TNF-α, IL-6, TLR2, and TLR4. Among the two fractions, Fraction 1 showed superior immunomodulatory effects across all measured parameters. These findings suggest that the polypeptide fractions, particularly Fraction 1, possess strong immune-stimulatory properties, likely mediated through the NF-κB signaling pathway, and hold potential for therapeutic applications aimed at boosting innate immune responses.

Additionally, a novel immunomodulatory protein, PEP 1b, isolated from *P. eryngii*, was found to activate macrophages and promote their polarization toward the M1 phenotype—a crucial step in initiating cellular immune responses. PEP 1b, with an Mw of 21.9 kDa, significantly increased the production of NO, IL-1β, IL-6, and TNF-α in RAW264.7 macrophages and upregulated inducible nitric oxide synthase (iNOS). Mechanistically, PEP 1b activated TLR4-dependent NF-κB and MAPK (JNK and ERK1/2) signaling pathways. Studies involving phosphorylation and inhibitors confirmed that these signaling events were essential for NO production and NF-κB nuclear translocation, ultimately leading to immune activation. These results highlight PEP 1b as a promising bioactive compound for enhancing innate immune responses (Hu et al., 2018).

A high Mw proteo-polysaccharide fraction, PN50G (~2,000 kDa), isolated from *Pleurotus nebrodensis*, was investigated for its bioactivity on RAW264.7 murine macrophage cells. At concentrations ranging from 5 to 20 μg/mL, PN50G significantly enhanced cell viability and pinocytosis. Additionally, treatments with 20 and 40 μg/mL PN50G notably increased the production of key immunomodulatory molecules, such as IL-6, NO, and IL-10. When co-treated with LPS, PN50G significantly reduced the pro-/anti-inflammatory cytokine ratios, including IL-6/IL-10, TNF-α/IL-10, and NO/IL-10, suggesting an anti-inflammatory bias. Moreover, macrophages treated with PN50G exhibited enhanced adherence to plastic surfaces and developed microvilli-like structures, indicative of morphological activation (Wang et al., 2014).

Another study identified and purified a novel polysaccharide (PN-S) from *Pleurotus nebrodensis*, which was shown to stimulate immune responses in RAW264.7 macrophages. PN-S promoted cell proliferation, phagocytosis, and progression through the S and G2/M phases of the cell cycle. Additionally, it significantly increased the production and gene expression of IL-6, NO, IFN-γ, and TNF-α in a dose-dependent manner, demonstrating its robust immunostimulatory potential through macrophage activation (Cui et al., 2015).

A β-D-glucan-rich polysaccharide fraction (BGP) extracted from *Pleurotus tuber-regium* was evaluated for its immunostimulatory properties in RAW264.7 macrophages. BGP treatment led to a significant increase in immune effector functions, such as inducible nitric oxide (iNO) production, TNF-α release, and phagocytic activity. No signs of toxicity or acute intoxication were observed in mice after BGP administration (Nworu et al., 2015).

In a study on polysaccharides from *Pleurotus albidus* (both basidiome and submerged culture), it was found that different extraction methods yielded structurally distinct β-glucans. Cold-water (BaCW) and exopolysaccharide (MyEX) extracts stimulated TNF-α and NO production but inhibited phagocytosis, while hot-water (BaHW) and endopolysaccharide (MyEN) extracts enhanced TNF-α, NO, IL-6 production, and phagocytosis. The hot alkali extract (BaHA) showed moderate effects. Structural analysis revealed that BaCW and MyEX are primarily β-(1,6)-branched β-(1,3)-glucans, BaHW and MyEN contain β-(1,3)-(1,2)-branched β-(1,6)-glucans, and BaHA is a mixture of β-(1,3)- and β-(1,6)-glucans. These findings position *P. albidus* as a promising source of immunomodulatory β-glucans (Castro-Alves et al., 2017).

A survey of 16 culinary medicinal mushroom species, focusing on hot-water extracts, revealed that *Pleurotus* species, particularly *P. eryngii* and *Pleurotus cystidiosus*, exhibited the highest β-glucan content, with levels around 18.19%. These species significantly enhanced macrophage viability, phagocytic activity, NO production, and NF-κB activation compared to other mushrooms, reinforcing *Pleurotus* species as potent sources of immune-boosting bioactive compounds (Abdullah et al., 2017).

### 10.4.2 Immune-Stimulatory Effect of *Pleurotus* sp. on DCs

DCs play a critical role in both innate and adaptive immunity, as they are the most potent antigen-presenting cells.

They are responsible for recognizing, capturing, processing, and presenting antigens to immune effector cells, influencing the type of immune response that is initiated. Immature DCs, which reside in peripheral tissues, have a strong capacity for endocytosing antigens. Upon exposure to various stimuli, these cells mature by binding to PRRs, such as TLRs and C-type lectins, which leads to increased expression of co-stimulatory molecules (CD40, CD80, CD86) and major histocompatibility complex (MHC) molecules. This maturation is also accompanied by the secretion of pro-inflammatory cytokines, essential for activating T cells and regulating immune responses (Li et al., 2016; Minato et al., 2016).

Minato et al. (2016) investigated the immunomodulatory effects of a β-1,3-branched β-1,6-glucan (PCPS) extracted from the fruiting bodies of *Pleurotus citrinopileatus*. This polysaccharide, with Mw of 450 kDa, was shown to significantly enhance the maturation of immature DCs. After 24 hours of treatment, PCPS increased the expression of surface markers like CD80, CD86, and HLA-DR. Furthermore, PCPS stimulated the production of CC-type chemokines such as CCL2 (MCP-1), CCL3 (MIP-1α), and CCL8 (MCP-2), as well as CXCL9 and CXCL10. The activation of DCs by PCPS was shown to involve the Dectin-1 receptor, triggering downstream Syk and Raf-1 signaling pathways. Additionally, PCPS activated TLRs, including TLR4 and TLR2, demonstrating its dual capacity to modulate both innate and adaptive immune responses. These findings highlight PCPS as a potent immunomodulatory agent with potential for use in functional food development or immunotherapy.

Expanding on their work, Minato et al. (2017) explored the pro-inflammatory effects of PCPS on THP-1 macrophages. The results showed that PCPS induced the secretion of TNF and increased the mRNA expression of both TNF and IL-1β. Interestingly, the inflammatory response to PCPS was modulated by external factors: co-treatment with 5′-GMP-Na, a nucleotide compound, enhanced TNF production, while vitamin $D_2$ suppressed the inflammatory response. These insights suggest that PCPS could serve as a valuable tool in therapeutic applications, particularly in modulating immune responses for immunotherapy or functional food development.

Li et al. (2015) evaluated the effects of *Pleurotus ferulae* water extract (PFWE) on bone marrow-derived DCs (BM-DCs). The results demonstrated that PFWE enhanced the expression of co-stimulatory markers like CD40, CD80, CD86, and MHC II, as well as IL-12 production in BM-DCs, partially through the TLR4 pathway. Moreover, PFWE improved BM-DCs' ability to stimulate $CD8^+$ T-cell proliferation and enhance antigen presentation. When combined with CpG-ODN, PFWE had an additive effect, further improving BM-DC maturation and function. This combination resulted in increased IFN-γ production and $CD8^+$ T-cell proliferation, suggesting that PFWE could potentiate antitumor immune responses and be used as an adjuvant for DC-based vaccines.

In a follow-up study, Li et al. (2016) further explored the synergistic effect of PFWE and CpG-ODN in DCs pulsed with human papillomavirus (HPV) E6 and E7 peptides. The combination significantly enhanced DC maturation and immune activation, leading to a substantial increase in antigen-specific $CD8^+$ T-cell responses. The PFWE + CpG-treated DCs induced the highest level of $CD8^+$ T-cell activation, indicating a strong enhancement of adaptive immune responses. This highlights the potential of PFWE in combination with CpG-ODN as a powerful adjuvant for improving the efficacy of vaccines.

Roopngam (2018) examined the immunomodulatory effects of *Pleurotus sajor-caju* β-glucan (PBG). The study showed that PBG enhanced the activation of human T-lymphocytes (both $CD4^+$ and $CD8^+$) and promoted IFN-γ release when co-cultured with human dendritic cells (HuDCs) pulsed with HPV16-E7 proteins. Both *in vitro* and *in vivo* analyses revealed that PBG boosted specific IgG neutralizing antibodies and enhanced T-cell responses. These results indicate that PBG has strong potential as an adjuvant for cervical cancer vaccines by stimulating both humoral and cellular immune responses, thereby improving vaccine efficacy.

### 10.4.3 Immune-Stimulatory Effect of *Pleurotus* sp. on Other Immune Cells

Roy et al. (2009) described the extraction and immunomodulatory effects of a water-soluble glucan derived from *Pleurotus florida*, cultivar Assam Florida. The glucan, obtained through hot-water extraction, has Mw of approximately 130 kDa. This polysaccharide exhibited a dose-dependent enhancement of NO production, with the peak NO production reaching 8.5 μM at a concentration of 80 μg/mL. Additionally, a concentration of 20 μg/mL of the glucan was found to maximize the proliferation of both splenocytes and thymocytes, indicating its potential in activating immune cells. These findings suggest that the glucan could serve as a promising candidate for further investigation as a natural immunomodulator.

In a follow-up study, the same team (Das et al., 2010) isolated a distinct water-soluble polysaccharide from the aqueous extract of the fruiting bodies of a somatic hybrid created by protoplast fusion between *P. florida* and *Volvariella volvacea*. Structural analysis revealed this polysaccharide to be a (1→6)-β-D-glucan with Mw of 185 kDa, distinguishing it from the polysaccharides found in either parent strain. This hybrid-derived glucan demonstrated significant immunostimulatory effects. When tested on peritoneal macrophages, it enhanced NO production, with an optimal level of 17.57 μM at 80 μg/mL. The proliferation of splenocytes peaked at 50 μg/mL, while thymocyte proliferation was maximized at 80 μg/mL. These results highlight the distinct and enhanced immunomodulatory effects of the hybrid-derived glucan compared to the parental strains, suggesting its potential for future research in immune modulation and health-related applications.

Nozaki et al. (2008) studied the immunomodulatory effects of acidic glycosphingolipids (AGLs) extracted from *Hypsizigus marmoreus* and *P. eryngii*. These AGLs stimulated the secretion of IFN-γ and IL-4 from T cells via CD11c⁺ DCs and promoted the proliferation of NK1.1⁺α/β TCR⁺ cells. Although less potent than α-GalCer, the AGLs modulated immune responses and influenced the Th1/Th2 balance, indicating their potential as immunomodulatory agents.

### 10.4.4 Immune-Stimulatory Effect of *Pleurotus* sp. in Mouse Models

Immunological research has predominantly been conducted through *in vitro* cell cultures or *in vivo* animal studies, with rodents, particularly mice, serving as common models for testing the effects of drugs and therapies. Fish models are also frequently employed to study innate immune responses and inflammatory diseases. However, significant differences between the immune systems of rodents, fish, and humans still exist, which can impact the effectiveness and predictability of clinical trials (Morsink et al., 2020).

In an earlier study, glucan derived from *P. ostreatus* demonstrated superior enhancement of delayed-type hypersensitivity (DTH) responses and phagocytic activity in mice compared to yeast glucan. This fungal glucan not only triggered a quicker immune response but also sustained its activation over time, leading to a significant rise in the phagocytic index. These findings suggest its potential as an effective immunomodulatory agent (Paulík et al., 1996). A similar trend was observed with *P. eryngii*, where supplementation with mushroom powder significantly potentiated cellular immune responses (Llauradó et al., 2013). In a separate investigation, polysaccharopeptides (PSP) enriched with β-D-glucans were extracted from the hot-water culture broth of *P. ostreatus* mycelium. A dose-response analysis revealed that administering PSP intraperitoneally at 25 mg/kg three times per week for five consecutive weeks led to a substantial increase in serum IFN-α levels, further highlighting its immunostimulatory potential (Refaie et al., 2009). Llauradó et al. (2016) examined the immunomodulatory and nutritional advantages of a cold-water extract from *P. ostreatus*. When administered to malnourished mice at a dose of 100 mg/kg body weight, the extract significantly increased phagocytic activity compared to control mice. This extract also promoted hematopoiesis, improved intestinal health, enhanced humoral immunity, and restored liver function, indicating broad systemic benefits. The study suggests that this cold-water extract could be developed into specialized foods or nutritional supplements with applications in immunotherapy.

A study by Jubori et al. (2018) assessed the immunomodulatory effects of a methanol extract from *P. ostreatus*. The extract was administered to male albino rats at a dose of 100 mg/kg body weight, leading to significant increases in white blood cell (WBC) and lymphocyte counts, as well as enhanced phagocytosis. These results indicated strong activation of innate immunity. Furthermore, the extract elevated key cytokines such as IL-2, TLR4, and MCP-1 while boosting immunoglobulins (IgG and IgM) and complement component C3, suggesting both humoral and cellular immune enhancement. These findings underscore the potential of *P. ostreatus* methanol extract for further exploration in immune health and therapeutic applications.

In another study, an aqueous extract (APTR) from *P. tuber-regium* was found to have immunomodulatory and immunorestorative effects. APTR was shown to be nontoxic at an oral dose of 45,000 mg/kg. *In vivo* studies in healthy mice revealed that oral supplementation with APTR resulted in a dose-dependent increase in spleen weight, indicating systemic immune activation. Moreover, APTR enhanced DTH responses and improved phagocytic activity. When administered before immunization with ovalbumin (OVA), APTR also elevated the production of antigen-specific antibodies, including IgG, IgG1, and IgG2a, demonstrating improved humoral immunity. Additionally, APTR countered dexamethasone-induced immunosuppression, thus restoring immune function (Nworu et al., 2015).

Morris et al. (2011) demonstrated that oral supplementation of *Pleurotus* fruiting body powder offered significant immune benefits in mice that had been immunosuppressed by cyclophosphamide. The supplementation led to recovery of bone marrow cellularity, increased WBC counts, and enhanced cell-mediated immune responses, showcasing the potential of *Pleurotus* to strengthen immune defenses in immunocompromised conditions.

In a recent study, a polysaccharide (CMP) from *P. citrinopileatus* mycelium was shown to have potent immunoregulatory effects in mice. CMP treatment increased lymphocyte proliferation, NK cell activity, immunoglobulin levels, and antioxidant enzyme activities, while decreasing markers of oxidative stress. Additionally, CMP improved immune organ indices, alleviated spleen tissue damage, and shifted the immune response toward a Th2 profile by raising IL-4 and IL-10 levels and lowering IL-12, IFN-γ, and TNF-α. Mechanistically, CMP activated the p62/Keap1/Nrf2 pathway, suggesting its potential as a natural immunomodulatory and antioxidant agent (Meng et al., 2023**a**).

### 10.4.5 Immune-Stimulatory Effect of *Pleurotus* sp. in Poultry Models

A study evaluated the immunomodulatory properties of purified β-glucan extracted from *P. florida* in broiler chickens. When chicks were fed 20 mg/kg of β-glucan for 15 days, there was a significant enhancement in their immune responses, including increased neutrophil superoxide production, lymphocyte proliferation, IL-2 secretion, and macrophage nitrite production compared to the control group. Additionally, the treated birds exhibited improved resistance to Ranikhet disease, highlighting the potential of β-glucan from mushrooms as a promising immunostimulant in poultry (Paul et al., 2012). In a follow-up study, supplementation with 30 mg/kg of β-glucan in the feed for 20 days further boosted immune function, with notable increases in nitrite production, lymphocyte proliferation (as indicated by a higher stimulation index), and enhanced phagocytic activity in intestinal intraepithelial lymphocytes (iIEL). Furthermore, the treated chickens showed a significant rise in cytokine levels, especially IFN-γ, at various post-treatment intervals (0, 7, and 14 days), underscoring the potent immunostimulatory effects of *P. florida*-derived β-glucan (Muthusamy et al., 2013).

### 10.4.6 Immune-Stimulatory Effect of *Pleurotus* sp. in Fish Models

With global fish production projected to reach around 210.7 million tons by 2030, it is crucial to adopt sustainable approaches to enhance fish health and productivity. One effective strategy is the dietary inclusion of bioactive compounds, which has shown promise in boosting fish health and resistance to disease. Among these, mushroom-derived supplements have demonstrated significant potential in modulating immune responses and promoting growth across various aquaculture species (Van Doan et al., 2016).

Ching et al. (2021) explored the immunomodulatory effects of β-glucan-enriched hot-water extract (HWE) from *Pleurotus pulmonarius* stalk waste when added to the diet of red hybrid tilapia (*Oreochromis* sp.). During a 30-day feeding trial, fish were fed HWE-supplemented diets and then exposed to PAMPs like LPS and poly(I:C) to simulate bacterial and viral infections, respectively. While the supplementation did not significantly improve growth parameters, it resulted in improved feed efficiency. Moreover, it had a pronounced impact on immune function, enhancing innate immune responses such as phagocytic activity in peripheral leukocytes. Fish in the HWE-fed group exhibited increased respiratory burst activity and upregulated expression of immune-related genes indicative of activated inflammatory and antiviral responses. Additionally, the fish showed enhanced bactericidal activity against *Streptococcus agalactiae*, further confirming the extract's ability to improve pathogen clearance. These results suggest that HWE from *P. pulmonarius* can effectively bolster both innate and adaptive immunity in tilapia, positioning it as a promising immunostimulant for aquaculture.

In another study, *Pangasius bocourti* were fed diets supplemented with 3 g/kg of *P. eryngii* mushroom, resulting in significant improvements in both specific growth rate (SGR) and feed conversion ratio (FCR) compared to the control group. The mushroom-supplemented fish also exhibited enhanced innate immune responses, including increased serum lysozyme activity, phagocytosis, and respiratory burst activity. When *Lactobacillus plantarum* was incorporated at a concentration of $10^8$ CFU/g, further improvements in FCR and SGR were observed. The combination of oyster mushroom and *L. plantarum* in a synbiotic diet led to the most substantial improvements in both growth performance and immune function. The synbiotic group demonstrated significantly higher SGR and lower FCR than the control and individual treatment groups. After being challenged with *Aeromonas hydrophila* at the end of the 90-day trial, all supplemented groups displayed significantly higher survival rates, with the highest survival seen in the synbiotic group. These findings underscore the potential of a synbiotic approach, combining *P. eryngii* and *L. plantarum*, to enhance both growth and immune responses in *P. bocourti* (Van Doan et al., 2016).

Kamilya et al. (2006) demonstrated that *Catla catla* could be successfully immunized against *A. hydrophila* using a formalin-killed whole-cell vaccine, with *P. florida*-derived glucan serving as an effective adjuvant. When compared to bovine lactoferrin, glucan significantly boosted immune responses, enhancing the vaccine's efficacy. In a subsequent study, the immunostimulatory effects of dietary glucan extracted from *P. florida* were evaluated in *C. catla*. Fish were fed diets supplemented with 0.5 g or 1 g/kg glucan for two weeks, followed by a control diet for four weeks. Both doses significantly improved superoxide production, phagocytosis, bactericidal activity, and lysozyme levels, especially in the 0.5 g/kg group. Additionally, both doses enhanced disease resistance, as indicated by increased survival rates following *A. hydrophila* challenge. These findings support the use of mushroom-derived glucan as a promising immunostimulant in aquaculture (Kamilya et al., 2008). Overall, dietary supplementation with *Pleurotus*-derived bioactive compounds significantly enhances immune responses, growth performance, and disease resistance in fish, highlighting their potential as effective and sustainable immunostimulants in aquaculture (Table 10.1). Further research is needed to optimize their application and explore additional species for broader aquaculture benefits.

### 10.4.7 Immune-Stimulatory Effect of *Pleurotus* sp. in Clinical Trials

In a clinical trial conducted by Tanaka et al. (2015), the consumption of oyster mushroom extract for 4 to 8 weeks did not cause any significant changes in liver or renal function or in blood lipid profile. However, it was found to enhance Th1-type immune responses, with a notable increase in plasma levels of IFN-γ, a cytokine that stimulates NK cells. Additionally, there was an increase in TNF-α, which supports tumor suppression. These findings suggest that oyster mushroom extract may have beneficial immunostimulatory effects without adverse health impacts (Figure 10.1).

## 10.5 CONCLUSION AND FUTURE DIRECTIONS

The genus *Pleurotus* represents a promising natural source of immunomodulatory agents, with numerous species

**TABLE 10.1**
**Immune-stimulatory Effects of *Pleurotus* Species in Various Experimental Models**

| Name | Country | Bioactive Compound/Extract | Experimental Model | Key Findings | References |
|---|---|---|---|---|---|
| *P. albidus* | Brazil | Hot-water polysaacharide (WFB) and endopoly-saccahride (culture) | RAW264.7 | ↑Phagocytosis, TNF-α, NO and IL-6 production | Castro-Alves et al., 2017 |
| *P. citrinopileatus* | Japan | β-1,3-branched β-1,6-glucan | Human dendritic cells | Induced DC maturation; ↑TNF, IL-1β, IL-6 and IL-12, and IL-10 expressions; ↑mRNA levels of chemokines; ↑Dectin-1, TLR4 and TLR2; activated Syk and Raf-1 pathways | Minato et al., 2016 |
| *P. citrinopileatus* | Japan | β-1,6-glucan | THP-1, peritoneal cells and spleen from BALB/c mice | ↑TNF and IL-1β mRNA expression | Minato et al., 2017 |
| *P. citrinopileatus* | China | Polysaccharide | Cyclophosphamide-induced immunosup-pressed mice | Enhanced immune response through p62/Keap1/Nrf2 pathway | Meng et al., 2023a |
| *P. citrinopileatus* | China | Polysaccharide | RAW264.7 | ↑TNF-α, IL-1β, and IL-6 through NF-κB pathway | Shi et al., 2024 |
| *P. cornucopiae* | Japan | Water extract | Volunteers of 34 to 64 years | Enhanced Th1-type immune responses | Tanaka et al., 2015 |
| *P. eryngii* | Thailand | Powder | *Pangasius bocourti* | ↑Serum lysozyme, phagocytosis and respiratory burst; improved specific growth rate and feed conversion ratio | Van Doan et al., 2016 |
| *P. eryngii* | China | Polypeptide | Ana-1 | ↑Phagocytosis, NO, $H_2O_2$, TNF-α, and IL-6 production; ↑TLR2 and TLR4expression | Sun et al., 2017 |
| *P. eryngii* | China | 3-O-methyl-galactose | RAW264.7 | ↑Phagocytosis, secretion of NO and cytokines through MAPKs, NF-κB pathways *via* TLR2 | Yan et al., 2019 |
| *P. eryngii* | China | A water-soluble polysaccahride | RAW264.7 | ↑NO, TNF-α, IL-1 and IL-6 through activation of p38, ERK, JNK in MAPKs and translocation of nuclear NF-κB | Xu et al., 2016 |
| *P. eryngii* | China | Protein | RAW264.7 | ↑NO, IL-1β, IL-6, iNOS and TNF-α through TLR4-NF-κB and MAPK signaling pathways | Hu et al., 2018 |
| *P. eryngii* | Brazil | Cold-and hot-water extracted polysaccahrides | THP-1 | ↑NO, IL-1β and IL-10 | Abreu et al., 2021 |
| *P. eryngii* | Norway | Water-soluble polysaccharide | D2SC/1 | ↑NO, TNF-α production in presence of IFN-γ | Ellefsen et al., 2021 |
| *P. ferulae* | China | Water extract | DCs from from C57BL/6 mice | ↑Expressions of CD40, CD80 and CD86; production of IL-12, IL-6 and TNF-α through TLR-4 pathway<br>↓PGE2 | Li et al., 2016 |
| *P. florida* | India | A water-soluble glucan | Peritoneal macrophages | ↑No production; splenocyte and thermocyte proliferation | Roy et al., 2009 |
| *P. florida* | India | β-glucan | Broiler chicken | ↑Superoxide anion, IL-2 and nitrite production | Paul et al., 2012 |
| *P. florida* | India | Glucan | *Catla catla* | ↑Superoxide anion, phagocytosis, bactericidal and lysosome activity | Kamilya et al., 2008 |

*(Continued)*

**TABLE 10.1** ***(Continued)***
**Immune-stimulatory Effects of *Pleurotus* Species in Various Experimental Models**

| Name | Country | Bioactive Compound/Extract | Experimental Model | Key Findings | References |
|---|---|---|---|---|---|
| *P. florida* | India | Glucan | *Catla catla* injected with formalin-killed *Aeromonas hydrophila* vaccine | ↑Antigen-specific proliferation, macrophage activating factor production and antibody production | Kamilya et al., 2006 |
| *P. florida* | India | β-glucan | Broilers | ↑Nitrite production, phagocytosis, lympho-proliferation | Muthusamy et al., 2013 |
| A somatic hybrid of *P. florida* and *Volvariella volvacea* | India | (1→6)-β-glucan | Peritoneal macrophages | ↑No production and splenocyte and thermocyte proliferation | Das et al., 2010 |
| *P. nebrodensis* | China | Glucan enriched polymer | RAW264.7 | ↑Viability, phagocytosis, production of iNOS, TNF-α, IL-6, IL-10; microvilli formation | Wang et al., 2014 |
| *P. nebrodensis* | China | Polysaccharide | RAW264.7 | ↑Proliferation, phagocytosis, IL-6, NO, TNF-α, iNOS, IFN-γ | Cui et al., 2015 |
| *P. ostreatus* | Iraq | Crude extract | Albino rats | ↑Phagocytosis; IL2, TLR4, MCP, IgG, IgM, C3 and C4 in serum level | Jubori et al., 2018 |
| *P. ostreatus* | China | Alkali-extracted polysaccharide | Peritoneal macrophage | ↑Phagocytosis, NO, iNOS, TNF-α, phosphorylation of p65 and Iκ-B degradation | Kong et al., 2014 |
| *P. ostreatus* strain CCEBI-3024 | Cuba | Aqueous extract | Female BALB/c malnourished mice | ↑Phagocytosis haemopoiesis, humoral immunity | Llauradó et al., 2016 |
| *P. ostreatus* | Cuba | Aqueous extract | RAW264.7 | ↑Phagocytosis, TNF-α, IL-6, NO production and iNOS expression; activated NF-κB promoter | Llauradó Maury et al., 2021 |
| *P. ostreatus* | India | Polysaccharide | RAW264.7 | ↑Phagocytosis, No, ROS, TNF-α | Devi et al., 2015 |
| *P. ostreatus* | China | Protein | RAW264.7 | ↑Lysosomal activity, pinocytic activity, NO production, $H_2O_2$ release | Mariga et al., 2014 |
| *P. ostreatus* | China | Selenium-enriched polysaccharide | RAW264.7 | ↑Proliferation, phagocytosis, CD80/CD86, NO, ROS, TNF-α, IL-1β, and IL-6 through NF-κB pathway | Wang et al., 2022 |
| *P. ostreatus* | China | Aqueous extract | RAW264.7 | ↑NO, ROS, TNF-α mediated through JNK/MAPK, Erk/MAPK, and NF-κB signaling pathways | Meng et al., 2023b |
| *P. pulmonarius* | Malaysia | Hot-water extract | Red hybrid tilapia challenged with PAMPs | ↑Phagocytosis, respiratory burst | Ching et al., 2021 |
| *P. pulmonarius* | Malaysia | Polysaccharide | THP-1 | ↑TNF-α, IL-1β, and IL-6 | Mokhtar et al., 2025 |
| *P. tuber-regium* | Nigeria | Hot-water extract | RAW264.7 | ↑Viability, phagocytosis, NO and TNF-α production | Nworu et al., 2015 |
| *P. giganteus, P. citrinopileatus, P. cystidiosus, P. eryngii, P. flabellatus, P. Florida* | Malaysia | Hot-water extract | RAW264.7 | ↑Viability, phagocytosis, NO production | Abdullah et al., 2017 |
| *Pleurotus* sp. strain (CCEBI-3024) | Cuba | Powder | Cyclophosphamide-treated mice | ↑White blood cell counts, stimulated cell-mediated immune responses | Morris et al., 2011 |
| *Pleurotus* sp. strain (CCEBI-3024) | Cuba | Hot-water extract | Murine peritoneal macrophages | Enhanced acid phosphatase activity | Llauradó et al., 2015 |

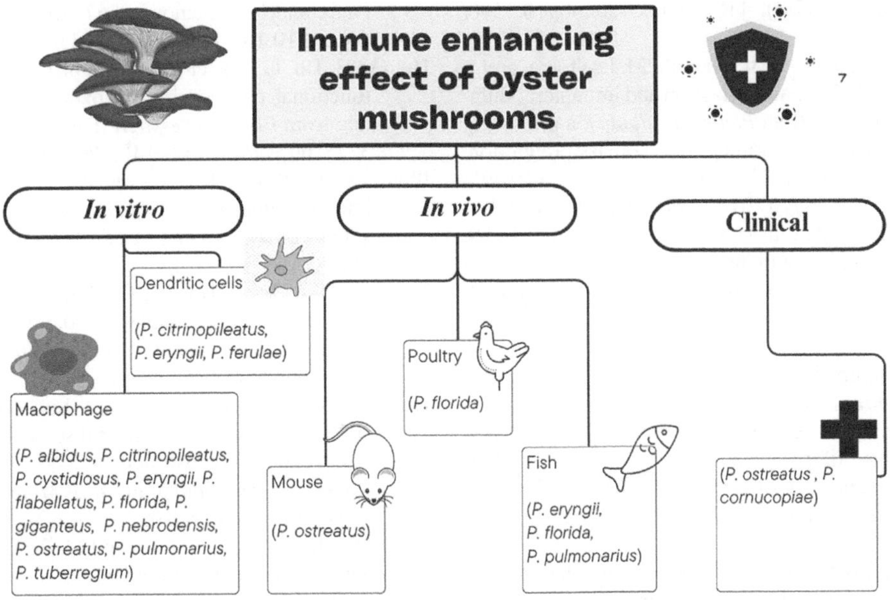


**FIGURE 10.1** Overview of immune-enhancing effects of different *Pleurotus* species.

demonstrating potent immunostimulatory activity across *in vitro*, *in vivo*, and clinical models. β-glucans, alkali-extracted polysaccharides, and selenium-enriched polymers from *P. ostreatus* have been shown to activate key immune mediators such as TNF-α, NO, and transcription factors like NF-κB and MAPK. Similarly, polysaccharides from *P. eryngii* with Mw ranging from 21.4 to 99.7 kDa enhance murine macrophage responses via TLR/NF-κB/MAPK signaling pathways. Proteins from king oyster mushrooms also share this immunostimulatory mechanism. Several other species—including *P. nebrodensis*, *P. tuber-regium*, *P. albidus*, *P. citrinopileatus*, *P. cystidiosus*, *P. ferulae*, and *P. sajor-caju*—along with hybrids like *P. florida* × *Volvariella volvacea*, have exhibited immune-enhancing potential *in vitro*. However, *in vivo* evidence remains limited, with notable outcomes from *P. ostreatus* (glucan and methanol extract) in murine models. Animal trials in poultry and aquaculture confirm enhanced disease resistance and innate immune responses upon supplementation with *Pleurotus* species. Furthermore, human clinical trials highlight their ability to promote Th1-type immune responses without metabolic side effects. Overall, the immunomodulatory potential of *Pleurotus* mushrooms is evident.

The future of research on the immunostimulatory effects of *Pleurotus* mushrooms lies in expanding investigations beyond commonly studied species like *P. ostreatus* and *P. eryngii* to include underexplored varieties such as *Pleurotus djamor*, *P. sajor-caju*, *Pleurotus sapidus*, and *Pleurotus flabellatus*, many of which may show promising bioactivity. A key focus should be on the isolation, structural characterization, and standardization of bioactive compounds such as β-glucans, proteo-polysaccharides, and immunoactive proteins, using advanced analytical tools. Elucidating the precise mechanisms through which these molecules act via pathways like TLRs, NF-κB, and MAPK will aid in developing targeted functional foods and adjuvants. Human clinical trials remain limited; therefore, large-scale, placebo-controlled studies are essential to establish efficacy, safety, and dosage. The effects of bio-fortification and agro-waste-based cultivation on compound yield and bioavailability, as well as the immunostimulatory effects of oyster mushrooms, should be systematically investigated to optimize their functional value. Additionally, evaluating whether bio-fortified mushrooms retain or enhance their immunostimulatory properties both *in vitro* and *in vivo* will bridge the gap between production and clinical application. Finally, translating research into practical applications requires interdisciplinary collaboration to integrate *Pleurotus*-based products into global food systems as affordable, effective, and sustainable immunonutraceuticals.

## REFERENCES

Abdullah, N., R. Abdulghani, S. M. Ismail, M. Hamdi, and Z. Abidin. 2017. Immune-stimulatory potential of hot water extracts of selected edible mushrooms. *Food and Agricultural Immunology* 28(3): 374–387. https://doi.org/10.1080/09540105.2017.1293011

Abreu, H., M. Zavadinack, F. R. Smiderle, T. R. Cipriani, L. M. C. Cordeiro, and M. Iacomini. 2021. Polysaccharides from *Pleurotus eryngii*: selective extraction methodologies and their modulatory effects on THP-1 macrophages. *Carbohydrate Polymers* 252: 117177. https://doi.org/10.1016/j.carbpol.2020.117177

Adams, S., D. Che, J. Hailong, et al. 2019. Effects of pulverized oyster mushroom (*Pleurotus ostreatus*) on diarrhea incidence, growth performance, immunity, and microbial composition in piglets. *Journal of the Science of Food and Agriculture* 99(7): 3616–3627. https://doi.org/10.1002/jsfa.9582

Calder, P. C., A. C. Carr, A. F. Gombart, and M. Eggersdorfer. 2020. Optimal nutritional status for a well-functioning immune system is an important factor to protect against viral infections. *Nutrients* 12(4): 1181. https://doi.org/10.3390/nu12041181

Castro-Alves, V. C., D. Gomes, N. Jr. Menolli, M. L. Sforça, and J. R. Nascimento. 2017. Characterization and immunomodulatory effects of glucans from *Pleurotus albidus*, a promising species of mushroom for farming and biomass production. *International Journal of Biological Macromolecules* 95: 215–223. https://doi.org/10.1016/j.ijbiomac.2016.11.059

Chen, S. N., F. H. Nan, M. W. Liu, M. F. Yang, Y. C. Chang, and S. Chen. 2023. Evaluation of immune modulation by β-1,3; 1,6 D-glucan derived from *Ganoderma lucidum* in healthy adult volunteers, a randomized controlled trial. *Foods* 12(3): 659. https://doi.org/10.3390/foods12030659

Ching, J. J., A. S. Shuib, N. Abdullah, et al. 2021. Hot water extract of *Pleurotus pulmonarius* stalk waste enhances innate immune response and immune-related gene expression in red hybrid tilapia *Oreochromis* sp. following challenge with pathogen-associated molecular patterns. *Fish and Shellfish Immunology* 116: 61–73. https://doi.org/10.1016/j.fsi.2021.06.005

Clark, R., and T. Kupper. 2005. Old meets new: the interaction between innate and adaptive immunity. *Journal of Investigative Dermatology* 125(4): 629–637. https://doi.org/10.1111/j.0022-202X.2005.23856.x

Cui, H. Y., C. L. Wang, Y. R. Wang, Z. J. Li, and Y. N. Zhang. 2015. The polysaccharide isolated from *Pleurotus nebrodensis* (PN-S) shows immune-stimulating activity in RAW264.7 macrophages. *Chinese Journal of Natural Medicines* 13(5): 355–360. https://doi.org/10.1016/S1875-5364(15)30026-1

Cumming, E., and C. Peters. 2024. Immune response to infection. *Anaesthesia and Intensive Care Medicine* 25(6): 439–443. https://doi.org/10.1016/j.mpaic.2024.04.003

Das, D., S. Mondal, S. K. Roy, et al. 2010. A (1→6)-beta-glucan from a somatic hybrid of *Pleurotus florida* and *Volvariella volvacea*: isolation, characterization, and study of immunoenhancing properties. *Carbohydrate Research* 345(7): 974–978. https://doi.org/10.1016/j.carres.2010.02.028

DeMaio, A., S. Mehrotra, K. Sambamurti, and S. Husain. 2022. The role of the adaptive immune system and T cell dysfunction in neurodegenerative diseases. *Journal of Neuroinflammation* 19: 251. https://doi.org/10.1186/s12974-022-02605-9

Devi, K. S., B. Behera, D. Mishra, and T. K. Maiti. 2015. Immune augmentation and Dalton's Lymphoma tumor inhibition by glucans/glycans isolated from the mycelia and fruit body of *Pleurotus ostreatus*. *International Immunopharmacology* 25(1): 207–217. https://doi.org/10.1016/j.intimp.2015.01.026

Ellefsen, C. F., C. W. Wold, A. L. Wilkins, F. Rise, and A. B. C. Samuelsen. 2021. Water-soluble polysaccharides from *Pleurotus eryngii* fruiting bodies, their activity and affinity for Toll-like receptor 2 and dectin–1. *Carbohydrate Polymers* 264: 117991. https://doi.org/10.1016/j.carbpol.2021.117991

Finn, O. J. 2012. Immuno-oncology: understanding the function and dysfunction of the immune system in cancer. *Annals of Oncology* 23(8): viii6–viii9. https://doi.org/10.1093/annonc/mds256

Ghosh, S., S. Khatua, A. Dasgupta, and K. Acharya. 2021. Crude polysaccharide from the milky mushroom, *Calocybe indica* modulates innate immunity of macrophage cells by triggering the MyD88-dependent TLR4/NF-κB pathway. *Journal of Pharmacy and Pharmacology* 73(1): 70–81. https://doi.org/10.1093/jpp/rgaa020

Gutierrez-Lanz, E. A., W. Y. Lee, and L. Pantanowitz. 2022. Lymphoid and hematopoietic systems (lymph nodes, thymus, spleen, bone marrow). In: *Normal Cytology*, ed. M. Lew, J. Pang, and L. Pantanowitz, 67–85. Springer, Cham. https://doi.org/10.1007/978-3-031-20336-7_7

Hu, Q., H. Du, G. Ma, et al. 2018. Purification, identification and functional characterization of an immunomodulatory protein from *Pleurotus eryngii*. *Food and Function* 9(7): 3764–3775. https://doi.org/10.1039/C8FO00604K

Ilham, S., C. Willis, K. Kim, et al. 2023. Cancer incidence in immunocompromised patients: a single-center cohort study. *BMC Cancer* 23(1): 33. https://doi.org/10.1186/s12885-022-10497-4

Janeway, C. A. Jr., P. Travers, M. Walport, and M. J. Shlomchik. 2001. Principles of innate and adaptive immunity. In: *Immunobiology: The Immune System in Health and Disease*, 5th ed. Garland Science, New York.

Jędrzejewski, T., M. Pawlikowska, J. Sobocińska, and S. Wrotek. 2023. COVID-19 and cancer diseases-the potential of *Coriolus versicolor* mushroom to combat global health challenges. *International Journal of Molecular Sciences* 24(5): 4864. https://doi.org/10.3390/ijms24054864

Jubori, S. M. A., M. M. Alrufae, and M. F. Darwesh. 2018. Assessment the immune stimulating effect of *Pleurotus ostreatus* crude extract of male albino rats. *Biochemical and Cellular Archives* 18(1): 1397–1401.

Kamilya, D., S. N. Joardar, C. B. Mal, and T. K. Maiti. 2008. Effects of a glucan from the edible mushroom (*Pleurotus florida*) as an immunostimulant in farmed Indian Major carp (*Catla catla*). *Israeli Journal of Aquaculture* 60: 37–45. https://doi.org/10.46989/001c.20471

Kamilya, D., T. K. Maiti, S. N. Joardar, and B. C. Mal. 2006. Adjuvant effect of mushroom glucan and bovine lactoferrin upon *Aeromonas hydrophila* vaccination in catla, *Catla catla* (Hamilton). *Journal of Fish Disease* 29(6): 331–337. https://doi.org/10.1111/j.1365-2761.2006.00722.x

Khatua, S., and K. Acharya. 2024. Insights into the potential of *Ganoderma* as an immune-stimulatory agent. In: *Ganoderma: Cultivation, Chemistry and Medicinal Applications Volume 2*, ed. K. Acharya, and S. Khatua, 127–147. CRC Press. https://doi.org/10.1201/9781003490258-7

Khatua, S., S. Chandra, and K. Acharya. 2021. Hot alkali extracted antioxidative crude polysaccharide from a novel mushroom enhances immune response via TLR mediated NF-κB activation: a strategy for full utilization of a neglected tribal food. *Journal of Food Biochemistry* 45(1): e13594. https://doi.org/10.1111/jfbc.13594

Khatua, S., J. Simal-Gandara, and K. Acharya. 2022. Understanding immune-modulatory efficacy *in vitro*. *Chemico-Biological Interactions* 352: 109776. https://doi.org/10.1016/j.cbi.2021.109776

Kong, F., F. E. Li, Z. He, et al. 2014. Anti-tumor and macrophage activation induced by alkali-extracted polysaccharide from *Pleurotus ostreatus*. *International Journal of Biological Macromolecules* 69: 561–566. https://doi.org/10.1016/j.ijbiomac.2014.05.045

Li, D., and M. Wu. 2021. Pattern recognition receptors in health and diseases. *Signal Transduction and Targeted Therapy* 6: 291. https://doi.org/10.1038/s41392-021-00687-0

Li, J., J. Li, A. Aipire, J. Luo, P. Yuan, and F. Zhang. 2016. The combination of *Pleurotus ferulae* water extract and CpG-ODN enhances the immune responses and antitumor efficacy of HPV peptides pulsed dendritic cell-based vaccine. *Vaccine* 34(31): 3568–3575. https://doi.org/10.1016/j.vaccine.2016.05.022

Li, J., X. Wang, W. Wang, et al. 2015. *Pleurotus ferulae* water extract enhances the maturation and function of murine bone marrow-derived dendritic cells through TLR4 signaling pathway. *Vaccine* 33(16): 1923–1933. https://doi.org/10.1016/j.vaccine.2015.02.063

Lima-Junior, D. S., N. Collins, and S-J. Han. 2022. Spleen, lymph nodes, and lymphoid tissues-Inflammatory system. In: *Macrophages in the Human Body: A Tissue Level* Approach, ed. N. O. S. Camara, and T. T. Braga, 67–105. Academic Press. https://doi.org/10.1016/B978-0-12-821385-8.00001-9

Liu, J. J., T. S. Huang, M. L. Hsu, et al. 2004. Antitumor effects of the partially purified polysaccharides from *Antrodia camphorata* and the mechanism of its action. *Toxicology and Applied Pharmacology* 201: 186–193. https://doi.org/10.1016/j.taap.2004.05.016

Llauradó, G., H. J. Morris, L. Ferrera, et al. 2015. *In-vitro* antimicrobial activity and complement/macrophage stimulating effects of a hot-water extract from mycelium of the oyster mushroom *Pleurotus* sp. *Innovative Food Science and Emerging Technologies* 30: 177–183. https://doi.org/10.1016/j.ifset.2015.05.002

Llauradó, G., H. J. Morris, Y. Lebeque, et al. 2013. Phytochemical screening and effects on cell-mediated immune response of *Pleurotus* fruiting bodies powder. *Food and Agricultural Immunology* 24(3): 295–304. https://doi.org/10.1080/09540105.2012.686988

Llauradó, G., H. J. Morris, Y. Lebeque, et al. 2016. Oral administration of an aqueous extract from the oyster mushroom *Pleurotus ostreatus* enhances the immunonutritional recovery of malnourished mice. *Biomedicine and Pharmacotherapy* 83: 1456–1463. https://doi.org/10.1016/j.biopha.2016.08.067

Llauradó Maury, G., H. J. Morris-Quevedo, A. Heykers, et al. 2021. Differential induction pattern towards classically activated macrophages in response to an immunomodulatory extract from *Pleurotus ostreatus* mycelium. *Journal of Fungi* 7(3): 206. https://doi.org/10.3390/jof7030206

Mariga, A. M., F. Pei, W. J. Yang, et al. 2014. Immunopotentiation of *Pleurotus eryngii* (DC. ex Fr.) Quel. *Journal of Ethnopharmacology* 153(3): 604–614. https://doi.org/10.1016/j.jep.2014.03.006

Marshall, J. S., R. Warrington, W. Watson, and H. L. Kim. 2018. An introduction to immunology and immunopathology. *Allergy Asthma and Clinical Immunology* 14(Suppl 2): Article no. 49. https://doi.org/10.1186/s13223-018-0278-1

Meng, M., Y. Sun, Y. Bai, et al. 2023a. A polysaccharide from *Pleurotus citrinopileatus* mycelia enhances the immune response in cyclophosphamide-induced immunosuppressed mice via p62/Keap1/Nrf2 signal transduction pathway. *International Journal of Biological Macromolecules* 228: 165–177. https://doi.org/10.1016/j.ijbiomac.2022.12.142

Meng, Q., Y. Niu, R. Wang, W. Niu, and L. Zhang. 2023b. Structural characterization and immunobiological activity of polysaccharides from *Astragalus* oyster mushroom. *Molecules* 28(13): 5280. https://doi.org/10.3390/molecules28135280

Michailidou, Z., M. Gomez-Salazar, and V. I. Alexaki. 2022. Innate immune cells in the adipose tissue in health and metabolic disease. *Journal of Innate Immunity* 14(1): 4–30. https://doi.org/10.1159/000515117

Minato, K. I., L. C. Laan, A. Ohara, and I. van Die. 2016. *Pleurotus citrinopileatus* polysaccharide induces activation of human dendritic cells through multiple pathways. *International Immunopharmacology* 40: 156–163. https://doi.org/10.1016/j.intimp.2016.08.034

Minato, K. I., A. Ohara, and M. Mizuno. 2017. A proinflammatory effect of the β-glucan from *Pleurotus cornucopiae* mushroom on macrophage action. *Mediators of Inflammation* 2017: 8402405. https://doi.org/10.1155/2017/8402405

Mishra, V. S. Tomar, P. Yadav, and M. P. Singh. 2021. Promising anticancer activity of polysaccharides and other macromolecules derived from oyster mushroom (*Pleurotus* sp.): an updated review. *International Journal of Biological Macromolecules* 182: 1628–1637. https://doi.org/10.1016/j.ijbiomac.2021.05.102

Mitra, S., S. Paul, S. Roy, et al. 2022. Exploring the immune-boosting functions of vitamins and minerals as nutritional food bioactive compounds: a comprehensive review. *Molecules* 27(2): 555. https://doi.org/10.3390/molecules27020555

Mokhtar, M., C. H. Leow, N. F. Mokhtar, et al. 2025. Immunomodulatory effects of bioactive polysaccharides from *Pleurotus pulmonarius* on LPS-stimulated THP-1 human macrophages. *Food Bioscience* 63: 105729. https://doi.org/10.1016/j.fbio.2024.105729

Morris, H. J., G. Llauradó, A. Gutiérrez, et al. 2011. Immunomodulating properties of *Pleurotus* sp. fruiting bodies powder on cyclophosphamide treated mice. In: *Proceedings of the 7th International Conference on Mushroom Biology and Mushroom Products (ICMBMP7)*, 324–333.

Morsink, M. A. J., N. G. A. Willemen, J. Leijten, R. Bansal, and S. R. Shin. 2020. Immune organs and immune cells on a chip: an overview of biomedical applications. *Micromachines (Basel)* 11(9): 849. https://doi.org/10.3390/mi11090849

Muthusamy, G., S. N. Joardar, I. Samanta, D. P. Isore, B. Roy, and T. K. Maiti. 2013. β-glucan from edible mushroom (*Pleurotus florida*) enhances mucosal immunity in poultry. *Advances in Animal and Veterinary Sciences* 1(4): 116–119.

Nozaki, H., S. Itonori, M. Sugita, et al. 2008. Mushroom acidic glycosphingolipid induction of cytokine secretion from murine T cells and proliferation of NK1.1 alpha/beta TCR-double positive cells *in vitro*. *Biochemical and Biophysical Research Communications* 373(3): 435–439. https://doi.org/10.1016/j.bbrc.2008.06.047

Nworu, C. S., S. A. Ihim, F. B. Okoye, C. O. Esimone, M. U. Adikwu, and P. A. Akah. 2015. Immunomodulatory and immunorestorative activities of β-D-glucan-rich extract and polysaccharide fraction of mushroom, *Pleurutus tuberregium*. *Pharmaceutical Biology* 53(11): 1555–1566. https://doi.org/10.3109/13880209.2014.991838

Oloke, J. K., and E. A. Adebayo. 2015. Effectiveness of immunotherapies from oyster mushroom (*Pleurotus* species) in the management of immunocompromised patients. *International Journal of Immunology* 3(2–1): 8–20. https://doi.org/10.11648/j.iji.s.2015030201.12

Paul, I., D. P. Isore, S. N. Joardar, et al. 2012. Orally administered β-glucan of edible mushroom (*Pleuratus florida*) origin upregulates innate immune response in broiler. *Indian Journal of Animal Sciences* 82(7): 745–748.

Paulík, S., J. Svrcek, A. Mojzisová, Z. Durove, Benísek, and M. Húska. 1996. The immunomodulatory effect of the soluble fungal glucan (*Pleurotus ostreatus*) on delayed hypersensitivity and phagocytic ability of blood leucocytes in mice. *Zentralbl Veterinarmed B* 43(3): 129–135. https://doi.org/10.1111/j.1439-0450.1996.tb00297.x

Poon, M. M. L., and D. L. Farber. 2020. The whole body as the system in systems immunology. *iScience* 23(9): 101509. https://doi.org/10.1016/j.isci.2020.101509

Pradeu, T., B. P. H. J. Thomma, S. E. Girardin, and B. Lemaitre. 2024. The conceptual foundations of innate immunity: taking stock 30 years later. *Immunity* 57(4): 613–631. https://doi.org/10.1016/j.immuni.2024.03.007

Raman, J., K. Y. Jang, Y. L. Oh, et al. 2020. Cultivation and nutritional value of prominent *Pleurotus* spp.: an overview.

*Mycobiology* 49(1): 1–14. https://doi.org/10.1080/12298093.2020.1835142
Refaie, F. M., A. Y. Esamt, A. S. Daba, and S. M. Taha. 2009. Characterization of polysaccharopeptides from *Pleurotus ostreatus* mycelium: assessment of toxicity and immunomodulation *in vivo*. *Micrologia Aplicada International* 21(2): 67–75.
Roopngam, P. E. 2018. Increased response of human tlymphocytes by dendritic cells pulsed with HPV16E7 and *Pleurotus sajor-caju* β-glucan (PBG). *Iranian Journal of Immunology* 15(4): 246–255. https://doi.org/10.22034/IJI.2018.39394
Roy, S. K., D. Das, S. Mondal, et al. 2009. Structural studies of an immunoenhancing water-soluble glucan isolated from hot water extract of an edible mushroom, *Pleurotus florida*, cultivar Assam Florida. *Carbohydrate Research* 344: 2596–2601. https://doi.org/10.1016/j.carres.2009.09.010
Sharma, Y., M. Arora, and K. Bala. 2024. The potential of immunomodulators in shaping the future of healthcare. *Discovery Medicine* 1: Article no. 37. https://doi.org/10.1007/s44337-024-00029-3
Shi, L., Y. Liu, Z-F. Zhang, Y-X. Li, Y-M. Wu, and Y-M. Lu. 2024. Structural characterization and immunomodulatory effects of polysaccharide PCP-1 from *Pleurotus citrinopileatus*. *Biosynthesis Nutrition Biomedical* 76(3–4): 2300093. https://doi.org/10.1002/star.202300093
Singh, M. R., K. Yadav, N. D. Chaurasiya, and D. Singh. 2022. Immune system and mechanism of immunomodulation. In: *Plants and Phytomolecules for Immunomodulation*, ed. N. S. Sangwan, M. A. Farag, and L. V. Modolo. Springer, Singapore. https://doi.org/10.1007/978-981-16-8117-2_1
Sollome, J., and R. C. Fry. 2015. Environmental contaminants and the immune system: a systems perspective. In: *Systems Biology in Toxicology and Environmental Health: From the Genome to the Epigenome*, ed. R. C. Fry, 171–186. Academic Press. https://doi.org/10.1016/B978-0-12-819968-8.00009-3
Son, J., Y. Hwang, E. M. Hong, et al. 2024. Effects of dietary yeast β-1,3/1,6-D-glucan on immunomodulation in RAW 264.7 cells and methotrexate-treated rat models. *International Journal of Molecular Sciences* 25(20): 11020. https://doi.org/10.3390/ijms252011020
Sun, Y., X. Hu, and W. Li. 2017. Antioxidant, antitumor and immunostimulatory activities of the polypeptide from *Pleurotus eryngii* mycelium. *International Journal of Biological Macromolecules* 97: 323–330. https://doi.org/10.1016/j.ijbiomac.2017.01.043
Tanaka, A., M. Nishimura, Y. Sato, H. Sato, and J. Nishihira. 2015. Enhancement of the Th1-phenotype immune system by the intake of Oyster mushroom (Tamogitake) extract in a double-blind, placebo-controlled study. *Journal of Traditional and Complementary Medicine* 6(4): 424–430. https://doi.org/10.1016/j.jtcme.2015.11.004
Tieu, S., A. Charchoglyan, L. Wagter-Lesperance, et al. 2022. Immunoceuticals: harnessing their immunomodulatory potential to promote health and wellness. *Nutrients* 14(19): 4075. https://doi.org/10.3390/nu14194075
Van Doan, H., S. Doolgindachbaporn, and A. Suksri. 2016. Effects of Eryngii mushroom (*Pleurotus eryngii*) and *Lactobacillus plantarum* on growth performance, immunity and disease resistance of Pangasius catfish (*Pangasius bocourti*, Sauvage 1880). *Fish Physiology and Biochemistry* 42(5): 1427–1440. https://doi.org/10.1007/s10695-016-0230-6
van Steenwijk, H. P., A. Bast, and A. de Boer. 2021. Immunomodulating effects of fungal beta-glucans: from traditional use to medicine. *Nutrients* 13: 1333. https://doi.org/10.3390/nu13041333
Vetvicka, V., O. Gover, M. Karpovsky, et al. 2019. Immune-modulating activities of glucans extracted from *Pleurotus ostreatus* and *Pleurotus eryngii*. *Journal of Functional Foods* 54: 81–91. https://doi.org/10.1016/j.jff.2018.12.034
Wang, C., H. Cui, Y. Wang, et al. 2014. Bidirectional immunomodulatory activities of polysaccharides purified from *Pleurotus nebrodensis*. *Inflammation* 37(1): 83–93. https://doi.org/10.1007/s10753-013-9714-z
Wang, D., J. Wang, H. Liu, M. Liu, Y. Yang, and S. Zhong. 2022. The main structural unit elucidation and immunomodulatory activity *in vitro* of a selenium-enriched polysaccharide produced by *Pleurotus ostreatus*. *Molecules* 27(8): 2591. https://doi.org/10.3390/molecules27082591
Wang, R., C. Lan, K. Benlagha, et al. 2024. The interaction of innate immune and adaptive immune system. *MedComm (2020)* 5(10): e714. https://doi.org/10.1002/mco2.714
Weyand, C. M., and J. J. Goronzy. 2016. Aging of the immune system. mechanisms and therapeutic targets. *Annals of the American Thoracic Society* 13(Suppl 5): S422–S428. https://doi.org/10.1513/AnnalsATS.201602-095AW
Xu, D., H. Wang, W. Zheng, et al. 2016. Charaterization and immunomodulatory activities of polysaccharide isolated from *Pleurotus eryngii*. *International Journal of Biological Macromolecules* 92: 30–36. https://doi.org/10.1016/j.ijbiomac.2016.07.016
Yan, J., Y. Meng, M. Zhang, et al. 2019. A 3-O-methylated heterogalactan from *Pleurotus eryngii* activates macrophages. *Carbohydrate Polymers* 206: 706–715. https://doi.org/10.1016/j.carbpol.2018.11.063

# 11 *Pleurotus*: A Solution for Other Ailments

*Madhumita Karmakar, Arghya Naskar, and Krishnendu Acharya**
Molecular and Applied Mycology and Plant Pathology Laboratory, Centre of Advanced Study, Department of Botany, University of Calcutta, Kolkata, West Bengal, India

## 11.1 INTRODUCTION

Edible mushrooms have garnered global recognition for their exceptional nutraceutical properties. Valued for their low-calorie content, high-quality protein, minimal fat and cholesterol levels, and richness in essential vitamins, they are widely regarded as a "superfood." Beyond their nutritional profile, edible mushrooms are a rich source of both primary and secondary metabolites that contribute to their therapeutic potential (Hamza et al., 2024). Mushrooms have also been esteemed as gourmet cuisine worldwide since ancient times due to their distinctive taste and nuanced flavour (Khan et al., 2022). Research conducted over a period of years has already shown that several mushrooms possess both medicinal properties and a great deal of health benefits. The bioactive metabolites of mushrooms exhibit a variety of therapeutic properties, such as antioxidant, antimicrobial, anticancer, antidiabetic, anti-inflammatory, and antiviral activities, and thus they are useful in the treatment of human ailments (Hamza et al., 2024).

Recently, the phrase "Functional Food" has gained popularity, referring to culinary items that provide therapeutic advantages in addition to providing nutritional fullness. A diverse array of naturally occurring substances in functional foods, containing active components that contribute to physiological actions, warrants careful consideration for their effective application in health maintenance (Essa et al., 2023). Mushrooms serve as a prime example of functional food. As the global population continues to rise, so do the demands for food and medicine security. In this context, mushrooms emerge as a promising source of functional food and nutraceuticals (Das et al., 2021).

Among the various medicinal mushrooms, *Pleurotus*, which is extensively cultivated, possesses the additional benefit of substantial availability owing to its production and increasing popularity across the globe. The production of *Pleurotus* mushrooms alone accounts for around 25% of the total amount of mushrooms farmed across the world (Raman et al., 2021). Moreover, *Pleurotus* mushrooms are regarded as a prime example of a functional food because of their capacity to improve human health in many ways. The *Pleurotus* genus encompasses several species, such as *P. sajor-caju, P. ostreatus, P. florida, P. djamor, P. flabellatus, P. sapidus, P. highking, P. cystidiosus, P. eryngii, P. tuberregium, P. ulmarius, P. geesteranus, P. pulmonarius,* and *P. citrinopileatus*, among others, with certain members being noteworthy for their significant nutritional and medicinal properties (Jarial et al., 2024). In this chapter, we aim to summarise and review the health benefits and therapeutic reports of *Pelurotus* spp., specifically highlighting the anti-aging, hypolipidemic, anti-hypercholesterolemic, neuroprotective, and cardioprotective activities.

## 11.2 ANTI-AGING PROPERTY OF *PLEUROTUS* SP.

Aging is a complex physiological process characterized by the gradual accumulation of molecular and cellular damage, which heightens susceptibility to chronic diseases such as cancer, diabetes, neurodegenerative disorders, and cardiovascular conditions, all leading contributors to global mortality (Murray et al., 2015; Franceschi et al., 2018). Though not a disease itself, aging is intrinsically linked to degenerative changes, including collagen depletion, fibroblast dysfunction, and elevated matrix metalloproteinases (MMPs), which accelerate tissue deterioration (Varani et al., 2006; Dewanjee et al., 2024). A key driver of these changes is oxidative stress resulting from an imbalance between reactive oxygen/nitrogen species (ROS/RNS) and the body's antioxidant defenses, which damage DNA, proteins, and lipids, and promote premature aging (Dash et al., 2024).

Emerging research suggests that aging and its associated pathologies can be delayed or alleviated through dietary interventions, particularly the consumption of bioactive compounds like polysaccharides, phenolics, and peptides that possess strong antioxidant properties (Fontana and Partridge, 2015; Gao et al., 2017; Hook et al., 2018). Among natural sources, oyster mushrooms stand out for their rich composition of functional biomolecules including β-glucans, proteins, linoleic acid, and complex carbohydrates, which contribute to a range of health-promoting effects. These include antioxidant, hepatoprotective, antihypertensive, anti-hypercholesterolemic, anti-aging, antimicrobial, anti-obesity, and antidiabetic activities (Madhanraj et al., 2019; Elkhateeb, 2020; Das et al., 2021).

The antioxidant potential of *Pleurotus* sp. is particularly relevant in the context of the free radical theory of aging, which attributes cellular decline to accumulated oxidative damage. *In vitro* models of aging commonly utilize agents

* Corresponding author: krish_paper@yahoo.com

DOI: 10.1201/9781003527572-11

like hydrogen peroxide or ultraviolet radiation to induce oxidative stress and mimic premature cellular senescence (Chen et al., 2001; Chainiaux et al., 2002). In this context, Shin et al. (2018) investigated the anti-aging effects of ethanolic extracts from *Pleurotus citrinopileatus*, which demonstrated strong dose-dependent radical scavenging activity along with high phenolic and flavonoid content. In $H_2O_2$-induced human fibroblast models, the extract significantly improved cell viability, restoring it from 64.1% to 82.4%–89.0% with extract concentrations of 100–500 µg/mL. Additionally, the extract suppressed the oxidative stress-induced increase in MMP-1 and helped restore type I collagen expression, indicating its protective role against skin aging.

Complementing these *in vitro* findings, Li et al. (2019) evaluated the *in vivo* anti-aging potential of acetylated mycelial polysaccharides (AMPS) extracted from *P. djamor* using a D-galactose-induced aging model in rats. The study involved five groups: normal, model, positive control (ascorbic acid), and two test groups treated with low and high doses of AMPS. After six weeks, the high-dose AMPS group showed significantly enhanced antioxidant enzyme activities. Glutathione peroxidase activity increased by 3.5, 2.8, and 2.6-fold in the liver, kidney, and brain, respectively, compared to the model group. Similarly, superoxide dismutase activity rose by up to 1.5-fold across organs, and catalase activity in the kidney and brain exceeded even that of the ascorbic acid-treated group, rising by 3.13% and 6.36%, respectively. Together, these findings underscore the significant antioxidant and anti-aging potential of *Pleurotus* mushrooms.

## 11.3 HYPOLIPIDEMIC ACTIVITY OF *PLEUROTUS* SP.

Hyperlipidemia, characterized by elevated cholesterol, triglycerides, and low-density lipoprotein cholesterol (LDL-C) levels along with reduced high-density lipoprotein cholesterol (HDL-C), contributes to inflammation, oxidative stress, and atherosclerosis, which are major risk factors for cardiovascular diseases (Saklayen, 2018; Xiao et al., 2021; Saha and Khatua, 2024). While statins are commonly prescribed, they may cause side effects like muscle pain, liver/kidney issues, and drug interactions (Thompson et al., 2016; Ramkumar et al., 2016). Natural compounds offer safer and cost-effective alternatives with broad pharmacological potential (Amirullah et al., 2018). Lovastatin, a 3-Hydroxy-3-methylglutaryl coenzyme A (HMG-CoA) reductase inhibitor, is naturally produced by *Pleurotus* spp., with reported levels ranging from 50 to 5991 µg/g dry weight (Gunde-Cimerman et al., 1993, 1995). Studies have demonstrated that *P. ostreatus* can improve lipid profiles by lowering triglycerides and LDL while raising HDL when included in high-fat diets (Dos Santos et al., 2013). Additionally, its high fiber content may contribute to anti-hyperlipidemic effects (Khan et al., 2022).

In a study, Anandhi et al. (2013) investigated the lipid-lowering effects of *P. ostreatus* extract and its key bioactive compound, chrysin, in rats with Triton WR-1339-induced hypercholesterolemia. Five groups were assessed: normal control, hyperlipidemic control, and three treatment groups receiving lovastatin (10 mg/kg/d), *P. ostreatus* extract (500 mg/kg/d), or chrysin (200 mg/kg/d) for one week. By day 10, untreated hyperlipidemic rats showed a 2.75-fold rise in triglycerides. In contrast, treatment groups exhibited significant reductions in triglycerides, with chrysin reducing levels by 2.46-fold, *P. ostreatus* by 2.06-fold, and lovastatin by 2.25-fold. LDL-C and very low-density lipoprotein cholesterol (VLDL-C) levels also dropped notably in treated groups, while HDL-C increased. Chrysin showed superior lipid-lowering activity, even outperforming the mushroom extract and matching lovastatin's effect. These results suggest that chrysin, a natural compound from *P. ostreatus*, may offer a potent, low-toxicity alternative to conventional statins for managing hyperlipidemia.

Valappil et al. (2020) evaluated the hypolipidemic effects of *Pleurotus florida* in six groups of male Wistar rats ($n$ = 6 per group). Group I served as normal control, receiving distilled water. Hyperlipidemia was induced in the other five groups by intraperitoneal injection of Triton WR-1339 (100 mg/kg). Group II remained untreated (hyperlipidemic control), group III received atorvastatin (2.5 mg/kg), and groups IV, V, and VI were treated with aqueous ethanolic extracts of *P. florida* at doses of 500, 250, and 100 mg/kg, respectively. Treatments were administered orally one hour prior to Triton injection after a 12-hour fast. Blood samples collected 24 hours post-treatment showed that the hyperlipidemic control group had significantly elevated total cholesterol, triglycerides, and LDL cholesterol by 2.21-, 6.69-, and 2.44-fold, respectively, compared to normal, while HDL cholesterol decreased 1.8-fold. In contrast, groups treated with atorvastatin and *P. florida* extracts showed notable reductions in lipid levels. Group IV showed fold reductions in LDL, triglycerides, and total cholesterol of 2.02, 2.43, and 7.74, respectively, compared to the hyperlipidemic control. Groups V and VI exhibited similar dose-dependent lipid-lowering effects. Additionally, the ratio of HMG-CoA to mevalonate increased in these groups, indicating inhibition of HMG-CoA reductase activity, the enzyme critical for cholesterol synthesis. *P. florida* also demonstrated significant inhibition of lipid peroxidation, reducing oxidative stress markers by 1.51- to 1.81-fold across treated groups.

Similar lipid-lowering effects have been reported by Zhao et al. (2020), where *P. eryngii* polysaccharides decreased cholesterol, triglycerides, and LDL-C while increasing HDL-C in hyperlipidemic mice. Furthermore, exopolysaccharides from submerged cultures of *P. eryngii* suppressed macrophage foam cell formation by downregulating the CD36 scavenger receptor, which mediates oxidized LDL uptake and atherosclerotic lesion formation (Chen et al., 2014). Acidic exopolysaccharides from *P. eryngii* also showed preventive effects against hyperlipidemia in high-fat diet mice, comparable to simvastatin, highlighting their potential as natural therapeutic agents and functional foods for lipid disorders (Zhang et al., 2017).

Oyetayo et al. (2020) observed that diets supplemented with *P. tuberregium* tubers (5%–10%) lowered mevalonate, triglycerides, total cholesterol, LDL, atherogenic indices, and reactive oxygen species in hypercholesterolemic rats, comparable to atorvastatin treatment.

## 11.4 ANTI-HYPERCHOLESTEROLEMIC EFFECT OF *PLEUROTUS* SP.

Hypercholesterolemia is characterized by elevated cholesterol and lipid deposits in blood vessels, accelerating atherosclerotic lesion formation, especially in coronary arteries (Dos Reis et al., 2022). Accumulation of LDL cholesterol in the subendothelial space leads to oxidized LDL, which is highly atherogenic and damaging to vascular cells (Anandhi et al., 2013). This condition significantly raises the risk of heart disease, stroke, and related complications (Félix-Redondo et al., 2013).

Mushrooms, particularly *P. ostreatus*, have shown potential in mitigating these effects. Bobek et al. (1998) demonstrated that supplementing diets with 5% dry *P. ostreatus* inhibited VLDL synthesis and cholesterol absorption in rats with diet-induced hypercholesterolemia. This effect is attributed to the mushroom's ability to bind bile acids, thereby preventing mixed micelle formation in the intestine, and to inhibit HMG-CoA reductase, the key enzyme in cholesterol biosynthesis (Bobek and Galbavý, 1999). In male Chinchilla rabbits consuming a diet with 1% cholesterol, the addition of 10% dry *P. ostreatus* powder resulted in a 65% reduction in serum cholesterol levels. Furthermore, cholesterol content in the liver, heart, muscle, and aorta decreased by 60%, 47%, 25%, and 80%, respectively, compared to rabbits that did not receive the dietary inclusion of *P. ostreatus* (Bobek and Galbavý, 1999). Several other species of *Pleurotus* like *P. sapidus* (Abidin et al., 2017), *P. eryngii* (Abidin et al., 2017), and *P. pulmonarius* (Abidin et al., 2018), also exhibit inhibitory activity against HMG-CoA reductase, thereby potentially reducing endogenous cholesterol synthesis (Figure 11.1).

Alam et al. (2009) investigated the effects of oyster mushrooms on lipid profiles and organ function in hypercholesterolemic rats. Forty rats were divided into five diet groups: a control group (normal diet), a hypercholesterolemic (HC) group (1% cholesterol), and three groups fed 1% cholesterol plus 5% powder of *P. ostreatus* (HC+PO), *Pleurotus sajor-caju* (HC+PS), or *P. florida* (HC+PF) for 40 days. The HC+PO group showed the greatest reductions in triglycerides and total cholesterol, 45% and 37%, respectively, compared to the HC group. The HC+PS and HC+PF groups also exhibited notable decreases in these lipids, though to a lesser extent. These lipid-lowering effects were attributed to the presence of lovastatin in the mushrooms.

Anandhi et al. (2013) conducted a study on male Wistar rats (150–200 g) housed under a 12-hour light/dark cycle at 25°C. Hypercholesterolemia was induced by injecting Triton WR-1339 (300 mg/kg body weight) after 12 hours of fasting. Rats were divided into five groups ($n = 6$ each): normal control, hypercholesterolemic control (saline-treated), lovastatin-treated (10 mg/kg/d), *P. ostreatus* aqueous extract-treated (500 mg/kg/d), and chrysin-treated groups. Blood samples were collected on day 1 (before Triton), day 3 (48 hours post-Triton), and day 10 (after 7 days of treatment).

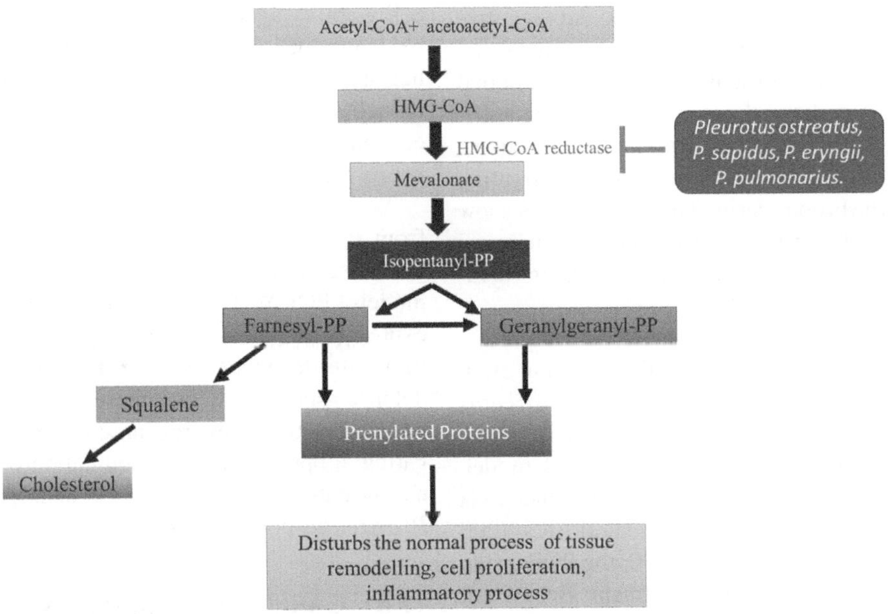


**FIGURE 11.1** Mevalonate pathway of cholesterol synthesis (Chatterjee et al., 2015) and inhibition of HMG-CoA reductase by different species of *Pleurotus* including *P. ostreatus* (Bobek et al., 1995; Gil-Ramirez et al., 2011; Abidin et al., 2017), *P. sapidus* (Abidin et al., 2017), *P. eryngii* (Abidin et al., 2017), *P. pulmonarius* (Abidin et al., 2018).

Triton injection significantly elevated total cholesterol in all hypercholesterolemic groups by day 3. By day 10, cholesterol levels remained high in the untreated group but were significantly reduced in lovastatin, *P. ostreatus*, and chrysin-treated groups, with *P. ostreatus* extract reducing serum cholesterol by 53.6%. These results suggest that *P. ostreatus* and its active compound chrysin effectively lower cholesterol levels comparable to lovastatin.

The inclusion of dried *P. ostreatus* powder at 5% of the diet resulted in a reduction of serum cholesterol in Wistar rats fed a 0.3% cholesterol diet by 31%–46% over periods of 8, 16, 28, and 52 weeks, respectively (Bobek et al., 1998). Similarly, 3% dietary inclusion of dried *P. eryngii* lowered serum cholesterol, triglycerides, and the area of atherosclerotic lesions (Mori et al., 2008). Another experiment revealed that aqueous extract of *P. tuber-regium* at a dosage of 200 mg/kg resulted in a 50% reduction in total cholesterol levels (Ikewuchi et al., 2013). Ramakrishnan et al. (2017) compared lovastatin content in different mushrooms and found that *P. ostreatus* produced the highest lovastatin levels under both solid-state and submerged fermentation conditions.

## 11.5 NEUROPROTECTIVE ACTIVITY OF *PLEUROTUS* SP.

Neurodegeneration refers to the progressive loss of neurons, leading to cognitive and motor impairments (Przedborski et al., 2003). Key causes include aging (Hou et al., 2019), genetic mutations (Conway et al., 2018), traumatic brain injury (Gavett et al., 2010), oxidative stress (Kim et al., 2015), central nervous system inflammation (Chitnis and Weiner, 2017), exposure to environmental toxins (Chin-Chan et al., 2015), and stress (Justice, 2018). These factors contribute to movement disorders like ataxia and cognitive decline seen in Alzheimer's, Parkinson's, Huntington's diseases, and multiple sclerosis (Yadav et al., 2020). Medicinal mushrooms are low in fat and calories but rich in nutrients and bioactive compounds such as vitamins, proteins, fibers, and carbohydrates, making them valuable in treating various diseases. Mushroom-derived polysaccharides show promise in slowing the progression of neurodegenerative disorders, positioning them as an important focus in therapeutic research (Yadav et al., 2020)

In this context, Yadav et al. (2020) reported neuroprotective effects of *Pleurotus giganteus*. Similarly, Zhang et al. (2016) investigated the polysaccharide from *P. ostreatus* (POP) on Alzheimer's disease (AD) in male Wistar rats (180–220 g). Rats were divided into normal, AD model, and POP-treated groups (15 rats each). AD was induced via intraperitoneal D-galactose (60 mg/kg) and oral $AlCl_3$ (200 mg/kg) for 60 consecutive days. On day 30, the POP group received 400 mg/kg polysaccharide treatment. Behavioral tests showed that AD rats exhibited aimless swimming, while normal and POP-treated rats actively searched for the hidden platform. AD rats had elevated acetylcholinesterase (AchE) levels, leading to reduced acetylcholine (Ach), essential for learning and memory. POP treatment lowered AchE levels compared to the AD group. Antioxidant enzyme activities (superoxide dismutase, glutathione peroxidase, and catalase) increased in the hippocampus, liver, and serum of POP-treated rats. Western blot analysis revealed that AD rats had higher expression of Alzheimer's markers such as amyloid precursor protein (APP), amyloid β peptide (Aβ), β-site APP cleaving enzyme 1 (BACE1), phospho-tau (p-tau), glycogen synthase kinase 3β (GSK3β), and lower levels of protein phosphatase 2A (PP2A). POP treatment reversed these proteins' expression patterns, decreasing APP, Aβ, BACE1, p-tau, and GSK3β, and increasing PP2A.

In another investigation, polysaccharides from *P. eryngii* (PEP) demonstrated neuroprotection against β-amyloid-induced toxicity in aged rats and cultured rat pheochromocytoma (PC12) cells. Long-term treatment increased cell survival, reduced intracellular calcium, and decreased β-amyloid-related cell death. PEP also lowered brain APP synthesis, possibly by reducing inducible nitric oxide synthase (iNOS) and cyclooxygenase-2 (COX-2) levels, suggesting modulation of calcium channels and inflammation as neuroprotective mechanisms (Zhang et al., 2020).

Liu et al. (2022) reported that *P. sajor-caju* polysaccharides alleviated oxidative stress, inhibited apoptosis in $H_2O_2$-damaged neuronal cells, and improved cognition in aging mice. Ćilerdžić et al. (2019) compared extracts of *P. ostreatus* and found hot water extract to be the most effective acetylcholinesterase inhibitor (~59.44% inhibition), only slightly less potent than galantamine (70.06%). It also inhibited tyrosinase moderately (35.48%), although less than α-kojic acid (93.98%).

Kushairi et al. (2020) evaluated hot water (PE-HWA) and ethanol (PE-ETH) extracts of *P. eryngii in vitro*, revealing that PE-ETH dose-dependently improved PC12 cell viability against $H_2O_2$-induced damage, reduced nitric oxide production in LPS-activated microglia, and promoted neurite outgrowth. PE-HWA showed no significant effect on cell viability but increased neurite-bearing cells at low concentrations, highlighting the potential of *P. eryngii* for neuro-health support.

Wang et al. (2023) isolated a polysaccharide (POP-W) from *P. ostreatus* composed mainly of mannose, glucose, galactose, and xylose. In $H_2O_2$-induced PC12 cell damage models, POP-W treatment increased anti-apoptotic Bcl-2 expression and decreased pro-apoptotic BAX and caspase-3 in a dose-dependent manner. It also enhanced antioxidant defense (superoxide dismutase, reduced glutathione) and reduced lactate dehydrogenase (LDH) release, while mitigating apoptosis through modulation of the PI3K/Akt signaling pathway, confirming the neuroprotective effects of *P. ostreatus* polysaccharides.

## 11.6 CARDIOPROTECTIVE ACTIVITIES OF *PLEUROTUS* SP.

Cardiovascular disease (CVD) remains the leading cause of death globally, despite advances in diagnosis and treatment. Its prevalence is rapidly increasing in low- and middle-income countries, driven by economic growth and unplanned urbanization, which elevate exposure to risk factors. Key contributors include tobacco use, physical inactivity, obesity, and poor nutrition, which are linked to hypertension, diabetes, and atherosclerosis. Additionally, environmental pollution plays a significant role in the onset and progression of CVD. The cardiovascular system is particularly sensitive to environmental hazards such as tobacco smoke, solvents, pesticides, and extreme noise and temperature (Cosselman et al., 2015). Since eliminating all causative factors of CVD is not always possible, some research focuses on prevention through biologically derived substances. Many mushrooms have long been used in traditional oriental medicine to manage cardiovascular and related conditions (Guillamón et al., 2010; Saha et al., 2024).

Sato et al. (2025) studied the effects of prolonged intake of *Pleurotus cornucopiae* (Tamogi-take) on age-related cardiac and vascular impairments in male C57BL/6N mice. Mice fed a diet containing this mushroom showed significant improvements in heart function, including preserved contractile and diastolic capacity, along with enhanced vascular endothelial function compared to controls. These cardioprotective effects were attributed to the antioxidant ergothioneine present in the mushrooms, which increased nuclear factor E2-related factor 2 (Nrf2) and heme oxygenase 1 (HO-1) protein expression, thereby reducing oxidative stress. The findings suggest that *P. cornucopiae* may be a promising dietary supplement to improve cardiovascular health and promote healthy aging.

Yan et al. (2015) reported the cardioprotective effects of a polysaccharide, PNPA, extracted from *Pleurotus nebrodensis* against myocardial ischemia-reperfusion (I/R) injury in rats. PNPA significantly reduced infarct size, oxidative stress, and myocardial cell death by enhancing antioxidant enzyme activities, including superoxide dismutase, catalase, and glutathione. Structurally, PNPA consists of a backbone with 1,3-linked D-glucopyranosyl and 1,3,6-linked D-galactopyranosyl residues, along with a terminal 1-linked D-mannopyranosyl residue. These findings highlight PNPA's potential as a therapeutic agent for cardiac I/R injury and emphasize the value of *Pleurotus* mushrooms in developing functional foods and natural pharmaceuticals.

## 11.7 CONCLUSION

Oyster mushrooms demonstrate considerable promise as both nutritional and therapeutic agents due to their rich bioactive compounds and health benefits. Current research supports their potential in managing various diseases, including cardiovascular, neurodegenerative, and metabolic disorders (Figure 11.2), as well as addressing malnutrition. However, to fully harness these benefits, more in-depth mechanistic studies are required to elucidate the molecular pathways involved in their bioactivities. Furthermore, integrating traditional knowledge with modern scientific research can accelerate the development of effective nutraceuticals and functional foods derived from *Pleurotus* mushrooms. Future efforts should also focus on increasing public awareness and accessibility to maximize their health impact, along with exploring sustainable cultivation and processing methods to support large-scale production. Overall, *Pleurotus* mushrooms hold great potential as

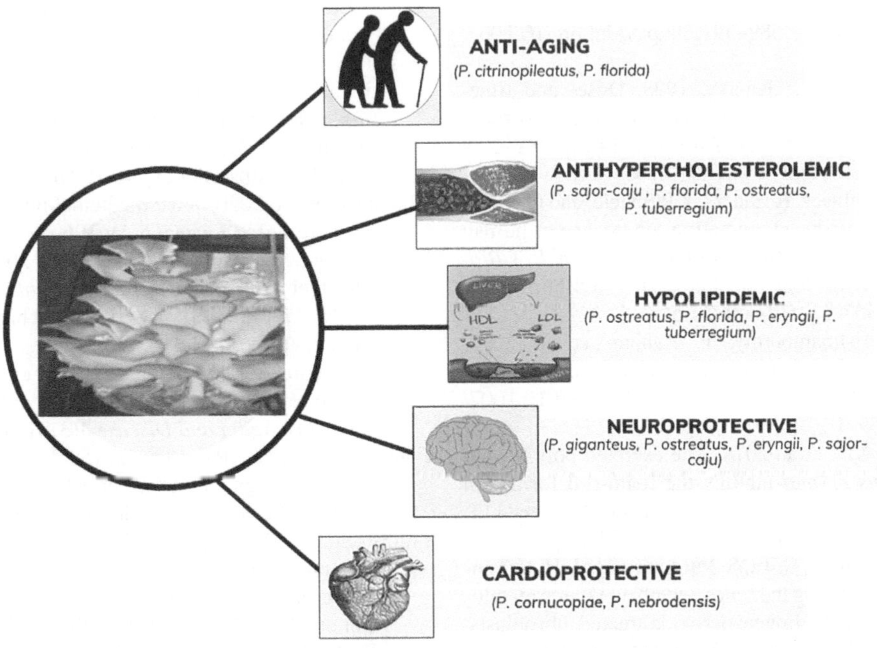


**FIGURE 11.2** Different species of *Pleurotus* showing a wide range of medicinal properties including anti-aging, anti-hypercholesterolemic, hypolipidemic, neuroprotective, and cardioprotective activities.

natural, affordable, and safe alternatives or complements to conventional therapies, paving the way for innovative strategies in health promotion and disease prevention.

## REFERENCES


Abidin, M. H. Z., N. Abdullah, and N. Z. Abidin. 2017. Therapeutic properties of *Pleurotus* species (oyster mushrooms) for atherosclerosis: a review. *International Journal of Food Properties* 20(6): 1251–1261. https://doi.org/10.1080/10942912.2016.1210162

Abidin, M. H. Z., N. Abdullah, and N. Z. Abidin. 2018. Anti-atherogenic potential of extracts from the gray oyster medicinal mushroom, *Pleurotus pulmonarius* (Agaricomycetes), *in vitro*. *International Journal of Medicinal Mushrooms* 20(3): 283–290. https://doi.org/10.1615/IntJMedMushrooms.2018025821

Alam, N., R. Amin, A. Khan, et al. 2009. Comparative effects of oyster mushrooms on lipid profile, liver and kidney function in hypercholesterolemic rats. *Mycobiology* 37(1): 37–42. https://doi.org/10.4489/MYCO.2009.37.1.037

Amirullah, N. A., N. Z. Abidin, and N. Abdullah. 2018. The potential applications of mushrooms against some facets of atherosclerosis: a review. *Food Research International* 105: 517–536. https://doi.org/10.1016/j.foodres.2017.11.023

Anandhi, R., T. Annadurai, T. S. Anitha, et al. 2013. Antihypercholesterolemic and antioxidative effects of an extract of the oyster mushroom, *Pleurotus ostreatus*, and its major constituent, chrysin, in triton WR-1339-induced hypercholesterolemic rats. *Journal of Physiology and Biochemistry* 69: 313–323. https://doi.org/10.1007/s13105-012-0215-6

Bobek, P., and Š. Galbavý. 1999. Hypocholesterolemic and antiatherogenic effect of oyster mushroom (*Pleurotus ostreatus*) in rabbits. *Food/Nahrung* 43(5): 339–342. https://doi.org/10.1002/(SICI)1521-3803(19991001)43:5<339::AID-FOOD339>3.0.CO;2-5

Bobek, P., M. Hromadová, and L. Ozdín. 1995. Oyster mushroom (*Pleurotus ostreatus*) reduces the activity of 3-hydroxy-3-methylglutaryl CoA reductase in rat liver microsomes. *Experientia* 51: 589–591. https://doi.org/10.1007/BF02128749

Bobek, P., L. Ozdín, and Š. Galbavý. 1998. Dose- and time-dependent hypocholesterolemic effect of oyster mushroom (*Pleurotus ostreatus*) in rats. *Nutrition* 14(3): 282–286. https://doi.org/10.1016/S0899-9007(97)00471-1

Chainiaux, F., J. P. Magalhaes, F. Eliaers, J. Remacle, and O. Toussaint. 2002. UVB-induced premature senescence of human diploid skin fibroblasts. *The International Journal of Biochemistry & Cell Biology* 34(11): 1331–1339. https://doi.org/10.1016/S1357-2725(02)00022-5

Chatterjee, S., P. Krishnamoorthy, P. Ranjan, et al. 2015. Statins and cognitive function: an updated review. *Current Cardiology Reports* 17: 1–8. https://doi.org/10.1007/s11886-014-0559-3

Chen, J., Y. Yong, X. Xia, et al. 2014. The excreted polysaccharide of *Pleurotus eryngii* inhibits the foam-cell formation via down-regulation of CD36. *Carbohydrate Polymers* 112: 16–23. https://doi.org/10.1016/j.carbpol.2014.05.068

Chen, Q. M., K. R. Prowse, V. C. Tu, S. Purdom, and M. H. K. Linskens. 2001. Uncoupling the senescent phenotype from telomere shortening in hydrogen peroxide-treated fibroblasts. *Experimental Cell Research* 265(2): 294–303. https://doi.org/10.1006/excr.2001.5182

Chin-Chan, M., J. Navarro-Yepes, and B. Quintanilla-Vega. 2015. Environmental pollutants as risk factors for neurodegenerative disorders: Alzheimer and Parkinson diseases. *Frontiers in Cellular Neuroscience* 9: 124. https://doi.org/10.3389/fncel.2015.00124

Chitnis, T., and H. L. Weiner. 2017. CNS inflammation and neurodegeneration. *The Journal of Clinical Investigation* 127(10): 3577–3587. https://doi.org/10.1172/JCI90609

Ćilerdžić, J., M. Galić, J. Vukojević, and M. Stajić. 2019. *Pleurotus ostreatus* and *Laetiporus sulphureus* (Agaricomycetes): possible agents against Alzheimer and Parkinson diseases. *International Journal of Medicinal Mushrooms* 21(3): 275–289. https://doi.org/10.1615/IntJMedMushrooms.2019030136

Conway, O. J., M. M. Carrasquillo, X. Wang, et al. 2018. ABI3 and PLCG2 missense variants as risk factors for neurodegenerative diseases in Caucasians and African Americans. *Molecular Neurodegeneration* 13: 1–12. https://doi.org/10.1186/s13024-018-0289-x

Cosselman, K. E., A. Navas-Acien, and J. D. Kaufman. 2015. Environmental factors in cardiovascular disease. *Nature Reviews Cardiology* 12(11): 627–642. https://doi.org/10.1038/nrcardio.2015.152

Das, A. K., P. K. Nanda, P. Dandapat, et al. 2021. Edible mushrooms as functional ingredients for development of healthier and more sustainable muscle foods: a flexitarian approach. *Molecules* 26(9): 2463. https://doi.org/10.3390/molecules26092463

Dash, U. C., N. K. Bhol, S. K. Swain, et al. 2024. Oxidative stress and inflammation in the pathogenesis of neurological disorders: mechanisms and implications. *Acta Pharmaceutica Sinica B* 15(1): 15–34. https://doi.org/10.1016/j.apsb.2024.10.004

Dewanjee, D., S. Ghosh, S. Khatua, and S. Rapior. 2024. *Ganoderma* in skin health care: a state-of-art review. In: *Ganoderma: Cultivation, Chemistry and Medicinal Applications, Volume 2*, ed. K. acharya, and S. Khatua, 79–101. CRC Press. http://doi.org/10.1201/9781003490258-5

Dos Reis, E. E., P. C. Schenkel, and M. Camassola. 2022. Effects of bioactive compounds from *Pleurotus* mushrooms on COVID-19 risk factors associated with the cardiovascular system. *Journal of Integrative Medicine* 20(5): 385. https://doi.org/10.1016/j.joim.2022.07.002

Dos Santos, L. F., A. L. Zanatta, V. T. Soccol, et al. 2013. Hypolipidemic and antiatherosclerotic potential of *Pleurotus ostreatus*, cultivated by submerged fermentation in the high-fat diet fed rats. *Biotechnology and Bioprocess Engineering* 18: 201–208. https://doi.org/10.1007/s12257-012-0561-9

Elkhateeb, W. A. 2020. What medicinal mushroom can do. *Chemistry Research Journal* 5(1): 106–118.

Essa, M. M., M. Bishir, A. Bhat, et al. 2023. Functional foods and their impact on health. *Journal of Food Science and Technology* 60(3): 820–834. https://doi.org/10.1007/s13197-021-05193-3

Félix-Redondo, F. J., M. Grau, and D. Fernández-Bergés. 2013. Cholesterol and cardiovascular disease in the elderly. Facts and gaps. *Aging and Disease* 4(3): 154.

Fontana, L., and L. Partridge. 2015. Promoting health and longevity through diet: from model organisms to humans. *Cell* 161(1): 106–118. https://doi.org/10.1016/j.cell.2015.02.020

Franceschi, C., P. Garagnani, C. Morsiani, et al. 2018. The continuum of aging and age-related diseases: common mechanisms but different rates. *Frontiers in Medicine* 5: 61. https://doi.org/10.3389/fmed.2018.00061

Gao, Y., Y. Wei, Y. Wang, F. Gao, and Z. Chen. 2017. *Lycium barbarum*: a traditional Chinese herb and a promising anti-aging agent. *Aging and Disease* 8(6): 778. https://doi.org/10.14336/AD.2017.0725

Gavett, B. E., R. A. Stern, R. C. Cantu, C. J Nowinski, and A. C McKee. 2010. Mild traumatic brain injury: a risk factor for neurodegeneration. *Alzheimer's Research and Therapy* 2: 18. https://doi.org/10.1186/alzrt42

Gil-Ramirez, A., C. Clavijo, M. Palanisamy, et al. 2011. Edible mushrooms as potential sources of new hypocholesterolemic compounds. In: *Proceedings of the 7th International Conference on Mushroom Biology and Mushroom Products*, 110–119.

Guillamón, E., A. García-Lafuente, M. Lozano, et al. 2010. Edible mushrooms: role in the prevention of cardiovascular diseases. *Fitoterapia* 81(7): 715–723. https://doi.org/10.1016/j.fitote.2010.06.005

Gunde-Cimerman, N., A. Plemenitaš, and A. Cimerman. 1993. *Pleurotus* fungi produce mevinolin, an inhibitor of HMG CoA reductase. *FEMS Microbiology Letters* 113(3): 333–337. https://doi.org/10.1111/j.1574-6968.1993.tb06536.x

Gunde-Cimerman, N., A. Plemenitaš, and A. Cimerman, 1995. A hydroxymethylglutaryl-CoA reductase inhibitor synthesized by yeasts. *FEMS Microbiology Letters* 132(1–2): 39–43. https://doi.org/10.1111/j.1574-6968.1995.tb07807.x

Hamza, A., A. Mylarapu, K. V. Krishna, and D. S. Kumar. 2024. An insight into the nutritional and medicinal value of edible mushrooms: a natural treasury for human health. *Journal of Biotechnology* 381: 86–99. https://doi.org/10.1016/j.jbiotec.2023.12.014

Hook, M., S. Roy, E. G. Williams, et al. 2018. Genetic cartography of longevity in humans and mice: current landscape and horizons. *Biochimica et Biophysica Acta (BBA)—Molecular Basis of Disease* 1864(9): 2718–2732. https://doi.org/10.1016/j.bbadis.2018.01.026

Hou, Y., X. Dan, M. Babbar, et al. 2019. Ageing as a risk factor for neurodegenerative disease. *Nature Reviews Neurology* 15(10): 565–581. https://doi.org/10.1038/s41582-019-0244-7

Ikewuchi, J. C., C. C. Ikewuchi, M. O. Ifeanacho, N. M. Igboh, and I. I. Ijeh. 2013. Moderation of hematological and plasma biochemical indices of sub-chronic salt-loaded rats by aqueous extract of the sclerotia of *Pleurotus tuberregium* (Fr) Sing's: implications for the reduction of cardiovascular risk. *Journal of Ethnopharmacology* 150(2): 466–476. https://doi.org/10.1016/j.jep.2013.09.002

Jarial, R. S., K. Jarial, and J. N. Bhatia. 2024. Comprehensive review on oyster mushroom species (Agaricomycetes): morphology, nutrition, cultivation and future aspects. *Heliyon* 10(5): e26539. https://doi.org/10.1016/j.heliyon.2024.e26539

Justice, N. J. 2018. The relationship between stress and Alzheimer's disease. *Neurobiology of Stress* 8: 127–133. https://doi.org/10.1016/j.ynstr.2018.04.002

Khan, N. S., P. Kumar, R. Taki, R. Giri, and Geetika. 2022. An overview and study on edible mushrooms: human health benefits and promoting quality of life. *Environment and Ecology* 40(2A): 540–553.

Kim, G. H., J. E. Kim, S. J. Rhie, and S. Yoon. 2015. The role of oxidative stress in neurodegenerative diseases. *Experimental Neurobiology* 24(4): 325. https://doi.org/10.5607/en.2015.24.4.325

Kushairi, N., C. W. Phan, V. Sabaratnam, S. Vidyadaran, M. Naidu, and P. David. 2020. Comparative neuroprotective, anti-inflammatory and neurite outgrowth activities of extracts of king oyster mushroom, *Pleurotus eryngii* (Agaricomycetes). *International Journal of Medicinal Mushrooms* 22(12): 1171–1181. https://doi.org/10.1615/IntJMedMushrooms.2020036938

Li, H., H. Zhao, Z. Gao, et al. 2019. The antioxidant and anti-aging effects of acetylated mycelia polysaccharides from *Pleurotus djamor*. *Molecules* 24(15): 2698. https://doi.org/10.3390/molecules24152698

Liu, M., Y. Dai, C. Song, J. Wang, Y. Liu, and Q. Wang. 2022. Structural characterization of a *Pleurotus sajor-caju* polysaccharide and its neuroprotection related to the inhibition of oxidative stress. *Nutrients* 14(19): 4047. https://doi.org/10.3390/nu14194047

Madhanraj, R., K. Ravikumar, M. R. Maya, et al. 2019. Evaluation of anti-microbial and anti-haemolytic activity of edible Basidiomycetes mushroom fungi. *Journal of Drug Delivery and Therapeutics* 9(1): 132–135. https://doi.org/10.22270/jddt.v9i1.2277

Mori, K., C. Kobayashi, T. Tomita, S. Inatomi, and M. Ikeda. 2008. Antiatherosclerotic effect of the edible mushrooms *Pleurotus eryngii* (Eringi), *Grifola frondosa* (Maitake), and *Hypsizygus marmoreus* (Bunashimeji) in apolipoprotein E–deficient mice. *Nutrition Research 28*(5): 335–342. https://doi.org/10.1016/j.nutres.2008.03.010

Murray, C. J., R. M. Barber, K. J. Foreman, et al. 2015. Global, regional, and national disability-adjusted life years (DALYs) for 306 diseases and injuries and healthy life expectancy (HALE) for 188 countries, 1990–2013: quantifying the epidemiological transition. *The Lancet* 386(10009): 2145–2191. https://doi.org/10.1016/S0140-6736(15)61340-X

Oyetayo, F. L., S. F. Akomolafe, and O. A. Ogundumi. 2020. Anti-hypercholesterolemic potential of diet supplemented with *Anchomanes difformis* and *Pleurotus tuberregium* tubers in high cholesterol fed rats. *Journal of Diabetes and Metabolic Disorders* 19: 1139–1155. https://doi.org/10.1007/s40200-020-00615-z

Przedborski, S., M. Vila, and V. Jackson-Lewis. 2003. Series introduction: neurodegeneration: what is it and where are we? *The Journal of Clinical Investigation* 111(1): 3–10. https://doi.org/10.1172/JCI17522

Ramakrishnan, M., C. Dubey, V. Tulasi, P. Kislai, and N. Manohar. 2017. Investigation of lovastatin, the anti-hypercholesterolemia drug molecule from three oyster mushroom species. *International Journal of Biomedical and Clinical Sciences* 2(4): 26–31.

Raman, J., K. Y. Jang, Y. L. Oh, et al. 2021. Cultivation and nutritional value of prominent *Pleurotus* spp.: an overview. *Mycobiology* 49(1): 1–14. https://doi.org/10.1080/12298093.2020.1835142

Ramkumar, S., A. Raghunath, and S. Raghunath. 2016. Statin therapy: review of safety and potential side effects. *Acta Cardiologica Sinica* 32(6): 631. https://doi.org/10.6515/ACS20160611A

Saha, A., P. Jana, and S. Khatua. 2024. A protective role of *Ganoderma* sp. on cardiovascular disease, atherosclerosis and hypertension. In: *Ganoderma: Cultivation, Chemistry and Medicinal Applications, Volume 2,* ed. K. Acharya, and S. Khatua, 102–126. CRC Press. http://doi.org/10.1201/9781003490258-6

Saha, S., and S. Khatua. 2024. Hypolipidemic and cholesterol lowering effects of *Ganoderma*. In: *Ganoderma: Cultivation, Chemistry and Medicinal Applications, Volume 1,* ed. K. Acharya, and S. Khatua, 189–214. CRC Press. http://doi.org/10.1201/9781003354789

Saklayen, M. G. 2018. The global epidemic of the metabolic syndrome. *Current Hypertension Reports* 20(2): 1–8. https://doi.org/10.1007/s11906-018-0812-z

Sato, M., D. Torigoe, Y. Kinoshita, et al. 2025. Long-term intake of Tamogi-take mushroom (*Pleurotus cornucopiae*) mitigates age-related cardiovascular dysfunction and extends healthy life expectancy. *npj Aging* 11(1): 1. https://doi.org/10.1038/s41514-024-00191-z

Shin, S. W., H. J. Koh, and E. S. Jung. 2018. Anti-skin aging effect of *Pleurotus citrinopileatus* extracts. *Journal of the Korean Society of Cosmetology* 8(3): 345–353.

Thompson, P. D., G. Panza, A. Zaleski, and B. Taylor. 2016. Statin-associated side effects. *Journal of the American College of Cardiology* 67(20): 2395–2410. https://doi.org/10.1016/j.jacc.2016.02.071

Valappil, G. P., S. Julius, and J. K. Krishnankutty. 2020. Hypolipidemic activity of *Pleurotus florida* against Triton WR 1339 induced hyperlipidemia. *Letters in Applied NanoBioScience* 10: 2107–2116. https://doi.org/10.33263/LIANBS102.21072116

Varani, J., M. K. Dame, L. Rittie, et al. 2006. Decreased collagen production in chronologically aged skin: roles of age-dependent alteration in fibroblast function and defective mechanical stimulation. *The American Journal of Pathology* 168(6): 1861–1868. https://doi.org/10.2353/ajpath.2006.051302

Wang, L., K. Li, Y. Cui, H. Peng, Y. Hu, and Z. Zhu. 2023. Preparation, structural characterization and neuroprotective effects against $H_2O_2$-induced oxidative damage in PC12 cells of polysaccharides from *Pleurotus ostreatus*. *Food Research International* 163: 112146. https://doi.org/10.1016/j.foodres.2022.112146

Xiao, P. T., Z. S. Xie, Y. J. Kuang, et al. 2021. Discovery of a potent FKBP38 agonist that ameliorates HFD-induced hyperlipidemia via mTOR/P70S6K/SREBPs pathway. *Acta Pharmaceutica Sinica B* 11(11): 3542–3552. https://doi.org/10.1016/j.apsb.2021.03.031

Yadav, S. K., R. Ir, R. Jeewon, M. Doble, et al. 2020. A mechanistic review on medicinal mushrooms-derived bioactive compounds: potential mycotherapy candidates for alleviating neurological disorders. *Planta Medica* 86(16): 1161–1175. https://doi.org/10.1055/a-1177-4834

Yan, B., L. Jing, and J. Wang. 2015. A polysaccharide (PNPA) from *Pleurotus nebrodensis* offers cardiac protection against ischemia–reperfusion injury in rats. *Carbohydrate Polymers* 133: 1–7. https://doi.org/10.1016/j.carbpol.2015.06.076

Zhang, C. J., J. Y. Guo, H. Cheng, et al. 2020. Protective effects of the king oyster culinary-medicinal mushroom, *Pleurotus eryngii* (Agaricomycetes), polysaccharides on β-amyloid-induced neurotoxicity in PC12 cells and aging rats, *in vitro* and *in vivo* studies. *International Journal of Medicinal Mushrooms* 22(4): 325–333. https://doi.org/10.1615/IntJMedMushrooms.2020033990

Zhang, C., J. Li, J. Wang, et al. 2017. Antihyperlipidaemic and hepatoprotective activities of acidic and enzymatic hydrolysis exopolysaccharides from *Pleurotus eryngii* SI–04. *BMC Complementary and Alternative Medicine* 17: 1–11. https://doi.org/10.1186/s12906-017-1892-z

Zhang, Y., X. Yang, G. Jin, X. Yang, Y. Zhang. 2016. Polysaccharides from *Pleurotus ostreatus* alleviate cognitive impairment in a rat model of Alzheimer's disease. *International Journal of Biological Macromolecules* 92: 935–941. https://doi.org/10.1016/j.ijbiomac.2016.08.008

Zhao, Y., X. Chen, Y. Zhao, et al. 2020. Optimization of extraction parameters of *Pleurotus eryngii* polysaccharides and evaluation of the hypolipidemic effect. *RSC Advances* 10(20): 11918–11928. https://doi.org/10.1039/C9RA10991A

# 12 *Pleurotus* Macrofungi-Assisted Nanoparticle Synthesis and Its Potential Applications

*Rusmita Buragohain*[1], *Qudsiya Y. Tamboli*[2], *Jibanjyoti Panda*[1], *Abdullah G. Al-Sehemi*[3], *Yugal Kishore Mohanta*[1], *and Sujogya Kumar Panda*[4*]

[1]Nano-Biotechnology and Translational Knowledge Laboratory, Department of Applied Biology, University of Science and Technology Meghalaya, Ri-Bhoi, Meghalaya, India,

[2]Department of Basic and Applied Science, MGM University, Aurangabad, Maharashtra, India

[3]Department of Chemistry, College of Science, King Khalid University, Abha, Saudi Arabia

[4]Centre for Biotechnology, Shiksha "O" Anusandhan (Deemed to be University), Bhubaneswar, Odisha, India

## 12.1 INTRODUCTION

There is a growing interest among experts all over the world in nanobiotechnology due to the immense potential implications that it could have on virtually every type of living organism (Bhardwaj et al., 2020**a**). In the year 1959, Richard Feynman presented the concept of miniaturization for the very first time. More importantly, he was the first person to provide hints regarding nanotechnology, which is a technique that involves materials with dimensions of approximately one hundred nanometers (Kumar et al., 2020a). There are several other fields that have profited tremendously from the ground-breaking discoveries that have been made possible by nanotechnology. These fields include nanomedicine, nano sensors, drug delivery systems, cosmetics, and electronics. As a result of the flexibility of Nanoparticles' (NPs') physical, chemical, and mechanical properties, as well as their higher performance in contrast to their larger alternatives, NPs have been driven to the forefront of technological breakthroughs. The idea behind green synthesis is to make NPs using biomolecules, enzymes, vitamins, and so on, that come from various living things, such as plants, microbes, algae, and so on (Patra et al., 2018). Microbes produce NPs by bioaccumulation, biomineralization, precipitation, and biosorption of harmful substances. Biosorption is a straightforward microbial detoxification pathway that results in NP production. A variety of NPs, including Ag, Au, ZnO, RGO, and others, have been thoroughly investigated by means of this environmentally friendly approach (Chopra et al., 2022). The high concentration of polyphenols, alkaloids, terpenoids, flavonoids, and so on in plant extracts may also set off the reduction reaction, resulting in the desired NPs (Sudheer et al., 2022).

There are a number of benefits associated with the utilization of microorganisms in green synthesis, which is sometimes referred to as bio-nanofactories. These benefits include cheap cost, environmental stability, biosorption, and safety (Bahrulolum et al., 2021). A powerful enzymatic machinery has been created by bacteria in order to facilitate the breakdown of the toxic chemicals that are produced during the formation of NPs. When it comes to biosynthesis, fungi are the most effective microbes to use since they secrete a large quantity of proteins and enzymes. These substances serve as reducing agents and contribute to the stability of NPs. This makes it possible to achieve a better level of productivity. Not only are well-formed mycelia simple to produce in a laboratory, but they also provide a larger surface area for interaction, which is not only practical but also inexpensive. This demonstrates that even after the synthesis steps have been completed, the downstream processing is quite straightforward (Hyde et al., 2024).

Many bioactive chemicals, like alkaloids, phenolics, proteins, amino acids, and others, are found in mushrooms. This is why they have been the subject of a lot of research about their health benefits and nutritional value. Metal NPs (MNPs) can be made from a lot of different biocompounds and parts, like mycelia, whole-mushroom powder, crude polysaccharides, enzymes, and proteins. According to Bhardwaj et al. (2020b), fungi have the ability to bioaccumulate metals, which enables them to accommodate metals into their cells and bind them to the walls of their cells to the greatest extent feasible. Fungi are reported to secrete a significant quantity of extracellular enzymes (Castro-Longoria et al., 2012; Bhardwaj et al., 2020b). According to Mirunalini et al. (2012), mycelial extracts release enzymes and metabolites and act as reducing agents in order to

* Corresponding author: sujogyapanda@soa.ac.in

DOI: 10.1201/9781003527572-12

**FIGURE 12.1** Worldwide distribution of *Pleurotus* species.

withstand the exposure to metal precursors that occurs during the creation of NPs.

"Myco-nanotechnology" is a relatively new field of nanotechnology that is seeing tremendous expansion. This field makes use of fungal biomass and the metabolites that are linked with it in order to produce NPs. Attributable to the existence of a wide variety of reducing enzymes and proteins, the use of micro or macro fungus for the production of NPs is the primary advantage utilized. As a result of the fact that fungi produce a greater variety of enzymes than bacteria do, the process of turning metal salts into NPs is rather straightforward (Panda et al., 2025).

"Oyster mushrooms" are a group of edible and nourishing mushrooms that belong to the genus *Pleurotus* (Wang et al., 2022). These mushrooms are known by their commonplace name. With the exception of Antarctica, oyster mushrooms are found in their natural habitat in tropical and subtropical forests, and as a result, they are not difficult to obtain (Chen et al., 2023). *Pleurotus* sp. found to be globally distributed, and some of the major geographical locations are presented in Figure 12.1. One of the primary functions of the fruiting bodies of oyster mushrooms is to absorb a variety of nutrients. These nutrients include proteins, amino acids, vitamin B (thiamine, riboflavin, and niacin), vitamin D, carbohydrates, and mineral salts (calcium, phosphorus, and iron). Additionally noteworthy are their antibacterial, anti-inflammatory, antifungal, and immunomodulatory properties (Panda et al., 2024).

In addition, they have the ability to reduce low levels of cholesterol and blood sugar (Iqbal et al., 2024). The components that are responsible for the formation of gels in *Pleurotus ostreatus*, namely β-1,3-D-glucan and pectin, are water-soluble and have the ability to bind with bile acids. They improve the clearance of plasma cholesterol and restrict the formation of cholesterol-bile micelles, cholesterol absorption, and endogenous production (Devi et al., 2024). By reducing the development and secretion of very low-density lipoproteins (VLDL), they further enhance the clearance of plasma cholesterol. In addition to other parameters, the type of fungus, its age, and its size all play a role in determining the nutrients that are present (Lok et al., 2007). Carbohydrates (40%–46% of the dry matter), protein (20%–25%), fiber (10%–21%), and amino acids (20%–41% of the dry matter) are all present in large proportions; although fat (10%–20% of the dry matter) is present in extremely low amounts. The fruiting bodies of fungi include many different micro- and macro-elements, including sodium, magnesium, phosphorus, calcium, potassium, iron, copper, and zinc, to name just a few of the elements that are present in high concentrations. *Pleurotus* mushrooms hold greater significance than the basidiomycetes used in commercial production (Bellettini et al., 2019). This is due to the fact that they possess superior nutritional, gastronomic, and therapeutic properties. It is possible to cultivate mushrooms on a wide variety of substrates with relative ease. At the present time, mushrooms hold a tremendous deal of potential for a wide range of applications, one of which is the manufacturing of MNPs (Owaid and Ibraheem, 2017).

Over the course of the past decade, researchers have shown a great deal of interest in the process of synthesizing NPs through the utilization of edible and therapeutic mushrooms. There are a number of macrofungi that are capable of synthesizing metallic NPs such as Au, Ag, Fe, and Pt, as well as non-metallic NPs such as Se and CdS (Šebesta et al., 2022). As a result of their ability to reduce metal ions into their zero-valent or nano forms, these macrofungi are considered to be environmentally beneficial biofactories (Sudheer et al., 2022).

In addition, NPs synthesized via biological routes often exhibit enhanced biological activity, greater stability, and extended shelf life. It is anticipated that flavin and fungal reductase play a critical role in the biosynthesis of these NPs. A deeper understanding of the precise mechanisms underlying NP biosynthesis by mushroom mycelia could enable the controlled synthesis of NPs with tailored size and morphology, thereby facilitating their application in diverse biomedical and environmental fields. In the course of microscopic and spectroscopic research of mushroom-mediated

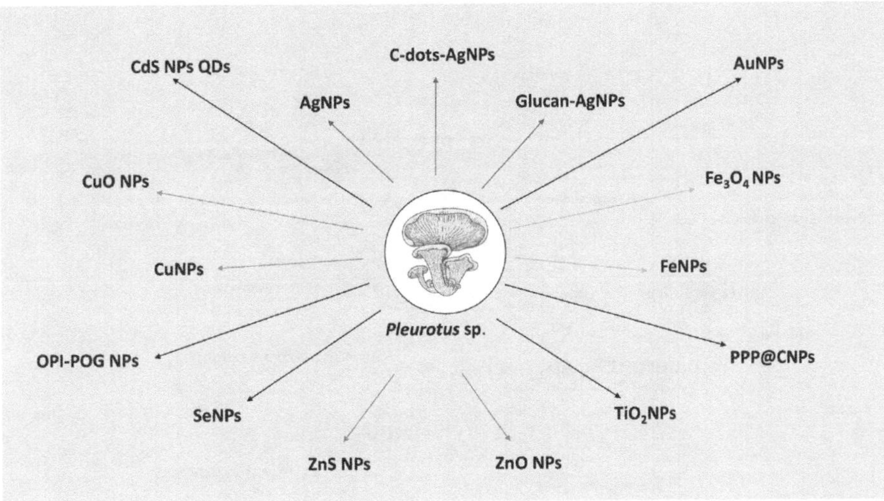


**FIGURE 12.2** Different NPs derived from *Pleurotus* mushroom.

NPs, a wide range of bioactive chemicals were discovered. These compounds include polysaccharides, enzymes, and proteins. These compounds serve as reducing agents and capping/stabilization agents, respectively, to facilitate the synthesis of these NPs. In the majority of instances, the NPs that are mediated by mushrooms are extremely crystalline, can be found in a variety of sizes, and can either be monodispersed or polydispersed (Mohanta et al., 2018).

Fungal cells, which resemble NP factories, are advantageous and helpful because they generate a lot of extracellular enzymes. In the process of making NP, these enzymes can operate as bio reducing (and stabilizing) agents. Furthermore, fungi produce superior NPs compared to bacteria. Microbial metabolites and enzymes that are generated by fungal cells are essential components in the production of MNPs, which reduce the toxicity of chemicals (Owaid and Ibraheem, 2017). The presence of harmful ions is not an unusual occurrence. When metal ions in solution are exposed to bacterial cells, they do not localize in a specific location; rather, they scatter uniformly across the environment that they are in. NPs that bind to the cell walls of bacteria increase the toxicity of the cells by producing a localized ion source through the continuous release of ions (McQuillan et al., 2011). Metal ions with a positive charge can quickly bond with the negatively charged surface of fungal cells. This is possible through electrostatic interactions and interactions that are unique to cell receptors. NPs can be produced by utilizing either intracellular or extracellular fungus. Both methods are viable strategies. Researchers have been interested in using fungal biomass for the large-scale synthesis of NPs because it has a higher capacity to bind metals than bacteria. It is possible to create a variety of metal nanoparticles (MNPs) using their oxides, nitrides, sulfides, and fungal biomasses (Gade et al., 2010) out of which, PtNPs, AgNPs, AuNPs, CuNPs, FeNPs, ZnNPs, and TiNPs are some examples of these types of NPs (Figure 12.2).

The NPs that are generated from fungal biomass possess distinctive optical, physical, and chemical properties that render them advantageous for a variety of chemical and medical applications (Rami et al., 2024). These applications include sensing, medicine, catalysis, and food packaging. The strong photostability, high quantum yield, and appropriate near-infrared (NIR) light-absorbing capacity are some of the qualities that characterize this material. The possibility that oyster mushrooms could induce myco-NPs through the utilization of wasted mushroom substrate (SMS) was first brought to light in the year 2007. As a result of this, *Pleurotus* mushrooms have gained medical significance in relation to a wide variety of bacteria that are associated with disease (Owaid, 2019).

Regarding the synthesis of MNPs, a large number of review publications have been explored and nevertheless, it is endeavored to address the link by conducting a literature study on the subject of the synthesis of various MNPs and their different applications, with a particular emphasis on NPs derived from *Pleurotus* mushrooms. Throughout the course of this chapter, it is intended to discuss a wide range of techniques and applications for the synthesis of NP derived from *Pleurotus* mushrooms. The creation of NPs by both intra- and extra-cellular methods is discussed in the first introductory section. Following that, a number of applications relating to catalysis, antioxidants, anticancer, and antibacterial properties, in addition to probable action mechanisms, were taken into consideration.

## 12.2 METHODOLOGY

*Pleurotus* spp. have emerged as a feasible source for nanomaterial synthesis due to their bioactive compounds, eco-friendly properties, and bioremediation potential. Bibliometric analysis clarifies research trends, impact, and scientific progress within this field. The principal terms featured in the study include *Pleurotus*-based nanomaterials, *Pleurotus* mushroom-mediated NPs, and green synthesis utilizing *Pleurotus*. The research employed the PubMed database, concentrating on a period of 10

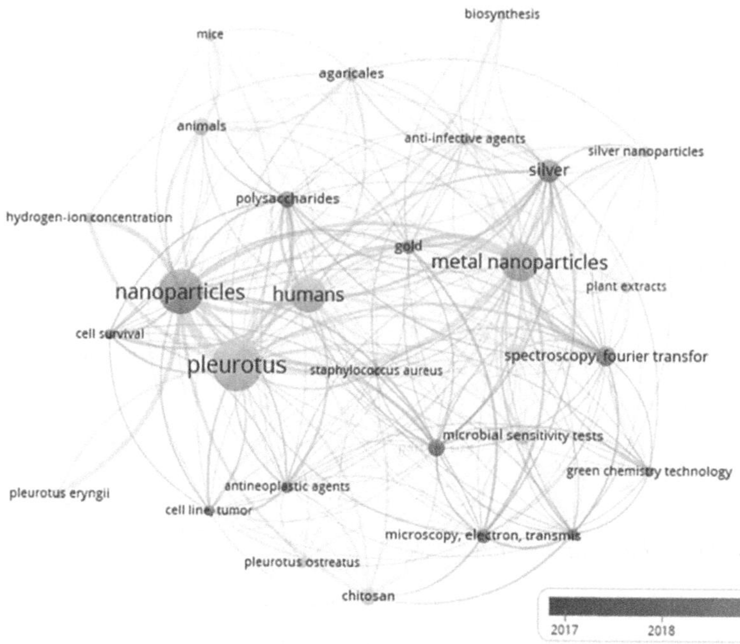


**FIGURE 12.3** Network visualization of *Pleurotus* on the basis of co-occurrence of nanoparticle and their applications.

to 15 years. Nonetheless, the majority of the research was recorded in recent years. A network visualization was generated based on the co-occurrence of the study (Figure 12.3).

## 12.3 GREEN SYNTHESIS OF METAL-BASED NPS MEDIATED BY GENUS *PLEUROTUS*

The bio-potential of the fungal cell wall is also thought to have a significant impact on metal ion absorption and reduction as well as MNP production. It is still not well understood how NPs are synthesized or which biological substances are involved in the process. As demonstrated in Figure 12.2, it is hypothesized that NPs mediated by fungi occur either in living organisms or in laboratory. In the *in vivo* approach, NPs are synthesized intracellularly, within fungal mycelia, where the majority of the hazardous transition metal ions are reduced to a non-toxic state. This approach allows for the intracellular production of NPs using washed mushroom mycelia (Osman et al., 2024). But before using the NPs for anything else, these should have to be separated from the mycelia. Alternatively, there are three ways to do NP synthesis *in vitro* utilizing extracts that do not contain any fungal cells. The first method involves using the fungal fermentation supernatant, which contains enzymes and extracellular proteins, to synthesize NPs. Biologically manufactured NPs (e.g., AuNPs) can be produced in the second way by releasing intracellular components into the media following cell wall disruption (Behzadi et al., 2017). Another option is to employ the watery liquid that is used to culture fungal mycelia in suspension to make NPs. These NPs are obtained by autolyzing the fungal cells and then dissolving their surface carbohydrates and membrane proteins into the watery liquid. Fungal mycelia must be washed and re-suspended in a sterile environment before this method may be used (Molnár et al., 2018).

Long term extensive studies on fungi and their potential impact on nanotechnology holds significant importance. There has been prior research suggesting that the NPs such as calcium, gold, silicon, iron, silver, lead, and gypsum could be synthesized by yeast, bacteria, and fungi. Due to their metal bioaccumulation properties, fungi have garnered a lot of attention for their ability to create MNPs. MNPs of oyster mushroom species were produced using mineral salts. Mycelia, polysaccharides, and proteins are the fungal materials used in NPs synthesis, because they can bioaccumulate metals and have a high metal tolerance. Moreover, fungi can extract metals from inside their cells and bind them to walls to a maximum. Mycelia from fungi offer superior holding ability in bioreactors, agitation, and high-flow pressure conditions compared to other microorganisms and plants. Fungus releases enzymes and metabolites necessary for survival when its mycelium comes into contact with the metal precursor (Mirunalini et al., 2012). The enormous synthesis of enzymes is a result of the vast quantities of extracellular enzymes secreted by fungi and their reduction helps with biomimetic mineralization, nanostructure, and the production of MNPs through pathways inside and outside of cells. Fungi extracts are also used as reducing agents and caps during synthesis. The NPs can be synthesized using either the mushroom fruiting bodies or the mycelium. *Pleurotus ostreatus* and other species can synthesize NPs intracellularly and extracellularly (Owaid, 2019). *Pleurotus ostreatus*, *Pleurotus sajor-caju*, and *Pleurotus djamor* var. *roseus* are among the species that can synthesize NPs extracellularly. Figure 12.4 illustrated the synthesis of MNPs obtained from *Pleurotus* and Table 12.1 exhibits different NPs obtained from *Pleurotus.*

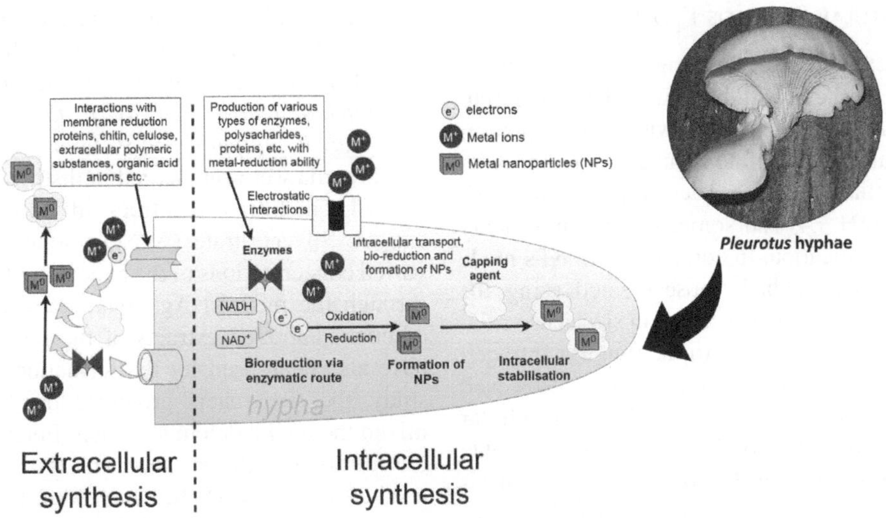


**FIGURE 12.4** Green synthesis of NPs using *Pleurotus* mushroom through extracellular and intracellular methods. Image is adopted from (Šebesta et al., 2022) under the distribution of common attributed license.

**TABLE 12.1**
***Pleurotus* Derived Synthesis of Different Metal and Metal Oxide NPs.**

| Species | NPs Type and Size (nm) | Chemical Used | Reaction Time (Hour) | Reducing Agent | Stabilizing Agent | Specific Temp. (°C) | Morphology | References |
|---|---|---|---|---|---|---|---|---|
| *Pleurotus citrinopileatus* | Ag, 6–10 | $AgNO_3$ | 24 | Mushroom extract, Nitrate | Mushroom extract | 60 | Spherical | Maurya et al., 2016 |
| *P. cornucopiae (citrinopileatus)* | Ag, 20–30 | $AgNO_3$ | 24 | AE | AE | 25 | Spherical | Owaid et al., 2015 |
| *P. cystidiosus* | Ag, 2–100 | $AgNO_3$ | 24 | AE | AE | 25 | ND | Madhanraj et al., 2017 |
| *P. cystidiosus* | Au, ND | $HAuCl_4$ | 24 | AE | AE | 29 | ND | Madhanraj et al., 2017 |
| *P. djamor* | ZnO, 70–80 | $Zn(NO_3)_2.5H_2O$ | 24 | ND | ND | RT | Spherical | Kumar et al., 2020c |
| *P. djamor* | $TiO_2$ | $TiCl_4$ | 20 min | AE | AE | RT | Spherical | Manimaran et al., 2021 |
| *P. eryngii* | Ag, 18.45 | $AgNO_3$ | 5 days | AE | AE | RT | Spherical | Acay and Baran, 2019 |
| *P. flabellatus* | Ag, 2–100 | $AgNO_3$ | 24 | AE | AE | 25 | ND | Madhanraj et al., 2017 |
| *P. flabellatus* | Au, ND | $HAuCl_4$ | 24 | AE | AE | 29 | ND | Madhanraj et al., 2017 |
| *P. florida* | Ag, 20 | $AgNO_3$ | Overnight; 72 | AE | AE | RT | Spherical | Sujatha et al., 2013 |
| *P. florida* | Au, 2–14 | $HAuCl_4$ | 1.5 | AE, glucan | Glucan | 70 | Spherical | Sen et al., 2013 |
| *P. florida* | Au, 20 | $HAuCl_4$ | 24 | AE | AE | RT | Spherical | Gurunathan et al., 2014 |
| *P. giganteus* | Ag, 5–25 | $AgNO_3$ | 3 days | AE | AE | 37 | Spherical | Debnath et al., 2019 |
| *P. ostreatus* | Ag, 4,28,50 | $AgNO_3$ | 24; 72; 1 | AE; mushroom broth | AE | 28; 75 | Spherical | Madhanraj et al., 2017 |
| *P. ostreatus* | Au, 22.9 | $HAuCl_4$ | 24 | AE | AE | 29 | Spherical | Madhanraj et al., 2017 |
| *P. ostreatus* | ZnS, 2–5 | $ZnCl_2$ | Overnight | mushroom | Mushroom Extract | 70 | Spherical, crystalline | Senapati and Sarkar, 2014 |
| *P. ostreatus* | Zn, 15 | $ZnS\text{-}N_3$ | 1 | AE | AE, Sodium azide | 4 | Uniform | Wu et al., 2010 |
| *P. platypus* | Ag, 0.56 μm | $AgNO_3$ | 72 | AE | AE | 37 | Spherical | Sujatha et al., 2013 |
| *P. pulmonarius* | Au, ND | $HAuCl_4$ | 24 | AE | AE | 29 | ND | Madhanraj et al., 2017 |
| *P. sajor-caju* | Ag, 5–50 | $AgNO_3$ | 48 | AE | AE | 25 | Spherical | Chaturvedi et al., 2020 |
| *P. sajor-caju* | Au, 16–18 | $HAuCl_4.3H_2O$ | Overnight | AE | AE | RT | Spherical | Chaturvedi et al., 2020 |

*Note:* AE—Aqueous Extract; ND-Not detected; RT—Room Temperature

### 12.3.1 Intracellular Synthesis Process

In this approach, enzyme exposure triggers ion transport, which leads to NP synthesis within the fungal cells (Mathur and Pillai, 2025). After being treated with a metal precursor, the mycelial cultures are left to incubate in darkness for a whole day. Mycelia are resuspended in phosphate-buffered saline (PBS) (pH 7.4) and sonicated to homogeneity for intracellular identification. In comparison to NPs made using the extracellular method, those created using the intracellular technique are much smaller in size. The variation in NPs size may be attributed to the fungus's intracellular nucleation processes during biosynthesis. However, compared to this intracellular route, the extracellular synthesis of MNPs is generally faster and more scalable (Narayanan and Sakthivel, 2010). Intracellular synthesis often complicates downstream processing and increases production costs due to the need for cell disruption and purification. Despite these challenges, this method remains valuable, particularly for applications involving composite film development, where intracellularly synthesized NPs can be directly integrated into biomaterial matrices (Khandel and Shahi, 2018).

### 12.3.2 Extracellular Synthesis Process

In the extracellular synthesis method, a fungal biomolecule aqueous filtrate is mixed with a metal precursor. This makes metal ions, which stick to the cells' surfaces. This approach is simple and inexpensive. Compared to the intracellular method, this technique is more effective because it does not require downstream processing. Therefore, the extracellular method primarily produces NPs. Extracellular metabolites of fungi are very important for their survival because they help them deal with changes in temperature, harmful substances (like metallic ions), and predators (Osemwegie et al., 2020). Additionally, this synthesis approach demonstrates the capacity to immobilize metallic ions in an appropriate carrier (Khandel and Shahi, 2018).

## 12.4 NPs Derived from *Pleurotus* Mushroom

### 12.4.1 Silver NPs (AgNPs) and Silver Composite NPs

AgNPs play an important role in biological and bio-medical system. These NPs can be formed through various techniques, including physical, chemical, and ionizing radiation methods, among others (Meher et al., 2024). However, these methods come with certain constraints, especially the chemicals used in the wet chemistry methods for producing AgNPs, which are not environmentally safe, overpriced, and can show highly toxic nature. Nonetheless, producing AgNPs through biosynthesis methods offers an excellent alternative, as it is more accessible, harmless, and ecologically sound compared to other synthesis methods (Kumar et al., 2020a). Several studies have reported the green synthesis of AgNPs using powdered basidiocarps and mycelia from different *Pleurotus* species, including *P. cornucopiae var. citrinopileatus*, *P. eous*, *P. florida*, *P. giganteus*, *P. ostreatus*, *P. platypus*, and *P. sajor-caju*. The basidiocarps and mycelia were submerged in distilled water, boiled, and filtered. The filtrate was freeze-dried to make the aqueous extract. Silver nitrate ($AgNO_3$) solution was mixed with varied concentrations of the aqueous extract to form AgNPs through the reduced $Ag^+$ ions to metallic silver (Ag). In addition, unboiled extracts of the mycelium of *Pleurotus* have also been used for the formation of AgNPs. In their study, the researchers grounded the fruiting bodies and mixed them with deionized water. Filter paper was used for the filtration of the mixture, and the resulting filtrate was used to produce AgNPs, which were found to be spherical in shape and the sizes ranged from 6 to 10 nm and the produced AgNPs were examined for their antibacterial ability against *Staphylococcus aureus* and *Escherichia coli* (Maurya et al., 2016).

Additionally, AgNPs were also synthesized using *P. tuber-regium* mushroom extract and 1 mM $AgNO_3$ solution. The two solutions were mixed and stirred for 2 hours at 90°C. The resulting AgNPs, which appeared as a black powder, had cubic and spherical shapes with an average size of 50 nm. Debnath et al. (2019) produced spherical AgNPs using 5 ml of aqueous extract of mushrooms, and 95 ml of a 1 mM silver nitrate ($AgNO_3$) solution was added to it, to reduce $Ag^+$ to metallic silver. The mixture was incubated for at 37°C for 3 days, leading to a color change from light yellow to yellowish-brown. The resulting AgNPs were crystalline and sizes ranging from 5 to 25 nm. The antibacterial activity of the AgNPs was evaluated against *E. coli*, *Bacillus subtilis*, *Pseudomonas aeruginosa*, and *S. aureus*. Similarly, several researchers have synthesized spherical AgNPs primarily, with sizes ranging from 2 to 100 nm, using a combination of mushroom extract with an $AgNO_3$ solution (Musa et al., 2018). Thulasinathan et al. (2023) produced C-dots-AgNPs to take advantage of the ability of the amine or phenolic hydroxyl groups on the aromatic rings of C-dots to directly reduce $Ag^+$ ions into metallic silver. The fruiting body of *Pleurotus* species were accumulated from wood decay and used to prepare water-soluble C-dots. The C-dot solution was mixed with 1 mM $AgNO_3$ to obtain the aimed C-dots-AgNPs. Glucan coated AgNPs were synthesized using *Pleurotus* spp., which were compared with free AgNPs to check antimicrobial ability. Different species of *Pleurotus* mushroom were picked, dried and grounded into powdered form and then cooled at 4°C (Irshad et al., 2024).

### 12.4.2 Cadmium Sulfide (CdS) NPs Quantum Dots (QDs)

Unlike conventional fluorescent organic dyes and green fluorescent proteins, CdS QDs offer advantages by overcoming issues such as spectral overlap, low signal intensity, and photobleaching (Borovaya et al., 2015). QDs have several remarkable properties, including high photostability, uniform shape, slow decay rates, precise emission spectra, large absorption cross-sections, and broad absorption spectra. The emission color of QDs is determined by their size and surface chemistry, and their chemical composition can be adjusted to emit across ultraviolet (UV), visible, or near-infrared (NIR) wavelengths. The growing interest in CdS QDs stems from their potential as luminescent probes and labels for biological imaging, disease diagnosis, and molecular histopathology. In 2015, the synthesis of luminescent CdS NPs was carried out using the mycelium of *P. ostreatus*, along with $CdSO_4$ and $Na_2S$. Concisely, a $CdSO_4$ solution was mixed with the mycelium and incubated for 10 days at 26°C, after which $Na_2S$ solution was added. The resulting NPs were spherical and their sizes ranged from 4 to 7 nm. Precisely, CdS QDs are highly valuable for finding the phenomenon of biomolecule interactions and cellular signaling pathways with the help of fluorescence microscopy (Borovaya et al., 2015).

### 12.4.3 Zinc Sulfide NPs (ZnSNPs) and Zinc Oxide NPs (ZnONPs)

The *P. ostreatus* extract was used to synthesize ZnSNPs, along with $ZnCl_2$ and $Na_2S$ solutions as the precursor substance. Primarily, small mushroom pieces were boiled and filtered. Varying concentrations of the resulting filtrate were then mixed with aqueous solutions of $ZnCl_2$ and $Na_2S$, and the mixtures were dried for 2 hours at 120°C. The filtrate acted as both a capping and stabilizing agent in the formation of spherical ZnSNPs. The resulting ZnSNPs were highly crystalline, ranging in size from 2.30 nm to 4.04 nm. The author observed that as the extract amount increased, the diameter of the ZnSNPs that are spherical, decreased (Senapati and Sarkar, 2014). *Pleurotus djamor* extract was used to make ZnONPs by mixing 20 ml of mushroom extract with 80 ml of a 5 mM Zn $(NO_3)_2.5H_2O$ solution. The mixture was continuously stirred for 24 hours at room temperature, and the color changed to light pink, indicating the formation of ZnONPs (Manimaran et al., 2021).

### 12.4.4 Titanium-Dioxide NPs ($TiO_2$ NPs)

The edible *P. djamor* mushroom was used to create $TiO_2$ NPs, which were then tested for their bactericidal and larvicidal qualities as well as their anticancer activities against A-549 (human lung carcinoma) cell line. First, 10 g of fresh *P. djamor* biomass were cut into small pieces after being cleaned for 10 minutes with deionized water. After adding the chopped biomass to 100 ml of double-deionized water, it was heated for 15 minutes at 60 °C before being filtered. Following two hours of stirring, 20 ml of the filtrate were combined with 80 ml of 5 mM $TiCl_4$ solution, and the mixture was left to stand at room temperature for 20 minutes until it turned brown. A wavelength of 345 nm was used to measure the extract's color intensity (Manimaran et al., 2021).

### 12.4.5 Gold NPs (AuNPs)

Using a photo-irradiation technique, AuNPs were created from the edible *P. florida* mushroom and evaluated for anticancer efficacy against K-562, HeLa, A-549, and MDA-MB cell lines. Initially, 5 g of fresh *P. florida* biomass were thoroughly washed with deionized water and subsequently cut into small pieces for further processing. After that, 500 ml of double-deionized water was added to these pieces, and they were agitated for half an hour. After an overnight incubation period, the mixture was filtered via filter paper. In the presence of sunshine, the resultant mushroom filtrate was utilized to convert $Au^+$ ions to Au, creating AuNPs with spherical to triangular forms and sizes ranging from 10 to 50 nm (Bhat et al., 2013).

*Pleurotus ferulae* polysaccharide (PFPS) derived AuNPs were produced by adding 10 mg/mL of PFPs with 0.01% chloroauric acid under burning hot conditions for 3 minutes, then 1% trisodium citrate was added and again reacted under very hot conditions for 10 minutes. This was then centrifuged for 20 minutes leading to the formation of PFPS-AuNPs, then collected and distilled water was added to it (Yuan et al., 2023). A glucan extracted from the edible mushroom *P. florida* (cultivar Assam Florida) was utilized to synthesize AuNPs via reduction of $HAuCl_4$. In this biosynthetic process, the glucan functioned both as a reducing and stabilizing agent. The resulting AuNPs were characterized using UV-Visible spectroscopy, high-resolution transmission electron microscopy (HR-TEM), X-ray diffraction (XRD), scanning electron microscopy (SEM), and Fourier-transform infrared spectroscopy (FT-IR). The findings indicated that the concentration of chloroauric acid significantly influenced the size distribution of the synthesized AuNPs (Sen et al., 2013).

### 12.4.6 Copper NPs (CuNPs) and Copper Oxide NPs (CuONPs)

Under gamma radiation exposure, homogeneous CuNPs were synthesized using aqueous fermented fenugreek powder and various polysaccharides, including citrus pectin, chitosan, sodium alginate, and pectin, along with fungal isolates of *P. ostreatus*. These natural biopolymers and fungal components acted as reducing and stabilizing agents, facilitating the green synthesis of CuNPs under irradiation conditions. Diameter of the CuNPs ranged from 25 to 36 nm. Because of their stability and small size, these CuNPs demonstrated antioxidant and antibacterial qualities, which made them useful for usage in industrial, pharmaceutical, medical, and cosmetic applications (Narayana et al., 2024). CuONPs were biosynthesized using the extract of *Pleurotus citrinopileatus* mushroom extract. The characterization of the biosynthesized CuONPs was carried out through spectral and microscopic methods, which confirmed their shape, size, structure and elemental composition. The CuONPS produced through mycosynthesis exhibited moderate antibacterial activity against *P. aeruginosa* (Manimaran et al., 2023b).

### 12.4.7 Iron (FeNPs) and Iron Oxide NPs (FeONPs)

FeNPs were synthesized inside the cells using the hyphae of *Pleurotus* species. The process involves the absorption of FeNPs through the fungal cell membrane, where Ferric ions ($Fe^{3+}$) are reduced to Ferrous ions ($Fe^{+2}$). This reduction process occurs during the uptake of iron by fungi (Mazumdar and Haloi, 2011). FeONPs were synthesized using an aqueous extract of *P. citrinopileatus*. The UV-vis spectroscopy of *P. citrinopileatus* derived FeONPs revealed a visible peak at 254 nm. XRD determined the crystalline shape of produced NPs. XRD showed that the produced NPs were pure crystals with a crystallite size of 14.5 nm. FTIR revealed the presence of Fe metallic ions. SEM–EDX and HR-TEM showed hexagonal and spherical agglomerations. The $Fe_3O_4$ NPs were enclosed in 3D honeycomb biochar, extracted from *P. eryngii*, to obtain $Fe_3O_4$@C nanocomposites. The reduction of $Fe_3O_4$ to the nanoscale enhanced ion mobility, while the honeycomb biochar, rich in pores and voids, helped alleviate the significant structural expansion of $Fe_3O_4$ NPs during lithiation process, thereby ensuring long-term cycle stability. Consequently, the $Fe_3O_4$@C nanocomposites demonstrated an impressive reversible capacity of 723 mAh $g^{-1}$ at 1.5 A $g^{-1}$ over 1000 charge-discharge cycles. This simple synthesis method could be adapted to create other transition metal oxide anode materials for next-generation lithium-ion batteries (He et al., 2022).

### 12.4.8 Selenium NPs (SeNPs)

Polysaccharide-protein complexes (PSPs), which were obtained from *P. tuber-regium* sclerotia, were used to create SeNPs. These NPs have low toxicity, excellent bioavailability, and anticancer qualities. SeNPs have been shown to inhibit the proliferation of human breast carcinoma MCF-7 cells by inducing cytotoxic effects specifically in cancer cells (Wong, 2012). In a dose-dependent way, PSP-SeNPs effectively increase the production of reactive oxygen species (ROS) and limit the development of MCF-7 human breast cancer cells. Through the cleavage of Poly (ADP-ribose) polymerase (PARP) and the activation of caspases, apoptosis is induced. The spherical PSP-SeNPs have a diameter of less than 50 nm. The production of SeNPs using *P. ostreatus* extract has been documented for its *in vitro* anticancer properties (Kaur et al., 2013).

### 12.4.9 Other NPs

#### 12.4.9.1 Oat Protein Isolate-P. *Ostreatus* β-Glucan Conjugate NPs (OPI-POG NPs)

The synthesis of β-carotene-OPI-POG conjugate NPs was carried out through emulsification and evaporation technique. β-carotene was encapsulated within these NPs (Lei et al., 2024).

#### 12.4.9.2 Pomegranate Peel Polyphenols-Encapsulated Chitosan NPs (PPP@CNPs)

Pomegranate peel polyphenols (PPP) were implanted in chitosan NPs and these PPP@CNPs were again embedded in *P. eryngii* polysaccharide (PEP) nanofibers with the help of electrospinning mechanism. The contamination of *E. coli* O157:H7 was controlled by the application of the nanofibers on the surface or pork and cucumber as there was no change in color and their structures were well maintained. The incorporation of PPP@CNPs to PEP nanofibers enhanced the size of the particle and the coarseness of the surface and upgraded the nanofibers' thermal stability (Cai et al., 2021).

## 12.5 VARIOUS POTENTIAL APPLICATIONS OF *PLEUROTUS* DERIVED NPS

Various types of MNPs derived from *Pleurotus* species are discussed above. These are beneficial in fields like medicine and industries (Figure 12.5).

### 12.5.1 Antimicrobial Properties

Over the last few decades, microbial infections have emerged as a significant global health concern, largely because of their constant evolution and their capacity to develop resistance, to current treatments. MNPs are recognized for their powerful antimicrobial properties against various microbes, including both Gram-negative and Gram-positive bacteria, and fungi. This activity is driven by their photodynamic effects and significant oxidative stress (Warnes et al., 2011). Additionally, when MNPS come into direct contact with the bacterial membrane, they cause the leakage of intracellular contents, disrupt the membrane's integrity, and lead to cell death. Many studies have been published on the bactericidal properties of MNPs (such as AgNPs, AuNPs, and ZnSNPs). However, only a limited number of studies explore the mechanisms underlying their antibacterial and antifungal activities. Furthermore, numerous studies in the literature suggest that the mechanisms affecting bacterial and fungal cells are largely similar (Nayak et al., 2024).

According to researchers, the antimicrobial activity of AgNPs is primarily attributed to the release of $Ag^+$ ions, which lead to protein denaturation, DNA damage, and inhibition of key enzymatic functions within microbial cells. Spherical AgNPs (20–25 nm) were produced from *P. florida* mushrooms by Bhat et al. (2011), and their antibacterial properties against *Salmonella typhi*, *Providencia alcalifaciens*, *S. aureus*, and *Proteus mirabilis* were assessed. In comparison to Gram-negative bacteria, the resultant NPs had more action against Gram-positive bacteria, with *P. mirabilis* showing the most noteworthy impact. A comparable study was conducted where AgNPs were synthesized using *P. pulmonarius*, *P. djamor*, and *Hypsizygus ulmarius* as starting materials. In 2014, Yehia and Al-Sheikh (2014) used a green synthesis approach to create spherical AgNPs

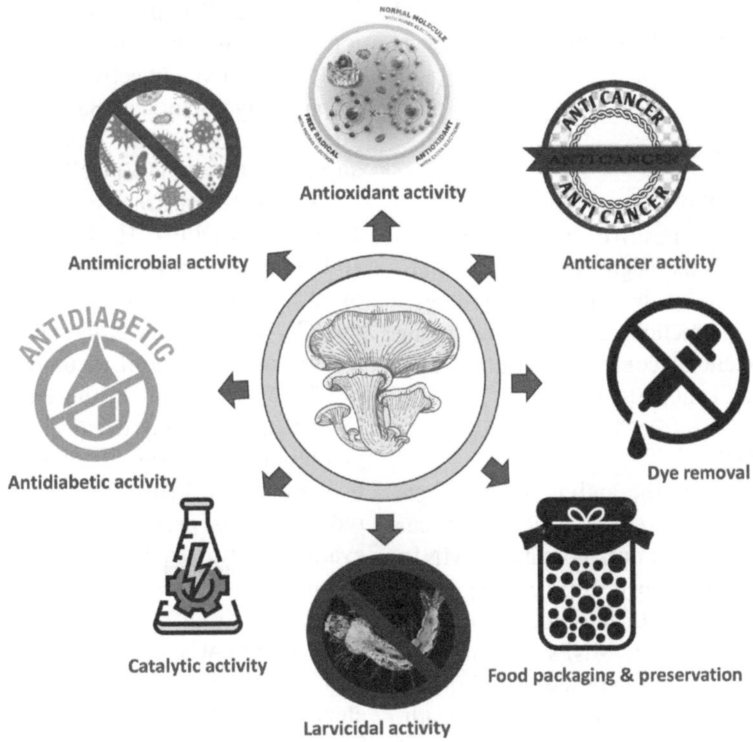


**FIGURE 12.5** Diverse applications of *Pleurotus* derived NPs.

(4–15 nm) from *P. ostreatus* and investigated their antifungal effectiveness against *C. albicans, C. glabrata, C. tropicalis, C. parapsilosis,* and *C. krusei*, among other species of *Candida*. AgNPs were more toxic to all *Candida* species (5–28 μg/mL) than amphotericin B (5–8 μg/mL) and fluconazole (13–33 μg/mL), according to the minimum inhibitory concentration (MIC) values. The bioactive white NPs made from *P. tuber-regium* extract and silver nitrate showed improved therapeutic efficacy against a variety of illnesses and ailments due to their compact size and capping capabilities. In 2018, Bawadekji et al. (2018) synthesized spherical AuNPs (~22.9 nm) using extract of *P. ostreatus* and tested their antimicrobial activity. The study found that the synthesized NPs exhibited notable toxic behavior against *C. albicans, P. aeruginosa*, and *S. aureus*; while no toxic nature was found against *E. faecalis, E. coli*, and *K. pneumoniae* (Bawadekji et al., 2018).

Acay and Baran (2019) reported the green synthesis of AgNPs using *P. eryngii* extract. They also tested the antibacterial capabilities of the product against a range of human pathogenic microbes, such as *Streptococcus pyogenes, E. coli, S. aureus, P. aeruginosa,* and *C. albicans*. For *S. aureus, E. coli, S. pyogenes, C. albicans,* and *P. aeruginosa,* the corresponding MIC were 0.035, 0.07, 0.018, 0.07, and 0.035 mg/L respectively. The AgNPs derived from *P. eryngii* extract were shown to have the potential to be a more beneficial alternative to existing silver nitrate and antibiotic therapies (Acay and Baran, 2019). Consequently, AgNPs were synthesized using *P. giganteus* and examined their microbial inhibition properties (Debnath et al., 2019). The $TiO_2$ NPs derived from *P. djamor* extract explained notable antibacterial effects against human pathogenic bacteria, with maximum zones of inhibition measured at 33±0.2 mm for *P. fluorescens*, and 32±0.4 mm for *S. aureus*, and 32 mm for *Corynebacterium diphtheriae* exhibiting significant inhibitory action (Manimaran et al., 2021). The ZnONPs produced from *P. djamor* illustrated the highest inhibition zones of 28.6 mm against *C. diphtheriae*, and 26.6 mm against *S. aureus,* 27 mm against *P. fluorescens*. The *P. ostreatus* derived $TiO_2$NPs exhibited maximum growth inhibition zone in *S. mutans* (14.3 mm), *E. coli* (14.1 mm), and *S. epidermidis* (13.3 mm) (Manimaran et al., 2023a). The antimicrobial efficacy of *Pleurotus cystidiosus*-produced CuONPs on human pathogens demonstrated a minimum zone of inhibition of approximately 25 mm observed in *Streptococcus mutans* and a maximum zone of inhibition of 32 mm against *Enterococcus faecalis* while other coliform bacteria such as *Enterobacter aerogenes* (30 mm), *E. coli* (28 mm) (Kamakshi et al., 2024). FeONPs produced from *P. citrinopileatus* shown antibacterial efficacy against both Gram-positive and Gram-negative bacteria. Greater antibacterial activity against *E. coli* (8.1 mm), *B. cereus* (7. 2 mm), and *K. pneumoniae* (3.1 mm) growth inhibitions has been demonstrated by *P. citrinopileatus*-FeONPs (Kumar et al., 2023). Glucan-AgNPs made from *Pleurotus sp.* exhibited exceptional antimicrobial activity against pathogenic microorganisms such as *S. aureus, E. coli* and *B. subtilis*. Amongst, the *Pleurotus* species, Glucan-AgNPs made from *P. ostreatus* showed highest inhibition zone against all the three microorganisms (Irshad et al., 2024).

## 12.5.2 Antimicrobial Mechanisms

### 12.5.2.1 Mechanical Damage

In this scenario, MNPs with a positive charge are able to interact electrostatically with the negatively charged components of the cell membrane, such as porins, peptidoglycans, and proteins. This interaction results in membrane damage, leakage of intracellular contents, and ultimately, inhibition of the cell (Wang et al., 2017).

### 12.5.2.2 ROS Imbalance

Another key mechanism behind the antimicrobial effects involves the induction of oxidative stress, which can occur either in the presence of light or in dark conditions. MNPs within the microbial cell can produce ROS, causing damage to proteins, DNA, enzymes, and ribosomes, ultimately resulting in cell death. In addition, MNPs can function as photo-absorbing materials when exposed to light, particularly near-infrared (NIR) light, leading to cell death. The photothermal effect occurs when electrons, excited to a higher energy state by light absorption, return to their ground state, releasing energy in the form of heat and molecular vibrations (Addae et al., 2014).

## 12.5.3 Free Radical Scavenging Ability

In the human body, an overproduction of reactive free radicals can occur due to factors like poor diet, emotional stress, smoking, and various health conditions (Dhanjal et al., 2020). MNPs demonstrated strong antioxidant effects in both internal and external cellular environments. Adebayo et al. (2012) utilized the β-carotene-linoleate model and the DPPH radical scavenging assay to synthesize metabolites from *P. pulmonarius* extract and assess their antioxidant activity. Their findings highlight the extract's potential in combating oxidative stress. Through hydrogen transfer, the study discovered that substances such glutathione, ascorbic acid, cysteine, tocopherol, polyhydroxy compounds, and aromatic amines identified in the metabolites helped to reduce and decolorize the violet 2,2-diphenyl-1-picrylhydrazyl (DPPH). Because of the phenolic components in the extract, the scientists observed that the metabolite exhibited inhibition levels of 75%, 80%, and 90% at a concentration of 2 mg/mL, which were comparable to butylated hydroxyanisole (BHA) and α-tocopherol, respectively. Further, Madhanraj et al. (2017) synthesized gold and silver NPs using an edible basidiomycete mushroom and evaluated their antioxidant properties through various radical scavenging tests. The results demonstrated that both AuNPs and AgNPs indicated notable antioxidant activity in a non-cellular environment. In 2020, Acay and Baran developed AgNPs from *P. eryngii* and evaluated their radical scavenging potential using various methods, including DPPH, β-carotene-linoleate model and ferrous ion chelation. They noted that at 10 mg/mL concentration, the antioxidant activities reached 85%, 77%, and 82%, respectively (Acay and Baran, 2020). Zinc is essential for guarding cells against oxidative stress and functions as an antioxidant. The ZnONPs formed by the assistance of *P. djamor* demonstrate significant free radical scavenging ability, with DPPH, $H_2O_2$, and 2,2'-azino-bis(3-ethylbenzothiazoline-6-sulfonic acid) (ABTS) scavenging abilities of 59%, 59.65%, and 59.30%, respectively. The corresponding $IC_{50}$ values for these activities are 428.35 μg/mL, 417.22 μg/mL, and 500 μg/mL (Kumar et al., 2020b).

There are two main mechanisms through which antioxidant activity may occur either the transfer of a hydrogen atom, or the transfer of a single electron. Donating a hydrogen atom helps balance or eliminate excessive free radicals. This procedure includes a series of assays to evaluate antioxidant potential, such as total oxyradical scavenging capacity, inhibition of induced low-density lipoprotein (LDL) oxidation, oxygen radical absorbance capacity (ORAC), and assessment of the ability to trap free radicals (Kumar et al., 2020b). As an alternative, single-electron transfer involves transferring a single electron to reduce substances such as metals, carbonyls, and radicals. As demonstrated by tests such as ferric reducing antioxidant power (FRAP) and DPPH radical testing, this reaction is frequently accompanied by a color shift when the substance is reduced. At the intracellular stage, MNPs are taken up by cells through endocytosis and help lower the levels of reactive oxygen species (ROS) produced by various probes, such as 2,7-dichlorodihydrofluorescein diacetate (DCFDA) (Wang et al., 2017).

## 12.5.4 Anticancer Potential

MNPs derived from fungi and other sources have shown impressive anti-cancer benefits in addition to their antibacterial and free radical scavenging capabilities. This is mostly because of their potent capacity to produce ROS in the presence of light or darkness. The anticancer effects of AgNPs (~136 nm) made from *Origanum vulgare* extract on human lung epithelial cells (A549) were examined by Sankar et al., in 2013. The findings demonstrated that AgNPs reduced cell viability in a dose-dependent manner, with 85% inhibition at 500 μg/mL. Bhat and colleagues produced AuNPs (12–15 nm, spherical) from *P. florida* extract by photo-irradiation and evaluated their anticancer effects on HeLa, K-562, MDA-MB, and A-549 cell lines. Across all cell lines, the resultant AuNPs showed a concentration-dependent lethal impact at concentrations between 10 and 30 μg/mL (Bhat et al., 2013). Gliga et al. (2014) investigated the toxicity of AgNPs toward human lung cells (BEAS-2B), focusing on how coating and particle size influenced their effects. Their findings revealed that AgNPs smaller than 10 nm exhibited the highest toxicity to BEAS-2B cells, likely due to their tendency to aggregate in the cell culture medium, their ability to localize within cells, and the process of cellular uptake, which led to the release of silver ions inside the cells. Furthermore, the confirmed that all AgNPs caused DNA damage in the BEAS-2B cells within 24 hours, highlighting their overall toxic impact. In the very same year, Yehia and Al-Sheikh (2014) utilized AgNPs synthesized from *P. ostreatus* (with sizes ranging from 4 to 15 nm and spherical in shape) as a potential anticancer treatment for MCF-7 cells. The results demonstrated that the AgNPs inhibited cell growth in a dose-dependent manner, with inhibition rates varying from 5% to 78% when the concentration ranged from 10 to 640 μg/mL. In the same target to cancer diagnostics, Ismail et al. (2015) studied the anticancer effects of AgNPs made from *P. ostreatus* extract on HepG2 and MCF-7 adenocarcinoma cell lines. According to the researchers, the production of ROS, which caused necrosis, apoptosis, and total cell death, was the main cause of the NPs deadly effects on cancer cells. Because of their high reactivity, ROS destroyed DNA and proteins oxidatively and interfered with mitochondrial function. Similarly, *P. djamor*-derived AgNPs were investigated as a possible anticancer therapy that targets PC3 cells in human prostate cancer (Raman et al., 2015). Regarding the toxicity, it was shown that MNPs are generally safe at low concentrations but can become toxic to healthy cells at higher doses. Several studies have highlighted the biosynthetic methods used to produce AgNPs for use as anticancer agents against various cell lines. Despite being synthesized through environmentally friendly approaches, AgNPs also exhibited some level of cytotoxicity against tumor cells. Although the widespread use of synthetic AgNPs has been documented, there remains a lack of sufficient research focused on managing their cytotoxic effects (Yehia and Al-Sheikh, 2014). Research on AgNPs derived from *P. eryngii* revealed significant cytotoxic effects on various cancer cell lines. The maximum inhibition observed was 73.46% against HeLa cells at a concentration of 60 μg/mL, 99.02% against PC-3 cells at 10 μg/mL, and 93.89% against MCF-7 cells at 20 μg/mL. The $IC_{50}$ values for these cells were 46.594, 2.185, and 6.169 μg/mL, respectively, after 24 hours of incubation (Beyth et al., 2015). In 2020, Chaturvedi and co-worker, investigated the cytotoxic effects of AgNPs and AuNPs synthesized from *P. sajor-caju* extract against the HCT-116 cancer line. The results illustrated that the feasibility of HCT-116 cells was minimized profoundly by treatment with the PS extract, AuNPs, and AgNPs, with values of $IC_{50}$ as 60, 80, and 50 μg/mL, respectively. The study found that the AgNPs produced by green synthesis exhibited the strongest antiproliferative activity compared to both the mushroom extract and AuNPs. This increased efficiency was attributed to the higher production of ROS, which induced oxidative stress, leading to irreversible damage to protein function and cellular integrity (Chaturvedi et al., 2020). The anticancer characteristics of $TiO_2$ NPs demonstrated prominent toxic nature against human lung cancer (A549) cell lines, with a maximum growth inhibition of 64% observed at 100 μg/mL concentration after 24 hours of exposure. The *in vitro* cytotoxicity assay of *P. ostreatus* derived $TiO_2$NPs showed an $IC_{50}$ value of 47.42 μg/mL against MG-63 cancer cells. The CuONPs, synthesized using *P. citrinopileatus* mushroom extract showed significant toxicity against the human colon cancer cell line (HT-29) at a concentration of 100 μg/mL (61.73%) after 24 hours of treatment (Manimaran et al., 2023**b**). The cancer fighting potential of *P. djamor*-derived ZnO NPs (ZnONPs) showed strong inhibitory activity against A549 cancer cells, showing a dose-dependent reaction with a lethal concentration that is required to kill 50% of the cell population ($LC_{50}$) of 42.26 μg/mL. The *P. citrinopileatus* derived FeONPs had considerable *in vitro* anticancer activity against the human lung cancer cell line A549, and their $IC_{50}$ value was determined to be 39.12 μg/mL (Kumar et al., 2023).

### 12.5.5 Larvicidal Effect

When $TiO_2$NPs made from *P. djamor* extract were administered to *Aedes aegypti* and *Culex quinquefasciatus* mosquitoes, histopathological examination showed significant harm. The mid- and hind-gut, muscles, and nerve ganglia of the brush border were severely damaged, and the caeca, digestive system, and cuticle and epithelial layer were completely collapsed. Both *A. aegypti* and *C. quinquefasciatus* larvae in their IV$^{th}$ instar shown significant larvicidal effects when treated with $TiO_2$ NPs; their respective $LC_{50}$ values were 5.88 μg/L and 4.84 μg/L, while their LC90 values were 22.80 μg/L and 19.33 μg/L. The treatment of *A. aegypti* larvae with ZnONPs resulted in significant morphological changes in their digestive system, including destruction of the midgut, damaged membranes, and severe injury to the brush border and cortex. There was also hyperplasia of the gut epithelial cells and alterations in the cytoplasmic contents. In *C. quinquefasciatus* larvae, the exposure led to complete decomposition of the abdominal regions, particularly affecting the caeca, midgut, and epithelial layer (Manimaran et al., 2021).

### 12.5.6 Antidiabetic Activity

The *in vitro* antidiabetic potential was assessed by evaluating the inhibition of α-amylase, and enzyme responsible for starch digestion. AgNPs synthesized from *P. giganteus* demonstrated strong α-amylase inhibitory activity, making them promising for the development of antidiabetic medications. The inhibition percentage increased as the concentration of the biosynthesized AgNPs was raised (Debnath et al., 2019).

### 12.5.7 Elimination of Dyes

El-Batal et al. (2015) used solid-state fermentation to separate fungal laccase from *P. ostreatus* in order to assess the chemical or dye leaching capabilities. With over 50% decolorization in just three hours, the researchers showed that this enzyme was efficient in decolorizing and breaking down a variety of dyes, including methyl orange, trypan blue, ramazol bright red, and ramazol brilliant yellow. The greatest notable decrease was seen in trypan blue and methyl orange. Furthermore, the laccase enzyme was employed in the production of AuNPs, demonstrating its potential for a range of industrial uses, including biological pretreatment procedures.

### 12.5.8 Catalytic Activity

When sodium borohydride is present, the AuNPs-glucan bioconjugates from *P. florida* work well as a heterogeneous catalyst to change 4-nitrophenol (4-NP) into 4-aminophenol (4-AP). AuNPs-glucan bioconjugates reduced 4-nitrophenol according to pseudo-first-order kinetics. Using AuNPs-glucan bioconjugates made with varying amounts of $HAuCl_4$, the impact of particle size and gold loading on the rate of 4-NP reduction was investigated. For the first time, a pure mushroom polysaccharide with a known structure is used to create catalytically active AuNPs (Sen et al., 2013).

Under exposure to UV-visible light, the photocatalytic degradation of methylene blue by $TiO_2$NPs was examined. Nearly 84.24% degradation was achieved quickly (between 30 and 180 minutes) because of the NPs' small size and strong dispersion. The photocatalytic activity of $TiO_2$NPs did not change significantly even after the photocatalyst was utilized three times. The $TiO_2$NPs that were made showed promise for cleaning up wastewater dyes because they got rid of the Methylene Blue dye in a water solution. So, this is the first time that our study shows that biosynthesis of $TiO_2$NPs using *P. ostreatus* extract can lead to great new organic and multipurpose goods. Investigations into the photocatalytic activity of CuONPs synthesized using *P. cystidiosus* revealed effective degradation of various textile dyes. After 90 minutes of exposure, the maximum degradation efficiency was observed for Congo red at 99.2%. In contrast, Malachite green exhibited moderate degradation, while Remazol blue showed relatively low degradation efficiency, indicating dye-specific variations in photocatalytic performance (Manimaran et al., 2023a).

### 12.5.9 Packaging and Storage of Food

Eco-friendly ZnNPs have the potential to be used in the food packaging and biomedical industries. The food industry could greatly benefit from the potential antibacterial properties of mushrooms, which have demonstrated activity against a range of pathogenic bacteria, such as *Listeria innocua, B. subtilis, S. faecalis, E. coli,* and *Micrococcus luteus*. Food contamination can be avoided and prolonged food preservation or storage is made possible by the use of MNPs (Senapati and Sarkar, 2014; Sharma and Dhanjal, 2016).

### 12.5.10 Miscellaneous Uses

β-Carotene-encapsulated OPI-POG NPs significantly alleviated cyclophosphamide (CTX)-induced immunosuppression in mice, suggesting their potential as an effective immunoprotective agent. This effect was linked to an increase in serum immunoglobulin levels, higher secretion of splenic cytokines, and improvements in spleen pathology. This was closely tied to elevated short-chain fatty acids (SCFAs) concentrations and shifts in the gut microbiota composition at the OUT level. Notably, *Erysipelotrichaceae* (OTU011) was reduced whereas *Akkermansia* (OTU052), *Bifidobacterium* (OTU055), and *Roseburia* (OTU436) were all more abundant. Therefore, by modifying the gut microbiota, a nanocarrier based on glycated proteins may be a potential method for administering hydrophobic chemicals to treat immunological diseases (Lei et al., 2024). Selenium (SeNPs) and ZnNPs made from extracellular extracts of *P. djamor* and *P. sajor-caju* uplifted the yield of fruit bodies on addition to straw and also improved the concentration of Se and Zn and the protein levels. These supplemented *Pleurotus* species, therefore, could be an invaluable nutraceutical (Debbarma et al., 2025). Glucan-AgNPs reduced infection caused by microorganisms more promptly and has less mutagenic and cytotoxic properties when compared with AgNPs and glucans. And its non-mutagenic and non-cytotoxic activities were increased as a result of the conjugation of glucans with AgNPs. Also, these NPs do not lead to the accumulation of silver in the tissues (Irshad et al., 2024). PFPS-AuNPs showed better activity in the dendritic cell maturation induction (phenotypic markers and functions), proliferation of robust $CD4^+$ and $CD8^+$ T cells and response to cellular system. They also increased the rate of absorption of PFPS and the immune activity, murine immunity, and spleen targeting efficiency is better and half-life is longer. PFPS-AuNPs also helped DC to migrate to lymph nodes that are draining and increased the activation of $CD4^+$ and $CD8^+$ $T_{cm}$ and $T_{em}$; and cellular immune responses of human papillomavirus DC-based vaccine, which are specific to the antigens, as adjuvant. These NPs, used to prepare HPV DC-based vaccine, notably restrained the TC-1 tumor growth by increased innate and adaptive responses of the immune system (Yuan et al., 2023).

## 12.6 FUTURE PROSPECTS AND CONCLUSION

In this extensive analysis, the chapter revealed at the potential of *Pleurotus* mushrooms as a tool for environmentally friendly NP synthesis. Therapeutic *Pleurotus* mushrooms have attracted considerable attention as a sustainable and biocompatible source for NPs synthesis, owing to their minimal toxicity risks and rich bioactive compound profile. There are two main environments where mycogenic production of NPs can take place: *in vivo* and *in vitro*. The *in vivo* processes the metal precursors into non-toxic MNPs by taking them in through the cell membrane. In the *in vitro* approach, mushroom components such as proteins, enzymes, polysaccharides, glucans, and broken cell walls assist in reducing the metal precursors. But, one of the biggest obstacles in green synthesis methods that use crude extracts, and the separation and purification of the important reducing agents is cost-effective. Mushroom based synthesis not only offers a less harmful technique to synthesize NPs, but it also gives the NPs important

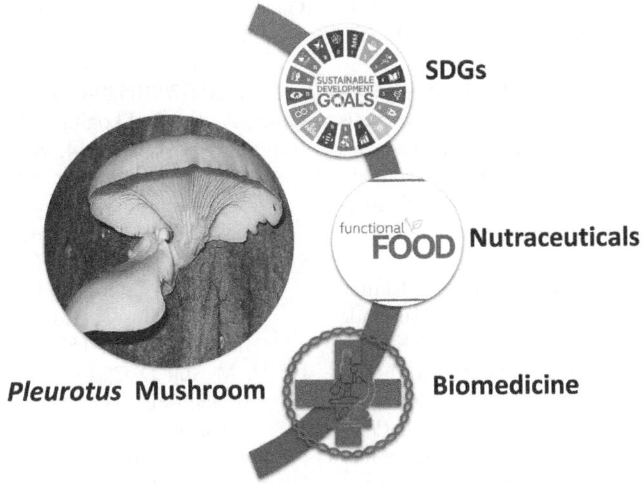


**FIGURE 12.6** Potential opportunities of *Pleurotus* derived NPs.

features, including improved stability and water dispersion. Mushroom polysaccharides and glucans act as caps, stopping NPs from growing and making them all the uniform size and shape. Anticancer, antibiotic, and antioxidant medicines, nutraceuticals are just a few of the biological domains that can benefit from these NPs along with many sustainable developments (Figure 12.6). Additionally, NPs with diverse physical properties were produced by varying the reaction conditions.

However, NPs with remarkable antibacterial and anticancer capabilities were created by modifying the parameters slightly, which led to NPs with UV-radiation, high-temperature, or brief exposure to gamma-radiation capabilities. The biomedical field uses them for drug delivery and biosensing, the industrial sector uses them for nutraceuticals, dye degradation and metal ion sensors, and the agricultural sector uses them to improve crop yields. Their remarkable physical and chemical properties make them useful in many other fields as well. The *Pleurotus* mushroom based synthesis of NPs and their applications have been the subject of substantial research. However, there is still much to be discovered, particularly regarding the precise pathways of mechanisms that contribute to the intra- and extracellular reduction of metal precursors. On top of that, artificial intelligence (AI) and computational models can help fill a gap in the discipline by allowing for finer-grained control over the desired NP features.

## REFERENCES


Acay, H., and M. Baran. 2019. Biosynthesis and characterization of silver nanoparticles using king oyster (*Pleurotus eryngii*) extract: effect on some microorganisms. *Applied Ecology and Environmental Research* 17(4): 9205–9214. https://doi.org/10.15666/aeer/1704_92059214

Acay, H., and Baran, M. 2020. Determination of antioxidant and cytotoxic activities of king oyster mushroom mediated AgNPs synthesized with environmentally friendly methods. *Medical Sciences and International Medical Journal* 9: 758. https://doi.org/10.5455/medscience.2020.02.014

Addae, E., X. Dong, E. McCoy, C. Yang, W. Chen, and L. Yang. 2014. Investigation of antimicrobial activity of photothermal therapeutic gold/copper sulfide core/shell nanoparticles to bacterial spores and cells. *Journal of Biological Engineering* 8(1): 11. https://doi.org/10.1186/1754-1611-8-11

Adebayo, E., J. Oloke, A. Adegunlola, and B. Olukemi. 2012. Phytochemical, antioxidant and antimicrobial assay of mushroom metabolite from *Pleurotus Pulmonarius*—LAU 09 (JF736658). *Journal of Microbiology and Biotechnology Research* 2: 366–374.

Bahrulolum, H., S. Nooraei, N. Javanshir, et al. 2021. Green synthesis of metal nanoparticles using microorganisms and their application in the agrifood sector. *Journal of Nanobiotechnology* 19(1): 86. https://doi.org/10.1186/s12951-021-00834-3

Bawadekji, A., A. Oueslati, A. Ali, and J. Basha. 2018. Biosynthesis of gold nanoparticles using *Pleurotus ostreatus* (Jacq. Ex. Fr.) Kummer extract and their antibacterial and antifungal activities. *Journal of Applied Environmental and Biological Sciences* 8(3): 142–147.

Behzadi, S., V. Serpooshan, W. Tao, et al. 2017. Cellular uptake of nanoparticles: journey inside the cell. *Chemical Society Reviews* 46(14): 4218–4244. https://doi.org/10.1039/c6cs00636a

Bellettini, M. B., F. A. Fiorda, H. A. Maieves, et al. 2019. Factors affecting mushroom *Pleurotus* spp. *Saudi Journal of Biological Sciences* 26(4): 633–646. https://doi.org/10.1016/j.sjbs.2016.12.005

Beyth, N., Y. Houri-Haddad, A. Domb, W. Khan, and R. Hazan. 2015. Alternative antimicrobial approach: nano-antimicrobial materials. *Evidence-Based Complementary and Alternative Medicine* 2015: 246012. https://doi.org/10.1155/2015/246012

Bhardwaj, K., A. Sharma, N. Tejwan, et al. 2020a. *Pleurotus* macrofungi-assisted nanoparticle synthesis and its potential applications: a review. *Journal of Fungi* 6(4): 351. https://doi.org/10.3390/jof6040351

Bhardwaj, K., D. Singh Dhanjal, A. Sharma, et al. 2020b. Conifer-derived metallic nanoparticles: green synthesis and biological applications. *International Journal of Molecular Sciences* 21(23): 1–22. https://doi.org/10.3390/ijms21239028

Bhat, R., R. Deshpande, S. Ganachari, D. Huh, and V. Abbaraju. 2011. Photo-irradiated biosynthesis of silver nanoparticles using edible mushroom *Pleurotus florida* and their antibacterial activity studies. *Bioinorganic Chemistry and Applications* 2011: 650979. https://doi.org/10.1155/2011/650979

Bhat, R., V. G. Sharanabasava, R. Deshpande, U. Shetti, G. Sanjeev, and A. Venkataraman. 2013. Photo-bio-synthesis of irregular shaped functionalized gold nanoparticles using edible mushroom *Pleurotus florida* and its anticancer evaluation. *Journal of Photochemistry and Photobiology B: Biology* 125: 63–69. https://doi.org/10.1016/j.jphotobiol.2013.05.002

Borovaya, M., Y. Pirko, T. Krupodorova, A. Naumenko, Y. Blume, and A. Yemets. 2015. Biosynthesis of cadmium sulphide quantum dots by using *Pleurotus ostreatus* (Jacq.) P. Kumm. *Biotechnology and Biotechnological Equipment* 29(6): 1156–1163. https://doi.org/10.1080/13102818.2015.1064264

Cai, M., G. Zhang, C. Li, X. Chen, H. Cui, and L. Lin. 2021. *Pleurotus eryngii* polysaccharide nanofiber containing pomegranate peel polyphenol/chitosan nanoparticles for control of *E. coli* O157:H7. *International Journal of Biological Macromolecules* 192: 939–949. https://doi.org/10.1016/j.ijbiomac.2021.10.069

Castro-Longoria, E., S. D. Moreno-Velázquez, A. R. Vilchis-Nestor, E. Arenas-Berumen, and M. Avalos-Borja. 2012. Production of platinum nanoparticles and nanoaggregates using *Neurospora crassa*. *Journal of Microbiology and Biotechnology* 22: 1000–1004. https://doi.org/10.4014/jmb.1110.10085

Chaturvedi, V. K., N. Yadav, N. K. Rai, et al. 2020. *Pleurotus sajor-caju*-mediated synthesis of silver and gold nanoparticles active against colon cancer cell lines: a new era of herbonanoceutics. *Molecules* 25(13): 3091. https://doi.org/10.3390/molecules25133091

Chen, Z., R. K. Kankala, L. Long, S. Xie, A. Chen, and L. Zou. 2023. Current understanding of passive and active targeting nanomedicines to enhance tumor accumulation. *Coordination Chemistry Reviews* 481: 215051. https://doi.org/10.1016/j.ccr.2023.215051

Chopra, H., S. Bibi, I. Singh, et al. 2022. Green metallic nanoparticles: biosynthesis to applications. *Frontiers in Bioengineering and Biotechnology* 10: 548. https://doi.org/10.3389/fbioe.2022.874742

Debbarma, S., S. Sharma, and A. Kalia. 2025. *Pleurotus* extract–mediated selenium and zinc nanoparticles exhibited improved yield of biofortified fruit bodies. *International Microbiology* 28(1): 107–118. https://doi.org/10.1007/s10123-024-00527-5

Debnath, G., P. Das, and A. K. Saha. 2019. Green synthesis of silver nanoparticles using mushroom extract of *Pleurotus giganteus*: characterization, antimicrobial, and α-amylase inhibitory activity. *BioNanoScience* 9(3): 611–619. https://doi.org/10.1007/s12668-019-00650-y

Devi, P. V., J. Islam, P. Narzary, D. Sharma, and F. Sultana. 2024. Bioactive compounds, nutraceutical values and its application in food product development of oyster mushroom. *Journal of Future Foods* 4(4): 335–342. https://doi.org/10.1016/j.jfutfo.2023.11.005

Dhanjal, D., S. Bhardwaj, R. Sharma, et al. 2020. Plant fortification of the diet for anti-ageing effects: a review. *Nutrients* 12: 3008. https://doi.org/10.3390/nu12103008

El-Batal, A. I., N. M. ElKenawy, A. S. Yassin, and M. A. Amin. 2015. Laccase production by *Pleurotus ostreatus* and its application in synthesis of gold nanoparticles. *Biotechnology Reports* 5: 31–39. https://doi.org/10.1016/j.btre.2014.11.001

Gade, A., A. Ingle, C. Whiteley, and M. Rai. 2010. Mycogenic metal nanoparticles: progress and applications. *Biotechnology Letters* 32(5): 593–600. https://doi.org/10.1007/s10529-009-0197-9

Gliga, A. R., S. Skoglund, I. O. Wallinder, B. Fadeel, and H. L. Karlsson. 2014. Size-dependent cytotoxicity of silver nanoparticles in human lung cells: the role of cellular uptake, agglomeration and Ag release. *Particle and Fibre Toxicology* 11(1): 11. https://doi.org/10.1186/1743-8977-11-11

Gurunathan, S., J. Han, J. H. Park, and J.-H. Kim. 2014. A green chemistry approach for synthesizing biocompatible gold nanoparticles. *Nanoscale Research Letters* 9(1): 248. https://doi.org/10.1186/1556-276X-9-248

He, D., M. Sun, D. Cao, G. He, and H. Chen. 2022. Rational design of nano-Fe3O4 encapsulated in 3D honeycomb biochar for enhanced lithium storage performance. *Nanotechnology* 33(3): 035401. https://doi.org/10.1088/1361-6528/ac2e76

Hyde, K. D., P. Baldrian, and Y. Chen, et al. 2024. Current trends, limitations and future research in the fungi? *Fungal Diversity* 125:1–71. https://doi.org/10.1007/s13225-023-00532-5

Iqbal, T., M. Sohaib, S. Iqbal, and H. Rehman. 2024. Exploring therapeutic potential of *Pleurotus ostreatus* and *Agaricus bisporus* mushrooms against hyperlipidemia and oxidative stress using animal model. *Foods* 13(5): 709. https://doi.org/10.3390/foods13050709

Irshad, A., R. Jawad, S. Sharif, et al. 2024. Bioengineering of glucan coated silver nanoparticles as dynamic biomedical compound; *in vitro* and *in vivo* studies. *Microbial Pathogenesis* 197: 107005. https://doi.org/10.1016/j.micpath.2024.107005

Ismail, A. F. M., M. M. Ahmed, and A. A. M. Salem. 2015. Biosynthesis of silver nanoparticles using mushroom extracts: induction of apoptosis in HepG2 and MCF-7 cells via caspases stimulation and regulation of BAX and Bcl-2 gene expressions. *Journal of Pharmaceutical and Biomedical Sciences* 5(1): 1–9.

Kamakshi, S., P. Anantha Lakshmi, R. Shenbhagaraman, P. Jenavio Maria Amirtham, R. Siva, G. Lakshmanan, and J. Selvakumari. 2024. Antibacterial and photo dye degradative ability of copper oxide nanoparticles from *Pleurotus cystidiosus*. *Nano Express* 5: 025029. https://doi.org/10.1088/2632-959X/ad560f

Kaur, G., A. Kalia, S. Kapoor, H. S. Sodhi, and P. K. Khanna. 2013. Scanning electron microscopy of *Pleurotus ostreatus* in response to inorganic selenium. In: *Proceedings of the Indian Mushroom Conference*, 17–18. Ludhiana, India.

Khandel, P., and S. K. Shahi. 2018. Mycogenic nanoparticles and their bio-prospective applications: current status and future challenges. *Journal of Nanostructure in Chemistry* 8(4): 369–391. https://doi.org/10.1007/s40097-018-0285-2

Kumar, H., K. Bhardwaj, K. Kuca, et al. 2020a. Flower-based green synthesis of metallic nanoparticles: applications beyond fragrance. *Nanomaterials* 10: 1–16. https://doi.org/10.3390/nano10040766

Kumar, H., K. Bhardwaj, E. Nepovimova, et al. 2020b. Antioxidant functionalized nanoparticles: a combat against oxidative stress. *Nanomaterials* 10(7): 1334. https://doi.org/10.3390/nano10071334

Kumar, M., B. Govindasamy, R. Devarajan, and S. Murugesan. 2020c. Biological applications of synthesized ZnO nanoparticles using *Pleurotus djamor* against mosquito larvicidal, histopathology, antibacterial, antioxidant and anticancer effect. *Journal of Cluster Science* 32: 1635–1647. https://doi.org/10.1007/s10876-020-01927-z

Kumar, M., D. Yanto, M. Govindasamy, B. Karunanithi, F. Alasmary, and R. Habab. 2023. Biological synthesis and characterization of iron oxide (FeO) nanoparticles using *Pleurotus citrinopileatus* extract and its biomedical applications. *Biomass Conversion and Biorefinery* 14: 1–11. https://doi.org/10.1007/s13399-023-04382-8

Lei, Z., H. Qiuhui, Z. Qiping, Z. Mingwen, and Z. Liyan. 2024. Oat protein isolate-*Pleurotus ostreatus* β-glucan conjugate nanoparticles bound to β-carotene effectively alleviate immunosuppression by regulating gut microbiota. *Food and Function* 15(4): 1867–1883. https://doi.org/10.1039/D3FO05158G

Lok, C-N., C-M. Ho, R. Chen, et al. 2007. Silver nanoparticles: partial oxidation and antibacterial activities. *Journal of Biological Inorganic Chemistry* 12(4): 527–534. https://doi.org/10.1007/s00775-007-0208-z

Madhanraj, R., M. Eyini, and P. Balaji. 2017. Antioxidant assay of gold and silver nanoparticles from edible Basidiomycetes mushroom fungi. *Free Radicals and Antioxidants* 7(2): 137–142. https://doi.org/10.5530/fra.2017.2.20

Manimaran, K., G. Balasubramani, C. Ragavendran, et al. 2020. Biological applications of synthesized ZnO nanoparticles using *Pleurotus Djamor* against mosquito larvicidal, Histopathology, antibacterial, antioxidant and anticancer effect. *Journal of Cluster Science* 32: 1635–1647. https://doi.org/10.1007/s10876-020-01927-z

Manimaran, K., S. Murugesan, C. Ragavendran, et al. 2021. Biosynthesis of $TiO_2$ nanoparticles using edible mushroom (*Pleurotus djamor*) extract: mosquito larvicidal, histopathological, antibacterial and anticancer effect. *Journal of Cluster Science* 32(5): 1229–1240. https://doi.org/10.1007/s10876-020-01888-3

Manimaran, K., D. H. Y. Yanto, F. C. Ardiati, et al. 2023a. Enhanced photocatalytic degradation, antimicrobial and anticancer efficiency of mycosynthesized TiO2 nanoparticles using *Pleurotus ostreatus* mushroom extract: an eco-friendly approach. *Journal of Environmental Chemical Engineering* 11(6): 111512. https://doi.org/10.1016/j.jece.2023.111512

Manimaran, K., D. H. Y. Yanto, C. Kamaraj, et al. 2023b. Eco-friendly approaches of mycosynthesized copper oxide nanoparticles (CuONPs) using *Pleurotus citrinopileatus* mushroom extracts and their biological applications. *Environmental Research* 232: 116319. https://doi.org/10.1016/j.envres.2023.116319

Mathur, P., and S. Pillai. 2025. https://doi.org/10.1016/j.fbr.2025.100445

Maurya, S., A. K. Bhardwaj, K. K. Gupta, et al. 2016. Green synthesis of silver nanoparticles using *Pleurotus* and its bactericidal activity. *Cellular and Molecular Biology* 62: 131. https://doi.org/10.4172/1165-158X.1000131

Mazumdar, H., and N. Haloi. 2011. A study on biosynthesis of iron nanoparticles by *Pleurotus* sp. *Journal of Microbiology and Biotechnology Research* 1(3): 39–49.

McQuillan, J., H. G. Infante, E. Stokes, and A. Shaw. 2011. Silver nanoparticle enhanced silver ion stress response in *Escherichia coli* K12. *Nanotoxicology* 6: 857–866. https://doi.org/10.3109/17435390.2011.626532

Meher, A., A. Tandi, S. Moharana, et al. 2024. Silver nanoparticle for biomedical applications: a review. *Hybrid Advances* 6: 100184. https://doi.org/10.1016/j.hybadv.2024.100184

Mirunalini, S., V. Arulmozhi, K. Deepalakshmi, and M. Krishnaveni. 2012. Intracellular biosynthesis and antibacterial activity of silver nanoparticles using edible mushrooms. *Notulae Scientia Biologicae* 4(4): 55–61. https://doi.org/10.15835/nsb448051

Mohanta, Y. K., D. Nayak, K. Biswas, et al. 2018. Silver nanoparticles synthesized using wild mushroom show potential antimicrobial activities against food borne pathogens. *Molecules* 23: 655. https://doi.org/10.3390/molecules23030655. ISSN: 1420–3049.

Molnár, Z., V. Bódai, G. Szakacs, et al. 2018. Green synthesis of gold nanoparticles by thermophilic filamentous fungi. *Scientific Reports* 8(1): 3943. https://doi.org/10.1038/s41598-018-22112-3

Musa, S. F., T. S. Yeat, L. Z. M. Kamal, et al. 2018. *Pleurotus sajor-caju* can be used to synthesize silver nanoparticles with antifungal activity against *Candida albicans*. *Journal of the Science of Food and Agriculture* 98(3): 1197–1207. https://doi.org/10.1002/jsfa.8573

Narayana, M. S., R. Nambi, C. Dastagiri, et al. 2024. *Pleurotus ostreatus* copper nanoparticles: *in vitro* and *in silico* evaluation of the antioxidant, antibacterial and antidiabetic activities. *Chemistry and Biodiversity* 22(5): e202402361. https://doi.org/10.1002/cbdv.202402361

Narayanan, K. B., and N. Sakthivel. 2010. Biological synthesis of metal nanoparticles by microbes. *Advances in Colloid and Interface Science* 156(1): 1–13. https://doi.org/10.1016/j.cis.2010.02.001

Nayak, D., A. K. Mishra, S. Malla, et al. 2024. Eco-friendly mycogenic synthesis of Iron Oxide nanoparticles using *Volvariella volvacea* and unveiling its potential as antibiofilm and anticancer agents. *ECS Journal of Solid State Science and Technology* 13: 087003. 2162–8777. https://doi.org/10.1149/2162-8777/ad658b

Osemwegie, O. O., C. O. Adetunji, E. A. Ayeni, et al. 2020. Exopolysaccharides from bacteria and fungi: current status and perspectives in Africa. *Heliyon* 6(6): e04205. https://doi.org/10.1016/j.heliyon.2020.e04205

Osman, A. I., Y. Zhang, M. Farghali, et al. 2024. Synthesis of green nanoparticles for energy, biomedical, environmental, agricultural, and food applications: a review. *Environmental Chemistry Letters* 22(2): 841–887. https://doi.org/10.1007/s10311-023-01682-3

Owaid, M. N. 2019. Green Synthesis of silver nanoparticles by *Pleurotus* (oyster mushroom) and their bioactivity: review. *Environmental Nanotechnology, Monitoring and Management* 12: 100256. https://doi.org/10.1016/j.enmm.2019.100256

Owaid, M. N., and I. J. Ibraheem. 2017. Mycosynthesis of nanoparticles using edible and medicinal mushrooms. *European Journal of Nanomedicine* 9(1): 5–23. https://doi.org/10.1515/ejnm-2016-0016

Owaid, M. N., J. Raman, H. Lakshmanan, et al. 2015. Mycosynthesis of silver nanoparticles by *Pleurotus cornucopiae* var. *citrinopileatus* and its inhibitory effects against *Candida* sp. *Materials Letters* 153: 186–190. https://doi.org/10.1016/j.matlet.2015.04.023

Panda, J., A. K. Mishra, P. C. Nath, et al. 2024. Wild edible mushrooms to achieve sustainable development goals: novel sources for food security, health, and well-being. *Food Bioscience* 60: 104277. https://doi.org/10.1016/j.fbio.2024.104277

Panda, J., A. Nongbet, S. K. Avula, et al. 2025. Macrofungi mediated nanoparticles for sustainable agriculture: recent advancement and future strategies. *Discover Sustainability* 6:375. https://doi.org/10.1007/s43621-025-01219-4

Patra, J. K., G. Das, L. F. Fraceto, et al. 2018. Nano based drug delivery systems: recent developments and future prospects. *Journal of Nanobiotechnology* 16(1): 71. https://doi.org/10.1186/s12951-018-0392-8

Raman, J., G. R. Reddy, H. Lakshmanan, et al. 2015. Mycosynthesis and characterization of silver nanoparticles from *Pleurotus djamor* var. *roseus* and their *in vitro* cytotoxicity effect on PC3 cells. *Process Biochemistry* 50(1): 140–147. https://doi.org/10.1016/j.procbio.2014.11.003

Rami, M. R., M. Meskini, and B. E. Sharafabad. 2024. Fungal-mediated nanoparticles for industrial applications: synthesis and mechanism of action. *Journal of Infection and Public Health* 17(10): 102536. https://doi.org/10.1016/j.jiph.2024.102536

Sankar, R., A. Karthik, A. Prabu, S. Karthik, K. S. Shivashangari, and V. Ravikumar. 2013. *Origanum vulgare* mediated biosynthesis of silver nanoparticles for its antibacterial and anticancer activity. *Colloids and Surfaces B: Biointerfaces* 108: 80–84. https://doi.org/10.1016/j.colsurfb.2013.02.033

Šebesta, M., H. Vojtková, V. Cyprichová, A. P. Ingle, M. Urík, and M. Kolenčík. 2022. Mycosynthesis of metal-containing nanoparticles—fungal metal resistance and mechanisms of synthesis. *International Journal of Molecular Sciences* 23(22): 14084. https://doi.org/10.3390/ijms24010304

Sen, I. K., K. Maity, and S. S. Islam. 2013. Green synthesis of gold nanoparticles using a glucan of an edible mushroom and study of catalytic activity. *Carbohydrate Polymers* 91(2): 518–528. https://doi.org/10.1016/j.carbpol.2012.08.058

Senapati, U. S., and D. Sarkar. 2014. Characterization of biosynthesized zinc sulphide nanoparticles using edible mushroom *Pleurotuss ostreatus*. *Indian Journal of Physics* 88(6): 557–562. https://doi.org/10.1007/s12648-014-0456-z

Sharma, D., and D. S. Dhanjal. 2016. Bio-nanotechnology for active food packaging. *Journal of Applied Pharmaceutical Science* 6(9): 220–226. https://doi.org/10.7324/JAPS.2016.60933

Sudheer, S., R. G. Bai, K. Muthoosamy, R. Tuvikene, V. K. Gupta, and S. Manickam. 2022. Biosustainable production of nanoparticles via mycogenesis for biotechnological applications: a critical review. *Environmental Research* 204: 111963. https://doi.org/10.1016/j.envres.2021.111963

Sujatha, S., S. Tamilselvi, K. Subha, and A. Panneerselvam. 2013. Studies on biosynthesis of silver nanoparticles using mushroom and its antibacterial activities. *International Journal of Current Microbiology and Applied Sciences* 2(12): 605–614.

Thulasinathan, B., V. Ganesan, P. Manickam, et al. 2023. Simultaneous electrochemical determination of persistent petrogenic organic pollutants based on AgNPs synthesized using carbon dots derived from mushroom. *Science of the Total Environment* 884: 163729. https://doi.org/10.1016/j.scitotenv.2023.163729

Wang, L., C. Hu, and L. Shao. 2017. The antimicrobial activity of nanoparticles: present situation and prospects for the future. *International Journal of Nanomedicine* 12: 1227–1249. https://doi.org/10.2147/IJN.S121956

Wang, S., X. Gao, Q. Sun, Y. Zhu, L. Qin, and B. Zhu. 2022. The phytochemical properties, pharmacological effects and traditional uses of *Actinidia eriantha* Benth.: a review. *Frontiers in Pharmacology* 13: 959900. https://doi.org/10.3389/fphar.2022.959900

Warnes, S., V. Caves, and C. Keevil. 2011. Mechanism of copper surface toxicity in *Escherichia coli* O157:H7 and *Salmonella* involves immediate membrane depolarization followed by slower rate of DNA destruction which differs from that observed for Gram-positive bacteria. *Environmental Microbiology* 14: 1730–1743. https://doi.org/10.1111/j.1462-2920.2011.02677.x

Wong, K. 2012. Preparation of highly stable selenium nanoparticles with strong anti-tumor activity using tiger milk mushroom. *Apoptosis* 134: 253–261.

Wu, H.-F., S. K. Kailasa, and L. Shastri. 2010. Electrostatically self-assembled azides on zinc sulfide nanoparticles as multifunctional nanoprobes for peptide and protein analysis in MALDI-TOF MS. *Talanta* 82(2): 540–547. https://doi.org/10.1016/j.talanta.2010.05.026

Yehia, R. S., and H. Al-Sheikh. 2014. Biosynthesis and characterization of silver nanoparticles produced by *Pleurotus ostreatus* and their anticandidal and anticancer activities. *World Journal of Microbiology and Biotechnology* 30(11): 2797–2803. https://doi.org/10.1007/s11274-014-1703-3

Yuan, P., L. Liu, A. Aipire, et al. 2023. Evaluation and mechanism of immune enhancement effects of *Pleurotus ferulae* polysaccharides-gold nanoparticles. *International Journal of Biological Macromolecules* 227: 1015–1526. https://doi.org/10.1016/j.ijbiomac.2022.11.277

# 13 Postharvest Management and Recipes of Cultivated *Pleurotus*

+-Debosmita Mukherjee[2], +-Moumita Parveen[1], Dipanwita Sahoo[1], Kankana Saha[1], and Narayan Chandra Mandal[1*]

[1] Mycology and Plant Pathology Laboratory, Department of Botany, Siksha Bhavana (Institute of Science), Visva Bharati, Santiniketan, West Bengal, India

[2] Directorate of Cinchona and Other Medicinal Plants, Mungpoo, West Bengal, India

+-Equal contribution

## 13.1 INTRODUCTION

People have always considered mushrooms to be a culinary marvel and delicacy, especially due to their unique flavour. More than two thousand types of mushrooms have been recorded worldwide; but only twenty-five types are widely considered as edible, and farmed commercially in a small number (Jayachandran et al., 2017; Cao et al., 2024). *Pleurotus* species or oyster mushroom is the most common type of mushroom cultivated worldwide and widely recognized for their exceptional nutritional value and environmental sustainability. They are low in fat and cholesterol; while being rich in essential nutrients such as proteins, complex carbohydrates, dietary fibre, vitamin D, and key minerals like potassium, phosphorus, selenium, and zinc (Sharma et al., 2017; Sawangwan et al., 2018). Unlike many plant-based foods, oyster mushrooms contain all essential amino acids, which make them a high-quality protein source suitable for diverse dietary preferences, including vegetarian and vegan diets (Bhagarathi et al., 2023). Notably, one of the distinguishing features of *Pleurotus* mushrooms is the presence of their prebiotic components, like- chitin, hemicellulose, and β-glucans (Aida et al., 2009; Sawangwan et al., 2018). These bioactive polysaccharides contribute to their functional food properties by promoting gut health, enhancing immune function, and exhibiting antibacterial, antioxidant, antiviral, and antiparasitic activities (Aida et al., 2009; Khatua et al., 2013). Additionally, *Pleurotus* mushrooms require minimal land, water, and energy inputs for cultivation, making them an efficient and eco-friendly food source compared to conventional crops and animal-based proteins (Kiiza et al., 2023).

Oyster mushrooms are the widely cultivated mushroom species all over the world, accounting for approximately 27% of total mushroom production (Jarial et al., 2024). China leads with 74% of worldwide oyster mushroom production. Countries like Italy, Poland, Netherlands, Romania, the Republic of Korea, Spain, Lithuania, and India also contribute enough to fulfil the market demand. Their commercial cultivation has expanded significantly due to their adaptability to various climates and their ability to grow on lignocellulosic agricultural waste materials such as straw, sawdust, and husks (Bhagarathi et al., 2023). Oyster mushrooms were originally cultivated in Germany during World War I as a survival food, and presently, it gained widespread acceptance for its excellent taste, aroma, and medicinal properties (El-Batal et al., 2015). The global demand for these mushrooms has surged due to increasing consumer interest in sustainable and functional foods. In India, oyster mushrooms rank second in production after button mushrooms, with their cultivation expanding rapidly due to favourable climatic conditions and growing market demand (Sharma et al., 2017). India's mushroom industry is experiencing an annual growth rate of over 15%, contributing significantly to domestic consumption and export markets. The country has become a key supplier of dried and processed oyster mushrooms, particularly for international markets such as the United States. Furthermore, Indian food industries are actively developing value-added products such as mushroom-enriched baked foods, dietary supplements, and protein-rich powder for the health-conscious citizen (Binsaleh et al., 2025).

One of the most promising applications of oyster mushrooms is their potential to serve as a sustainable and nutritious alternative to traditional protein sources. Due to their high protein content, well-balanced amino acid profile, and fibrous texture, *Pleurotus* mushrooms have been incorporated into meat analogues, offering a viable solution for reducing dependence on animal-based proteins. Unlike conventional plant-based meat substitutes, which often require extensive processing and artificial flavoring, oyster mushrooms naturally possess umami-rich compounds such as glutamic acid, aspartic acid, and ribonucleotides, which contribute to their meaty taste (Chirinang et al., 2009).

Their application in food fortification has also been widely studied, particularly in baked and processed foods. For example, substituting wheat flour with oyster mushroom powder in bread formulations has been found to improve protein and fibre content without significantly compromising

* Corresponding author: narayanchandra.mandal@visva-bharati.ac.in

DOI: 10.1201/9781003527572-13

sensory acceptability (Sangeeta et al., 2024). Similarly, mushroom-based protein blends are being developed for use in dairy alternatives, functional beverages, and nutraceutical formulations (Bulam et al., 2022). As consumer awareness of sustainable and plant-based diets continues to grow, *Pleurotus* mushrooms are expected to play a pivotal role in reshaping global food systems. Research on enhanced preservation technologies, such as edible coatings and bio-preservation techniques, aims to extend shelf-life while maintaining nutritional quality (Sun et al., 2023; Huo et al., 2023).

Despite their numerous advantages, oyster mushrooms are highly perishable, with a shelf-life of only a few days under ambient conditions. Their high moisture content (85%–95%) makes them particularly susceptible to microbial spoilage, enzymatic browning, and rapid textural degradation (Maray et al., 2018). Once harvested, oyster mushrooms undergo physiological and biochemical changes, leading to quality deterioration that affects their appearance, texture, flavour, and nutritional value (Sharma et al., 2007; Schill et al., 2021; Cao et al., 2024). Microbial contamination is one of the primary causes of spoilage in *Pleurotus* mushrooms. Several fungal pathogens, including *Aspergillus*, *Rhizopus*, *Penicillium*, *Cladobotryum*, and *Trichoderma* species, contribute to decay by producing mycotoxins and breaking down the mushroom's cellular structure. These fungi thrive in high-humidity environments and can cause mold growth, discoloration, and off-flavors, rendering the mushrooms unmarketable. Additionally, bacterial infections caused by *Bacillus*, *Pseudomonas*, *Enterobacter*, and *Enterococcus* species can lead to slimy texture, unpleasant odors, and soft rot. *Pseudomonas tolaasii*, in particular, is responsible for bacterial blotch disease, which causes brown discoloration and water-soaked lesions on the mushroom surface, significantly reducing its market value (Liang et al., 2024).

In addition to microbial spoilage, enzymatic browning is another major postharvest issue. This process is driven by polyphenol oxidase (PPO) and peroxidase (POD) enzymes, which catalyze the oxidation of phenolic compounds in mushroom tissues, leading to undesirable darkening. The oxidation reaction accelerates when mushrooms are bruised or exposed to high oxygen levels, making them appear aged and less appealing to consumers. Added to this, mechanical injuries sustained during harvesting, handling, and transportation further exacerbate browning, as damaged tissues become more susceptible to microbial invasion and moisture loss (Zhu et al., 2025).

Another key challenge in *Pleurotus* postharvest management is textural degradation due to rapid water loss. The mushrooms have a soft, spongy texture that is highly sensitive to dehydration. Even minor moisture loss can cause shrinkage, wrinkling, and hardening of the fruiting bodies. The absence of a protective cuticle, as found in other fruits and vegetables, makes oyster mushrooms particularly prone to desiccation when stored under low-humidity conditions. High respiration rates also contribute to tissue softening and breakdown, making proper storage conditions crucial for maintaining quality and extending shelf-life (Jafri et al., 2013; Ramteke et al., 2020).

These spoilage factors collectively impact the marketability of oyster mushrooms, leading to significant economic losses for growers and suppliers. To mitigate these issues, various postharvest preservation strategies have been explored, ranging from traditional drying and refrigeration to advanced processing techniques such as high-pressure processing (HPP), gamma irradiation, ultraviolet (UV) treatment, and modified atmosphere packaging (MAP) (Lu et al., 2012).

This book chapter focuses on the postharvest management and food applications of *Pleurotus* mushrooms. It explores various preservation methods, including both conventional and innovative techniques, and examines their roles. Additionally, the chapter highlights the growing potential of *Pleurotus* species in functional food development and meat analogue industries. By addressing the postharvest management techniques followed for oyster mushroom, this chapter aims to provide comprehensive insights into optimizing the utilization, food processing, and market value of *Pleurotus* mushrooms for the human consumption.

## 13.2 CULTIVATION OF *PLEUROTUS*

The nutritional, economic, and ecological advantages of mushroom cultivation, particularly that of *Pleurotus ostreatus* have positioned it as an essential element of modern sustainable agriculture. *Pleurotus ostreatus* is prized for its low fat content and rich nutrient profile, including dietary fiber, high-quality protein, essential vitamins, and minerals, making it a valuable addition to the global food market and a key component of a balanced diet. Moreover, its remarkable ability to grow on a wide array of substrates including agricultural and industrial wastes enables the conversion of lignocellulosic biomass into edible protein. This not only contributes to food security but also supports environmental sustainability by promoting efficient organic waste recycling (Thakur, 2020; Aditya et al., 2024).

According to recent production data, several oyster mushroom species are frequently grown in India inclusing *P. ostreatus*, *P. florida*, *P. sajor-caju*, *P. cystidiosus*, *P. tuberregium*, and *P. djamor*. The ideal temperature range for oyster mushroom growth is 22–30 °C. Numerous lignocellulosic substrates, such as weeds, wood, straw, husks, and plant remnants, can support their large-scale cultivation (Figure 13.1). In the pasteurization and composting procedures used on mushrooms, many kinds of straws are used. Furthermore, the length of the straw, the inoculum ratio during substrate inoculation, and the substrate quality significantly influence the biological efficiency of production (Sánchez et al., 2010).

## 13.3 NUTRITIONAL VALUE AND HEALTH BENEFITS OF *PLEUROTUS*

### 13.3.1 Nutritional Value

The protein deficit has been fulfilled in recent decades by edible mushrooms, which replaced a diet predominantly

**FIGURE 13.1** Cultivation of *Pleurotus ostreatus* (28 ± 2 °C).

composed of cereals, such as wheat, rice, and maize. In addition, mushrooms are regarded as a high-quality food because of their health advantages and nutritional content. Numerous investigations have evaluated the nutritional value of *Pleurotus*, primarily using dried fruiting bodies as the standard form for analysis. The fruiting body of fresh *Pleurotus* contains many distinct bioactive compounds, including phenolic compounds, flavonoids, hydroxycinnamic acids, hydroxybenzoic acids, lignans, tannins, stilbenes, and oxidized polyphenols, and it also contains 85%–95% moisture. Additionally, *Pleurotus* is fat-free, rich in protein, carbohydrate, minerals (potassium, iron, copper, zinc, and manganese), the vitamins $B_1$, $B_2$, $B_{12}$, C, D, E, and K, and dietary fibers. In fact, dietary *P. ostreatus* mushrooms rank as the second most significant mushroom in the world for food supplements. Consumers today are very interested in bioactive foods that have positive effects on boosting human health and lowering risks. Mushrooms are regarded as a functional food that offers both nutritional value and health advantages (Lesa et al., 2022). The following section discusses the most prevalent nutrients found in *Pleurotus*.

#### 13.3.1.1 Carbohydrate

The structural skeleton of mushrooms primarily consists of carbohydrates, which are crucial for preserving the high osmotic concentration and supporting the rapid metabolism rate while maintaining energy release. The percentage of carbohydrates in *Pleurotus* mushrooms ranges from 24.95% to 75.88%. Chitin and polysaccharides make up a significant portion of the nutrients found in mushrooms (Raman et al., 2021).

#### 13.3.1.2 Protein Contents

Edible oyster mushrooms are rapidly recognized as a promising source of novel proteins. The crude protein level of dietary oyster mushrooms is usually high. The protein content of mushrooms varied from 12% to 35% depending on the species. Several research articles conclude that the digestibility of oyster mushroom protein is as high as 72% to 83%. Since mushrooms contain several essential amino acids that are present in animals, they are highly significant for vegetarians (Raman et al., 2021).

#### 13.3.1.3 Lipid

The fat content of *Pleurotus* species is very low, ranging between 1.08% and 9.4%, and has an average value of 2.85%. The chief lipid constituents are: triglycerides (61.4%), phospholipids (55.3%), free fatty acids. The dominant fatty acids are oleic acid (79.4%), palmitic acid (14.3%), and linoleic acid (6.3%), which are of critical significance to human health. The richness of linoleic acid in *Pleurotus* enhances the nutritional significance, as this essential fatty acid cannot be synthesized by the human body and must be obtained through diet (Silva et al., 2002).

#### 13.3.1.4 Minerals

Oyster mushrooms are valued for their rich nutritional profile, which includes a wide range of vital minerals. According to scientific investigations, the health-promoting qualities of *Pleurotus* mushrooms are attributed to their significant levels of potassium (K), phosphorus (P), magnesium (Mg), calcium (Ca), and iron (Fe). The most prevalent mineral, potassium, is essential for maintaining fluid balance and neuronal function. Iron promotes oxygen transport and immunological function; whereas phosphorus and magnesium are essential for energy metabolism and bone health. Furthermore, trace metals, including copper (Cu), zinc (Zn), and selenium (Se), are present in significant concentrations and support enzymatic and antioxidant activities (Khan et al., 2012).

#### 13.3.1.5 Vitamins

According to references, the fruit bodies of mushrooms are rich in vitamins, particularly vitamins B (specifically thiamine, folic acid, pyridoxine, pantothenic acid, riboflavin, nicotinic acid, nicotinamide, and cobalamin), C, and $D_2$. Several studies have shown that *P. ostreatus* mushrooms are low in vitamins $B_{12}$, C, and $B_2$ and high in folacin, vitamin $B_1$, and vitamin $B_3$ (Effiong et al., 2024).

## 13.4 HEALTH BENEFITS AND MEDICINAL IMPORTANCE OF *PLEUROTUS* SP.

Mushrooms are considered "the ultimate health food" and have been utilized in traditional medicine all across the world since inception. The fruit body, mycelium, and

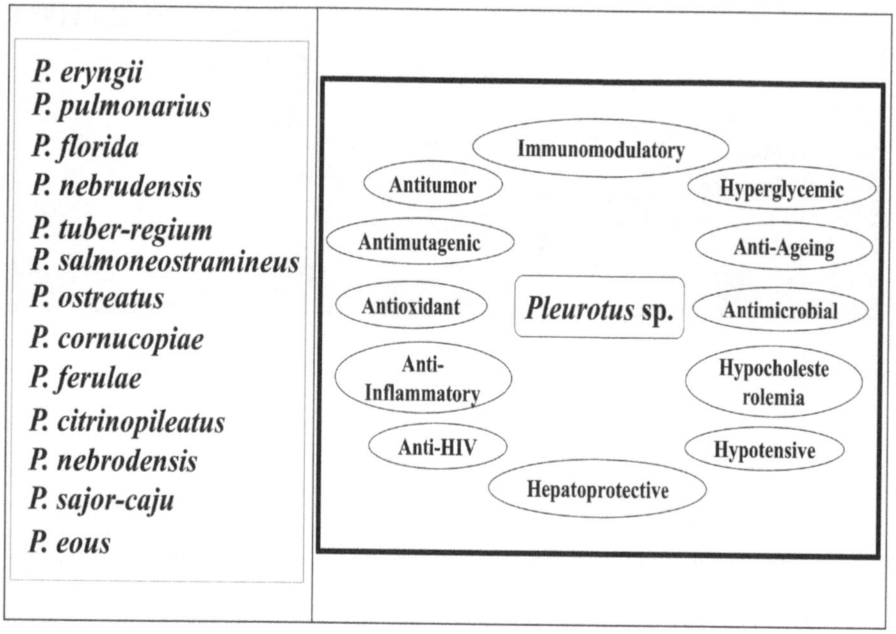


**FIGURE 13.2** Diffferent *Pleurotus* species with their numerous health-promoting effects.

extracts of edible macrofungi have been considered functional foods due to their potential to treat a variety of human diseases. The identification of these mushrooms as "Green Medicine" has shown promising results in lowering oxidative stress and improving general health due to their antioxidant components (Irfan et al., 2022). To evaluate their anti-inflammatory, immunomodulatory, antimicrobial, antiviral, hepatoprotective, anti-cancer, antihypercholesterolemic, antioxidative, antihypertensive, and anti-aging qualities, numerous studies have been carried out (Figure 13.2) (Adams et al., 2019). Positive aspects of the mushrooms' incorporation in dietary considerations are also presented by the analysis of their nutritional qualities (Medihi et al., 2024).

## 13.5 SPAWN SPOILAGE AND POSTHARVEST DETERIORATION OF *PLEUROTUS*

Cultivating *P. ostreatus* presents several challenges, with one of the most significant being its susceptibility to various diseases. Fungal pathogens, in particular, pose a major threat as they can contaminate the substrate or directly infect the mushrooms during growth. This can result in considerable losses in output and quality. To improve mushroom output and control illnesses, there is a growing interest in investigating sustainable and ecological solutions (Prasad et al., 2023). Abdel-Aziz et al. (2016) investigated the microbial ecology of oyster mushrooms across various stages from cultivation in bottles or on shelves to distribution in supermarkets and outdoor markets. Their findings raised significant hygiene concerns, as these conditions may contribute to the risk of foodborne disease outbreaks. Notably, both pre- and post-distribution samples showed a high prevalence of bacterial families such as Pseudomonadaceae and Enterobacteriaceae, indicating potential microbial contamination throughout the supply chain. Scientists reported that *Pseudomonas fluorescens*, *Ewingella americana*, and *Pantoea beijingensis* were the cause of oyster mushroom rotting (Tajalipour et al., 2014; Sharma et al., 2007). Our observations conclusively demonstrated that spawn spoilage in *P. ostreatus* is primarily mediated by *Rhizopus delemar*, along with species of *Penicillium* and *Aspergillus*. Additionally, bacterial involvement in spawn deterioration was identified, with *Citrobacter freundii*, *Enterobacter hormaechei*, and *Enterobacter cloacae* being the predominant species implicated (Mukherjee et al., 2025).

Mushrooms have a very short shelf life (5–7 days at 1–4 °C) due to their high respiration rate and lack of natural barriers to prevent water loss and microbial invasion. During the postharvest period, their quality rapidly declines, leading to moisture loss, discoloration, texture degradation, off-flavors, nutrient loss, and microbial spoilage caused by bacteria and fungi. These issues have resulted in substantial economic losses for farmers. To address this, it is crucial to preserve mushrooms in a way that prevents bacterial and fungal contamination. As a result, various postharvest preservation techniques have been developed and applied (Lyn et al., 2020).

Due to their exceptional perishability, the postharvest management of *Pleurotus* mushrooms is particularly challenging and demands meticulous care. Unlike many other vegetables, mushrooms lack a protective outer layer, making them highly susceptible to microbial contamination (especially fungal: Table 13.1), enzymatic degradation, and moisture loss. Therefore, efficient postharvest handling is crucial to minimize financial losses, extend shelf life, and preserve their quality (Guo et al., 2023).

**FIGURE 13.3** Various types of recipes prepared using *Pleurotus*.

**TABLE 13.1**
**Spoilage Causing Organisms of Different *Pleurotus* sp.**

| Name of Different *Pleurotus* sp. | Causal Organism for Spoilage | References |
|---|---|---|
| *Pleurotus ostreatus* | *Fusarium oxysporium* | Omokaro and Ogechi, 2013; Suada et al., 2015; Mwangi, 2016; Agbagwa et al., 2020; Sobowale et al., 2022; Mukherjee et al., 2025 |
| | *Aspergillus flavus* | |
| | *Aspergillus niger* | |
| | *Rhizopus oryzae* | |
| | *Trichoderma harzianum* | |
| | *Trichoderma asperellum* | |
| | *Penicillium frequentans* | |
| | *Mucor* spp. | |
| | *Rhizopus delemer* RH1 | |
| *Pleurotus sajor- caju* | *Aspergillus* sp. | Fadahunsi et al., 2013; Bhargav et al., 2023 |
| | *Fusarium chlamydosporum* | |
| | *Phoma exigua* | |
| | *Aspergillus niger* | |
| | *Aspergillus fumigatus* | |
| | *Penicillium chrysogenum* | |
| | *Fusarium pallidoroseum* | |
| | *Botryodiplodia theobromae* | |
| | *Rhizopus Stolonifer* | |
| *P. pulmonarius* | *Rhizopus arrhizus* | Hassan and Ibrahim, 2022; Sobowale et al., 2022 |
| | *Mucor hiemalis* | |
| | *Penicillium glabrum* | |
| | *Trichoderma longibrachiatum* | |
| | *Trichoderma pleuroticola* | |
| | *Aspergillus flavus* | |

**Why do we need mushrooms to be preserved?**

- To prevent rapid texture softening, which affects their quality and consumer appeal
- To avoid loss of essential nutrients such as proteins and carbohydrates during storage
- To inhibit the formation of brown discoloration, which results from enzymatic browning and degrades visual quality
- To minimize the depletion of soluble bioactive compounds that contribute to flavor and health benefits
- To ensure food security and safety, as mushrooms are highly perishable and prone to microbial spoilage

### 13.5.1 Excessive Moisture Content and Susceptibility to Fungal Contamination

The high-water content of *Pleurotus* (about 85%–95%) makes them prone to rapid desiccation when dried at room temperature. *Pleurotus flabellatus*, for instance, loses about 32% of its water content in 24 hours at 22–28 C. This moisture loss leads to wilting, shrinkage, and texture degradation, rendering the mushrooms unmarketable (Silva et al., 2025). Additionally, the high water activity accelerates microbial growth, resulting in rapid spoilage. Without proper storage and preservation, mushrooms quickly become slimy, discolored, and unsuitable for consumption within just a few days after harvest (Eswaran and Ramabadran, 2000).

### 13.5.2 Quick Deterioration of Quality

Exposure to oxygen triggers PPO, which accelerates the browning process in mushrooms. As cell walls begin to break down, structural distortion occurs, leading to a loss of firmness and a mushy texture. Inadequate storage conditions further exacerbate the problem, causing rapid moisture loss. This results in weight reduction and a significant decline in market value (Li et al., 2021).

### 13.5.3 Restricted Shelf-Life

*Pleurotus* mushrooms have a short shelf-life of two to three days at room temperature because of their high metabolic activity and environmental sensitivity. The shelf-life, even when refrigerated, is only 7–14 days (Singh et al., 2010). Several preservation techniques, including drying, refrigeration, MAP, and chemical treatments, must be used to avoid waste (Schill et al., 2021).

### 13.5.4 Preventing Economic Losses

Farmers, traders, and suppliers suffer significant financial losses as a result of inadequate postharvest management. Due to spoiling, a significant portion of produced mushrooms never make it to the market. Profitability is increased by using effective postharvest methods to keep mushrooms fresh and marketable for extended periods (Raman et al., 2021).

### 13.5.5 Ensure Safety and Quality

Foodborne diseases can result from the contamination of fresh mushrooms by bacteria and fungi such as *Pseudomonas* sp., *Escherichia coli*, and *Aspergillus* sp. The microbial load can be reduced, and consumer safety is ensured by appropriate storage, packaging, and sanitation practices (Silva et al., 2025).

### 13.5.6 Retaining Medicinal and Nutritional Properties

*Pleurotus* mushrooms are high in proteins, minerals (potassium, selenium), vitamins (B-complex, D), and bioactive substances such as β-glucans and polysaccharides that have immune-stimulating, anti-inflammatory, and antioxidant qualities (Zheng et al., 2025). The nutritional

**TABLE 13.2**
**Postharvest Preservation Techniques and the Purpose of Application**

| Traditional Methods | Advanced Methods | Common Purpose |
|---|---|---|
| Cold Storage (1–4 °C) | Vacuum-sealing | • Extend the shelf-life |
| | Chemical treatment | • Enhance nutritional value |
| Drying (Hot Air) | High electric field (HEF) treatment | • Maintain texture and flavor |
| | Irradiation | • Reduce respiration rates |
| Freezing (−18 °C) | Cold Plasma Technology | • Long-term preservation |
| | Modified Atmosphere Packaging (MAP) | • Value addition and preservation |
| Blanching | Chitosan Coating | |
| | High-Pressure Processing (HPP) | |
| Pickling and Fermentation | Ozone Treatment | |
| | Ultrasound treatment | |
| | Electrolyzed Water | |
| Biopreservation: Probable application of lactic acid bacteria (LAB) | | |

and therapeutic benefits of these bioactive substances are diminished if stored improperly since they break down rapidly (Raman et al., 2021).

Based on the above factors it is very much necessary to prevent mushroom spoilage. By properly treating mushrooms with a variety of chemicals and physical treatments, microbial spoilage can be prevented. There are two different postharvest preservation strategies to prevent mushroom spoilage: Traditional preservation methods and advanced preservation techniques (Table 13.2).

## 13.6 POSTHARVEST PRESERVATION STRATEGIES AND MANAGEMENT

### 13.6.1 Traditional Preservation Methods

#### 13.6.1.1 Drying

For controlling microbial and biochemical reactions, drying is a good method where moisture is removed from the product. As water content is reduced, the shelf-life increases. Perhaps drying is an old-aged methodology till it is effective in the case of mushrooms, where the water content is reduced by 10%–12% to protect mushrooms from microbial attacks. Now many drying mechanisms like sun drying, air drying, and mechanical drying play a vital role. There are several drying methods:

(a) Sun drying (Cheap but needs regular sunshine and produces an unattractive product)
(b) Air drying (Cross-flow drying at 60 °C for approximately 8 hours. Through-flow drying at 60 °C for approximately 6 hours to dry to ~5% moisture)
(c) Vacuum shelf drying (Done at temperature of below 35° C, this process maintains texture and flavor more than hot air drying) (Wang et al., 2024).

#### 13.6.1.2 Freezing and Storage at Low Temperature

A popular technique for maintaining the freshness and quality of edible mushrooms is postharvest cold storage. To lower metabolic rates and delay spoiling, edible mushrooms should be rapidly stored in a refrigerated environment after harvest, ideally between 0 and 4 °C. Applying low temperature reduces physiological metabolic rates, enzymatic activity, the intensity of respiration, and the growth and reproduction of microorganisms. The texture and nutritional value of mushrooms can be preserved and microbial development inhibited by freezing them. It is advised to freeze edible mushrooms for extended periods, either slowly or quickly (Li et al., 2015).

#### 13.6.1.3 Blanching

In this preservation method, mushrooms are blanched by submerging them in boiling water for a specified duration, followed by rapid cooling or freezing to halt enzymatic activity and extend shelf life. Blanching is used to render enzymes inactive, which might cause decrease in the nutritional value of mushrooms include the enzymes gluconase and PPO, which cause the mushrooms to brown and reduce their sugar and protein content (Farag et al., 2025).

### 13.6.2 Advanced Preservation Techniques

#### 13.6.2.1 Chemical Treatment

Chemical treatments that use preservatives like sodium nitrate and sodium bicarbonate are frequently used. However, they have negative impacts on health that can lead to high blood pressure, tumors, and cancer. Citric acid and potassium sorbate are two examples of chemicals used to prevent microbial development and in spoilage prevention. While certain chemicals used in mushroom preservation can be beneficial, improper handling may leave behind residues that could trigger allergic reactions or other health issues in sensitive individuals. Chemical fungicides and insecticides are occasionally employed in mushroom farming to manage infections. However, because of the possible harmful leftovers that could endanger human health, their usage is usually discouraged (Jafri et al., 2013).

#### 13.6.2.2 Irradiation

Utilizing non-thermal physical processes, irradiation eliminates bacteria and pests. One safe, sustainable, and efficient way to extend the shelf-life of mushrooms is by irradiation. The procedure preserves the original flavor, color, and nutrition of mushrooms worldwide and does not leave behind any hazardous residues. However, the impact of irradiation on mushroom nutrients depends on several factors, including the mushroom species, treatment method, radiation dose, and the type of radiation used. The most prevalent irradiation sources are UV light, electron beams, and gamma rays. Food that has been exposed to radiation up to 10 kGy is still safe to eat (Dawadi et al., 2022).

#### 13.6.2.3 MAP

Several studies have been carried out to slow down the deterioration of mushroom quality after harvest, including the use of MAP. It depends on several variables, including the respiration rate of the packaged product and the permeability of the film to gases ($O_2$ and $CO_2$), to alter the composition of the surrounding gas. Freshly sliced *Agaricus bisporus*, *Lactarius deliciosus*, and *Lentinus edodes* have all been packaged using MAP with a gas mixture of $O_2$ and $CO_2$. Chemical treatments have also been used in the past to increase the shelf-life from 4 to 6 days. These involve dipping the oyster mushroom in various solutions and then packaging the samples in MAP with a gas mixture of $O_2$ (1.5%) and $CO_2$ (20%) (Xiao et al., 2011; Lyn et al., 2020).

#### 13.6.2.4 Chitosan Coating

Among various preservation techniques, edible coatings have gained significant attention for their effectiveness in reducing water loss and limiting gas exchange in *Pleurotus eryngii*. These coatings also serve as carriers for functional ingredients with antioxidant and antimicrobial properties, further enhancing postharvest protection. In recent years, a wide range of edible coatings has been developed and successfully applied to fruits and vegetables to extend shelf life. Chitosan, a natural and biodegradable polymer, is one such material that has shown promise as an edible coating for mushrooms. It helps slow down quality deterioration during storage. However, its effectiveness in preserving *P. eryngii* is somewhat limited, likely due to its poor solubility and relatively weak antioxidant and antimicrobial activities (Liu et al., 2016).

#### 13.6.2.5 HPP

HPP enhances the safety and shelf-life of plant-based foods by physically inactivating enzymes and microorganisms, while preserving bioactive compounds with minimal impact on nutritional value and sensory qualities. Like other plant-based foods, mushrooms also require physical modification through such technologies to expand their range of applications. In the food sector, HPP is a noteworthy physical processing technique with a lot of promise. It can preserve the quality of fresh foods that have little impact on nutrition and flavor, like mushrooms. It has been evaluated how high pressure affects the texture, color, and yield of mushrooms. When compared to HPP-treated mushrooms alone, vacuum-treated mushrooms exhibit superior color values and resemble traditional blanched mushrooms. Additionally, pressure-treated mushrooms have a similar yield to blanched mushrooms but with better firmness (Sangeeta et al., 2024).

#### 13.6.2.6 Ozone Treatment

PPO enzyme is the main factor causing browning. As a potent antibacterial and reducing agent, ozone can aid in delaying the enzyme's activity and improving the quality as well as the shelf-life of *Pleurotus* (Anjaly et al., 2020).

#### 13.6.2.7 Ultrasound Treatment

Ultrasound or ultrasonic treatment is one of the latest non-thermal technologies used to extend the shelf-life of fresh produce during storage. For highly perishable fruits like peaches (*Prunus persica*), this method, along with other postharvest treatments, has shown promise in prolonging freshness and preserving quality. Numerous studies have demonstrated that acoustic waves are an external factor that significantly affects plants' biological index and can either stimulate or inhibit development. On the other hand, research has been done on the use of acoustic waves in perishable crop storage. By applying acoustic waves with different exposure times, the fresh harvest of white oyster mushrooms (*Pleurotus florida*) can be improved without sacrificing quality (Paclibare et al., 2019).

#### 13.6.2.8 Cold Plasma (CP) Technology

The food processing industry has adopted CP, an innovative non-thermal technique, because of its short processing time, lack of toxicity, lack of residue generation, and ease of experimental setup. Reactive species (reactive nitrogen species, reactive oxygen species), free electrons, ions, and ultraviolet photons make up CP, which is known as a partially ionized gas. The lipids and proteins found in microbial (spoilage-causing organisms) cells are severely oxidized by these extremely reactive plasma species causing harm to the cell membrane, causing cell death, which leads to the shelf-life extension of mushrooms (Shabani et al., 2021).

#### 13.6.2.9 High Electric Field (HEF) Treatment

A new physical processing technique being investigated for food preservation is HEF treatment. This method has several benefits, including minimal energy consumption, little heat generation, nondosage residue, and easy operation. By altering the structure, dipole moment, and binding of enzymes to the reaction matrix, HEF has been shown in numerous investigations to dramatically reduce enzyme activity. Additionally, modifying the EF strength can have the impact of postponing tissue disintegration and drip loss during postharvest storage (Chang et al., 2023). Recent reports have shown that HEF treatment can reduce the activity of lipoxygenase and PPO in *P. ostreatus*, thereby potentially preserving its sensory and nutritional qualities during storage. This decrease in enzyme activity eventually increases the shelf-life of mushrooms by 1.5 times by preventing oxidative damage and browning during storage (Chang et al., 2023).

#### 13.6.2.10 Nanocoatings

The biopolymer known as nano-chitosan is produced by deacetylating the chitin presents in crab exoskeletons and then processing it into nanoparticles that are less than 100 nanometers in size. By increasing surface area and reactivity, this nanoscale form greatly boosts antibacterial efficacy against microbes. Therefore, as compared to regular chitosan, nano-chitosan has better antibacterial and antioxidant properties, which makes it a useful ingredient for food safety and preservation (Falahati et al., 2025).

#### 13.6.2.11 Vacuum-Sealing

By drastically lowering oxidation and microbiological spoiling, vacuum-sealing of mushrooms in oxygen-depleted conditions extends their shelf-life. However, it can result in an excessive buildup of moisture, which could lead to the growth of anaerobic microorganisms (Silva et al., 2025).

#### 13.6.2.12 Electrolyzed Water

Water with active oxygen compounds created by electrolyzing neutral salt solutions is known as electrolyzed water, and it has superior bacteriostatic, disinfecting, and cleaning properties. It primarily breaks down microbial cells and internal structures, which inhibits the growth of harmful bacteria on edible mushroom surfaces by producing active oxygen compounds and modifying the environment's acidity and alkalinity (Cao et al., 2024).

#### 13.6.2.13 Bio-Preservation

Utilizing natural or regulated microbiota and their antimicrobial metabolites to increase the shelf-life and safety of food products is known as bio-preservation. This procedure uses beneficial microorganisms, including lactic acid bacteria (LAB), or their metabolites, to prevent the growth of pathogenic and spoilage-causing microbes (Gerez et al., 2009; Garcha et al., 2012). A natural, secure, and eco-friendly alternative to artificial chemical preservatives, bio-preservation is becoming increasingly popular as a food preservation technique (Abouloifa et al., 2021; Shehata et al., 2024). The practical use of microbes and their natural products for food preservation has been a widely recognized method throughout human history. Utilizing microorganisms and their byproducts, many workers have frequently reported that food has a long shelf-life and doesn't spoil or harm the environment. Probiotic use may be a superior way to preserve foods and extend their shelf-life more safely (Mandal and Mandal, 2023). Probiotics were defined as "live microorganisms which when administered in adequate amounts confer a health beneficiary effect on the host" by the ICMR-DBT committee and the FAO/WHO group (Ganguly et al., 2011). Aflatoxin, produced by *Aspergillus* species, is a highly toxic and carcinogenic secondary metabolite that can be successfully detoxified by *Limosilactobacillus fermentum* LAB212, according to a report by Mandal et al. (2024a). They showed that the LAB212 strain could considerably lower aflatoxin contamination in groundnuts across several formulations through a variety of *in vitro* tests. According to the study, this strain has potent aflatoxin degradation abilities, which makes it a viable option for bioremediation in applications about food safety. Additionally, potential of LAB212 as a bio-preservative was highlighted, suggesting that it might be used as an environmentally acceptable substitute for traditional chemical preservatives in the mitigation of aflatoxin contamination in food products to reduce the food borne disease using appropriate formulation in future (Mandal and Mandal, 2023; Mandal et al., 2024b).

**Desirable properties of *Lactobacillus*:**

- **Lactic Acid Production**
- **Probiotic Benefits**
- **Antimicrobial Activity**
- **Enzymatic Activity**
- **Flavour and Aroma Development**
- **Safety and non-pathogenicity**
- **Immune System Modulation**
- **Adhesion to Intestinal Mucosa**
- **Stress Tolerance in the gut**
- **Nutrient Production**

The potential of the genus *Lactobacillus* makes it essential for modern food improvements. Recently, a variety of techniques have been employed to produce dairy products that are resistant to mycotoxicological contamination and can lower dairy product contamination. The LAB have drawn more interest in recent years for use in mushroom bio preservation due to the content of mushrooms as well as possible antifungal activity and GRAS (Generally Recognized as Safe) status. LAB as bio-preservatives would be a blooming approach to lessen the chance of deteriorating mushrooms while also improving their nutritional value (Mukherjee et al., 2025).

## 13.7 RECIPES OF CULTIVATED *PLEUROTUS*

*Pleurotus* mushrooms are highly versatile in culinary applications, offering excellent nutritional value and functional food qualities. Their incorporation into diverse dishes supports gut microbiome balance, boosts immune function, and contributes to cardiovascular and renal health. As sustainable ingredients suitable for vegetarian and vegan diets, they are gaining prominence in functional food development. Ongoing studies are exploring their potential in fermented products, gluten-free formulations, and as alternative protein sources, reinforcing their role in future-ready, health-oriented food systems (Table 13.3) (Aishah et al., 2013; Fallea and Venturella, 2025).

## 13.8 CONCLUSION AND FUTURE ASPECTS

The potential for improving postharvest management and expanding the culinary applications of *Pleurotus* species is vast, with significant implications for food security, sustainability, and the functional food industry. As demand for nutrient-dense and environmentally friendly food sources rises, oyster mushrooms are emerging as an ideal alternative to both plant- and animal-based proteins. Their rich nutritional content, rapid growth cycle, and ability to thrive on agricultural waste make them a sustainable solution for addressing global food challenges while reducing environmental impact. From the nutritional perspective, high protein content, balanced amino acid profile, and fibrous texture make them a natural substitute for meat. Their bioactive components, including β-glucans, polysaccharides, secondary metabolites, and phenolic compounds also offer immunomodulatory, anti-inflammatory, anti-microbial, and cholesterol-lowering benefits.

However, maintaining the texture and overall quality of mushrooms after harvest remains a significant challenge due to their high moisture content and perishable nature; often restricting the market availability and consumer demands. In conventional practices, physical and chemical treatments have been applied to prevent dehydration and microbial contamination. These practices are either not cost-effective or may hamper the texture and nutritional value of the mushroom. In this context, application of selective probiotic LAB can be a healthier substitute. This approach can increase the shelf-life by forming a bacterial coating on the surface, and its metabolites will help in value addition to the *Pleurotus* mushrooms. Notably, the composite form of probiotic-prebiotic (synbiotic) will make a perfect nutraceutical with the ability to modulate gut environment toward a healthy one, deciphering maximum utilization of food components and improving the host growth, physiology, and immunity. On the other hand, fortification of available mushroom products with potent probiotics can be another option to incorporate health-beneficiating metabolites and improve the shelf-life and quality of the ready-to-eat food.

Advancements in biotechnology and food science are aspects to optimize *Pleurotus* production, postharvest management and utilization. Techniques such as precision

**TABLE 13.3**
**Different Cultivated *Pleurotus* Recipes**

| Name of the Dishes | Countries Being Major Consumers | Ingredients | Recipe | Reference |
|---|---|---|---|---|
| *Pleurotus* Stir-Fry | China | *Pleurotus*<br>Olive oil<br>Garlic<br>Onion<br>Red bell pepper<br>Soy sauce<br>Black pepper | Heat olive oil<br>↓<br>Saute garlic and onions<br>↓<br>Add *Pleurotus*, salt and pepper (cook for 5 minutes)<br>↓<br>Add soy sauce and serve | Hess et al., 2018; Siddique et al., 2002 |
| *Pleurotus* Soup | France | *Pleurotus*<br>Vegetable Broth<br>Onion<br>Garlic<br>Coconut milk<br>Thyme<br>Black pepper<br>Salt | Saute onions and garlic<br>↓<br>Add chopped *Pleurotus* (cook for 5 minutes)<br>↓<br>Add vegetable broth (boil for 10 minutes)<br>↓<br>Add thyme, salt and pepper<br>↓<br>Add coconut milk (cook for another 5 minutes) and serve | Saed et al., 2022 |
| *Pleurotus* burger patties | Philippines | *Pleurotus*<br>Quinoa<br>Breadcrumbs<br>Onion<br>Garlic<br>Paprika<br>Salt<br>Black pepper<br>Egg<br>Water | Saute onions and garlic<br>↓<br>Add mushrooms and cook<br>↓<br>Mix with quinoa, breadcrumbs, spices, and egg<br>↓<br>Patties formation<br>↓<br>Refrigerate<br>↓<br>Saute each side and serve | Tabaldo-Tucar and Solar, 2019 |
| *Pleurotus* crackers | Malaysia | Mushroom powder<br>Tapioca flour<br>Salt<br>Sugar<br>Iced water<br>Palm oil | Mushroom powder mixed with tapioca flour<br>↓<br>Dough formation and shaped into a cylindrical form (steamed for 90 min at 100 °C); refrigerated overnight<br>↓<br>Slice and deep fry using palm oil | Yahya et al., 2017 |
| *Pleurotus* Bread | China<br>Italy | Wheat flour<br>*Pleurotus* powder<br>Salt<br>Baking powder<br>Olive oil<br>Warm water | Mix all the ingredients<br>↓<br>Add olive oil and warm water<br>↓<br>Dough (keep it for 1 h)<br>↓<br>Loaf formation and bake at 180 °C for 30 min | Adeoye et al., 2019 |
| *Pleurotus* Pasta | Italy | *Pleurotus*<br>Whole wheat pasta<br>Olive oil<br>Garlic<br>Red pepper flakes<br>Tomatoes<br>Parsley<br>Salt<br>Black pepper Grated Parmesan cheese | Cook pasta<br>↓<br>Add garlic and red pepper flakes in oil<br>↓<br>Add *Pleurotus* (sauté for 5 minutes)<br>↓<br>Add cherry tomatoes, salt, and black pepper<br>↓<br>Mix with pasta | Sathvara et al., 2024 |

**TABLE 13.3 *(Continued)***
**Different Cultivated *Pleurotus* Recipes**

| Name of the Dishes | Countries Being Major Consumers | Ingredients | Recipe | Reference |
|---|---|---|---|---|
| *Pleurotus* risotto | Italy | Arborio rice<br>Vegetable broth<br>*Pleurotus* White wine<br>Onion<br>Garlic<br>Olive oil<br>Cheese<br>Thyme<br>Salt<br>Black pepper | Saute onions and garlic<br>↓<br>Add *Pleurotus* (cook for 5–10 minutes)<br>↓<br>Add Arborio rice (cook for another 5 minutes)<br>↓<br>Pour in white wine and add vegetable broth with some salt and pepper | Sathvara et al., 2024 |
| *Pleurotus* Tacos | Mexico | *Pleurotus*<br>Cumin<br>Smoked paprika<br>Chili powder<br>Lime juice<br>Olive oil<br>Onion<br>Corn tortillas<br>Cilantro Tomatoes | Saute onions<br>↓<br>Add shredded *Pleurotus* along with the spices and cook for 5 minutes<br>↓<br>Assemble the mixture in tortillas or taco shell | Wong, 2017 |

fermentation and microbial engineering can enhance their nutritional profile, flavour, and shelf-stability, expanding their applications in the alternative protein sector. Use of bio-preservatives, smart packaging technologies, and artificial intelligence-driven monitoring systems can also reduce postharvest losses by ensuring optimal storage conditions. As consumer preferences shift toward ethical, sustainable, and health-conscious foods, *Pleurotus* mushrooms have the potential to reshape global food systems. By integrating innovative postharvest management, these mushrooms can play a key role in reducing food waste, increasing market reach, and contributing to a more sustainable and resilient food industry.

## REFERENCES


Abdel-Aziz, S. M., M. M. Asker, A. A. Keera, and M. G. Mahmoud. 2016. Microbial food spoilage: control strategies for shelf life extension. In: *Microbes in Food and Health*, ed. N. Garg, S. Abdel-Aziz, and A. Aeron. Springer, Cham. https://doi.org/10.1007/978-3-319-25277-3_13

Abouloifa, H., S. Gaamouche, Y. Rokni, et al. 2021. Antifungal activity of probiotic *Lactobacillus* strains isolated from natural fermented green olives and their application as food bio-preservative. *Biological Control* 152: 104450. https://doi.org/10.1016/j.biocontrol.2020.104450

Adams, S., D. Che, J. Hailong, et al. 2019. Effects of pulverized oyster mushroom (*Pleurotus ostreatus*) on diarrhea incidence, growth performance, immunity, and microbial composition in piglets. *Journal of the Science of Food and Agriculture* 99(7): 3616–3627. https://doi.org/10.1002/jsfa.9582

Adeoye, B. K., K. A. Adeleye, I. F. Ani, A. R. Akinlade, E. O. Ngozi, and N. C. Ajuzie. 2019. Quality evaluation of bread enriched with mushroom powder. *Journal of Environmental Science, Toxicology and Food Technology* 13(6): 11–18.

Aditya, N., R. S. Jarial, K. Jarial, J. N. Bhatia. 2024. Comprehensive review on oyster mushroom species (Agaricomycetes): morphology, nutrition, cultivation and future aspects. *Heliyon* 10(5): e26539. https://doi.org/10.1016/j.heliyon.2024.e26539

Agbagwa, S. S., E. C. Chuku, and C. G. Mbah. 2020. Evaluation of yield and spoilage Moulds of *Pleurotus ostreatus* grown on sawdust, wood ash and cassava bran. *International Journal of Research and Innovation in Applied Science* 5(8): 137–140.

Aida, F. M. N. A., M. Shuhaimi, M. Yazid, and A. G. Maaruf. 2009. Mushroom as a potential source of prebiotics: a review. *Trends in Food Science and Technology* 20(11–12): 567–575. https://doi.org/10.1016/j.tifs.2009.07.007

Aishah, M. S., and W. I. Wan Rosli. 2013. The effect of addition of oyster mushroom (*Pleurotus sajor-caju*) on nutrient composition and sensory acceptation of selected wheat-and rice-based products. *International Food Research Journal* 20(1): 183–188.

Anjaly, S., A. Khanashyam, B. Balasubrahmanyam, and B. K. Yadav. 2020. Potentials of ozone pretreatment in prolonging the freshness of oyster mushrooms (*Pleurotus Florida*). *Malaysian Journal of Medicine and Health Sciences* 16(13): 119–120.

Bhagarathi, L. K., G. Subramanian, and P. N. B DaSilva. 2023. A review of mushroom cultivation and production, benefits and therapeutic potentials. *World Journal of Biology Pharmacy and Health Sciences* 15(2): 1–56. https://doi.org/10.30574/wjbphs.2023.15.2.0333

Bhargav, P., D. Gupta, R. S. Jarial, N. Sharma, and R. Thakur. 2023. Fungal and bacterial contaminants associated with the spoilage of mushroom spawn. *International Journal of Plant and Soil Science* 35(6): 85–93. https://doi.org/10.9734/ijpss/2023/v35i62842

Binsaleh, N. K., A. S. Bazaid, H. Barnawi, et al. 2025. Chemical characterization, anticancer, antioxidant and anti-obesity activities with molecular docking studies of *Pleurotus ostreatus* biomass exposed to moist heat. *Journal of Food Measurement and Characterization* 19: 2476–2495. https://doi.org/10.1007/s11694-025-03125-9

Bulam, S., N. Üstün, and A. Pekşen. 2022. Oyster mushroom (*Pleurotus ostreatus*) as a healthy ingredient for sustainable functional food production. *Mantar Dergisi* 13(3): 131–143. https://doi.org/10.30708/mantar.1192063

Cao, Y., L. Wu, Q. Xia, K. Yi, and Y. Li. 2024. Novel post-harvest preservation techniques for edible fungi: a review. *Foods* 13(10): 1554. https://doi.org/10.3390/foods13101554

Chang, C.-K., C.-Y. Hou, K.-C. Cheng, et al. 2023. Exploring the potential of alternative current electric field treatment for preservation and the tissue softening mechanism in *Pleurotus ostreatus*. *Scientia Horticulturae* 321: 112284. https://doi.org/10.1016/j.scienta.2023.112284

Chirinang, P., and K.-O. Intarapichet. 2009. Amino acids and antioxidant properties of the oyster mushrooms, *Pleurotus ostreatus* and *Pleurotus sajor-caju*. *Science Asia* 35(2009): 326–331.

Dawadi, E., P. B. Magar, S. Bhandari, S. Subedi, S. Shrestha, and J. Shrestha. 2022. Nutritional and post-harvest quality preservation of mushrooms: a review. *Heliyon* 8(12): e12093. https://doi.org/10.1016/j.heliyon.2022.e12093

Effiong, M. E., C. P. Umeokwochi, I. S. Afolabi, and S. N. Chinedu. 2024. Assessing the nutritional quality of *Pleurotus ostreatus* (oyster mushroom). *Frontiers in Nutrition* 10: 1279208. https://doi.org/10.3389/fnut.2023.1279208

El-Batal, A. I., M. E. Nora, A. S. Yassin, and M. A. Amin. 2015. Laccase production by Pleurotus ostreatus and its application in synthesis of gold nanoparticles. *Biotechnology Reports* 5: 31–39. https://doi.org/10.1016/j.btre.2014.11.001

Eswaran, A., and R. Ramabadran. 2000. Studies on some physiological, cultural and postharvest aspects of oyster mushroom, *Pleurotus eous* (Berk.) sacc. *Tropical Agricultural Research* 12: 360–374.

Fadahunsi, I., D. Ayansina, and A. Okunrotifa. 2013. Biocontrol of mushroom spoilage fungi and aflatoxin evaluation during storage. *Nature and Science* 11(7): 7–13.

Falahati, M., P. Ariaii, Z. Khoshkhoo, G. Asadi, and S. E. Hosseini. 2025. Nano-chitosan-Aloe Vera Coating with Tomato Seed Protein Hydrolyzate for Preserving Button Mushroom (*Agaricus bisporus*) Quality. *Food Biophysics* 20: 45. https://doi.org/10.1007/s11483-024-09924-9

Fallea, A., and G. Venturella. 2025. Health foods made from medicinal mushrooms: innovative recipes created by an Italian chef. *International Journal of Medicinal Mushrooms* 27(6): 81–91. https://doi.org/10.1615/IntJMedMushrooms.2025058296

Farag, M. A., N. Kamal, H. S. Hamezah, et al. 2025. The role of microorganisms and microbial enzymes in commercial fermented mushroom production: a comprehensive review of their action mechanisms, quality attributes and health benefits. *Food Production, Processing and Nutrition* 7(1): 8. https://doi.org/10.1186/s43014-024-00278-w

Ganguly, N. K., S. K. Bhattacharya, B. Sesikeran, et al. 2011. ICMR-DBT guidelines for evaluation of probiotics in food. *Indian Journal of Medical Research* 134(1): 22–25.

Garcha, S., and N. K. Natt. 2012. *In situ* control of food spoilage fungus using *Lactobacillus acidophilus* NCDC 291. *Journal of Food Science and Technology* 49: 643–648. https://doi.org/10.1007/s13197-011-0482-1

Gerez, C. L., M. I. Torino, G. Rollán, and G. F. de Valdez. 2009. Prevention of bread mould spoilage by using lactic acid bacteria with antifungal properties. *Food Control* 20(2): 144–148. https://doi.org/10.1016/j.foodcont.2008.03.005

Guo, Y., X. Chen, P. Gong, et al. 2023. Advances in postharvest storage and preservation strategies for *Pleurotus eryngii*. *Foods* 12(5): 1046. https://doi.org/10.3390/foods12051046

Hassan, A. A., and M. T. Ibrahim. 2022. Isolation, morphological and molecular identification of the pathogenic and competitors fungi associated with the edible mushroom *Pleurotus* sp. and control them. *IOP Conference Series: Earth and Environmental Science* 1060: 012118. https://doi.org/10.1088/1755-1315/1060/1/012118

Hess, J., Q. Wang, T. Gould, and J. Slavin. 2018. Impact of *Agaricus bisporus* mushroom consumption on gut health markers in healthy adults. *Nutrients* 10(10): 1402. https://doi.org/10.3390/nu10101402

Huo, J., M. Zhang, D. Wang, A. S. Mujumdar, B. Bhandari, and L. Zhang. 2023. New preservation and detection technologies for edible mushrooms: a review. *Journal of the Science of Food and Agriculture* 103(7): 3230–3248. https://doi.org/10.1002/jsfa.12472

Irfan, H., A. Saleem, A. Ali, and M. Akmal. 2022. Medicinal importance of *Pleurotus* species (Oyster mushroom): a review. *World Journal of Fungal and Plant Biology* 3(1): 01–12. https://doi.org/10.5829/idosi.wjfpb.2012.3.1.303

Jafri, M., A. Jha, D. S. Bunkar, and R. C. Ram. 2013. Quality retention of oyster mushrooms (*Pleurotus florida*) by a combination of chemical treatments and modified atmosphere packaging. *Postharvest Biology and Technology* 76: 112–118. https://doi.org/10.1016/j.postharvbio.2012.10.002

Jarial, R. S., K. Jarial, and J. N. Bhatia. 2024. Comprehensive review on oyster mushroom species (Agaricomycetes): morphology, nutrition, cultivation and future aspects. *Heliyon* 10(5): e26539. https://doi.org/10.1016/j.heliyon.2024.e26539

Jayachandran, M., J. Xiao, and B. Xu. 2017. A critical review on health promoting benefits of edible mushrooms through gut microbiota. *International Journal of Molecular Sciences* 18(9): 1934. https://doi.org/10.3390/ijms18091934

Khan, M. A., and M. Tania. 2012. Nutritional and medicinal importance of *Pleurotus* mushrooms: an overview. *Food Reviews International* 28(3): 313–329. https://doi.org/10.1080/87559129.2011.637267

Khatua, S., S. Paul, and K. Acharya. 2013. Mushroom as the potential source of new generation of antioxidant: a review. *Research Journal of Pharmacy and Technology* 6(5): 496–505.

Kiiza, A., and P. Byandusya. 2023. Environmental, economic, and social benefits of cultivating the oyster mushroom *Pleurotus ostreatus* (Jacq.,) p. Kumm. In a community organization in southwest Uganda. *International Multidisciplinary Scientific GeoConference: SGEM* 23(5.1): 369–377. https://doi.org/10.5593/sgem2023/5.1/s21.47

Lesa, K. N., M. U. Khandaker, F. M. R. Iqbal, et al. 2022. Nutritional value, medicinal importance, and health-promoting effects of dietary mushroom (*Pleurotus ostreatus*). *Journal of Food Quality* 2022(1): Article ID 2454180. https://doi.org/10.1155/2022/2454180

Li, R., Q. Zheng, J. Lu, et al. 2021. Chemical composition and deterioration mechanism of *Pleurotus tuoliensis* during postharvest storage. *Food Chemistry* 338: 127731. https://doi.org/10.1016/j.foodchem.2020.127731

Li, Z., T. Song, C. Feng, L. Hao, and J. Shi. 2015. Effect of different storage temperature on hypobaric storage quality of *Pleurotus eryngii*. *Transactions of the Chinese Society of Agricultural Engineering* 31(3): 332–338. https://doi.org/10.3969/j.issn.1002-6819.2015.03.045

Liang, Y., K. Luo, B. Wang, B. Huang, P. Fei, and G. Zhang. 2024. Inhibition of polyphenol oxidase for preventing browning in edible mushrooms: a review. *Journal of Food Science* 89(11): 6796–6817. https://doi.org/10.1111/1750-3841.17322

Liu, J., C.-G. Meng, X.-C. Wang, Y. Chen, J. Kan, and C.-H. Jin. 2016. Effect of protocatechuic acid-grafted-chitosan coating on the postharvest quality of *Pleurotus eryngii*. *Journal of Agricultural and Food Chemistry* 64(38): 7225–7233. https://doi.org/10.1021/acs.jafc.6b02468

Lu, S., and Q. Xia. 2012. Effects of combined treatments with modified-atmosphere packaging on shelf-life improvement of food products. In: *Progress in Food Preservation*, ed. R. Bhat, A. K. Alias, and G. Paliyath, 67–109. Wiley-Blackwell. https://doi.org/10.1002/9781119962045.ch4

Lyn, F. H., Z. A. M. Adilah, M. A. R. Nor-Khaizura, B. Jamilah, and Z. A. N. Hanani. 2020. Application of modified atmosphere and active packaging for oyster mushroom (*Pleurotus ostreatus*). *Food Packaging and Shelf Life* 23: 100451. https://doi.org/10.1016/j.fpsl.2019.100451

Mandal, S., A. D. Bhowmik, A. Chattopadhyay, and N. C. Mandal. 2024a. *Limosilactobacillus fermentum* LAB212 effectively ameliorates toxigenicity of aflatoxin. *The Nucleus* 67(3): 569–578. https://doi.org/10.1007/s13237-024-00488-0

Mandal, S., S. Barman, A. Sen, and N. C. Mandal. 2024b. Mitigating aflatoxin contamination in food product using *Limosilactobacillus fermentum* LAB212. *Food Bioscience* 61: 104750. https://doi.org/10.1016/j.fbio.2024.104750

Mandal, S., and N. C. Mandal. 2023. Formulation of food grade *Limosilactobacillus fermentum* for antifungal properties isolated from home-made curd. *Scientific Reports* 13(1): 20371. https://doi.org/10.1038/s41598-023-45487-4

Maray, A. R. M., M. K. Mostafa, and A. E. D. M. El-Fakhrany. 2018. Effect of pretreatments and drying methods on physico-chemical, sensory characteristics and nutritional value of oyster mushroom. *Journal of Food Processing and Preservation* 42(1): e13352. https://doi.org/10.1111/jfpp.13352

Medihi, N. I., Z. A. Haiyee, R. Sukor, and S. Raseetha. 2024. Exploring the functional properties and nutritional values of colored oyster mushrooms species (*Pleurotus*, Agaricomycetes): a review. *International Journal of Medicinal Mushrooms* 26(6): 25–38. https://doi.org/10.1615/IntJMedMushrooms.2024053563

Mukherjee, D., A. Mishra, and N. C. Mandal. 2025. Isolation and characterization of spawn contaminating microbes and their inhibitory impact on mycelial growth of *Pleurotus ostreatus*. *Journal of Mycopathological Research* 63(1): 159–168. https://doi.org/10.1016/j.micres.2019.126393

Mwangi, R. W. 2016. Biocontrol of green mould disease of oyster mushroom (*Pleurotus ostreatus*) using *Bacillus amyloliquefaciens*. *Journal of Environmental Sustainability Advancement Research* 2016(2): 49–54.

Omokaro, O., and A. A. Ogechi. 2013. Cultivation of mushroom (*Pleurotus ostreatus*) and the microorganisms associated with the substrate used. *E-Journal of Science and Technology* 8(4): 49–59.

Paclibare C. O., J. A. Lavarias, A. M. Tuates Jr, C. G. G. Somera, and R. G. Peneyra. 2019. Influence of acoustic waves on the shelf life quality of white oyster mushroom (*Pleurotus florida*). *International Journal on Advanced Science Engineering and Information Technology* 9(6): 2022–2027. https://doi.org/10.18517/ijaseit.9.6.8365

Prasad, P., and A. Roy. 2023. Critical analysis of therapeutic effectiveness of *Pleurotus ostreatus*. *Pharmaceutical and Biosciences Journal* 11(4): 20–26.

Raman, J., K.-Y. Jang, Y.-L. Oh, et al. 2021. Cultivation and nutritional value of prominent *Pleurotus* spp.: an overview. *Mycobiology* 49(1): 1–14. https://doi.org/10.1080/12298093.2020.1835142

Ramteke, A. Y., A. Nayak, A. Sagar, S. K. Kesari, and P. W. Ramteke. 2020. Recent advances in mushrooms preservation. *International Journal of Chemical Studies* 8(2): 2376–2381. https://doi.org/10.22271/chemi.2020.v8.i2aj.9107

Saed, B., M. El-Waseif, H. Fahmy, et al. 2022. Physicochemical and sensory characteristics of instant mushroom soup enriched with Jerusalem artichoke and cauliflower. *Foods* 11(20): 3260. https://doi.org/10.3390/foods11203260

Sánchez, C. 2010. Cultivation of *Pleurotus ostreatus* and other edible mushrooms. *Applied Microbiology and Biotechnology* 85(5): 1321–1337. https://doi.org/10.1007/s00253-009-2343-7

Sangeeta, S., D. Sharma, S. Ramniwas, R. Mugabi, J. Uddin, and G. A. Nayik. 2024. Revolutionizing mushroom processing: innovative techniques and technologies. *Food Chemistry: X* 23: 101774. https://doi.org/10.1016/j.fochx.2024.101774

Sathvara, A. B., and C. D. Afuwale. 2024. A comprehensive review on mushrooms as a nutraceutical superfood. *International Journal of Contemporary Microbiology* 10(2): 22–33. https://doi.org/10.37506/3038td36

Sawangwan, T., W. Wansanit, L. Pattani, and C. Noysang. 2018. Study of prebiotic properties from edible mushroom extraction. *Agriculture and Natural Resources* 52(6): 519–524. https://doi.org/10.1016/j.anres.2018.11.020

Schill, S., B. Stessl, N. Meier, A. Tichy, M. Wagner, and M. Ludewig. 2021. Microbiological safety and sensory quality of cultivated mushrooms (*Pleurotus eryngii*, *Pleurotus ostreatus* and *Lentinula edodes*) at retail level and post-retail storage. *Foods* 10(4): 816. https://doi.org/10.3390/foods10040816

Shabani, E., A. S. Lavasani, and N. Helali. 2021. The study of cold plasma on extending of shelf life of Oyster mushroom. *International Journal of Bio-Inorganic Hybrid Nanomaterials* 10(3): 143–154.

Sharma, S. R., S. Kumar, and V. P. Sharma. 2007. Diseases and competitor moulds of mushrooms and their management. In: [*Solan, Himachal Pradesh: National Research Centre for Mushroom (N.R.C.M 2007)*], Indian Council of Agricultural Research.

Sharma, V. P., S. K. Annepu, Y. Gautam, M. Singh, and S. Kamal. 2017. Status of mushroom production in India. *Mushroom Research* 26(2).

Shehata, M. G., T. Alsulami, N. M. Abd El-Aziz, et al. 2024. Biopreservative and anti-mycotoxigenic potentials of *Lactobacillus paracasei* MG847589 and its bacteriocin in soft white cheese. *Toxins* 16(2): 93. https://doi.org/10.3390/toxins16020093

Siddique, S. S., S. Gupta, S. Gupta, and D. Kumar. 2002. Impact of cooking and processing on mushroom nutrient retention. In: *Mushroom Magic*, ed. D. Kumar, S. Gupta, M. J. Ansari, et al., 174–190. CRC Press.

Silva, M., M. Vida, A. C. Ramos, F. J. Lidon, F. H. Reboredo, and E. M. Gonçalves. 2025. Storage temperature effect on quality and shelf-life of *Hericium erinaceus* mushroom. *Horticulturae* 11(2): 158. https://doi.org/10.3390/horticulturae11020158

Silva, S. O., S. M. G. da Costa, and E. Clemente. 2002. Chemical composition of *Pleurotus pulmonarius* (Fr.) Quél., substrates and residue after cultivation. *Brazilian Archives of Biology and Technology* 45: 531–535. https://doi.org/10.1590/S1516-89132002000600018

Singh, P., H.-C. Langowski, A. A. Wani, and S. Saengerlaub. 2010. Recent advances in extending the shelf life of fresh *Agaricus* mushrooms: a review. *Journal of the Science of Food and Agriculture* 90(9): 1393–1402. https://doi.org/10.1002/jsfa.3971

Sobowale, A. A., L. C. Uzoma, A. O. Aduramigba-Modupe, and B. A. Bamkefa. 2022. Fungitoxicity of *Trichoderma longibrachiatum* (Rifai) metabolites against *Fusarium oxysporum*, *Aspergillus niger* and *Aspergillus tamarii*. *American Journal of Plant Sciences* 13(7): 984–993. https://doi.org/10.4236/ajps.2022.137065

Suada, I. K., I. Sudarma, B.-S. Kim, J.-Y. Cha, and S. Ohga. 2015. Fungal contaminant threaten Oyster mushroom (*Pleurotus ostreatus* (Jacq. ex Fr.) Kummer) cultivation in Bali. *Journal of the Faculty of Agriculture, Kyushu University* 60(2): 309–313. https://doi.org/10.5109/1526343

Sun, J., R. Ren, L. Yao, J. Li, L. Tong, J. Yuan, and D. Wang. 2023. Effect of combined chitosan and hyperbranched poly-L-lysine based coating on prolonging the shelf life of oyster mushroom (*Pleurotus ostreatus*). *Foods* 13(1): 77. https://doi.org/10.3390/foods13010077

Tabaldo-Tucar, M., and M. N. J. B. Solar. 2019. Choose your patty: the sensory characterization and consumer acceptance of burger patties with oyster mushroom (*Pleurotus ostreatus*) enrichment. *Journal of Physics: Conference Series* 1254(1): 012017. https://doi.org/10.1088/1742-6596/1254/1/012017

Tajalipour, S., N. Hassanzadeh, H. K. Jolfaee, A. Heydari, and A. Ghasemi. 2014. Biological control of mushroom brown blotch disease using antagonistic bacteria. *Biocontrol Science and Technology* 24(4): 473–484. https://doi.org/10.1080/09583157.2013.873113

Thakur, M. P. 2020. Advances in mushroom production: key to food, nutritional and employment security: a review. *Indian Phytopathology* 73(3): 377–395. https://doi.org/10.1007/s42360-020-00244-9

Wang, D., Y. Wang, R. Pandiselvam, D. Su, and H. Xu. 2024. Comparative analysis of drying methods on *Pleurotus eryngii*: impact on drying efficiency, nutritional quality, and flavor profile. *Food and Bioprocess Technology* 17(1): 4598–4616. https://doi.org/10.1007/s11947-024-03402-3

Wong, K. 2017. Investigating the utilization of mushrooms in beef-based products for improved health. Master thesis, University of Massachusetts Amherst. https://scholarworks.umass.edu/masters_theses_2/487

Xiao, G., M. Zhang, L. Shan, Y. You, and V. M. Salokhe. 2011. Extension of the shelf-life of fresh oyster mushrooms (*Pleurotus ostreatus*) by modified atmosphere packaging with chemical treatments. *African Journal of Biotechnology* 10(46): 9509–9517. https://doi.org/10.5897/AJB08.974

Yahya, F. A. R. I. D. A. H., N. N. M. Yusof, and C. K. Chen. 2017. Effect of varying ratios of oyster mushroom powder to tapioca flour on the physicochemical properties and sensory acceptability of fried mushroom crackers. *Malaysian Applied Biology* 46(1): 57–62.

Zheng, N., Y.-M. Ma, H.-Y. Lei, et al. 2025. Progress of postharvest preservation technology of edible mushroom. *Food and Medicine Homology* 2(1): 9420028. https://doi.org/10.26599/FMH.2025.9420028

Zhu, J., J. Sun, H. Li, Pet al. 2025. Underlying the early signaling and gene expression of *Pleurotus ostreatus* mycelia during injury response. *World Journal of Microbiology and Biotechnology* 41(3): 80. https://doi.org/10.1007/s11274-025-04296-9

# 14 Global Market Prospects and Opportunities for Oyster Mushrooms

[1,2,3]*M. A. Oke**, [1,2,4]*E. A. Adebayo*, [1,2]*N. A. Ajisope*, [1,2]*O. A. Okunade*, and [3]*B. M. Olowe*

[1]Department of Pure and Applied Biology, Ladoke Akintola University, Ogbomoso, Nigeria

[2]Microbiology and Nanobiotechnology Laboratory, LAUTECH, Ogbomoso, Nigeria

[3]Department of Biological Sciences, Bamidele Olumilua University of Education, Science and Technology, Ikere-Ekiti, Nigeria

[4]Department of Biological Science, Ajayi Crowther University, Oyo, Nigeria

## 14.1 INTRODUCTION

The world continues to grapple with numerous challenges arising from the unchecked growth of the human population. One of the most pressing issues is the massive volume of waste generated daily, which is increasingly released into the environment, posing serious threats to both human health and ecological balance (García-Ramón et al., 2023). In addition, rising population levels contribute to growing unemployment and social insecurity. As the demand for jobs and sustainable income sources increases, the failure to meet these needs can result in significant social and economic instability (Minhaj and Imran, 2021). Moreover, the need for nutritious and affordable food intensifies with population growth. A lack of access to a balanced diet can lead to a range of health problems and nutritional deficiencies among affected populations (Tripathi et al., 2019). In response to these global pressures, nature offers numerous sustainable solutions, one of which is the cultivation and utilization of mushrooms. The global mushroom market was valued at $46.1 billion in 2020 and is projected to grow at a compound annual growth rate (CAGR) of 9.5%, potentially reaching $86.6 billion by 2025 (Jarial et al., 2024).

Out of the approximately 15,000 identified mushroom species worldwide, about 2,000 are edible, 700 possess therapeutic value, and only 100 are cultivated commercially (Jarial et al., 2024). Among the various edible fungi, *Pleurotus* species (commonly known as oyster mushrooms) have emerged as a highly promising resource with remarkable nutritional, medicinal, environmental, and economic benefits. As such, they are the third most cultivated mushrooms globally and the most widely produced species in Africa (Adebayo and Oloke, 2017; Jarial et al., 2024). China leads global production, contributing nearly 74% of the total output, followed by extensive cultivation in Southeast Asia, India, Europe, and Africa (Jarial et al., 2024).

One of the most remarkable features of *Pleurotus* mushrooms is their ability to convert a wide variety of lignocellulosic biomass and agro-industrial wastes into edible biomass. Their saprophytic nature enables them to thrive on organic substrates such as straw, sawdust, and crop residues (Getachew et al., 2019; Besufekad et al., 2020). They play a key role in recycling agricultural, forestry, and food-processing waste, making them vital agents in sustainable agriculture and circular economy initiatives (Sánchez, 2010; Tesfay et al., 2020; Sganzerla et al., 2022). With an estimated 998 million tons of agricultural and organic waste generated annually, the ability of *Pleurotus* species to utilize this biomass offers a scalable, eco-friendly waste management solution (Raman et al., 2021). Besides, *Pleurotus* mushrooms are highly adaptable and can grow under a wide range of environmental conditions. They flourish in temperatures between 10 °C and 30 °C, relative humidity levels of 55%–70%, and pH values ranging from 6.0 to 8.0, making them particularly suitable for cultivation in both rural and urban environments with minimal technological input (Shen et al., 2014; Argaw et al., 2023).

There are approximately 40 known species of *Pleurotus* distributed across tropical and temperate regions (Golak-Siwulska et al., 2018), with 26 species specifically documented in the scientific literature (Bernardi et al., 2013; Bumanlag et al., 2018; Raman et al., 2021). Many of them have been recognized for their food and medicinal value for over a thousand years (Tolera and Abera, 2017; Muszyńska et al., 2018; Argaw et al., 2023). They offer a low-cost, high-nutrient food source that is particularly valuable in addressing poverty and malnutrition. These mushrooms are rich in protein, carbohydrates, fiber, essential fatty acids, minerals, mycocellulose, and vitamins such as riboflavin and niacin (Argaw et al., 2023). Additionally, their low calorie and fat content make them an ideal functional food for health-conscious consumers (Törős et al., 2022; Juárez-Hernández et al., 2023). Their culinary versatility also contributes to their popularity. They are commonly used in soups, salads, stir-fries, and pasta dishes and can even serve as a meat substitute in vegetarian and vegan diets due to their mild

* Corresponding author: oke.morenikeji@bouesti.edu.ng

DOI: 10.1201/9781003527572-14

flavor and meat-like texture (Siddiq et al., 2018; Salehi, 2019; Tagkouli et al., 2021; Sangeeta et al., 2024).

Beyond nutrition, *Pleurotus* species possess notable therapeutic and environmental applications. They are used in mycoremediation and biofermentation processes to detoxify polluted soils and water, thanks to their enzymatic capacity to break down pollutants. Postharvest mushroom substrates also hold potential as biofertilizers, animal feed additives, and even sources of bioenergy, further extending the ecological and economic value of their cultivation (Raman et al., 2021; Törős et al., 2022; Juárez-Hernández et al., 2023).

*Pleurotus* mushrooms present a multifaceted opportunity to address some of the most critical global challenges—ranging from food and nutrition insecurity to environmental degradation and unemployment. Their global cultivation, ease of production, high adaptability, and numerous applications in food, medicine, and environmental sustainability make them one of nature's most valuable gifts. This chapter explores these diverse benefits of *Pleurotus* species, highlighting their roles in global production, consumption, market dynamics, ecological sustainability, and livelihood generation.

## 14.2 DIVERSITY AND TAXONOMY

Oyster mushrooms, scientifically classified under the genus *Pleurotus*, are commonly known as wood-decaying fungi and are predominantly found in tropical forest ecosystems. These fungi are well-regarded for their ability to grow on a variety of agricultural residues, making them highly adaptable and suitable for sustainable cultivation practices. Taxonomically, *Pleurotus* belongs to the Kingdom Fungi, Phylum Basidiomycota, Class Agaricomycetes, Order Agaricales, Family Pleurotaceae. The name *Pleurotus* is derived from the Greek word "pleuro," referring to the lateral attachment of the stipe (stem) to the pileus (cap)—a characteristic feature of the genus. These mushrooms are primarily saprophytic, feeding on dead organic matter, although some species may occasionally exhibit parasitic behavior. Oyster mushrooms are commonly identified by their habitat and cap coloration (Juárez-Hernández et al., 2023).

*Pleurotus* species predominantly inhabit both temperate and tropical forest ecosystems. In nature, they are commonly found growing on decomposing logs and, less frequently, on the dried trunks of deciduous and coniferous trees (Jarial et al., 2024). These mushrooms thrive on a wide variety of agricultural waste materials, including fallen branches, decaying tree remnants, and moist wooden logs. Their ability to colonize and break down lignocellulosic matter is facilitated by their potent lignin-degrading enzymes (particularly laccase and versatile peroxidases), which rank them among the most effective wood-decaying fungi in terrestrial ecosystems globally (Raman et al., 2021; Juárez-Hernández et al., 2023). Owing to these properties, *Pleurotus* species are extensively cultivated worldwide and are recognized as one of the most commercially valuable edible mushrooms (Knop et al., 2015). These fungi display remarkable diversity in their morphological features, which include variations in cap (pileus) shape and size, stem (stipe) length and position, color, texture, and structural patterns depending on species and environmental conditions (Jarial et al., 2024).

Morphologically, oyster mushrooms are characterized by a fleshy, spatula- or fan-shaped cap (pileus), and a short to elongated stipe that may be lateral, central, or absent altogether (Shukla and Jaitly, 2011; Argaw et al., 2023). The presence of pronounced ridges and furrows on the underside of the pileus enhances their visual appeal and contributes to their widespread popularity. Their reproductive spores, found within the gills, are smooth, cylindrical, and germinate readily on standard mycological media within 48 to 96 hours of incubation (Argaw et al., 2023).

**TABLE 14.1**
**Structural Accounts of *Pleurotus* Species (Jarial et al., 2024)**

| Parameters | Features |
|---|---|
| **Shape** | |
| Cap | Shell or fan-shaped, with a broad, convex to flat cap |
| Spore | Elongated, ellipsoid, and cylindrical |
| **Size** | |
| Cap | Diameter (5–25 cm) |
| Stem | Length (2–10 cm), Thickness (1–5 cm) |
| Spore | Approximately 5–10 $\mu$m long |
| **Color** | |
| Cap | It could vary depending on the strain, but it is generally found in shades of white, gray, or tan and exhibits blue, purple, yellow, or pink hues |
| Gills | White when young, whirling cream to light yellow when matured with hints of pink or lilac |
| Stem | Whitish to cream-colored or pale gray. Can also develop brown spots with age |
| Flesh | White to creamy |
| Spore | White to light yellow or pale yellow |

**TABLE 14.1 *(Continued)***
**Structural Accounts of *Pleurotus* Species (Jarial et al., 2024)**

| Parameters | Features |
|---|---|
| **Stem** | |
| General Features | Short, stubby, and lateral, attached to the cap eccentrically. Varies in length and thickness |
| Surface | Smooth, while sometimes fibrous or silky |
| Color | Whitish to cream-colored or pale gray but also can develop brown spots with age |
| Surface | Smooth and sometimes fibrous or silky |
| **Surface** | |
| Cap | Smooth and often soft, with some wrinkles or ridges, dry in some cases, and can become slightly slimy when dumped |
| Spore | Smooth |
| Stem | Smooth and sometimes fibrous or silky |
| **Texture** | |
| Cap | Smooth and often somewhat wrinkled |
| Flesh | Thick, firm, and succulent |
| **Others** | |
| Gills attachment | Decurrent |
| Gills arrangement | Compact/closely spaced |
| Odor | Mild to slightly fragrant, often with a hint of anise or almond |
| Habitat | Saprotrophic grows on dead or decaying hardwood trees, logs or stumps |
| Climate | Tropical, subtropical, and temperate regions contingent upon species type |
| Cultivating season | Can occur throughout the year depending upon species type |
| Distribution | Widely distributed across the globe, particularly in temperate and subtropical regions |

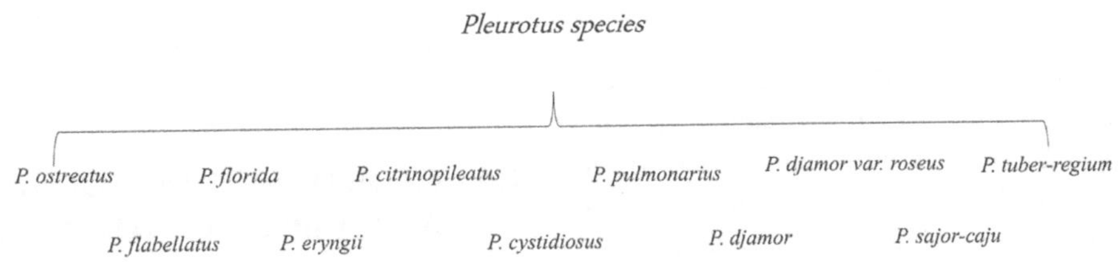


**FIGURE 14.1** Some common commercially cultivated *Pleurotus* species.

The basidiocarps of *Pleurotus* mushrooms exhibit a range of colors and shapes, often resembling a shell or oyster. Cap colors vary widely and may include white, cream, gray, yellow, pink, or light brown, depending on the species and growing conditions (Singh and Singh, 2011; Raman et al., 2021), as outlined in Table 14.1.

Approximately 40 species of *Pleurotus* are widely distributed across tropical and temperate regions. However, only 26 of these species have been successfully cultivated using a variety of lignocellulosic waste materials worldwide (Golak-Siwulska et al., 2018; Barh et al., 2019). Several of the most prominent and commercially cultivated *Pleurotus* species are illustrated in Figure 14.1 (Zakia and Srivastava, 1962; Suguimoto et al., 2001; Mandeel et al., 2005; Islam et al., 2009; Singh and Singh, 2011; Bernardi et al., 2013; Bumanlag et al., 2018; Golak-Siwulska et al., 2018; Chitra et al., 2018; Jarial et al., 2024). Despite this, many species within the genus remain unidentified or underexplored, particularly in biodiverse regions. A significant number of *Pleurotus* species have been reported across Asia, as documented by Bao et al. (2004).

The wide range of substrates used for cultivating *Pleurotus* species (combined with diverse agroclimatic conditions) has led to considerable morphological variability within the genus (Kong, 2004). Their fruiting bodies typically exhibit a shell- or spatula-like shape and can present in various colors, including white, cream, gray, yellow, pink, light brown, and even blue, depending on the species and environmental factors. A notable characteristic of some *Pleurotus* species is the emission of a distinctive, sweet aroma resembling anise or licorice (Jarial et al., 2024).

## 14.3 GLOBAL CULTIVATION OF OYSTER MUSHROOM

Oyster mushrooms are among the most widely cultivated edible fungi globally, ranking third only to the white button mushroom in terms of production volume. They account for approximately 27% of total global mushroom cultivation, reflecting their widespread popularity and economic significance. Large-scale cultivation is prevalent across Southwest Asia, India, Europe, and Africa, with a rapid surge observed particularly in Asian countries due to favorable climatic conditions, low production costs, easy access to agricultural residues, and high yield potential. Historically, the commercial cultivation of oyster mushrooms can be traced back to 1917, when Flack in Germany utilized wood logs and stumps as substrates. This cultivation method was later adopted and modified by Block, Taso, and Hou in the United States. In India, initial trials began in the 1960s, with commercial cultivation gaining momentum by the mid-1970s. At present, several countries including Italy, Poland, the Netherlands, Romania, the Republic of Korea, Spain, Lithuania, and India have established themselves as leading producers of oyster mushrooms. Nonetheless, China dominates global production, contributing approximately 74% to the total output (Jarial et al., 2024).

The genus *Pleurotus* demonstrates remarkable adaptability to various agroclimatic conditions and substrates. It thrives within a wide temperature range (10–30 °C), tolerates pH levels from 6.0 to 8.0, and requires high humidity for optimal development (Shen et al., 2014; Bumanlag et al., 2018). The ideal environmental conditions for robust mycelial growth include temperatures between 25 and 30 °C and relative humidity levels of 75%–90% (Ragupathi et al., 2016). However, studies have shown that elevated temperatures combined with low humidity can significantly inhibit mycelial development (Thakur, 2020). Seasonal analysis has demonstrated that optimal growth and biological efficiency occur during the cooler months, particularly from December to February, with reduced productivity observed during warmer months such as May and June (Uddin et al., 2011; Chitra et al., 2018; Deora et al., 2021).

The cultivation of *Pleurotus* mushrooms not only offers nutritional and medicinal value but also serves as an ecologically sustainable practice. These fungi possess robust ligninolytic capabilities typical of white-rot fungi, enabling them to effectively decompose lignocellulosic biomass. Their ability to grow on a wide range of biodegradable materials, including agricultural and forestry residues, makes them valuable agents in waste bioconversion. This process supports sustainable agriculture and forestry practices by promoting organic waste recycling, enhancing soil fertility, and contributing to nutrient cycling within ecosystems (Sganzerla et al., 2022; Mane et al., 2007).

### 14.3.1 Cultivation Techniques

The successful cultivation of *Pleurotus* species (oyster mushrooms) involves a sequence of well-defined stages, each requiring precision and appropriate environmental conditions. The initial and most critical step is the development of pure mycelium from the selected mushroom species, which ensures consistency and quality in subsequent cultivation phases. Over 200 types of agricultural and agro-industrial waste materials have been reported as suitable substrates for edible mushroom cultivation (Poppe, 2000). However, the choice of substrate and cultivation parameters largely depends on the specific *Pleurotus* species being grown. Each species exhibits unique growth requirements, influencing the composition of substrates, inoculation methods, and environmental conditions for optimal production.

The general stages involved in oyster mushroom cultivation include:

Spawn Preparation—This involves inoculating a sterilized grain or other carrier material with pure mycelium to produce mushroom spawn, which acts as the planting material.

Substrate Preparation—Selected agricultural residues (e.g., straw, sawdust, corn cobs) are processed, moistened, and pasteurized or sterilized to eliminate competing microorganisms.

Spawning—The prepared substrate is inoculated with spawn under aseptic conditions and packed into bags, bottles, or trays, depending on the cultivation system.

Incubation—The inoculated substrate is incubated in a dark, warm, and humid environment to promote mycelial colonization. Once colonization is complete, fruiting is induced by adjusting light, humidity, and ventilation.

Harvesting—Mature fruiting bodies are harvested at the appropriate developmental stage, typically before the caps fully flatten, to ensure quality and shelf life (Argaw et al., 2023).

## 14.4 GLOBAL PRODUCTION AND MARKET ANALYSIS OF OYSTER MUSHROOMS

Mushrooms are broadly classified into edible, medicinal, poisonous, and multi-functional categories (Cheung, 2010; Bijla, 2023; El Sheikha and Hu, 2018). Among the edible and medicinal varieties, oyster mushrooms have emerged as a highly valuable crop, playing a pivotal role in addressing numerous global challenges. From combating food insecurity and environmental degradation to reducing unemployment and enhancing health and economic well-being, mushroom cultivation has evolved into a profitable agro-industry with multifaceted benefits (Ojha and Shukla, 2023; Singh et al., 2024). The rising demand for highly nutritious, low-fat, and cholesterol-free foods also serve as a primary market driver (Thakur, 2020; Rath, 2023; Ogwu et al., 2025). Alongside this, increased investments in research and development aim to enhance production efficiency, develop improved strains, and ensure the safety and quality of mushrooms for human consumption and export (Zhang et al., 2014; De Cianni et al., 2023; Sangeeta et al., 2024).

In recent decades, oyster mushrooms have gained substantial traction, particularly in India and across Asia. The *Pleurotus* genus includes the largest number of commercially cultivated species, prized for their ability to grow year-round with minimal inputs. Compared to white button mushrooms, oyster mushrooms offer a more profitable alternative owing to their low capital investment, reduced labor requirements, and impressive yield potential (Nongthombam et al., 2021). Their simplicity of cultivation, especially in rural areas with access to raw agro-waste, has made them an ideal crop for small- and medium-scale farmers (Masevhe et al., 2016). Another key factor contributing to their rising popularity is their adaptability. Oyster mushrooms can thrive in a wide range of climatic conditions, making them sustainable and environmentally friendly (Rana et al., 2015; Chauhan et al., 2024).

Reflecting their growing global significance, the oyster mushroom market was valued at USD 62.91 billion in 2024 and is projected to reach USD 90.73 billion by 2029, driven by a CAGR of 7.6%. This growth is largely fueled by increasing consumer interest in functional foods and natural remedies. Despite its promise, oyster mushroom cultivation faces certain limitations. High operational costs, the need for technical expertise, and challenges such as pest attacks, climatic variations, and animal interference can affect consistent production. Additionally, a portion of the population experiences allergic reactions or side effects such as dryness, itching, or rashes from prolonged consumption, posing a minor hurdle to market growth (Market Data Forecast, 2024).

## 14.5 GLOBAL CONSUMPTION

The global demand for mushrooms, particularly oyster mushrooms, has surged due to their medicinal and nutraceutical value. The Asia-Pacific region, led by China, dominates both production and consumption, driven by strong domestic, medical, and commercial demand. In North America, the market is rapidly expanding as health-conscious lifestyles and interest in alternative medicine boost oyster mushroom consumption (Market Data Forecast, 2024).

## 14.6 BENEFICIAL OPPORTUNITIES OF OYSTER MUSHROOM

### 14.6.1 Employment Opportunity

Mushroom cultivation, particularly of *Pleurotus* species, is an accessible and inclusive agricultural practice suitable for individuals of all ages and genders. Women form a significant proportion of mushroom farmers. It requires minimal land, low investment, a short cultivation cycle, and utilizes agricultural waste as a substrate, making it a sustainable alternative to traditional farming. In India, most mushroom growers are small-scale or marginal farmers with limited capital, and oyster mushroom farming offers them a cost-effective way to enhance income and food security (Bijla, 2023; Chauhan et al., 2024).

As the third most widely cultivated mushroom globally, *Pleurotus* spp. are easy to grow and have high market demand. Their production not only generates employment and reduces poverty but also provides an effective way for farmers to diversify income sources, build economic resilience, and improve livelihoods (Chauhan et al., 2024).

### 14.6.2 Nutritional Value

Oyster mushrooms are important due to their exceptional culinary and nutritional value, and they contribute enormously to environmental sustainability (Aditya et al., 2023a). They are rich in proteins, containing the essential amino acids needed by the body for daily functions, as well as vitamins, minerals, and fibers, and they serve as a source of antioxidants that protect the human body against damage triggered by free radicals. Oyster mushrooms comprise a large array of vitamins that play major roles in energy metabolism, nerve function, and skin health. They contain minerals such as phosphorus, copper, potassium, and selenium (Adebayo et al., 2014a, 2014b; Aditya et al., 2022). Potassium is important for supporting a healthy heart and muscles. At the same time, selenium acts as an antioxidant, while copper helps in iron absorption and the formation of red blood cells. A unique feature of oyster mushrooms is their ability to produce ergosterol, making them a natural plant-based source of vitamin D, which supports bone formation, the immune system, and overall body function. Table 14.2 depicts the nutritional composition of some *Pleurotus* spp.

#### 14.6.2.1 Proteins and Amino Acids

*Pleurotus* species are excellent sources of high-quality protein, containing all essential amino acids, making them especially beneficial for vegetarian diets (Raman et al., 2020). Among edible mushrooms, oyster mushrooms are recognized for their relatively high protein content, averaging 2–3 g/100 g fresh weight. This protein contributes significantly to tissue repair, immune system support, and overall health (Aditya et al., 2021a, 2021b, 2021c).

The protein content in *Pleurotus* mushrooms varies depending on the species, substrate used for cultivation, and agro-climatic conditions (Elkanah et al., 2022). On a dry weight basis, values range from 9.29 to 37.4 g/100 g (Ritota and Manzi, 2019). For example, *Pleurotus* ostreatus has shown protein levels of 28.85% in fresh samples (Tolera and Abera, 2017); while Korean varieties reported 28.57% and up to 30.4% according to FAO (1973) and Lee et al. (2018). *Pleurotus eous* contains around 24.10%, and *Pleurotus djamor* var. *roseus* ranges between 23.91 and 35.5 g/100 g (Alam et al., 2008; Kortei and Wiafe-Kwagyan, 2015; Jegadeesh et al., 2018). The lowest protein content was observed in *Pleurotus eryngii*, reported at 11.95 g/100 g (Akyuz and Kirbağ, 2010).

#### 14.6.2.2 Carbohydrates

*Pleurotus* mushrooms contain carbohydrate, ranging between 24.95 and 75.88% (Koutrotsios et al., 2014; Raman et al., 2020). In addition, carbohydrate content varied depending on species type (Alam et al., 2008; Ahmed et al., 2009; Adebayo et al., 2014a; Raman et al., 2020).

#### 14.6.2.3 Dietary Fiber

Oyster mushrooms are a valuable source of dietary fiber, making them a beneficial addition to a balanced diet. On average, they provide about 1 to 2 g of fiber/100 g of fresh

**TABLE 14.2**
**Nutritional Composition (As % of Dried Mushrooms) of Some *Pleurotus* Species**

| *Pleurotus* spp. | Moisture (%) | Proteins (%) | Carbohydrates (%) | Fats (%) | Ash (%) | Fiber (%) | References |
|---|---|---|---|---|---|---|---|
| *P. citrinopileatus* | 88.9 | 30.0 | 42.50 | 3.90 | 7.65 | 20.78 | Singh and Singh, 2011 |
| *P. cystidiosus* | 91.1 | 15.6 | 55.92 | 2.05 | 6.30 | 20.05 | Raman et al., 2021 |
| *P. djamor* | 86.8 | 24.1 | 45.59 | 4.73 | 9.84 | 15.91 | Bumanlag et al., 2018 |
| *P. djamor var. roseus* | 79.5 | 35.5 | 44.75 | 1.72 | 5.90 | 14.60 | Jegadeesh et al., 2018 |
| *P. eryngii* | 91.0 | 11.9 | 39.85 | 7.50 | 4.89 | 28.29 | Gąsecka et al., 2016 |
| | 88 | 20 | 53 | 2.8 | 7.5 | 7.5 | Sardar et al., 2017 |
| | 88 | 18.8 | 57 | 2.3 | 5.5 | 10 | Sardar et al., 2022 |
| *P. flabellatus* | 91.0 | 21.6 | 57.40 | 1.80 | 10.7 | 11.90 | Mshandete and Cuff, 2008 |
| *P. florida* | 87.5 | 20.5 | 42.83 | 2.31 | 9.02 | 11.50 | Ahmed et al., 2009 |
| *P. ostreatus* | 90.7 | 18.3 | 71.25 | 2.58 | 7.82 | 14.31 | Jacinto-Azevedo et al., 2021 |
| *P. pulmonarius* | 78.8 | 20.3 | 34.00 | 2.62 | 7.33 | 9.00 | Silva et al., 2002 |
| *P. sajor-caju* | 87.0 | 24.6 | 39.82 | 2.29 | 8.28 | 10.90 | Rajak et al., 2011 |
| *P. tuber-regium* | 87.1 | 22.1 | 63.03 | 1.06 | 2.97 | 10.86 | Ijeh et al., 2009 |

weight (Aditya et al., 2021a, 2021b, 2021c). The exact fiber content may vary depending on the mushroom's size, maturity, and cultivation conditions. Fiber plays a key role in maintaining digestive health and helps reduce blood cholesterol levels. *Pleurotus* species are particularly rich in various types of fiber, contributing to their potential in regulating blood glucose and lowering cholesterol (Xu et al., 2012; Okolo et al., 2016; Raman et al., 2020).

#### 14.6.2.4 Lipids

*Pleurotus* species are known for their low lipid content, making them a heart-healthy dietary option. Despite their low-fat levels, they contain beneficial fatty acids such as triglycerides, sterols, sterol esters, phospholipids, linoleic acid, and oleic acid (Raman et al., 2020). These lipid components are considered favorable for cardiovascular health, as they pose minimal risk of plaque formation in blood vessels. However, lipid content can vary between species. According to Musieba et al. (2013), these variations are species-specific, and a survey by Lavelli et al. (2018) reported fat content in *Pleurotus* species ranging from 0.9% to 7.5%.

#### 14.6.2.5 Vitamins and Minerals

Oyster mushrooms are rich in essential vitamins and minerals, making them a valuable component of a balanced diet. They contain notable amounts of vitamin C, which acts as a powerful antioxidant, along with vitamins B1 (thiamine) and B2 (riboflavin), crucial for energy metabolism, cell growth, and repair. Additionally, they are an excellent dietary source of folic acid (vitamin B9), a vital nutrient that the human body cannot synthesize and must obtain through food (Sopanrao et al., 2010; Okolo et al., 2016; Raman et al., 2020). *Pleurotus* species also provide a broad spectrum of minerals, including potassium, phosphorus, zinc, iron, sodium, and selenium. Potassium plays a key role in regulating heart rhythm, fluid balance, nerve function, and blood pressure, while iron is critical for oxygen transport and overall cellular function (Adebayo et al., 2014b; Adebayo et al., 2017). Reported mineral concentrations include 4.22–7.70 μg/g for zinc, 5.5–13.4 μg/g for iron, and selenium levels ranging between 0.011 and 0.512 mg/100 g (Tang et al., 2016; Golak-Siwulska et al., 2018; Kikuchi, 2025).

### 14.6.3 Therapeutic Potentials

Oyster mushrooms are recognized for their wide-ranging medicinal properties, largely due to their rich content of bioactive compounds (Figure 14.2; Table 14.3). Their consumption enhances the nutritional quality of the diet, supplying essential vitamins and minerals that contribute to improved overall health. One of the key therapeutic attributes of oyster mushrooms is their immune-boosting ability, particularly linked to the presence of β-glucans. These polysaccharides stimulate immune cells and enhance the body's defense mechanisms (Golak-Siwulska et al., 2018; Aditya and Bhatia, 2023; Aditya et al., 2023b, 2023c). Additionally, they possess strong anti-inflammatory properties, which may help manage chronic inflammatory conditions. Oyster mushrooms also contribute to cardiovascular health by potentially lowering cholesterol levels, partly through the inhibition of cholesterol absorption. Furthermore, several studies have indicated their antitumor and anticancer activities, attributed to specific bioactive compounds found in their fruiting bodies. Another significant property of oyster mushrooms is their high antioxidant potential. These natural antioxidants combat oxidative stress, protect cellular integrity, and support overall wellness (Aditya et al., 2023a; Tiwari et al., 2023). Thus, incorporating oyster mushrooms into the diet not only enhances nutrition but also offers promising preventive health benefits.

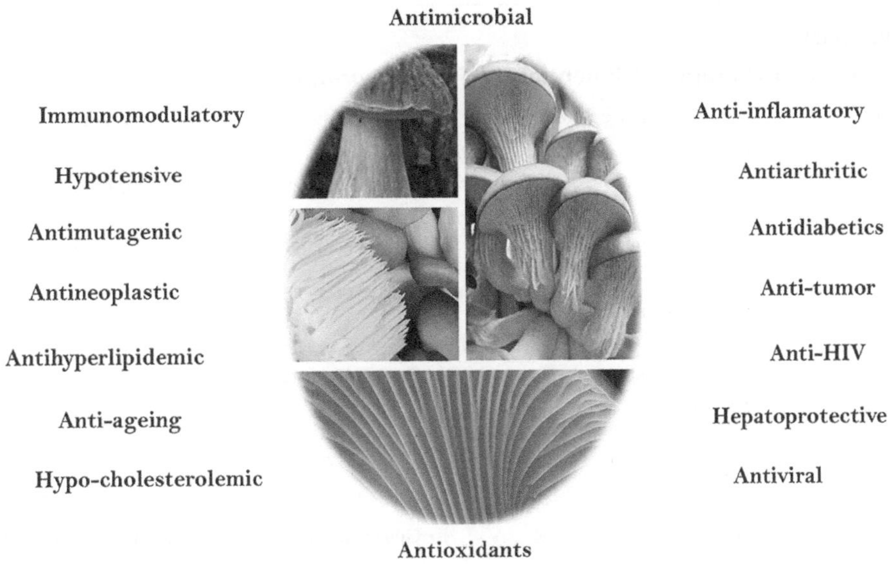


**FIGURE 14.2** Medicinal properties of oyster mushrooms.

**TABLE 14.3**
**Bioactive Components and Therapeutic Potentials of Some *Pleurotus* Species**

| Bioactive Composition | Therapeutic Potentials | References |
|---|---|---|
| | *P. ostreatus* | |
| β-glucans | Anticancer | Abidin et al., 2017;Golak-Siwulska et al., 2018;Wjabn et al., 2020;Gutef et al., 2020;Mishra et al., 2021; Binsaleh et al., 2025 |
| α-glucan | Antitumour | |
| Proteins | Immunomodulatory | |
| Polysaccharides | Antihypercholesterolemic | |
| Lectins | Anti-atherogenic | |
| Polysaccharide peptides and polysaccharide-protein complex | Antiviral<br>Antibacterial | |
| lectin | Anti-oxidative | |
| Proteoglycans | Hypoglycemic | |
| Heteroglycan | Anti-inflammatory | |
| Lovastatin | Anti-arthritic | |
| Lovastatin | Anti-atopic dermatitis | |
| Chrysin | Anticataractogenic | |
| Lovastatin | Antinociceptive | |
| Pleuran | | |
| Vitamin C | | |
| | *P. geesteranus* | |
| Polysaccharide-protein | Anticancer | Binsaleh et al., 2025 |
| | *P. citrinopileatus* | |
| Glycoprotein | Anticancer | Mishra et al., 2021;Binsaleh et al., 2025 |
| | *P. nebrodensis* | |
| Nebrodeolysin | Anticancer | Cui et al., 2014, 2016;Gargano et al., 2024;Binsaleh et al., 2025 |
| Polysaccharides | Antibacterial | |
| Proteins | | |
| | *P. cornucopiae* | |
| Laccase | Inhibition of the hepatitis C virus and hepatoma | Mishra et al., 2021;Binsaleh et al., 2025 |
| | *P. djamor* | |
| Polysaccharides | Anticancer | Manimaran et al., 2021 |
| Ribonuclease | Antibacterial | |
| Ubiquinone-9 | Antioxidants | |

*(Continued)*

**TABLE 14.3** *(Continued)*
**Bioactive Components and Therapeutic Potentials of Some *Pleurotus* Species**

| Bioactive Composition | Therapeutic Potentials | References |
|---|---|---|
| | *P. citrinopileatus* | |
| lectin | Antitumour | Younis et al., 2019;Wang et al., 2021 |
| | *P. cornucopiae* | |
| Polysaccharides<br>β-Glucan | Immunomodulatory<br>Antitubercular,<br>Anti-inflammatory | Minato et al., 2017;Costanzo et al., 2022 |
| | *P. sajor-caju* | |
| Lovastatin<br>β-glucans | Antihypercholesterolemic<br>Hypoglycemic<br>Anti-arthritic<br>Anti-rheumatoid | Patel et al., 2012b;Chen et al., 2015 |
| | *P. florida* | |
| Lovastatin<br>Lectins<br>Polysaccharides | Antihypercholesterolemic<br>Anti-oxidative<br>Anti-inflammatory | Gunawardena et al., 2014; Pandimeena et al., 2015; Corrêa et al., 2016 |
| | *P. cornucopiae* | |
| D-mannitol<br>Peptides | Anti-atherogenic<br>Antihypertensive | Abidin et al., 2017 |
| | *P. eryngii* | |
| Ergothioneine<br>Proteins | Anti-atherogenic<br>Antifungal<br>Antidepressant | Abidin et al., 2017;Park et al., 2021 |
| | *P. nebrodensis* | |
| Proteins | Antiviral | Hassan et al., 2015 |
| | *P. abalonus* | |
| Polysaccharides | Antiviral | Chen et al., 2015 |
| | *P. citrinopileatus* | |
| Lectins | Antiviral | Hassan et al., 2015 |
| | *P. eous* | |
| Fatty acids esters<br>β-glucans | Antibacterial<br>Antinociceptive | Suseem et al., 2013 |
| | *P. cystidiosus* | |
| Ergosterol<br>Heteroglycan | Antifungal<br>Antioxidant | Panda et al., 2017;Garcia et al., 2020;Atawong and Kriangkrai, 2024 |
| | *P. citrinopileatus* | |
| Phenols | Anti-oxidative<br>Antibacterial | Chomcheon et al., 2013 |
| | *P. tuber-regium* | |
| Lectins<br>Polysaccharides | Anti-oxidative<br>Antibacterial | Adebayo et al., 2018;Okolo and Orisakwe, 2021; Sheng et al., 2023 |
| | *P. abalonus* | |
| Polysaccharide-peptide | Hypoglycemic | Hassan et al., 2015 |
| | *P. ostreatoroseus* | |
| Polysaccharides | Anti-inflammatory | Gunawardena et al., 2014;Corrêa et al., 2016 |

#### 14.6.3.1 Antioxidant

Oxidative stress plays a central role in the development of numerous degenerative diseases in humans (Adebayo et al., 2014a, 2018; Aditya et al., 2022). Oyster mushrooms have demonstrated significant potential to mitigate oxidative damage due to their rich composition of natural antioxidants. These include phenolic compounds, flavonoids, selenium, and vitamins, all of which contribute to their potent free radical scavenging activities (Chang and Miles, 1989; Kumar, 2020; Bassi et al., 2024; Sultana et al., 2024).

A notable study by Zhang et al. (2012) evaluated the antioxidant properties of two polysaccharide fractions, namely PSPO-1a and PSPO-4a, isolated from the fruiting bodies of *P. ostreatus*. These fractions exhibited strong radical scavenging activities, particularly against 2,2-diphenyl-1-picrylhydrazyl (DPPH) and superoxide anions, with PSPO-1a displaying greater efficacy than PSPO-4a.

In another investigation, Mitra et al. (2013) explored the nitric oxide (NO) scavenging ability of *P. ostreatus*, linking its high antioxidant potential to the presence of β-glucans. Furthermore, other species, such as *Pleurotus porrigens* and *Pleurotus florida*, have also shown strong DPPH radical scavenging activity (Wong and Chye, 2009), reinforcing the genus *Pleurotus* as a valuable natural source for developing antioxidant-rich functional foods and dietary supplements.

#### 14.6.3.2 Antimicrobial

*Pleurotus* species have been reported to exhibit significant antimicrobial activities, primarily attributed to the presence of a wide array of bioactive myco-constituents (Adebayo et al., 2018; Bhardwaj et al., 2020). Their ability to act against various pathogenic microorganisms has been demonstrated in several studies. Notably, Ahmed et al. (2022) reported that methanolic extracts of *Pleurotus* showed the strongest antimicrobial inhibition, followed by acetone and petroleum ether extracts, with inhibitory effects increasing proportionally with the extract concentration.

In another study, Mirunalini et al. (2012) investigated the mycosynthesis of silver nanoparticles (AgNPs) using *P. ostreatus*, which showed remarkable antimicrobial activity, particularly against Gram-positive bacteria, as evidenced by significant zones of inhibition. These findings are supported by several earlier studies that utilized different extraction media to assess the antimicrobial potential of *Pleurotus* species (Cowan, 1999; Wang et al., 2000; Wolff et al., 2008; Akyuz et al., 2010; Mirunalini et al., 2012; Devi et al., 2024). Collectively, this highlights the promising role of *Pleurotus* mushrooms as natural sources of antimicrobial agents, with potential applications in food preservation, pharmaceuticals, and biomedicine.

#### 14.6.3.3 Anticancer

Oyster mushrooms possess a range of bioactive compounds that exhibit notable anticancer properties. Research has shown that extracts from oyster mushrooms can inhibit the proliferation of cancer cells, particularly in the liver. For instance, studies have demonstrated that these extracts can suppress the growth of hepatocellular carcinoma cells, highlighting their therapeutic potential (Mishra et al., 2022). In addition, other investigations have revealed the efficacy of oyster mushroom extracts against hematological malignancies. Specifically, they have been found to induce cytotoxic effects and promote apoptosis in leukemia cells, suggesting their role as a promising natural agent for leukemia treatment (Devi et al., 2024; Jayaprakash et al., 2024; Sreedharan et al., 2025). These anticancer activities are attributed to the presence of polysaccharides (particularly β-glucans), phenolic compounds, and other secondary metabolites that exert immunomodulatory, antioxidant, and antiproliferative effects.

#### 14.6.3.4 Immune-Boosting

Oyster mushrooms contain bioactive compounds that have been shown to significantly boost the immune system. Their extracts stimulate the production of white blood cells, which play a crucial role in defending the body against infections. These mushrooms exhibit strong immunomodulatory properties, largely attributed to their rich composition of polysaccharides, proteoglycans, and other secondary metabolites. For instance, Selegean et al. (2009) reported the immunostimulatory effects of polysaccharides isolated from *P. ostreatus*, which enhanced immune responses in broiler chickens infected with infectious bursal disease virus, particularly four weeks post-hatching. Similarly, Sarangi et al. (2006) isolated three neutral proteoglycans from oyster mushrooms that exhibited both immunomodulatory and anticancer properties.

#### 14.6.3.5 Anti-inflammatory

Oyster mushrooms are rich in bioactive compounds with notable anti-inflammatory properties. Inflammation is a key contributor to chronic diseases such as heart disease, cancer, and arthritis, and numerous studies have demonstrated the ability of *Pleurotus* species to mitigate inflammatory responses in both *in vitro* and *in vivo* models (Bassi et al., 2024). For example, the anti-inflammatory effects of *Pleurotus sajor-caju* were evident in an experimental rat model, where oral administration of its aqueous and methanolic extracts (500–1000 mg/kg) significantly reduced inflammation and arthritis symptoms (Patel et al., 2012a). Similarly, β-glucans isolated from *P. ostreatus* were shown to enhance treatment efficacy by modulating inflammatory responses (Rovenský et al., 2011).

In another study by Choi et al. (2013), ethanolic extracts of *P. eryngii* administered orally to mice led to a dose-dependent reduction in dermatitis. The treatment also lowered serum levels of immunoglobulin E and thymus and activation-regulated chemokine (TARC), key markers of allergic inflammation. Although some cytokines and skin lesions were still present, the overall findings confirmed the potential of oyster mushrooms in reducing inflammation-associated conditions.

#### 14.6.3.6 Blood-Sugar Lowering and Cardioprotective

Oyster mushrooms contain bioactive compounds that help lower blood-sugar levels, making them beneficial for individuals with diabetes or prediabetes (Agrawal et al., 2010; Agunloye and Oboh, 2022; Devi et al., 2024). These mushrooms also support cardiovascular health. Their extracts have been shown to reduce cholesterol levels and improve the function of blood vessels, contributing to better heart health (Dicks and Ellinger, 2020; Effiong et al., 2024; Devi et al., 2024).

#### 14.6.3.7 Skin-Protective

Oyster mushrooms contain bioactive compounds, including antioxidants such as polyphenols, flavonoids, and vitamin C, which help protect the skin from damage caused by ultraviolet radiation and environmental pollutants. These compounds neutralize free radicals, reduce oxidative stress, and support skin cell regeneration, thereby contributing to healthier, more resilient skin (Bassi et al., 2024; Devi et al., 2024; Sultana et al., 2024).

### 14.6.4 Environmental Safety and Sustainability

Oyster mushroom production is environmentally sustainable, as it utilizes agricultural waste as a cultivation substrate. The technology not only helps to recycle biomass but also converts the waste into mushroom compost, a valuable organic fertilizer that enhances soil fertility for other crops. The wood-degrading ability of *Pleurotus* spp. contributes to the development of healthy soil ecosystems (Pavlik et al., 2023). Moreover, oyster mushrooms play a role in environmental remediation by removing waste efficiently and economically, thereby promoting sustainability.

Environmental contamination by heavy metals, non-biodegradable pollutants that persist and magnify through the food chain, poses a serious health risk. *Pleurotus* species have demonstrated the ability to bioaccumulate these toxic elements, suggesting their potential in bioremediation. However, this also highlights the importance of regulating substrate sources, as heavy metal accumulation in mushrooms can transfer to humans through consumption (Kapahi and Sachdeva, 2017; Innes, 2023).

### 14.6.5 Economic Value

Mushroom production holds significant potential as a holistic solution for revitalizing both global and national economies. Mushrooms are rich in high-quality protein, making them an essential component of the human diet, particularly in underdeveloped regions where cereal-based diets dominate. The proteins found in mushrooms are of excellent quality, containing a wide array of essential amino acids (Sadler, 2003; Chauhan et al., 2024). As such, incorporating mushrooms into diets could address key issues related to food insecurity, malnutrition, and protein deficiency, thereby helping to uplift the economic status of countries burdened by these challenges (Singh et al., 1995; Dunkwal et al., 2007; Chauhan et al., 2024).

Oyster mushrooms are affordable and accessible to both affluent and economically disadvantaged populations, making them an ideal food source for improving general nutrition and quality of life. They are also rich in non-starchy carbohydrates, dietary fiber, minerals, and vitamins (notably C and B complex), while containing no cholesterol and only trace amounts of fat. Their delicate flavor allows them to enhance a variety of savory dishes (Minato et al., 2017).

The production and consumption of *Pleurotus* species contribute significantly to individual well-being due to their high nutritional value and digestibility. Their role in promoting health and preventing various diseases has been well established, attributed to their diverse array of bioactive compounds (Patel et al., 2012a, 2012b; Chauhan et al., 2024). The therapeutic applications of these bioactive further reinforce the health-promoting status of Pleurotus mushrooms (Lee et al., 2012), and the resulting health improvements can positively influence a nation's economic productivity.

One of the pressing global issues affecting economic development is inadequate employment. Oyster mushroom cultivation offers a viable alternative livelihood, especially for youth and women. It requires minimal land, low capital investment, easily accessible inputs, and has a fast cultivation cycle, making it suitable for adoption at the household level. This provides an opportunity to generate meaningful income for individuals and communities (Thakur, 2020).

Moreover, oyster mushroom production contributes to environmental sustainability and enhanced agricultural productivity. The cultivation of *Pleurotus* species relies heavily on the recycling of diverse agricultural wastes (Tiram, 2013; Mahari et al., 2020; Chauhan et al., 2024). These mushrooms efficiently bioconvert agro-wastes that would otherwise pose environmental hazards, transforming them into valuable biomass (Adebayo and Martinez-Carrera, 2015; Mahari et al., 2020; Fufa et al., 2021). This process not only mitigates waste pollution but also produces biofertilizer from spent mushroom substrates, which can restore soil fertility and boost crop yields (Patel et al., 2012b; Chauhan et al., 2024).

## 14.7 CONCLUSION

*Pleurotus* species (oyster mushrooms) are among the most widely cultivated edible mushrooms, ranking third in global production. These lignocellulose-degrading fungi are highly valued for their broad acceptability and global consumption. In recent years, they have garnered increasing international attention due to their diverse health-promoting properties and substantial economic value. Nutritionally, *Pleurotus* mushrooms are rich sources of essential vitamins, minerals, dietary fiber, cholesterol-free content, low caloric value, and high-quality proteins, making them a beneficial dietary component across all age groups. Their unique flavor profile and palatability further enhance their appeal as a versatile culinary ingredient.

Beyond nutrition, *Pleurotus* species are known for their bioactive compounds, which possess a range of therapeutic properties, including immunomodulatory, antioxidant, anti-inflammatory, antimicrobial, and anticancer activities. Given these benefits, promoting the cultivation of edible mushrooms such as *Pleurotus* can create significant livelihood opportunities, particularly for marginalized communities and smallholder farmers, including youth and women, by providing a sustainable and profitable enterprise.

Furthermore, increased mushroom production can contribute meaningfully to the national economy through food security, value-added product development, and rural employment. However, to fully realize these benefits, it is essential to develop and adopt improved, scalable cultivation techniques aimed at enhancing yield and quality. The future of oyster mushroom cultivation is promising—offering multifaceted economic, environmental, and social benefits for both producers and society at large.

## REFERENCES

Abidin, M. H. Z., N. Abdullah, and N. Z. Abidin. 2017. Therapeutic properties of *Pleurotus* species (oyster mushrooms) for atherosclerosis: a review. *International Journal of Food Properties* 20(6): 1251–1261. https://doi.org/10.1080/10942912.2016.1210162

Adebayo, E. A., and D. Martinez-Carrera. 2015. Oyster mushrooms (*Pleurotus*) are useful for utilizing lignocellulosic biomass. *African Journal of Biotechnology* 14(1): 52–67. https://doi.org/10.5897/AJB2014.14249

Adebayo, E. A., D. Martínez-Carrera, P. Morales, M. Sobal, H. Escudero, M. E. Meneses, and M. Bonilla. 2018. Comparative study of antioxidant and antibacterial properties of the edible mushrooms *Pleurotus levis*, *P. ostreatus*, *P. pulmonarius* and *P. tuber-regium*. *International Journal of Food Science and Technology* 53(5): 1316–1330. https://doi.org/10.1111/ijfs.13712

Adebayo, E. A., and J. K. Oloke. 2017. Oyster mushroom (*Pleurotus* species); a natural functional food. *The Journal of Microbiology, Biotechnology and Food Sciences* 7(3): 254. https://doi.org/10.15414/jmbfs.2017/18.7.3.254-264

Adebayo, E. A., J. K. Oloke, D. A. Aina, and T. C. Bora. 2014a. Antioxidant and nutritional importance of some *Pleurotus* species. *Journal of Microbiology, Biotechnology* and *Food Sciences* 3(4): 289–294.

Adebayo, E. A., J. K. Oloke, M. A. Azeez, I. O. Omomowo, and T. C. Bora. 2014b. Assessment of the genetic diversity among ten genotypes of *Pleurotus* (oyster mushroom) using nutrient and mineral compositions. *Scientia Horticulturae* 166: 59–64. https://doi.org/10.1016/j.scienta.2013.12.010

Aditya, B., R. S. Jarial, and J. N. Bhatia. 2021a. Entrepreneurship opportunities through waste management. In: *Entrepreneurship Emerging Trends and Challenges*, ed. S. Sruthi, P. Rengarajan, 101–111. Nitya Publications, Bhopal, India.

Aditya, B., R. S. Jarial, and J. N. Bhatia. 2021b. Waste management - need of the hour. In: *Emerging Issues in Commerce, Economics, Information Technology and Management*, ed. A. K. Sarkar, 107–119. Red'Shine Publications Pvt. Ltd. Lunawada, Gujrat, India.

Aditya, B., R. S. Jarial, and J. N. Bhatia. 2021c. Empowering rural women through mushroom farming - a profitable and sustainable enterprise. In: *Women Empowerment in India*, 55–63. Scentics Publications, New Delhi, India.

Aditya, B., R. S. Jarial, and K. Jarial. 2022. Commercial cultivation of the Elm Oyster mushroom *Hypsizygus ulmarius* (Agricomycetes) on different substrates and its medicinal benefits. *International Journal of Medicinal Mushrooms* 24(12): 87–93. https://doi.org/10.1615/IntJMedMushrooms.2022045380

Aditya, B., R. S. Jarial, and J. N. Bhatia. 2023a. Performance of various agroforestry wastes for the cultivation of elm oyster mushroom *Hypsizygus ulmarius* (Agaricomycetes) in India and its biochemical constituents. *International Journal of Medicinal Mushrooms* 25(08): 55–62. https://doi.org/10.1615/IntJMedMushrooms.2023049037

Aditya, B., R. S. Jarial, and K. Jarial. 2023b. Influence of physiological parameters on mycelial growth and cultural characteristics of blue oyster mushroom [*Hypsizygus ulmarius*] (Bull.: Fr.) Redhead]. *Indian Journal of Ecology* 50(01): 151–156. https://doi.org/10.55362/IJE/2023/3869

Aditya, B., R. S. Neeraj, K. J. Jarial, and J. N. Bhatia. 2023c. Oyster mushrooms: versatile mushrooms with potential health benefits. *The Science World* 3(08): 2128–2133. https://doi.org/10.5281/zenodo.8290368

Agrawal, R. P., A. Chopra, G. S. Lavekar, et al., 2010. Effect of oyster mushroom on glycemia, lipid profile and quality of life in type 2 diabetic patients. *Australian Journal of Medical Herbalism* 22(2): 50–54.

Agunloye, O. M., and G. Oboh. 2022. Blood glucose lowering and effect of oyster (*Pleurotus ostreatus*)-and shiitake (*Lentinus subnudus*)-supplemented diet on key enzymes linked diabetes and hypertension in streptozotocin-induced diabetic in rats. *Food Frontiers* 3(1): 161–171. https://doi.org/10.1002/fft2.111

Ahmed, H. A., M. S. Youssef, and T. A. Magraby. 2022. Antimicrobial and antioxidant activities of *Pleurotus ostreatus* as edible oyster mushroom in Egypt. *Journal of Environmental Studies and Sciences* 28: 2434.

Ahmed, S. A., J. A. Kadam, V. P. Mane, S. S. Patil, and M. M. V. Baig. 2009. Biological efficiency and nutritional contents of *Pleurotus florida* (Mont.) Singer cultivated on different agro-wastes. *Nature and Science of Sleep* 7(1): 44–48.

Akyuz, M., and S. Kirbağ. 2010. Effect of various agro-residues on nutritive value of *Pleurotus eryngii* (DC. ex Fr.) Quel. var. ferulae Lanzi. *Journal of Agricultural Sciences* 16: 83–88. https://doi.org/10.1501/Tarimbil_0000001125

Akyuz, M., A. Onganer, P. Erecevıt, and S. Kırbag. 2010. Antimicrobial activity of some edible mushrooms in the eastern and southeast Anatolia region of Turkey. *Gazi University Journal of Science* 23(2): 125–130.

Alam, N., R. Amin, A. Khan, et al. 2008. Nutritional analysis of cultivated mushrooms in Bangladesh–*Pleurotus ostreatus, Pleurotus sajor-caju, Pleurotus florida* and *Calocybe indica. Mycobiology* 36(4): 228–232. https://doi.org/10.4489/MYCO.2008.36.4.228

Argaw, B., T. Tesfay, T. Godifey, and N. Asres. 2023. Growth and yield performance of oyster mushroom (*P. ostreatus* (Jacq.: Fr.) Kummer) using waste leaves and sawdust. *International Journal of Agronomy* 2023(1): 8013491. https://doi.org/10.1155/2023/8013491

Atawong, P., and W. Kriangkrai. 2024. In vitro study on skin-related bioactivity of Pleurotus cystidiosus extract. Doctoral dissertation, Naresuan University.

Bao, D., S. Kinugasa, Y. Kitamoto. 2004. The biological species of oyster mushrooms (*Pleurotus spp.*) from Asia based on mating compatibility tests. *Journal of Wood Science* 50: 162–168. https://doi.org/10.1007/s10086-003-0540-z

Barh, A., V. P. Sharma, S. K. Annepu, S. Kamal, S. Sharma, and P. Bhatt. 2019. Genetic improvement in *Pleurotus* (oyster mushroom): a review. *3 Biotech* 9(9): 322. https://doi.org/10.1007/s13205-019-1854-x

Bassi, S., M. Benvenuti, S. Mirata, et al. 2024. Enhanced antioxidant and anti-inflammatory activity of the extracts of *Pleurotus ostreatus* edible mushroom grown on *Lavandula angustifolia* residues. *Food Bioscience* 60: 104382. https://doi.org/10.1016/j.fbio.2024.104382

Bernardi, E., E. Minotto, and J. S. D. Nascimento. 2013. Evaluation of growth and production of *Pleurotus* sp. in sterilized substrates. *Arquivos do Instituto Biológico* 80: 318–324. https://doi.org/10.1590/S1808-16572013000300009

Besufekad, Y., A. Mekonnen, B. Girma, et al. 2020. Selection of appropriate substrate for production of oyster mushroom (*Pleurotus ostreatus*). *Genetics* 11(1): 15–25. https://doi.org/10.5897/JYFR2019.0187

Bhardwaj, K., A. Sharma, and N. Tejwan. 2020. *Pleurotus* macrofungi assisted nanoparticle synthesis and its potential applications a review. *Journal of Fungi* 6: 351. https://doi.org/10.3390/jof6040351

Bijla, S. 2023. Status of mushroom production: global and national scenario. *Mushroom Research* 32(2): 91–98.

Binsaleh, N. K., A. S. Bazaid, H. Barnawi, B. Alharbi, A. Alsolami, A. Y. Babsail, and H. Qanash. 2025. Chemical characterization, anticancer, antioxidant and anti-obesity activities with molecular docking studies of *Pleurotus ostreatus* biomass exposed to moist heat. *Journal of Food Measurement and Characterization* 19: 1–20. https://doi.org/10.1007/s11694-025-03125-9

Bumanlag, C. P. B., S. P. Kalaw, R. M. R. Dulay, and R. G. Reyes. 2018. Optimum conditions for mycelia growth and basidiocarp production of *Pleurotus djamor* on corn-based media. *International Journal of Biology, Pharmacy and Allied Sciences* 7(4): 558–575.

Chang, S. T., and P. G. Miles. 1989. *Edible Mushrooms and Their Cultivation*. CRC Press, Inc., Florida.

Chauhan, A., S. Negi, R. Kaur, M. Chauhan, R. Dhanai, and P. S. Negi. 2024. Cultivation of oyster mushroom could be a viable option for doubling the farmer's income-an overview. *International Journal of Economic Plants* 11(4): 387–393. https://doi.org/10.23910/2/2024.5725a

Chen, R. R., Z. K. Liu, F. Liu, and T. B. Ng. 2015. Antihyperglycaemic mechanisms of an aceteoside polymer from rose flowers and a polysaccharide–protein complex from abalone mushroom. *Natural Product Research* 29(6): 558–561. https://doi.org/10.1080/14786419.2014.952230

Cheung, P. C. 2010. The nutritional and health benefits of mushrooms. *Nutrition Bulletin* 35(4): 292–299. https://doi.org/10.1111/j.1467-3010.2010.01859.x

Chitra, K., R. Venkatesh, K. Dhanalakshmi, P. T. Sharavanan, C. Bali Sasikumar, and K. Karthikeyani Vijayakumari. 2018. Production and economic analysis of oyster mushroom (*Pleurotus florida*). *International Journal of Current Microbiology and Applied Sciences* 7(9): 379–383.

Choi, J. H., H. G. Kim, S. W. Jin, et al. 2013. Topical application of *Pleurotus eryngii* extracts inhibits 2, 4-dinitrochlorobenzene-induced atopic dermatitis in NC/Nga mice by the regulation of Th1/Th2 balance. *Food and Chemical Toxicology* 53: 38–45. https://doi.org/10.1016/j.fct.2012.11.025

Chomcheon, P., B. Kheawkum, P. Sriwiset, S. Dulsamphan, and C. Dulsamphan. 2013. Antibacterial activity of crude extracts from edible mushrooms *Pleurotus citrinopileatus* and *Tricholoma crassum* Berk. *Thai Journal of Pharmaceutical Sciences* 37: 107–111.

Corrêa, R. C. G., T. Brugnari, A. Bracht, R. M. Peralta, and I. C. Ferreira. 2016. Biotechnological, nutritional and therapeutic uses of *Pleurotus* spp. (Oyster mushroom) related with its chemical composition: a review on the past decade findings. *Trends in Food Science and Technology* 50: 103–117. https://doi.org/10.1016/j.tifs.2016.01.012

Costanzo, V., M. Gilhen-Baker, D. Beresford-Kroeger, and Roviello. 2022. Tree-inhabiting polypore fungi as sources of a cornucopia of bioactive compounds. *Future Microbiology* 17(12): 899–902. https://doi.org/10.2217/fmb-2022-0098

Cowan, M. M. 1999. Plant products as antimicrobial agents. *Clinical Microbiology Reviews* 12(4): 564–582. https://doi.org/10.1128/cmr.12.4.564

Cui, H., C. Wang, Y. Wang, Z. Li, Y. Zhang, M. Chen, and F. Li. 2014. *Pleurotus nebrodensis* polysaccharide induces apoptosis in human non-small cell lung cancer A549 cells. *Carbohydrate Polymers* 104: 246–252. https://doi.org/10.1016/j.carbpol.2014.01.001

Cui, H., S. Wu, Y. Sun, T. Wang, Z. Li, M. Chen, and C. Wang. 2016. Polysaccharide from *Pleurotus nebrodensis* induces apoptosis via a mitochondrial pathway in HepG2 cells. *Food and Function* 7(1): 455–463. https://doi.org/10.1039/c5fo00884k

De Cianni, R., L. Pippinato, and T. Mancuso. 2023. A systematic review on drivers influencing consumption of edible mushrooms and innovative mushroom-containing products. *Appetite* 182: 106454. https://doi.org/10.1016/j.appet.2023.106454

Deora, A., S. S. Sharma, P. Kumari, V. Dahima, S. Kumar, and M. Rohith. 2021. Cultivation of Kabul Dhingri (*Pleurotus eryngii*) mushroom by standardizing protocols in subtropical zones of world. *Scientific Reports* 11(1): 14692. https://doi.org/10.1038/s41598-021-94038-2

Devi, P. V., J. Islam, P. Narzary, D. Sharma, and F. Sultana. 2024. Bioactive compounds, nutraceutical values and its application in food product development of oyster mushroom. *Journal of Future Foods* 4(4): 335–342. https://doi.org/10.1016/j.jfutfo.2023.11.005

Dicks, L., and S. Ellinger. 2020. Effect of the intake of oyster mushrooms (*Pleurotus ostreatus*) on cardiometabolic parameters—a systematic review of clinical trials. *Nutrients* 12(4): 1134. https://doi.org/10.3390/nu12041134

Dunkwal, V., S. Jood, and S. Singh. 2007. Physico-chemical properties and sensory evaluation of *Pleurotus sajor caju* powder as influenced by pre-treatments and drying methods. *British Food Journal* 109(9): 749–759. https://doi.org/10.1108/00070700710780715

Effiong, M. E., C. P. Umeokwochi, I. S. Afolabi, and S. N. Chinedu. 2024. Assessing the nutritional quality of *Pleurotus ostreatus* (oyster mushroom). *Frontiers in Nutrition* 10: 1279208. https://doi.org/10.3389/fnut.2023.1279208

El Sheikha, A. F., and D. M. Hu. 2018. How to trace the geographic origin of mushrooms? *Trends in Food Science and Technology* 78: 292–303. https://doi.org/10.1016/j.tifs.2018.06.008

Elkanah, F. A., M. A. Oke, and E. A. Adebayo. 2022. Substrate composition effect on the nutritional quality of *Pleurotus ostreatus* (MK751847) fruiting body. *Heliyon* 8(11): e11841. https://doi.org/10.1016/j.heliyon.2022.e11841

Food and Agricultural Organization. Energy and protein requirements, report of a joint FAO/WHO ad-hoc Committee. Rome (Italy): Food and Agriculture Organization; 1973. (Food Nutrition Report Series No. 52).

Fufa, B. K., B. A. Tadesse, and M. M. Tulu. 2021. Cultivation of *Pleurotus ostreatus* on agricultural wastes and their combination. *International Journal of Agronomy* 2021(1): 1465597. https://doi.org/10.1155/2021/1465597

Garcia, K., C. J. Garcia, R. Bustillos, and R. M. Dulay. 2020. Mycelial biomass, antioxidant, and myco-actives of mycelia of abalone mushroom *Pleurotus cystidiosus* in liquid culture. *Journal of Applied Biology and Biotechnology* 8(2): 94–97. https://doi.org/10.7324/JABB.2020.80215

García-Ramón, F., M. Malnati-Ramos, J. Rios-Mendoza, et al. 2023. Avocado Hass peel from industrial by-product: effect of extraction process variables on yield, phenolic compounds and antioxidant capacity. *Frontiers in Sustainable Food Systems* 7: 1255941. https://doi.org/10.3389/fsufs.2023.1255941

Gargano, M. L., G. Venturella, G. Balenzano, V. Ferraro, F. Cirlincione, and G. Mirabile. 2024. *Pleurotus nebrodensis* (Basidiomycota), a rare endemic mushroom of Sicily: current and future issues. *Italian Botanist* 17: 55–68. https://doi.org/10.3897/italianbotanist.17.123048

Gąsecka, M., M. Mleczek, M. Siwulski, and P. Niedzielski. 2016. Phenolic composition and antioxidant properties of *Pleurotus ostreatus* and *Pleurotus eryngii* enriched with selenium and zinc. *European Food Research and Technology* 242: 723–732. https://doi.org/10.1007/s00217-015-2580-1

Getachew, A., A. Keneni, and M. Chawaka. 2019. Production of oyster mushroom (*Pleurotus ostreatus*) on substrate composed from wheat straw, waste paper and cotton seed waste. *International Journal of Microbiology and Biotechnology* 4(2): 38–44. https://doi.org/10.11648/j.ijmb.20190402.12

Golak-Siwulska, I., A. Kałużewicz, T. Spiżewski, M. Siwulski, and K. Sobieralski. 2018. Bioactive compounds and medicinal properties of Oyster mushrooms (*Pleurotus* sp.). *Folia Horticulturae* 30(2): 191–201. https://doi.org/10.2478/fhort-2018-0012

Gunawardena, D., L. Bennett, K. Shanmugam, K. King, R. Williams, D. Zabaras, and G. Münch. 2014. Anti-inflammatory effects of five commercially available mushroom species determined in lipopolysaccharide and interferon-γ activated murine macrophages. *Food Chemistry* 148: 92–96. https://doi.org/10.1016/j.foodchem.2013.10.015

Gutef, A. H., A. A. Al-Attraqchi, A. T. Tawfeeq, and H. B. Saheb. 2020. Evaluation of antibacterial potential of fruiting body extracts of *Pleurotus ostreatus in vitro* and *in vivo* study. *Drug Invention Today* 14(6): 854–863.

Hassan, M. A. A., R. Rouf, E. Tiralongo, T. W. May, and J. Tiralongo. 2015. Mushroom lectins: specificity, structure and bioactivity relevant to human disease. *International Journal of Molecular Sciences* 16(4): 7802–7838. https://doi.org/10.3390/ijms16047802

Ijeh, I. I., I. A. Okwujiako, P. C. Nwosu, and H. I. Nnodim. 2009. Phytochemical composition of *Pleurotus tuber regium* and effect of its dietary incorporation on body/organ weights and serum triacylglycerols in albino mice. *Journal of Medicinal Plants Research* 3(11): 939–943.

Innes, C. 2023. *Pleurotus* spp. as Agents of Mycoremediation: a review. An undergraduate honors thesis, Portland State University.

Islam, M. Z., M. H. Rahman, and F. Hafiz. 2009. Cultivation of oyster mushroom (*Pleurotus flabellatus*) on different substrates. *International Journal of Sustainable Crop Production* 4(1): 45–48.

Jacinto-Azevedo, B., N. Valderrama, K. Henríquez, M. Aranda, and P. Aqueveque. 2021. Nutritional value and biological properties of Chilean wild and commercial edible mushrooms. *Food Chemistry* 356: 129651. https://doi.org/10.1016/j.foodchem.2021.129651

Jarial, R. S., K. Jarial, and J. N. Bhatia. 2024. Comprehensive review on oyster mushroom species (Agaricomycetes): morphology, nutrition, cultivation and future aspects. *Heliyon* 10(5): e26539. https://doi.org/10.1016/j.heliyon.2024.e26539

Jayaprakash, B., A. R. Suresh, R. Thiruvengadam, et al. 2024. Evaluation of oyster mushroom (*Pleurotus ostreatus*)-derived anthraquinone on the induction of apoptosis and suppression of MMP-2 and MMP-9 expression in breast cancer cells. *International Journal of Medical Sciences* 21(6): 1016. https://doi.org/10.7150/ijms.93334

Jegadeesh, R., H. Lakshmanan, J. Kab-Yeul, V. Sabaratnam, and N. Raaman. 2018. Cultivation of pink oyster mushroom *Pleurotus djamor* var. *roseus* on various agro-residues by low-cost technique. *Journal of Mycopathological Research* 56(3): 213–220.

Juárez-Hernández, E. O., M. L. Pérez-Zavala, M. Román-Reyes, J. E. Barboza-Corona, and K. L. Macías-Sánchez. 2023. Overview of *Pleurotus* spp., edible fungi with various functional properties. *International Food Research Journal* 30(5): 1074–1092. https://doi.org/10.47836/ifrj.30.5.01

Kapahi, M., and S. Sachdeva. 2017. Mycoremediation potential of *Pleurotus* species for heavy metals: a review. *Bioresources and Bioprocessing* 4(1): 32. https://doi.org/10.1186/s40643-017-0162-8

Kikuchi, T. 2025. Ergostane-type steroids from mushrooms of *Pleurotus* genus. *Journal of Natural Medicines* 79: 1–14. https://doi.org/10.1007/s11418-024-01872-5

Knop, D., O. Yarden, and Y. Hadar. 2015. The ligninolytic peroxidases in the genus *Pleurotus*: divergence in activities, expression, and potential applications. *Applied Microbiology and Biotechnology* 99: 1025–1038. https://doi.org/10.1007/s00253-014-6256-8

Kong, W. S. 2004. Descriptions of commercially important *Pleurotus* species. In: *Mushroom Growers' Handbook 1: Oyster Mushroom Cultivation,* ed. K. W. Choi, 54–61. MushWorld, Seoul.

Kortei, N. K., and M. Wiafe-Kwagyan. 2015. Comparative appraisal of the total phenolic content, flavonoids, free radical scavenging activity and nutritional qualities of *Pleurotus ostreatus* (EM-1) and *Pleurotus eous* (P-31) cultivated on rice (*Oryzae sativa*) straw in Ghana. *Journal of Advances in Biology and Biotechnology* 3(4): 153–164. https://doi.org/10.9734/JABB/2015/16522

Koutrotsios, G., K. C. Mountzouris, I. Chatzipavlidis, and G. I. Zervakis. 2014. Bioconversion of lignocellulosic residues by *Agrocybe cylindracea* and *Pleurotus ostreatus* mushroom fungi–assessment of their effect on the final product and spent substrate properties. *Food Chemistry* 161: 127–135. https://doi.org/10.1016/j.foodchem.2014.03.121

Kumar, K. 2020. Nutraceutical potential and processing aspects of oyster mushrooms (*Pleurotus* species). *Current Nutrition and Food Science* 16(1): 3–14. https://doi.org/10.2174/1573401314666181015111724

Lavelli, V., C. Proserpio, F. Gallotti, M. Laureati, and E. Pagliarini. 2018. Circular reuse of bio-resources: the role of Pleurotus spp. in the development of functional foods. *Food and Function* 9(3): 1353–1372. https://doi.org/10.1039/C7FO01747B

Lee, K., U. Sim, Y. Choi, and J. Lee. 2018. Nutritional compositions and antioxidant activities of frequently consumed mushrooms in Korea. *The Korean Society of Food Science and Nutrition* 47(11): 1178–1184. https://doi.org/10.3746/jkfn.2018.47.11.1178

Lee, K. H., S. L. Morris-Natschke, X. Yang, et al. 2012. Recent progress of research on medicinal mushrooms, foods, and other herbal products used in traditional Chinese medicine. *Journal of Traditional and Complementary Medicine* 2(2): 1–12.

Mahari, W. A. W., W. Peng, W. L. Nam, H. Yang, X. Y. Lee, Y. K. Lee, and S. S. Lam. 2020. A review on valorization of oyster mushroom and waste generated in the mushroom cultivation industry. *Journal of Hazardous Materials* 400: 123156. https://doi.org/10.1016/j.jhazmat.2020.123156

Mandeel, Q. A., A. A. Al-Laith, and S. A. Mohamed. 2005. Cultivation of oyster mushrooms (*Pleurotus* spp.) on various lignocellulosic wastes. *World Journal of Microbiology and Biotechnology* 21: 601–607. https://doi.org/10.1007/s11274-004-3494-4

Mane, V. P., S. S. Patil, A. A. Syed, and M. M. V. Baig. 2007. Bioconversion of low quality lignocellulosic agricultural waste into edible protein by *Pleurotus sajor-caju* (Fr.) Singer. *Journal of Zhejiang University Science B* 8: 745–751. https://doi.org/10.1631/jzus.2007.B0745

Manimaran, K., G. Balasubramani, C. Ragavendran, D. Natarajan, and S. Murugesan. 2021. Biological applications of synthesized ZnO nanoparticles using *Pleurotus djamor* against mosquito larvicidal, histopathology, antibacterial, antioxidant and anticancer effect. *Journal of Cluster Science* 32: 1635–1647. https://doi.org/10.1007/s10876-020-01927-z

Market Data Forecast. 2024. From https://www.marketdataforecast.com/market-reports/oyster-mushroom-market. Accessed 20 September 2024.

Masevhe, M. R., N. J. Taylor, and P. Soundy. 2016. Alternative substrates for cultivating oyster mushrooms (*Pleurotus ostreatus*). *South African Journal of Plant and Soil* 33(2): 97–103. https://doi.org/10.1080/02571862.2015.1079932

Minato, K. I., A. Ohara, and M. Mizuno, 2017. A proinflammatory effect of the β-glucan from *Pleurotus cornucopiae* mushroom on macrophage action. *Mediators of Inflammation* 2017(1): 8402405. https://doi.org/10.1155/2017/8402405

Minhaj, N., and M. Imran. 2021. Impact of population growth and unemployment on Pakistan economy. *RADS Journal of Business Management* 3(2): 194–200.

Mirunalini, S., V. Arulmozhi, K. Deepalakshmi, and M. Krishnaveni. 2012. Intracellular biosynthesis and antibacterial activity of silver nanoparticles using edible mushrooms. *Notulae Scientia Biologicae* 4(4): 55–61. https://doi.org/10.15835/nsb448051

Mishra, V., S. Tomar, P. Yadav, and M. P. Singh. 2021. Promising anticancer activity of polysaccharides and other macromolecules derived from oyster mushroom (*Pleurotus* sp.): an updated review. *International Journal of Biological Macromolecules* 182: 1628–1637. https://doi.org/10.1016/j.ijbiomac.2021.05.102

Mishra, V., S. Tomar, P. Yadav, S. Vishwakarma, and M. P. Singh. 2022. Elemental analysis, phytochemical screening and evaluation of antioxidant, antibacterial and anticancer activity of *Pleurotus ostreatus* through *in vitro* and *in silico* approaches. *Metabolites* 12(9): 821. https://doi.org/10.3390/metabo12090821

Mitra, P., S. Khatua, and K. Acharya. 2013. Free radical scavenging and NOS activation properties of water soluble crude polysaccharide from *Pleurotus ostreatus*. *Asian Journal of Pharmaceutical and Clinical Research* 6(3): 67–70.

Mshandete, A. M., and J. Cuff. 2008. Cultivation of three types of indigenous wild edible mushrooms: *Coprinus cinereus*, *Pleurotus flabellatus* and *Volvariella volvocea* on composted sisal decortications residue in Tanzania. *African Journal of Biotechnology* 7(24): 4551–4562.

Musieba, F., S. Okoth, R. K. Mibey, S. Wanjiku, and K. Moraa. 2013. Proximate composition, amino acids and vitamins profile of *Pleurotus citrinopileatus* singer: an indigenous mushroom in Kenya. *American Journal of Food Technology* 8(3): 200–206.

Muszyńska, B., A. Grzywacz-Kisielewska, K. Kała, and J. Gdula-Argasińska. 2018. Anti-inflammatory properties of edible mushrooms: a review. *Food Chemistry* 243: 373–381. https://doi.org/10.1016/j.foodchem.2017.09.149

Nongthombam, J., A. Kumar, Ladli, B. Manikanta, M. Madhushekhar and S. Patidar. 2021. A review on study of growth and cultivation of oyster mushroom. *Plant Cell Biotechnology and Molecular Biology* 22(5–6): 55–65.

Ogwu, M. C., E. A. Kosoe, and S. C. Izah, 2025. Future trends and research directions: emerging uses and trends in mushroom consumption utilization. In: *Bioactive Compounds in Edible Mushrooms. Reference Series in Phytochemistry*, ed. S. C. Izah, M. C. Ogwu, and M. Akram, 1–16. Springer, Cham. https://doi.org/10.1007/978-3-031-52642-8_45-1

Ojha, J., and S. K. Shukla. 2023. Mushroom cultivation: a lucrative agribusiness venture. *Indian Farming* 73(4): 33–35.

Okolo, K. O., and O. E. Orisakwe. 2021. *In vitro* antioxidants and hepatoprotective effects of *Pleurotus tuber-regium* on carbon tetrachloride–treated rats. *Journal of Basic and Clinical Physiology and Pharmacology* 32(2): 67–78. https://doi.org/10.1515/jbcpp-2020-0034

Okolo, K. O., I. M. Siminialayi, and O. E. Orisakwe. 2016. Protective effects of *Pleurotus tuber-regium* on carbon-tetrachloride induced testicular injury in Sprague Dawley rats. *Frontiers in Pharmacology* 7: 480. https://doi.org/10.3389/fphar.2016.00480

Panda, B. C., P. Maity, A. K. Nandi, M. Pattanayak, D. K. Manna, S. Mondal, and S. S. Islam. 2017. Heteroglycan of an edible mushroom *Pleurotus cystidiosus*: structural characterization and study of biological activities. *International Journal of Biological Macromolecules* 95: 833–842. https://doi.org/10.1016/j.ijbiomac.2016.11.121

Pandimeena, M., M. Prabu, R. Sumathy, and R. Kumuthakalavalli. 2015. Evaluation of phytochemicals and in vitro antiinflammatory, anti-diabetic activity of the white oyster mushroom, *Pleurotus florida*. *International Research Journal of Pharmaceutical and Applied Sciences* 5(1): 16–21.

Park, Y. S., S. Jang, H. Lee, S. Kang, H. Seo, S. Yeon, and C. W. Yun. 2021. Identification of the antidepressant function of the edible mushroom *Pleurotus eryngii*. *Journal of Fungi* 7(3): 190. https://doi.org/10.3390/jof7030190

Patel, P., D. Patel, and N. Patel. 2012a. Experimental investigation of anti-rheumatoid activity of *Pleurotus sajorcaju* in adjuvant-induced arthritic rats. *Chinese Journal of Natural Medicines* 10(4): 269–274. https://doi.org/10.1016/S1875-5364(12)60054-5

Patel, Y., R. Naraian, and V. K. Singh. 2012b. Medicinal properties of *Pleurotus* species (oyster mushroom): a review. *World Journal of Fungal and Plant Biology* 3(1): 1–12.

Pavlik, M., M. Dzurenko, A. Kiiza, J. Akanyijuka, and P. Byandusya. 2023. Environmental, economics, and social benefits of cultivating the oyster mushroom *Pleurotus ostreatus* (JACQ.,). P. Kumm. in a community organization in southwest Uganda. 23rd SGEM International Multidisciplinary Scientific GeoConference 2023. https://doi.org/10.5593/sgem2023/5.1/s21.47

Poppe, J. 2000. Use of agricultural waste materials in the cultivation. *Mushroom Science* 15(1): 3–23.

Ragupathi, V., S. Kumerasan, S. Selvaraju, V. Karthikeyan, and C. V. Karthikeyan. 2016. Optimizing the growth conditions and adopting new methods growing oyster and milky mushrooms in same conditions. *International Journal of Herbal Medicine* 4(3): 1–4.

Rajak, S., S. C. Mahapatra, and M. Basu. 2011. Yield, fruit body diameter and cropping duration of oyster mushroom (*Pleurotus sajor caju*) grown on different grasses and paddy straw as substrates. *European Journal of Medicinal Plants* 1(1): 10–17.

Raman, J., K. Y. Jang, Y. L. Oh, et al. 2021. Cultivation and nutritional value of prominent *Pleurotus* spp.: an overview. *Mycobiology* 49(1): 1–14. https://doi.org/10.1080/12298093.2020.1835142

Raman, J., H. Lakshmanan, K. Y. Jang, M. Oh, Y. L. Oh, and J. H. Im, 2020. Nutritional composition and antioxidant activity of pink oyster mushrooms (*Pleurotus djamor* var. *roseus*) grown on a paddy straw substrate. *Journal of Mushroom* 18(3): 189–200. https://doi.org/10.14480/JM.2020.18.3.189

Rana, N., D. Vaidya, S. Sharma, and N. Chauhan. 2015. Chemical profile and amino acids composition of edible mushroom.

*International Journal of Agriculture, Environment and Biotechnology* 8(3): 675. https://doi.org/10.1177/026010600701800407

Rath, S. S. 2023. Strengthening economic and nutrition security through mushroom value chain management model in Odisha.

Ritota, M., and P. Manzi. 2019. *Pleurotus* spp. cultivation on different agri-food by-products: example of biotechnological application. *Sustainability* 11(18): 5049. https://doi.org/10.3390/su11185049

Rovenský, J., M. Stančíkova, K. Švík, K. Bauerová, and J. Jurčovičová. 2011. The effects of β-glucan isolated from *Pleurotus ostreatus* on methotrexate treatment in rats with adjuvant arthritis. *Rheumatology International* 31: 507–511. https://doi.org/10.1007/s00296-009-1258-z

Sadler, M. 2003. Nutritional properties of edible fungi. *Nutrition Bulletin* 28(3): 305–308. https://doi.org/10.1046/j.1467-3010.2003.00354.x

Salehi, F. 2019. Characterization of different mushrooms powder and its application in bakery products: a review. *International Journal of Food Properties* 22(1): 1375–1385. https://doi.org/10.1080/10942912.2019.1650765

Sánchez, C. 2010. Cultivation of *Pleurotus ostreatus* and other edible mushrooms. *Applied Microbiology and Biotechnology* 85: 1321–1337. https://doi.org/10.1007/s00253-009-2343-7

Sangeeta, S. D. Sharma, S. Ramniwas, R. Mugabi, J. Uddin, and G. A. Nayik. 2024. Revolutionizing Mushroom processing: innovative techniques and technologies. *Food Chemistry: X* 23: 101774. https://doi.org/10.1016/j.fochx.2024.101774

Sarangi, I., D. Ghosh, S. K. Bhutia, S. K. Mallick, and T. K. Maiti. 2006. Anti-tumor and immunomodulating effects of *Pleurotus ostreatus* mycelia-derived proteoglycans. *International Immunopharmacology* 6(8): 1287–1297. https://doi.org/10.1016/j.intimp.2006.04.002

Sardar, H., M. A. Ali, M. A. Anjum, et al. 2017. Agro-industrial residues influence mineral elements accumulation and nutritional composition of king oyster mushroom (*Pleurotus eryngii*). *Scientia Horticulturae* 225: 327–334. https://doi.org/10.1016/j.scienta.2017.07.010

Sardar, H., M. A. Anjum, S. Hussain, et al. 2022. Deciphering the role of moringa leaf powder as a supplement in the cotton waste substrate for the growth and nutrition of king oyster mushroom. *Scientia Horticulturae* 293: 110694. https://doi.org/10.1016/j.scienta.2021.110694

Selegean, M., M. V. Putz, and T. Rugea. 2009. Effect of the polysaccharide extract from the edible mushroom *Pleurotus ostreatus* against infectious bursal disease virus. *International Journal of Molecular Sciences* 10(8): 3616–3634. https://doi.org/10.3390/ijms10083616

Sganzerla, W. G., S. D. Todorov, and A. P. da Silva. 2022. Research trends in the study of edible mushrooms: nutritional properties and health benefits. *International Journal of Medicinal Mushrooms* 24(5): 1–18. https://doi.org/10.1615/IntJMedMushrooms.2022043738

Shen, Y., M. Gu, Q. Jin, et al. 2014. Effects of cold stimulation on primordial initiation and yield of *Pleurotus pulmonarius*. *Scientia Horticulturae* 167: 100–106. https://doi.org/10.1016/j.scienta.2013.12.021

Sheng, L., X. Sun, C. Mo, M. Hao, X. Wei, and A. Ma. 2023. Relationship between antioxidant enzymes and sclerotial formation of *Pleurotus tuber-regium* under abiotic stress. *Applied Microbiology and Biotechnology* 107(4): 1391–1404. https://doi.org/10.1007/s00253-022-12358-3

Shukla, S., and A. K. Jaitly. 2011. Morphological and biochemical characterization of different oyster mushroom (*Pleurotus* spp.). *Journal of Phytology* 3(8): 18–20.

Siddiq, M., R. Ravi, and A. Sami. 2018. Edible mushrooms: production, processing, and quality. In: *Handbook of Vegetables and Vegetable Processing*, ed. N. K. Sinha, 701–725. Wiley-Blackwell. https://doi.org/10.1002/9780470958346.ch32

Silva, S. O., S. M. G. D. Costa, and E. Clemente. 2002. Chemical composition of *Pleurotus pulmonarius* (Fr.) Quél., substrates and residue after cultivation. *Brazilian Archives of Biology and Technology* 45: 531–535. https://doi.org/0.1590/S1516-89132002000600018

Singh, K., S. Sharma, P. Kumar, and D. Singh. 2024. Exploring the potential of oyster mushroom cultivation as remunerative agri-enterprise in Punjab. *Indian Journal of Ecology* 51(6): 1341–1344. https://doi.org/10.55362/IJE/2024/4406

Singh, M. P., and V. K. Singh. 2011. Yield performance and nutritional analysis of *Pleurotus citrinopileatus* on different agrowastes and vegetable wastes. In: *Proceedings of the 7th International Conference on Mushroom Biology and Mushroom Products (ICMBMP7)*, pp. 4–7.

Singh, S., C. G. Kumar, and S. Singh. 1995. Production, processing and consumption patterns of mushrooms. *Indian Food Industry* 14: 38–47.

Sopanrao, P. S., A. S. Abrar, T. S. Manoharrao, and B. M. M. Vaseem. 2010. Nutritional value of *Pleurotus ostreatus* (Jacq: Fr) Kumm cultivated on different lignocellulosic agro-wastes. *Innovative Romanian Food Biotechnology* (7): 66–76.

Sreedharan, P. L., M. Kishorkumar, E. G. Neumann, and S. S. Kurup. 2025. The emerging role of oyster mushrooms as a functional food for complementary cancer therapy. *Foods* 14(1): 128. https://doi.org/10.3390/foods14010128

Suguimoto, H. H., A. M. Barbosa, R. F. Dekker, and R. J. Castro-Gomez. 2001. Veratryl alcohol stimulates fruiting body formation in the oyster mushroom, *Pleurotus ostreatus*. *FEMS Microbiology Letters* 194(2): 235–238. https://doi.org/10.1111/j.1574-6968.2001.tb09475.x

Sultana, S., M. Shamsuzzaman, M. A. S. Miah, et al. 2024. Assessment of proximate composition, mineral element profile and antioxidant properties of the edible oyster mushroom grown in Bangladesh. *Discover Food* 4(1): 126. https://doi.org/10.1007/s44187-024-00216-2

Suseem, S. R., and A. M. Saral. 2013. Analysis on essential fatty acid esters of mushroom pleurotus eous and its antibacterial activity. *Asian Journal of Pharmaceutical and Clinical Research* 6(1): 188–191.

Tagkouli, D. G., S. Bekiaris, M. E. Pantazi, G. Anastasopoulou, A. Koutrotsios, A. Mallouchos, and N. Kalogeropoulos. 2021. Volatile profiling of *Pleurotus eryngii* and *Pleurotus ostreatus* mushrooms cultivated on agricultural and agro-industrial by-products. *Foods* 10(6): 1287. https://doi.org/10.3390/foods10061287

Tang, C., P. C. X. Hoo, L. T. H. Tan, et al. 2016. Golden needle mushroom: a culinary medicine with evidenced-based biological activities and health promoting properties. *Frontiers in Pharmacology* 7: 474. https://doi.org/10.3389/fphar.2016.00474

Tesfay, T., T. Godifey, R. Mesfin, and G. Kalayu. 2020. Evaluation of waste paper for cultivation of oyster mushroom (*Pleurotus ostreatus*) with some added supplementary materials. *AMB Express* 10(1): 15. https://doi.org/10.1186/s13568-020-0945-8

Thakur, M. P. 2020. Advances in mushroom production: key to food, nutritional and employment security: a review. *Indian Phytopathology* 73: 377–395. https://doi.org/10.1007/s42360-020-00244-9

Tiram, C. 2013. Effect of different drying techniques on the nutritional values of oyster mushroom (*Pleurotus sajor-caju*). *Sains Malaysiana* 42(7): 937–941.

Tiwari, H., R. Naresh, R. Bhatt, K. Y. Aditya, D. D. Manisha, and S. K. Kataria. 2023. Underutilized nutrient rich millets:

challenges and solutions for India's food and nutritional security: a review. *International Journal of Plant and Soil Science* 35(02): 45–56. https://doi.org/10.9734/IJPSS/2023/V35i22758.

Tolera, K. D., and S. Abera. 2017. Nutritional quality of Oyster mushroom (*Pleurotus ostreatus*) as affected by osmotic pretreatments and drying methods. *Food Science and Nutrition* 5(5): 989–996. https://doi.org/10.1002/fsn3.484

Törős, G. H., H. El-Ramady, and J. Prokisch. 2022. Edible mushroom of *Pleurotus* spp.: a case study of oyster mushroom (*Pleurotus ostreatus* L.). *Environment, Biodiversity and Soil Security* 6(2022): 51–59. https://doi.org/10.21608/jenvbs.2022.117554.1161

Tripathi, A. D., R. Mishra, K. K. Maurya, R. B. Singh, and D. W. Wilson. 2019. Estimates for world population and global food availability for global health. In: *The Role of Functional Food Security in Global Health*, pp. 3–24. Academic Press. https://doi.org/10.1016/B978-0-12-813148-0.00001-3

Uddin, M. N., S. Yesmin, M. A. Khan, M. Tania, M. Moonmoon, and S. Ahmed. 2011. Production of oyster mushrooms in different seasonal conditions of Bangladesh. *Journal of Scientific Research 3*(1): 161–161. https://doi.org/10.3329/jsr.v3i1.6130

Wang, H. X., and T. B. Ng. 2000. Isolation of a novel ubiquitin-like protein from *Pleurotus ostreatus* mushroom with anti-human immunodeficiency virus, translation-inhibitory, and ribonuclease activities. *Biochemical and Biophysical Research Communications* 276(2): 587–593. https://doi.org/10.1006/bbrc.2000.3540

Wang, Q., L. L. Niu, H. Liu, Y. R. Wu, M. Li, and Q. Jia. 2021. Structural characterization of a novel polysaccharide from *Pleurotus citrinopileatus* and its antitumor activity on H22 tumor-bearing mice. *International Journal of Biological Macromolecules* 168: 251–260. https://doi.org/10.1016/j.ijbiomac.2020.12.053

Wjabn, J., S. M. Handunnetti, C. A. Wanigatunge. G. H. Fernando, D. T. U. Abeytunga, and T. S. Suresh. 2020. Anti-inflammatory activity of *Pleurotus ostreatus*, a culinary medicinal mushroom, in Wistar rats. *Evidence-based Complementary and Alternative Medicine* 2020: 6845383–6845383. https://doi.org/10.1155/2020/6845383

Wolff, E. R. S., E. Wisbeck, M. L. L. Silveira, R. M. M. Gern, M. S. L. Pinho, and S. A. Furlan. 2008. Antimicrobial and antineoplasic activity of *Pleurotus ostreatus*. *Applied Biochemistry and Biotechnology* 151(2): 402–412. https://doi.org/10.1007/s12010-008-8208-1

Wong, J. Y., and F. Y. Chye. 2009. Antioxidant properties of selected tropical wild edible mushrooms. *Journal of Food Composition and Analysis* 22(4): 269–277. https://doi.org/10.1016/j.jfca.2008.11.021

Xu, W., J. J. H. Huang, and P. C. K. Cheung. 2012. Extract of *Pleurotus pulmonarius* suppresses liver cancer development and progression through inhibition of VEGF-induced PI3K/AKT signaling pathway. *PLoS ONE* 7(3): e34406. https://doi.org/10.1371/journal.pone.0034406

Younis, A. M., M. M. Abdel-Aziz, and M. Yosri. 2019. Evaluation of some biological applications of *Pleurotus citrinopileatus* and *Boletus edulis* fruiting bodies. *Current Pharmaceutical Biotechnology* 20(15): 1309–1320. https://doi.org/10.2174/1389201020666190904162403

Zakia, B., and H. C. Srivastava. 1962. Studies on the cultivation of *Pleurotus* sp. on paddy straw. *Food Science* 11(12): 363–365.

Zhang, Y., L. Dai, X. Kong, and L. Chen. 2012. Characterization and *in vitro* antioxidant activities of polysaccharides from *Pleurotus ostreatus*. *International Journal of Biological Macromolecules* 51(3): 259–265. https://doi.org/10.1016/j.ijbiomac.2012.05.003

Zhang, Y., W Geng, Y. Shen, Y. Wang, and Y. C. Dai. 2014. Edible mushroom cultivation for food security and rural development in China: bio-innovation, technological dissemination and marketing. *Sustainability 6*(5): 2961–2973. https://doi.org/10.3390/su6052961

# Index

## Q



## R



## S



## T



## U



## V



## W



## X



## Z